VACUUM
MICROELECTRONICS

VACUUM MICROELECTRONICS

Edited by

Wei Zhu
Bell Laboratories - Lucent Technologies

A WILEY-INTERSCIENCE PUBLICATION

JOHN WILEY & SONS

New York · Chichester · Weinheim · Brisbane · Singapore · Toronto

Library of Congress Cataloging-in-Publication Data is available.

ISBN 0-471-32244-X

Printed in the United States of America

10 9 8 7 6 5 4 3 2 1

CONTENTS

▰▰▰▰ PREFACE

Vacuum microelectronics is the science and technology of building micrometer-scale devices that operate with ballistic electrons in vacuum. It is a field that has been experiencing tremendous growth in recent years. As a technology, it starts with some unmatched advantages: electrons in vacuum can travel far faster with less energy dissipation than in any semiconductor. This enables faster modulation and higher electron energies than are possible with semiconductor structures, so vacuum microelectronic devices (also sometimes called cold cathode devices or field emission devices) can operate at higher frequencies and higher power in a wider temperature range, as well as in high radiation environments. Potential applications include flat panel field emission displays, miniaturized microwave power amplifier tubes, advanced sensors, atomic resolution storage, and electron sources for microscopes, ion guns, mass spectrometers, cathode-ray tubes, x-ray generators, high energy accelerators, and electron beam lithography. In the last several years, field emission displays have moved beyond the research laboratory to actual prototypes and commercial products.

This book is designed to provide a comprehensive and up-to-date treatment of this rapidly evolving field. It brings together a group of respected experts to provide an in-depth coverage of the science and technologies involved in vacuum microelectronics. The book focuses on field emission from cold cathodes, which is currently the most common and also most promising source of electrons for vacuum microelectronics. The book strives to connect the fundamental and practical aspects of the field so that a broad base of readers can find the book useful. To this end, the book goes from the physics of field emission to fabrication of cathodes to discussions of the two major device applications for the technology, namely field emission displays and microwave power amplifier tubes.

In Chapter 1, Utsumi presents a historical account of the evolution of vacuum device technologies. His insightful observations about the rich and colorful history and unique perspectives about the future of the technology give readers a clear overview of how the field has developed. In Chapter 2, Kochanski, Zhu and Goren review the technological state of the art by examining the scientific and technical problems that remain to be solved before large scale commercialization is possible. By discussing the potentials and challenges of vacuum microelectronics in the context of real devices, they present a thoughtful assessment of the promise and reality of the technology.

On theoretical issues, Jensen presents a broad and representative review of the complex and diverse theoretical work on field emission in Chapter 3. He gradually proceeds from basic theory to more advanced topics, and the in-depth coverage will make this chapter an excellent introduction to newcomers in the field and a useful reference to those already acquainted with the subject. In Chapter 4, Spindt, Brodie, Holland and Schwoebel introduce the all-important Spindt cathode that essentially laid the foundation for present-day vacuum microelectronics. As pioneers of the technology, the authors describe in detail the fabrication technology and operating characteristics of this microfabricated, gated, field emitter array (FEA). They also present many of the very significant application opportunities that exist for this enabling cathode technology. In Chapter 5, Shaw and Itoh discuss the silicon field emitter arrays and their fabrication, emission characteristics and various circuit elements that are made possible with the semiconducting FEA material. The practical issues concerning the stability and reliability of the Si FEAs are also carefully examined. In Chapter 6, Zhu, Baumann and Bower describe the emerging new cathodes by reviewing field emission properties of carbon-based field emitters, including diamond, diamond-like carbon and carbon nanotube materials, and discussing physical mechanisms involved and potential device applications that are being explored. Three non-carbon based cathodes – surface conduction emitters (SCEs), ferroelectric emitters and metal-insulator-metal (MIM) emitters – are also briefly reviewed.

For device applications, Busta describes field emission displays, currently the most active application for vacuum microelectronics, in Chapter 7. The basic display structure and operation principles, as well as major issues related to structural fabrication, encapsulation, gettering and panel addressing, are presented in detail. In Chapter 8, Murphy and Kodis discuss the application of field emitter arrays to microwave power tubes. The authors describe the potential advantages of such devices over conventional vacuum tubes and semiconductor devices and walk the readers through key device concepts, physical principles that underlie the issues, and simple mathematical analyses that describe them.

A book of this magnitude and diversity would not have been possible without the direct and indirect assistances of many other people. It is my great pleasure to work with a highly motivated group of people who provided the essential ingredients of diligence and perseverance to accomplish this task. To begin with, I want to express my gratitude to Takao Utsumi, who not only agreed to pen the first chapter of the book, but also provided important help in the initiation phase of this project. Without his guidance and advice, this project might not have gotten off the ground.

I want to thank the chapter authors for their well-written chapters. They are respected experts and leaders in the field and have spent an incredible amount of time and energy in developing these chapters. It is their devotion and contributions, as well as their cooperation and support, that made this book possible.

Each of the chapters was reviewed by experts in the field, who devoted precious hours of free time to improving the book with their critical comments and invaluable suggestions. Thanks go to Akintunde Ibitayo (Tayo) Akinwande of MIT, Christopher Bower of Lucent Technologies, Heinz Busta of Sarnoff, Thomas Felter of Lawrence Livermore National Laboratory, Richard Fink of SI Diamond Technology, Kevin Jensen of Naval Research Laboratory, Gregory Kochanski of Lucent Technologies, Lawrence Pan of Candescent Technologies, Jonathan Shaw of Naval Research Laboratory, Lynwood Swanson of FEI, Alec Talin of Motorola, Dorota Temple of MCNC and Paul VonAllmen of Motorola.

I also want to take this opportunity to express my respect to the late Henry Gray. He agreed to write a chapter for this book and had started working on it before he fell ill and passed away in July, 1999. He was full of ideas and always inspired people to try new things. I benefited a great deal from many conversations with him. I will always remember his knowledge, enthusiasm, energy and devotion to this field.

I want to acknowledge my colleagues here at Bell Labs, Gregory Kochanski, Sungho Jin, Christopher Bower and John Graebner for many years of fruitful collaborations in field emission research. In particular, I am grateful to Greg for agreeing to take up the primary authorship of the technology overview chapter on short notice and delivering it on a tight schedule. I benefited greatly over the years from his profound understanding of the field and enormous talent and resourcefulness in identifying the fundamental physics of the problems and formulating brilliant and workable solutions.

Today, vacuum microelectronics has evolved into a highly stimulating and focused interdisciplinary field that is full of far-reaching scientific and technological promises, and huge efforts are being made to commercialize this technology. Technological advances in device design and fabrication are constantly occurring. Cathode lifetime, stability and performance are continually improving. Structural characterization and property measurements of new emitter materials are being extensively conducted. A good understanding of emission theories relevant to practical cathode/device operating environments is beginning to evolve. Moreover, nanotechnology is providing new stimulus and promises further excitement, as nanoscale materials and fabrication tools are increasingly being applied to building vacuum microelectronic devices. It is, therefore, with great enthusiasm for the future of vacuum microelectronics that I present this book.

April, 2001   WEI ZHU
Murray Hill, NJ

LIST OF CONTRIBUTORS

Baumann, P.K., AIXTRON AG, Kackertstr. 15-17, 52072, Aachen, Germany

Bower, C.A., Bell Laboratories, Lucent Technologies, 600 Mountain Avenue, Murray Hill, NJ 07974, USA

Brodie, I., SRI International, 333 Ravenswood Avenue, Menlo Park, CA 94025, USA

Busta, H.H., Sarnoff Corporation, 201 Washington Road, Princeton, NJ 08543, USA

Goren, Y., Teledyne Electronic Technologies, 11361 Sunrise Park Drive, Rancho Cordova, CA 95742, USA

Holland, C.E., SRI International, 333 Ravenswood Avenue, Menlo Park, CA 94025, USA

Itoh, J., National Institute of Advanced Industrial Science and Technology, Tsukuba, Japan

Jensen, K.L., Naval Research Laboratory, 4555 Overlook Avenue SW, Washington D.C. 20375, USA

Kochanski, G.P., Bell Laboratories, Lucent Technologies, 600 Mountain Avenue, Murray Hill, NJ 07974, USA

Kodis, M.A., Jet Propulsion Laboratory, California Institute of Technology, Pasadena, CA 91125, USA

Murphy, R.A., Lincoln Laboratory, Massachusetts Institute of Technology, 244 Wood Street, Lexington, MA 02420, USA

Schwoebel, P.R., SRI International, 333 Ravenswood Avenue, Menlo Park, CA 94025, USA

Shaw, J., Naval Research Laboratory, 4555 Overlook Avenue SW, Washington D.C. 20375, USA

Spindt, C.A., SRI International, 333 Ravenswood Avenue, Menlo Park, CA 94025, USA

Utsumi, T., LEEPL Corporation, 4155 Akiya, Yokosuka, 240-0105, Japan

Zhu, W., Bell Laboratories, Lucent Technologies, 600 Mountain Avenue, Murray Hill, NJ 07974, USA

CHAPTER 1

Historical Overview

TAKAO UTSUMI

LEEPL Corporation, 4155 Akiya, Yokosuka 240-0105, Japan

1.1. INTRODUCTION

When semiconductor transistors were invented by Bardeen, Brattain, and Shockley in the 1950s [1], and integrated circuits were subsequently developed in the 1960s [2,3], people generally thought that the time of using vacuum tubes was over. They were large, fragile, and inefficient. They required a vacuum to operate and used a cathode heated to over $1000°C$ to generate the electrons. However, the emergence of vacuum microelectronic (VME) devices has generated renewed interests in the exploitation of vacuum tubes for many new applications. Vacuum microelectronics, a new inter-disciplinary science and technology, deals with vacuum devices of micrometer dimensions that are made by microfabrication techniques developed for the semiconductor industry. To help understand the evolution of this technology, I will discuss some historical developments that occurred in this field.

1.2. SHOULDERS' PROPOSAL

Historically, there were two important technological breakthroughs that made the birth of modern vacuum microelectronics possible. The first was the microfabrication technology developed for on-chip large scale integration of micron-sized solid-state devices. If a vacuum triode that was manually machined and assembled 40 years ago could be operated at frequencies as high as 4 GHz [4], it would not be difficult to imagine that the new microfabrication techniques, which could produce devices with dimensions three orders of magnitude smaller, would enable much faster speeds for VME devices.

The second breakthrough was the successful use of field emitters as the source of electrons. Field emission is a quantum mechanical tunneling phenomenon in which electrons escape from a solid surface into vacuum. In contrast to the commonly used thermionic emission from hot filaments, field emission occurs at room

Vacuum Microelectronics, Edited by Wei Zhu.
ISBN 0-471-32244-X. 2001 John Wiley & Sons, Inc.

temperature from unheated "cold" cathodes under an electric field. Field emission offers several attractive characteristics, including instantaneous response to field variation, resistance to both temperature fluctuation and radiation, and a nonlinear, exponential current–voltage relationship in which a small change of voltage can induce a large change of emission current.

However, field emission requires a very large electric field, more than 10 MV/cm, to obtain a reasonable current. A traditional way to obtain this high field was to use a very sharp needle, the tip curvature of which was of the order of a few hundred nanometers, obtainable by wet etching. Even so, high voltages of the order of a few thousand volts were necessary to draw a meaningful current.

The advent of microfabrication technology changed the situation dramatically. It allowed an electrode to be fabricated in the very close vicinity (e.g., a few micrometers) of an emitting cathode, thus significantly lowering the operating voltage. The curvature of the emitting tips could be made much smaller than that of etched needles, leading to more effective field concentrations at the tips. Additionally, it became possible to create large arrays of emitters to achieve high emission currents and current densities. It certainly appeared that with the microfabrication technology, vacuum microelectronics could reap the benefits of both conventional vacuum and solid-state devices.

About 40 years ago, Shoulders of Stanford Research Institute (SRI) presented a device concept in an article entitled "Microelectronics using electron-beam-activated machining techniques" [5]. At that time, the microfabrication technology that we know today had not come to existence, and Fairchild Corporation had just announced its 4-transistor monolithic chips with a newly invented planar process [6]. However, Shoulders, with tremendous vision and insight, proposed "to devise vacuum tunnel effect devices of micron sizes with switching times in 10^{-10} s that (i) operate at 50 V, (ii) have high input impedance, (iii) are insensitive to temperature effects up to 1000°C, (iv) are insensitive to ionizing radiation effects up to the limits of the best known dielectric materials, and (v) have a useful lifetime of many hundreds of years."

The proposed device structure was exactly what the microfabrication technology was developed to create. The powerful combination of the field emission cold cathode with the microfabrication techniques would seem to be capable of overcoming most of the drawbacks associated with the traditional vacuum tubes. It is an irony, though, that while semiconductor electronics advanced rapidly soon after the time of Shoulders' publication, vacuum microelectronics took 40 years to get off the ground. Nevertheless, his power of technological imagination was truly gigantic.

1.3. GROUNDBREAKING WORK

Field emission was traditionally studied within the discipline of surface science. The early history of surface science was full of studies of cathodes, which led to many important discoveries. For example, Langmuir's adsorption isotherm for gas–solid surface interactions [7], which essentially laid the foundation of modern surface science, was a direct result of his studies on heated tungsten filaments. The experiment conducted by Davisson and Germer in 1927 [8], which established the wave nature of electrons, was originally designed to investigate the role of positive ion bombardment in electron emission from oxide-coated nickel cathodes used in vacuum triode amplifiers. Later, this work also led to the discovery of low energy electron diffraction (LEED) [9].

Armed with the "new" quantum mechanics, Fowler and Nordheim calculated a relationship between the emission current density and the applied electric field in 1928 based on electron tunneling through a simple two-dimensional (2-D) triangular barrier [10,11]. However, experimental verifications of the theory were not easy because of the need for very high electric fields. It was also difficult at the time to obtain field emission under controlled and reproducible conditions, due to the perturbing effects of invisible microstructural defects and impurities on the emitter surfaces. Catastrophic vacuum arcs often occurred between electrodes at high voltages. In fact, how high a voltage could be applied between electrodes without causing a vacuum arc had become a critical design issue at that time for all types of vacuum devices ranging from cathode ray tubes to high-energy accelerators. These vacuum-breakdown phenomena and associated physical mechanisms were topics of intensive research in the 1950s–1960s [12–15].

As indicated earlier, it was necessary to employ a sharp needle for obtaining a sufficiently high electric field for electron emission. In 1937, Müller etched a fine tungsten wire and placed it at the center of a spherical glass vessel, the inside surface of which was coated with a fluorescent material [16]. He observed a symmetric pattern on the fluorescent screen produced by field-emitted electrons that traveled radially from the tip to the screen. This was essentially the field emission microscope (FEM) that we know today. Using this FEM setup, Good and Müller [17] and Dyke and Dolan [18] examined and verified the Fowler–Nordheim equation over a wide range of currents and voltages. Because of the strong dependency of current density on the local electric field and the work function of the emitter surface, both of which were sensitive to the atmoic structure and any adsorbates at the surface, this field emission apparatus also became an excellent tool for studying surface phenomena such as crystallographic structure, chemisorption, surface diffusion, and impurity effects. The FEM was further explored to achieve atomic resolutions, because a single adatom on the tip surface could induce local field enhancement and result in increased emission current, thus creating a bright spot on the screen. However, this field enhancement distorted the equipotential lines in the vicinity of the adatom. Because of this field distortion and the initial transverse velocities of electrons emitted from the Fermi level, FEM did not possess atomic resolutions [19]. Such resolutions were realized much later through field ion microscopy (FIM) [20] and scanning tunneling microscopy (STM) [21].

It gradually became clear during the course of extensive studies on field emission in the 1950s–1970s that etched needles had many limitations when they were used as cathodes in miniaturized devices. Among them, the necessity of using high voltages (several thousands of volts) and maintaining a high vacuum (10^{-9} torr or better) were the most problematic. Additionally, the etched tips had a limited lifetime due to the sputtering damage from high-energy ions created by electron impact ionization. Nonetheless, etched needles have been successfully deployed in many devices such as microwave amplifiers, high-resolution cathode ray tubes, electron microscopes, electron beam lithography, and flash X-ray photography [18].

1.4. INVENTION OF SPINDT CATHODE

By the mid-1960s, excitement and enthusiasm for Shoulders' concept of microminiaturized vacuum tubes incorporating a field emission electron source had waned considerably, mainly because technical difficulties encountered in using etched needles as electron emitters in such devices had stalled any momentum. Then, Spindt of SRI, who was hired by Shoulders to carry on the work, displayed unyielding persistence in pursuing the device concept and finally succeeded in fabricating the field emitters that Shoulders had dreamed of. The first publication of his work appeared in 1968 [22], and the emitters, now known as the Spindt cathode, consisted of a multilayer structure of Mo gate/SiO_2 insulator/Mo cathode cones on a Si substrate fabricated by thin film vacuum deposition techniques. The technology allowed the emitters to be fabricated in arrays of up to many thousands tips at a packing density of tens of millions per square centimeter. An extensive review of the general properties of Spindt cathodes, also commonly referred to as field-emitter arrays (FEAs), was published in 1976 [23].

Very briefly, the operating gate voltage in a FEA was typically around 100 V and could be reduced to a few tens of volts by further shrinking the device dimension. Densely packed FEAs of the order of 10^7 emitters/cm^2 could be fabricated. The lifetime of these emitters could exceed more than 10 000 h, and the excessive noise from a single emitter could be reduced statistically by increasing the number of emitters. Because of these outstanding properties, devices built on FEAs became attractive. Two of the most notable developments were the exploitation of flat panel field emission displays [24] and high frequency vacuum tube devices [25]. The impact of the work by Spindt and his colleagues at SRI was extensive and global. Many researchers have since worked on the refinement, characterization, and applications of the Spindt emitters. An article titled "Vaccum Microelectronics" co-authored by Brodie and Spindt in 1992 [26] documented major progresses achieved since the invention of the Spindt cathode.

There were variations to the cone-shaped Spindt metal emitters, both geometrically, such as wedge and thin film edge emitters, and materials-wise, such as carbides and semiconductors. In particular, Gray of Naval Research Laboratory fabricated a structure called "vacuum field effect transistor" based on Si-FEAs in 1986 [27]. The structure consisted of emitters, gates, and collectors on the same planar surface of a silicon wafer and exhibited both voltage and power gains with a transit time of only 5 ps from the emitter to the collector. This was a fine example of the use of state-of-the-art silicon fabrication technology in building advanced vacuum microelectronic devices [28,29]. In addition to the great variety of available fabrication techniques, the $I–V$ characteristics of semiconductor-FEAs could be made different from those of metal-FEAs. This opened up new application fronts, including light-sensitive FEAs, electron velocity-saturated FEAs, noble metal-silicided tip-FEAs, and monolithically integrated FEAs with silicon and optoelectronic integrated circuits [30–32].

By the middle of 1987, the surging worldwide interest in FEAs and their applications necessitated the formation of an international forum for researchers to gather and exchange information. A nine-member international steering committee was formed with Gray and Spindt serving as co-chairmen to organize a conference. In the summer of 1988, the first International Vacuum Microelectronics Conference (IVMC) was held in Williamsburg, VA, and the conference scope covered all aspects of the field, including fundamental emission physics and device physics; theory, simulation, and modeling; materials, processing, fabrication, and micromachining; FEAs and multielectrode structures; field emission displays (FEDs), high frequency devices, and other applications. This was truly an exciting event, symbolizing the dawn of a new era for vacuum microelectronics, a term coined at the conference. Since then, the IVMCs have been held annually at locations around the world.

1.5. FIELD EMITTER ARRAYS

The first 6 years following the 1st IVMC (1988–1993) could be called the "Years of the Spindt Cathode," because attention was almost entirely devoted to the characterization and understanding of Spindt-type FEAs and their practical applications. Researchers probed all types of issues in an effort to optimize the technology so that one can "draw the most current at the least applied voltage from the smallest device structure" [33]. There were many notable breakthroughs reported in this period that had far-reaching implications for the technology. For example, Van Veen of Philips Research Labs reported the largest current obtained from a single Mo tip: 850 μA at a gate voltage of 205 V [34]. Spindt reported the highest current of 6 mA drawn from a 12 tip Mo-FEA, which was equivalent to a current density of 320 A/cm^2 or 500 μA/tip [35]. He further found that when the number of tips increased, the average emission current per tip actually decreased due to variations in tip geometry. A 5000-tip FEA yielded a total current of 100 mA, which corresponded to $\sim$20 μA/tip.

Bozler et al. of MIT Lincoln Laboratory fabricated the smallest-dimension Mo-FEAs at the time using interferometric lithography, featuring a gate-to-tip distance of 0.08 μm, a tip-to-tip distance of 0.32 μm, and a tip density of 10^9/cm^2 [36]. They measured a current of 1 μA (1 nA/tip or 1 A/cm^2) at 25 V from an uncesiated 10 × 10 μm array of 900 tips, a record low gate voltage for FEAs. The cesiated array achieved 1 μA at an even lower voltage of 10 V, and the maximum current density reached 1600 A/cm^2, again a record for FEAs.

Schowebel et al. [37] reported that treating the emitters in a hydrogen plasma was effective in removing surface contaminants and enhancing the emission uniformity. This emitter pre-conditioning has become a standard procedure in the processing of FEAs. Gomer [19], Spindt et al. [38], and Goodhue et al. [39] did extensive work in understanding the nature of emission noise from FEAs. They identified that emission predominantly originated from nano-scale protrusions on the tip surface, and the change of atomic structures at these protrusions because of enhanced surface diffusion in high electric fields led to discrete jumps in emission current, and hence the occurrence of flip-flop or flicker noises.

Perhaps the most exciting development in this period was the successful demonstration of a 6″ flat panel FED based on Mo-FEAs by Researchers at LETI of France [40]. The practical difficulty of achieving uniform emission over the entire panel was overcome by the incorporation of underlayer in-series resistors below the emitters [41]. The brightness, clarity, stability, and uniformity of the display clearly established FED as a viable technology for desk-top and many other applications, and created the much-needed momentum for further advancing the technology.

1.6. NEW CATHODE MATERIALS

In the subsequent IVMC years (1994–2000), there has been a noticeable shift in research activities. Besides the strong push in FEA-based FEDs by companies such as Pix-Tech and Candescent Technologies, researchers have been increasingly turning their attentions to the search for new cathode materials. The main motivation is to find materials or structures that are more robust and manufacturing-friendly than the FEAs. Of the many reported cathode materials, diamond and carbon nanotube emitters have attracted the most interest. As an example, nearly one-third of the conference papers were on diamond related work at the 10th IVMC (Kyongju, Korea) in 1997.

Diamond, with its negative electron affinity (NEA) on hydrogen-terminated surfaces [42] and the ability of depositing it in thin film forms at low pressures by chemical vapor deposition [43,44], has been shown to exhibit low emission threshold fields of 3–40 V/μm for a current density of 10 mA/cm^2. These values compare favorably to those required for metal or semiconductor FEAs. Similarly, the one-dimensional carbon nanotubes [45], which are nanometers in diameter (1–30 nm) and micrometers in length (1–20 μm), are also low-field emitters. Emission currents as high as 1 μA from single nanotubes [46] and current densities as high as 4 A/cm^2 from multiple nanotubes [47] have been observed. It is interesting to note that the emitting structure of nanotubes has been speculated to consist of a single atomic chain of 10–100 carbon atoms that are pulled out from an open graphene sheet at the nanotube end by the strong local electric field [48]. This may prove to be the ultimate atomic-scale field emitter structure that researchers have been searching for. Active development efforts on diamond- and carbon nanotube-based FEDs and lighting elements are ongoing at a number of companies, including SI Diamond Technology, Samsung, and ISE Electronics.

Some of the other important developments reported in this period include the achievement of noiseless emission current from a single tip in a MOSFET integrated Si-FEA [49,50], the never-ending quest for noiseless field emitting materials [51,52], the fundamental understanding of field emitted super-coherent electron beams [53], the monochromatic electron emission from superconductors [54], and surface-emitting cold cathode devices based on porous polycrystalline silicon films [55,56]. For high frequency device applications, researchers at NEC successfully demonstrated a cold cathode X-band miniaturized traveling wave tube [57]. The tube had a 27.5 W output power and 19.5 dB gain at 10.5 GHz, which was realized by an emission current of 60 mA or a current density of 10.6 A/cm^2 from a 14 520-tip metal FEA. In the United States, programs were established to develop cold cathode based klystrodes with the target performance of 50 W output power, 10 dB gain, and 50% efficiency operating at 10 GHz, but with limited success [58,59]. Russian researchers reported a number of novel cold cathode rf devices, such as super-miniaturized reflecting klystrons and low voltage backward-wave tubes [60].

1.7. FUTURE

By glancing at Table 1.1 that compares the characteristics of vacuum microelectronic devices with semiconductor devices, it becomes quite clear that vacuum is a vastly superior transport medium compared to solids. The electron velocity in vacuum can approach the speed of light (3×10^{10} cm/s), while the saturation velocity in solid-state devices is typically limited to 10^7 cm/s by collisions with optical and acoustic phonons. An electron moves ballistically in vacuum, while in solids the electronic

TABLE 1.1. Characteristics of Solid-State and VME Devices

Properties	Solid-State Devices	VME Devices
Current density	10^4–10^5 A/cm^2	$\sim 2 \times 10^3$ A/cm^2
Voltage	>0.1 V	>10 V
Structure	Solid/solid interface	Solid/vacuum
Electron transport		
Medium	Solid	Vacuum
Ballistic	<0.1 μm, Low temp.	100% Ballistic
Coherence	Length <0.1 μm	Length $\gg$ 0.1 μm
	$t < 10^{-13}$ s at RT	$t \gg 10^{-13}$ s
Lens effect	Difficult	Easy
Noise		
Thermal noise	Random motion of carriers	Comparable
Flicker noise	Surface/interface effects	Worse
Shot noise	Fluctuation in generation/ recombination rates of carriers	Comparable
Electron energy	<0.3 eV	Several to 1000 eV
Cutoff frequency	<20 GHz (Si) <100 GHz (GaAs)	<100–500 GHz
Power	Small	Large
Radiation hardness	Poor	Excellent
Temperature sensitivity	$-30 - +50°$C	<500°C
Fabrication/materials	Well established (Si), established (GaAs)	Not well established
Applications	Microprocessors, memory devices, optoelectronic devices, rf devices	Flat panel displays, microwave power tubes, electron/ion sources, e-beam lithography, e-beam memories, and excitation devices

ballistic motion is rarely sustained for more than 0.1 μm, even in a two-dimensional quantum state at very low temperatures. Similarly, the coherence of electron waves in vacuum can be preserved for a long distance, while in solids it can not be maintained further than 0.1 μm. Therefore, various forms of electron-beam devices can be built by taking advantage of vacuum as the electron transport medium, and electro-optical components such as lenses, reflectors, and deflectors can also be made.

However, the largest current density that is ultimately achievable in FEAs is limited by space charge effects and other structural parameters, and can be estimated at about 2200 A/cm^2. In contrast, current densities in a solid-state device can be as high as 10^4–10^5 A/cm^2, because the equal numbers of electrons and ions in a solid-state device can neutralize the negative charge of the electrons. Vacuum devices are based on only one type of charged carrier, either negative or positive, while solid-state devices can have two carriers of opposite charges, electrons and holes. This prevents vacuum microelectronics from making energy-efficient CMOS-type devices.

For a vacuum triode, the maximum current density of FEAs will limit the cutoff frequency to less than 20 GHz based on a simple unity-current-gain frequency calculation. This limitation is due to a low transconductance that, in turn, is limited by both the capacitance and the space charge current density. The highest speed at which solid-state devices of Si and GaAs can be operated at is about 20 GHz and 100 GHz, respectively. However, if FEAs are incorporated into linear beam devices such as klystrons, traveling wave tubes (TWT), and backward wave oscillator (BWO) tubes instead of the simple triode structure, the operating frequency of vacuum microelectronic devices can be increased to 100–500 GHz.

As far as current noise is concerned, among the three main types of noise, thermal noise, shot noise, and flicker noise, the first two types of noise are similar for both vacuum and solid-state devices. However, flicker noise is usually much greater in vacuum devices, because it originates from surface and interface effects. Under normal FEA operating conditions, it is difficult to maintain atomic stability at vacuum/solid interfaces because of complicated physical or chemical reactions such as gas adsorption, impurity migration, and ion impact. In contrast, the sources of flicker noise in solids are impurities and defects, which are relatively few and are contained inside the crystals. Fortunately, the noise from FEAs can be statistically reduced when the number of tips is greater than $\sim$1000. An extremely clean operating environment or the use of some highly stable emitting materials [61] will also produce much lower noise.

It is important to understand that different applications call for different schemes of optimization for device characteristics. For flat panel display applications, the current density rarely needs to exceed more than 0.01 A/cm^2. Instead, it is more essential to find a low cost solution to uniformly produce large area FEAs that can be operated at low voltages. In this regard, the recent trend of exploring new cathode materials such as carbon-based emitters is encouraging. It is conceivable that FEDs, owning to their inherent advantages of brighter image, wider viewing angle, less energy consumption, and higher resolution, will someday be able to compete effectively with liquid crystal displays and other technologies in the flat panel display market. It is also probably vital for the future of this technology that researchers look beyond the current display and microwave devices and start to think about next generation, functional "beam" devices for memory, switching, and other applications. Only by doing so, we will be able to allow the unique capability of vacuum microelectronics in creating "beam" devices to complement solid-state technology in bringing further advanced products to the marketplace.

Dedication

The field of vacuum microelectronics has made great strides since the 1st IVMC conference was organized by the late Henry Gray in 1988. He had devoted his entire energy to promote this new branch of science and technology throughout the world. He strongly believed in international cooperation for advancing the field and always encouraged newcomers, no matter young, old, native, or foreign, to do research in this area. Beside his own technical contributions, he played a pivotal role in organizing and building the vacuum microelectronics community. He died on July 21, 1999 at the age of 63. As a close friend, I dedicate this chapter to him.

REFERENCES

1. J. Bardeen and W. H. Brattain, *Phys. Rev.* **74**, 230 (1948).
2. R. N. Noyce, US Patent 2,981,877 (1959).
3. J. S. Kilby, *IEEE Trans. Electron Devices* **ED-23**, 648 (1976).
4. J. W. Gewartowski and H. A. Watson, *Principles of Electron Tubes*, D. Van Norstrand, Co.: Princeton, NJ, p. 229, 1965.
5. K. R. Shoulders, *Adv. Comp.* **2**, 135 (1961).
6. G. E. Moore, *Proc. SPIE* **2437**, 2 (1995).
7. I. Langmuir, *Trans. Faraday Soc.* **17**, 1 (1921).
8. C. J. Davisson and L. H. Germer, *Phys. Rev.* **30**, 705 (1927).
9. R. K. Gehrenbeck, Electron diffraction: Fifty years ago, *Physics Today*, p. 34, January 1978.
10. R. H. Fowler and L. W. Nordheim, *Proc. Roy. Soc. (London), A* **119**, 173 (1928).
11. L. Nordheim, *Proc. Roy. Soc. (London), A* **121**, 626 (1928).
12. W. P. Dyke, J. K. Trolan, E. E. Martin, and J. P. Barbour, *Phys. Rev.* **91**, 1043 (1953).
13. L. H. Germer and W. S. Boyle, *J. Appl. Phys.* **27**, 32 (1956).
14. I. Brodie, *J. Appl. Phys.* **35**, 2324 (1964).
15. T. Utsumi, *J. Appl. Phys.* **38**, 2989 (1967).
16. E. W. Müller, *Ergeb. Exakt. Naturwiss* **27**, 290 (1953).
17. R. H. Good and E. W. Müller, *Handbuch der Physik* **21**, 178 (1956).
18. W. P. Dyke and W. W. Dolan, *Advances in Electronics and Electron Physics*, Vol. 8, Academic Press: New York, p. 89, 1956.
19. R. Gomer, *Field Emission and Field Ionization*, American Institute of Physics: New York, 1993.

20. E. W. Müller, *Phys. Rev.* **102**, 618 (1956).

21. G. Binning, H. Rohrer, C. Gerber, and E. Weibel, *Phys. Rev. Lett.* **50**, 120 (1983).

22. C. Spindt, *J. Appl. Phys.* **39**, 3504 (1968).

23. C. Spindt, I. Brodie, L. Humphrey, and E. R. Westerberg, *J. Appl. Phys.* **47**, 5248 (1976).

24. R. Meyer, A. Ghis, P. Ramband, and F. Muller, Development of matrix array of cathode emitters on a glass substrate for flat display applications, in *Proc. 1st IVMC*, Williamsburg, VA, p. 10, 1988,

25. I. Brodie and C. A. Spindt, *Appl. Surf. Sci.* **2**, 149 (1979).

26. I. Brodie and C. A. Spindt, Vacuum microelectronics, in *Advances in Electronics and Electron Physics, Vol. 83* (P. W. Hawkes, Ed.), Academic Press: New York, p. 1, 1992.

27. H. F. Gray, G. J. Campisi, and R. F. Greene, A vacuum field effect transistor using silicon field emission arrays, in *Technical Digest of IEDM*, Washington, D.C. p. 776, 1986.

28. R. N. Thomas, R. A. Wickstrum, D. K. Schroder, and H. C. Nathanson, *Solid State Electron* **17**, 155 (1974).

29. A. M. E. Hoeberechts, Field emission devices, US Patent 4,095,133 (1978).

30. J. R. Anthur, Jr., *J. Appl. Phys.* **36**, 3221 (1965).

31. J. L. Shaw and H. F. Gray, Emission limiting in silicon field emitter, in *Proc. 11th IVMC*, Asheville, N.C., p. 146, 1998.

32. H. S. Uh, B. G. Park, and J. D. Lee, Formation of Mo silicide on poly-Si field emitter arrays, in *Proc. 10th IVMC*, Kingju, Korea, p. 371, 1997.

33. T. Utsumi, *IEEE Trans. Electron Devices* **38**, 2276 (1991).

34. G. N. A. van Veen, *J. Vac. Sci. Technol. B* **12**, 655 (1994).

35. C. A. Spindt, C. E. Holland, and R. D. Stowell, *Appl. Surf. Sci.* **16**, 268 (1983).

36. C. O. Bozler, C. T. Harris, S. Rabe, D. D. Rathman, M. A. Hollis, and H. I. Smith, *J. Vac. Sci. Technol. B* **12**, 629 (1994).

37. P. R. Schowebel, C. A. Spindt, I. Brodie, and C. E. Holland, in *Proc. 7th IVMC*, Grenoble, France, p. 378, 1994.

38. C. A. Spindt, I. Brodie, L. Humphrey, and E. R. Westerberg, *J. Appl. Phys.* **47**, 5248 (1976).

39. W. D. Goodhue, P. M. Nitishin, C. T. Harris, C. O. Bozler, D. D. Rathman, G. D. Johnson, and M. A. Hollis, *J. Vac. Sci. Technol. B* **12**, 693 (1994).

40. J. L. Grand-Clement, Technology, applications and market analysis for field emission displays (FEDs), in *Proc. 6th IVMC*, Newport, RI, p. 3, 1993.

41. A. Ghis, R. Meyer, P. Rambaud, F. Levy, and T. Leroux, *IEEE Trans. Electron Devices* **38**, 2320 (1991).

42. F. J. Himpsel, J. A. Knapp, J. A. Van Vechten, and D. E. Eastman, *Phys. Rev. B* **20**, 624 (1979).

43. B. V. Spitsyn, L. L. Bouilov, and B. V. Derjaguin, *J. Cryst. Growth* **52**, 219 (1981).

44. S. Matsumoto, Y. Sato, M. Kame, and N. Setaka, *Jpn. J. Appl. Phys.* **21**, L183 (1982).

45. S. Iijima, *Nature* **354**, 56 (1992).

46. K. A. Dean and B. R. Chalamala, *Appl. Phys. Lett.* **76**, 375 (2000).

47. W. Zhu, C. Bower, O. Zhou, G. Kochanski, and S. Jin, *Appl. Phys. Lett.* **75**, 873 (1999).

48. A. G. Rinzler, J. H. Hafner, P. Nikolaev, L. Lou, S. G. Kim, D. Tomanek, P. Nordlander, D. T. Colbert, and R. E. Smalley, *Nature* **269**, 1550 (1995).

49. S. Kaneman, K. Ozawa, K. Ehara, T. Hirano, and J. Itch, MOSFET-structured Si field emitter tip, in *Proc. 10th IVMC*, Kingju, Korea, p. 34, 1997.

50. J. Itoh, T. Hirano, and S. Kanemaru, *Appl. Phys. Lett.* **69**, 1577 (1996).

51. A. Nagashima, N. Tejima, Y. Jamou, T. Kawai, and C. Oshima, *Surf. Sci.* **357 & 358**, 307 (1996).

52. M. Terai, N. Hasegawa, M. Okusawa, S. Otani, and C. Oshima, *Appl. Surf. Sci.* **130–132**, 876 (1998).

53. K. Nagaoka and C. Oshima, *Jpn. Surf. Sci.* **17**, 738 (1996).

54. K. Nagaoka, T. Yamashita, S. Uchiyama, M. Yamada, H. Fujii, and C. Oshima, *Nature* **396**, 557 (1998).

55. N. Koshida, T. Okazaki, X. Sheng, and H. Koyama, *Jpn. J. Appl. Phys.* **34**, 705 (1995).

56. T. Komoda, X. Sheng, and N. Koshida, *J. Vac. Sci. Technol. B* **17**, 1076 (1999).

57. H. Makishima, H. Imura, M. Takahashi, H. Fukui, and A. Okamoto, Remarkable improvements of microwave electron tubes through the development of the cathode materials, in *Proc. 10th IVMC*, Kingju, Korea, p. 194, 1997.

58. C. P. Spindt, C. E. Holland, P. R. Schwoebel, and I. Brodie, Field-emitter-array development for microwave application (II), in *Proc. 10th IVMC*, Kingju, Korea, p. 200, 1997.

59. S. G. Bandy, C. A. Spindt, M. A. Hollis, W. D. Palmer, B. Goplen, and E. G. Wintucky, Application of gated field emitter arrays in microwave amplifier tubes, in *Proc. 10th IVMC*, Kingju, Korea, p. 132, 1997.

60. Y. V. Gulyaev and N. I. Sinitsyn, *IEEE Trans. Electron Devices* **36**, 2742 (1989).

61. J. Brodie, Fluctuation phenomena in field emission from molybdenum, in *Proc. 2nd IVMC*, Bath, England, p. 89, 1989.

CHAPTER 2

Technological Overview

GREGORY P. KOCHANSKI AND WEI ZHU

Bell Laboratories, Lucent Technologies, 600 Mountain Avenue, Murray Hill, New Jersey 07974-0636

YEHUDA GOREN

Teledyne Electronic Technologies, 11361 Sunrise Park Drive, Rancho Cordova, California 95742-6587

2.1. INTRODUCTION

Since the 1960s, there has been a substantial interest in field emitters and vacuum microelectronics (VME) that, paradoxically, has been stimulated by the growth and evolution of solid-state technologies. Robust semiconductor markets have paid for technologies that can be applied to field emitters and vacuum devices, and have created demands for other components, such as displays, that are hard to make from silicon.

One might think that VME should rival solid-state technology because vacuum, as a medium, offers unique and unsurpassed characteristics. For instance, the limiting carrier velocity in vacuum is the speed of light, which is far better than those in silicon (Si) or gallium arsenide (GaAs). Similarly, carrier mobility in vacuum is virtually infinite, higher than in any semiconductor. However, the reality is that vacuum technology, after more than 30 years of development, is just barely off the ground, while solid-state technology dominates the entire landscape. Why? We will see that advantages and disadvantages of vacuum devices are strongly application dependent, and VME is suitable for a few interesting areas. We will look at this technology from three perspectives:

1. What is the best plausible performance?
2. What are the problems that limit the technology?
3. What are potential solutions to these problems?

We will avoid discussing cost, despite its obvious importance to technical success, because it depends so much on factors outside the scope of this book: plant investment, maturity of processing tools, and details of design and manufacturing processes.

Vacuum Microelectronics, Edited by Wei Zhu.
ISBN 0-471-32244-X. 2001 John Wiley & Sons, Inc.

2.2. PROMISE AND REALITY

Let's consider the switching speed of a device as an example. In a silicon field effect transistor (FET), this is commonly limited by the transit time of the charge carriers between the source and drain. The saturation velocity for electrons is about 3×10^7 cm/s in silicon because of optical and acoustic phonon scattering. In vacuum, however, the carrier velocity can reach 3×10^{10} cm/s before saturation. That is why even a manually assembled triode vacuum tube built in the early 1950s with large mechanical tolerances could operate at a frequency of 4 GHz [1].

Of course, the above comparison is over simplified and application dependent. It takes energy to accelerate the electrons, and that energy is limited by the applied voltages. Those applied voltages are decreasing, as silicon technology advances and device dimensions shrink. If a chip runs on 1.5 V, the electrons won't get the energy to accelerate above 7.3×10^7 cm/s, no matter how perfect the medium in which they travel. The limiting velocity in vacuum is then just about twice the limit in silicon, which would reduce transit times by about 25%. Even if all emitter problems were solved, one would get only a moderate increase in device performance. We see here that, even under the most generous assumptions, VME is unlikely to prosper in high-density applications like processors and memory.

High power devices are a different story. The applied voltages are intentionally higher, so there is now enough energy to accelerate electrons in vacuum to velocities much higher than the saturation velocity in any semiconductor. These faster electrons will reduce the transit time between the emitter and gate, thus allowing higher frequency operation. Further, the absence of electron–phonon scattering in vacuum leads to less energy dissipation and higher efficiency. Overall, one could expect substantially better device performance.

As a second example, it is often claimed that vacuum devices can better withstand high temperatures, because they don't suffer from leakage current that comes from electrons excited across the semiconductor band gap. This leakage increases rapidly with temperature and effectively prevents useful operation of silicon devices above 200°C. However, real devices are not that simple.

A VME device has components other than vacuum and emitters: it has walls, insulators, and interfaces. Many things happen as the temperature increases. Gases that are physisorbed or chemisorbed on the walls can desorb into the vacuum space. The gas molecules can become ionized, eventually eroding the emitter structure. Likewise, atoms diffuse more readily at high temperatures. VME devices tend to be more affected by diffusion than do solid state devices for two reasons: (1) They depend on atomically sharp tips to concentrate the electric field over their emitters, and just a small amount of diffusion can modify and destroy the delicate tip structure, and (2) unlike semiconductor devices, they have solid/vacuum interfaces, and diffusion is faster on such interfaces than in the bulk because surface atoms have fewer bonds and are, therefore, more easily moved by thermal fluctuations. At sufficiently high temperatures, these processes can easily make the device unusable. The actual performance of a VME device at elevated temperatures will be basically an experimental question, controlled by defects and contamination as well as details of the design. One simply cannot make sweeping statements about such a performance comparison.

On the positive side, vacuum devices are extremely radiation-resistant, partly because the active medium is vacuum, and unlike semiconductors, vacuum cannot be directly damaged by particle bombardment. Radiation-induced defects in the solid parts of the devices also turn out to be largely irrelevant. For instance, a collision that would dislodge an atom out of its position to make a vacancy-interstitial pair would be serious in the crystalline silicon of a FET. The defect would create states deep in the band gap that would scatter and trap electrons and substantially degrade the device properties. A similar collision occurring in a metal or an insulator in a VME device would have little effect. Most practical insulators are amorphous anyhow and have many states deep in the gap already, so an extra defect has little practical consequence.

Building useful field emitters requires producing uniform tips with a diameter of about 50 nm or less. At first glance, this requires lithographic details smaller and more precise than anything that is commercially feasible. This requirement makes field emitter arrays seem completely impractical or prohibitively expensive. Yet, while the uniformity of field emitters is still a major concern, there are techniques that can make reasonably uniform emitters that are considerably smaller than conventional lithography. Spindt emitters (see Chapter 4), silicon tips sharpened by etching or oxidation (see Chapter 5), and various edge emitters (see Chapters 4 and 7) all make the construction of FEAs possible by taking advantage of favorable material properties or growth conditions to produce extremely sharp tips. The recent advance in nanotechnology also offers the possibility of building structures with atomic and molecular precisions. In particular, the discovery of carbon nanotubes (see Chapter 6) presents a promising route to produce atomically identical nanometer-sized tips, potentially solving one of the most difficult problems in the field.

2.3. CASE STUDIES

We will highlight the problems and potentials of field emitters and VME by discussing two examples: field emission flat panel displays and microwave power amplifier tubes. There are other applications, such as lithography [2], microscopy [3], and data storage [4], but displays and microwave tubes seem most likely to have substantial commercial impact, and they also provide a way to discuss technical challenges in the context of real devices. Both are in direct competition with existing technologies, and we will compare these with alternative technologies where possible.

2.3.1. Flat Panel Displays

Currently, the hottest application for field emitters is in building flat panel displays. A field emission display (FED) is an array of a million or more electron guns, each exciting a single phosphor pixel. The electron guns are typically triodes, each with a gate and cathode, and they are addressed in a square matrix, where the gate electrodes form the rows and the cathodes the columns of the matrix. The electron guns are approximately a millimeter from the phosphor screen (the anode). An FED also contains ceramic spacers to keep atmospheric pressure from crushing the anode down onto the cathode, along with getters for maintaining a clean vacuum, and external driver electronics.

Several companies have demonstrated displays or even reached production [5]. The potential market for flat panel displays is large, as displays are often the largest single cost in many consumer electronic devices. Better displays can add a lot of value to computers and especially portable electronics.

We will see that the biggest technical challenges in building displays are not directly connected with the field emitters, but instead are related to the pillars that separate the anode from the cathode and cleanliness issues. Field emitters are not going to be made in garage shops. The cleanliness requirements for building a real device are stringent, primarily to minimize outgassing, and, therefore, cathode contamination.

2.3.1.1. FED Advantages. The largest technical advantage of FEDs is that they are more efficient than their competitors. Overall, FEDs end up using perhaps half of the power of the currently dominant technology: a color active matrix liquid crystal display (AMLCD). This is a significant selling point for any battery powered device, because it immediately doubles the battery life or reduces the device's weight. The weight reduction can be substantial, as the battery is often 20–50% of the weight of a portable device, and the display is often where most of the energy goes. At a constant battery lifetime, most devices would become noticeably lighter.

Color LCDs consist of a lamp and an array of light valves. The lamp is typically a normal fluorescent tube, and it converts electrical energy into photons with an efficiency that is comparable to the electron-excited phosphor screen in FEDs. However, the LCD light valve reduces the output intensity by a factor of 10 or more for color active matrix displays. Light is lost in LCDs because polarizers absorb 50% of the incident light, color filters absorb or waste roughly 2/3 of the light, and transistors and wiring block up to 50% of the light. In addition, of course, power is wasted on any pixels that are turned off.

In an FED, the light comes almost directly out. The phosphors are typically a white powder, so a bare phosphor screen would reflect far too much room light. One needs to put an absorptive filter in front of the phosphor screen to reduce the reflected light. Normal practice on cathode ray tubes (CRTs) is to use a neutral density filter that reduces the transmitted light by about a factor of five [6], although improved performance has been achieved by reducing the absorption at wavelengths that the phosphors emit and increasing it elsewhere. For instance, the front glass would have three transmission bands, near the peaks of the red, blue, and green phosphor emissions. With this type of customized filters, losses can be smaller, approaching an attenuation factor of two. One can presumably optimize the filters further, because each pixel could have its own filter, tuned to the color of the phosphor on that specific pixel [7].

FEDs also have a faster response time than LCDs, even AMLCDs. In an LCD, it takes more than 30 ms for the liquid crystal to reorient in response to a change in the applied electric field, and the light output changes along with the molecular orientation. This response time is noticeable and can smear rapidly changing videos. FEDs, on the other hand, can be built with refresh times less than 20 ms, and should be able to respond as well to video as a high-quality CRT.

2.3.1.2. Phosphor Efficiency. Actually meeting our assumption that the phosphor efficiency in the FEDs matches the efficiency of the fluorescent tube inside the LCDs entails solving some difficult problems. Specifically, the display must be designed to operate with 5 kV or more between the anode and emitter electrodes. Such high operating voltages are needed because phosphor efficiencies increase with voltage, as does the operating life [8]. While high voltage operation is necessary to achieve good efficiency, the design must become more complicated to prevent arcing, and a focusing electrode may become necessary as the high voltage forces the anode to be further from the emitters. At low voltages (<1 kV), too large a fraction of the electron beam's energy is lost outside the phosphor (e.g., in an aluminum film deposited on the phosphor). Also, at low voltages, most of the electron–hole pairs in the phosphor are created close to the particle's surface, and they are likely to recombine on some surface defects without emitting any light. On the other hand, electron–hole pairs that are created deep in the bulk of the phosphor particle by a high energy incident electron are unlikely to encounter the surface and will probably recombine radiatively on some intentionally introduced dopant atoms. The overall result is fewer photons per joule when the beam's energy is insufficient to penetrate to the center of the phosphor particles.

An additional problem at low voltages is that the electrons can stimulate desorption of atoms at the surface of phosphor particles. With a low voltage beam, the electrons will deposit the beam power near the surface. Worse yet, higher current densities are necessary to achieve the same brightness as a high-voltage beam. As a result, surface damage at the phosphor particles can be substantial. In ZnS-based phosphors, this process can lead to darkening of the phosphor, presumably due to the accumulation of defects and excess zinc at the grain surface. Any desorbed atoms cause lifetime problems for the displays by spoiling the vacuum and contaminating the emitters. This process can be controlled by proper surface treatments of the phosphors, e.g., coating the phosphors with a thin layer of oxide [9], which minimizes damage to the phosphor surface and prevents sulfur desorption.

2.3.1.3. Pillar Challenges. A display needs pillars that hold the anode and cathode apart against atmospheric pressure. Pillars are also necessary to ensure that the cathode–anode gap is stable if the front of the display is touched. These pillars are probably the worst manufacturing problem in a display (because they provide surfaces which connect the anode and the emitter), and electrical breakdown along a surface is far easier, especially in vacuum, than through the solid.

If we consider an insulator in vacuum and throw in a few electrons, the insulator's surface will generally become charged. Contrary to expectations, the sign of the charge is often positive, because incoming electrons can knock electrons off the insulator, a process known as secondary emission. If, on average, there is more than one outgoing electron per incoming electron, the insulator will pick up a positive charge. The positively charged surface can then attract more electrons. This process doesn't run away when we consider an isolated object, because the positive charge will eventually become large enough to prevent the secondary electrons from leaving, and the system will reach a steady state.

However, if we put the insulator between electrodes and establish a continuous voltage gradient across the insulator, the secondary electrons can always hop downhill toward the anode. This can lead to a runaway process where the insulator becomes positively charged, which makes the voltage gradients near the negative electrode (i.e., the cathode) stronger. The stronger gradients attract more electrons to the insulator and can also enhance field emission from the cathode, leading to further increases in the positive charge on the insulator. The process can lead to the formation of an arc along the surface of the pillar long before it would breakdown through the bulk [10,11]. Even before it causes an arc, the charge on the pillars can deflect the electrons in their vicinity and make the pillars optically more obvious to the viewer, especially in displays that have widely spaced pillars.

In general, to reduce charging, the pillars should be designed so that materials that emit relatively few secondary electrons are exposed to bombardment. An ideal pillar design would have exactly one electron leaving the vicinity of each incident electron impact. Designs with fewer than one secondary electron per primary are the next best thing, as the runaway process that leads to an arc does not occur: when the surface charges negatively, it repels incoming electrons, inhibiting further charging. However, material choices alone cannot provide a complete solution, because the pillars need to be extremely resistive, and most suitable materials (i.e., insulators or semiconductors) emit many secondary electrons per primary. Over a broad range of voltages that are unfortunately relevant to field emission display operation (e.g., 500–5000 V), most suitable pillar materials typically emit 2–5 secondary electrons per incident electron.

The charging problem can be tackled by several techniques, e.g., controlled conductivity of the pillars to prevent charge buildup [12] or control of pillar geometry and secondary electron emission coefficients [13–15]. In a practical system, the secondary electron emission will never be exactly 1 for 1, so it becomes necessary to make the pillars slightly conductive to allow excess charge to flow away. This conductivity needs to be carefully controlled, and it needs to be as small as possible, because any conductivity through the pillars is a direct drain on the anode power supply, which reduces the display's efficiency. In order to prevent charging effects, the current flowing through the pillars from the power supply needs to be comparable to the current of primary electrons that hits the pillar minus the current of secondaries that escape. Consequently, care needs to be taken to minimize the number of electrons that hit the pillars and to choose materials that minimize charging, otherwise the power wasted in the pillars could be a substantial fraction of the total display power.

A rough-surfaced pillar will also help to minimize the current that must flow through the pillar. If most of the primary electrons impact deep in trenches or holes, then many of their secondary electrons will not be able to escape. They will hit the sides on the way out at an energy that is typically too low for them to generate secondaries of their own. If the secondary electrons can be trapped locally (within a few microns), then they don't contribute to the charge redistribution on the scale of the electrode spacing that leads to breakdown. Given only the smallest amount of conductivity in the pillar material, the secondaries will then be able to recombine with the positive charges left at the primary impact site.

Pillars are subject to energetic electron impacts, and this adds some further complications to the design. The materials must be chosen to minimize outgassing under electron bombardment, and they also must be chosen to maintain (or at least develop) stable surface properties. It does no good to control the surface conductivity or the secondary emission at the point when the device leaves the factory, unless one can be assured that the properties will remain in acceptable ranges over the device's lifetime.

Even with pillar charging controlled, electrical breakdown of the display is still a possibility. Arcs can be initiated by runaway spike growth on an emitter, or by a loose particle that becomes charged and rattles around inside the display. Because of this possibility, the display needs to be designed so that small arcs remain localized, and do only local damage. Electronic solutions to the problem have been explored by the authors, based on the realization that an arc will generate a puff of gas that will travel at thermal velocities, about 1000 m/s. This gas puff (and the entrained bits and droplets of electrode materials) can cause damage on nearby pixels via ion bombardment, heating, and contamination, but the gas puff takes approximately 100 ns to arrive at the neighboring pixels. This is enough time for the driver electronics to shut down emission on nearby pixels. Careful design of electrodes and interelectrode capacitances can also passively reduce electron emission from pixels near an arc. With electron emission shut down, other pixels are less likely to be damaged by the gas puff and debris from the arc.

2.3.1.4. Cathode Issues

2.3.1.4.1. Uniformity. For flat panel displays, the current density requirement from a field emitter cathode is quite modest. Approximately 1–10 mA/cm^2 is sufficient to excite high voltage phosphors. The main problems are uniformity and lifetime. In an ideal world, we could build individual field emitters to perform identically, but the physics of field emission seems to forbid this brute-force approach to uniformity. In field emission, the current depends very sensitively on the electric field at the surface of the emitters. Electrons escape from the emitter via a quantum mechanical tunneling process, and the rate of escape depends exponentially on the barrier width and height [16]. The dependence on electric field is so strong (often doubling current for a 10% increase in electric field) that a single atom can dominate the current from a field emitter, and a single emitter can dominate the current from a device. The resulting emission nonuniformity degrades the display resolution and compromises the picture quality. In addition, the few dominant, "hot" emitters are vulnerable to destruction due to ion bombardment, leading to arcing and catastrophic failure of the device.

How, then, can one hope to build a practical device, when the properties of individual emitters, as manufactured, can vary by orders of magnitude? First of all, one depends on the statistics of large numbers of emitters. Field emitters are small (nanometers to a few microns) and cheap, so one can afford to put thousands to millions on a single pixel of a display. It seems to require at least 10^6 active emission sites per square centimeter to achieve high resolutions (>70 dpi).

Second, one can use emitters, such as single-wall carbon nanotubes, that are atomically identical. If all the nanotubes were separate, and if their length could be controlled, then uniformity would be enforced by quantum mechanics. In practice, single-wall nanotubes tend to form bundles, and the details of bundle geometry, and especially bundle terminations, are hard to control. However, one is at least removing one source of nonuniformity by using identical emitters.

Third, one can use a controlled aging process that selectively erodes the best few emitters. This tends to happen naturally, as the best few emitters suffer most from ion bombardment due to the presence of more energetic electrons and, therefore, more ions in their vicinity. Better emitters will also tend to have a larger density of adsorbates near their tips because mobile adsorbates are attracted to high-field regions. So if chemical reactions that erode the tips are possible, they will also happen more on the emitters with the highest fields and the highest electron densities. FEAs tend to become more stable [17], and presumably more uniform as they age. Unfortunately, since this process works by damaging the best emitters, one ends up with a uniform array of mediocre emitters. Figures of merit such as the electric field required to reach a useful current density (e.g., 10 mA/cm^2) always decline during the aging process.

Fourth, one can add a resistance in series with the emitter [18,19]. The resistance converts very steep exponentially rising current from the emitter into an effective system $I\text{--}V$ characteristic that is not so steep and can be controlled primarily by the resistor's value. While this technique works well and is valuable, adding the resistance can substantially raise the required driver voltage when the resistance needs to be large enough to cover up severe nonuniformities. For the best emitters, more of the gate–emitter voltage can be dropped across the resistor than across the vacuum gap. The technique of using a series capacitance [20] will also raise the required driver voltage, but avoids the manufacturing difficulties of producing the precisely matched, large value resistors that are required for uniform emission. A series impedance can also limit the current prior to runaway and subsequent failure due to arcing. Unfortunately, the higher driver voltages that are needed will increase the cost of the driver chips, which is already a substantial part of the total cost of the display module.

Fifth, there is a possibility that field emitters can be operated in a space-charge limited regime. Ideally, this provides a natural feedback loop on the emission rate. As more electrons are emitted, they form a cloud of negative charge in front of the emitters. This negative charge then reduces the field at the emitter surface and returns the emission rate to its nominal value. This is not a hard current limit, but the current will increase relatively slowly, with the voltage dependence changing from a steep exponential to a shallow $V^{3/2}$ dependence [21,22]. However, direct operation of a space-charge limited field emitter requires large currents because the electric fields are large near the emitter tip. At the required fields, electromigration in the emitter as well as Joule heating and the electrostatic stress near the tip are all close to values at which the emitter will self-destruct. We note that it might be possible to use a multiple layer gate structure to decelerate electrons into a low field/low velocity region where the charge density might become large enough at a practical level of current per emitter to provide a space charge limit to the emission current. However, we are not aware of any such designs or models, much less experiments.

And, finally, one has to keep the device clean and free of adsorbates. Unlike a hot cathode that keeps itself clean by thermally evaporating off contaminants [23], field emitters are cold, and the adsorbates remain on the emitters. Even small amounts of contamination can have a dramatic effect on the emitter performance. For example, in his absorption studies with the field emission and field ion microscope, Holscher found that, by adding CO to a 30 nm radius clean tungsten emitter operating at 400 V, the current dropped from 40 nA to 30 pA upon addition of one monolayer [24]. To put this effect in context, a 10^{-9} torr vacuum will deposit a monolayer of adsorbate in 20 min (although there may be processes that remove adsorbates from the tips, such as field ionization or sputtering). Electronegative atoms like sulfur and oxygen [25] should generally reduce emission, but there are other adsorbates that can cause substantial increase in current [21,26–28]. In any event, any difference in adsorbate density will then translate into nonuniformity in emission current.

Sulfur is an interesting case, because it is expected to be released by electron-stimulated decomposition of ZnS-based phosphors, which are the most efficient and desirable phosphors available. One of the authors (Kochanski, with R. F. Bell) has measured Zn and S desorption from P2 phosphor (ZnS:Cu) in UHV with an electron beam to mimic a display (2 kV, 0.5 mA/cm^2). They measured an upper limit to the partial pressure of sulfur at 10^{-12} torr, and an upper limit of 2×10^{-14} torr for Zn or S$_2$. These pressures imply that a display in operation will take at least 300 h to deposit a monolayer of S on the emitter surface. As a result, there seems to be no strong reason to avoid ZnS-based phosphors in FEDs. In a practical example, researchers at Candescent Technologies reported stable, reliable operation in the presence of sulfur from ZnS-based phosphors [29].

Of course, there is nothing wrong with an adsorbate-covered tip, so long as the adsorbate doesn't corrode the underlying structure and is itself stable. It may be possible to design a display so that the emitter surface is intentionally saturated with a particular, tightly bonded adsorbate. However, it is not at all clear how one could, in practice, design getters and materials to pump away all but the preferred adsorbate, maintain an adequate vacuum, and control the amount of the preferred adsorbate, all at the same time.

2.3.1.4.2. Lifetime. Emitter lifetime is primarily limited by three mechanisms: (1) ion bombardment and sputtering, (2) surface chemistry, and (3) contamination. In any but a perfect vacuum, the electrons from field emitters have a chance of ionizing atoms of the residual gas inside the device. These ions will be accelerated back towards the emitters, and some small fraction will strike the active sites of field emitters and damage them [30,31]. The probability of an ionizing electron–ion collision can be calculated by the Bethe–Bohm approximation [32], which gives a probability that peaks at a few tens of electron volts, and gradually decreases at higher energies. On the other hand, the sputtering yield, which is the number of atoms that the incoming ion will remove from the emitter, increases linearly with the ion energy [33]. The product of the two is roughly independent of energy at all interesting energies (above 100 eV).

While an accurate calculation requires consideration of the focusing effects of the field geometry near both the emitter and the gate, a crude estimate of the maximum allowed pressure at which ion bombardment damage becomes important can be obtained by simply considering a uniform array of emitters in a diode configuration. We find that the damage rate, D (in monolayers per hour), the rate at which the topmost layer of atoms in the emitter is removed, is

$$D \approx 2 \times 10^{-3} \left(\frac{P}{10^{-10}\,\text{torr}} \right) \left(\frac{j}{1\,\text{A cm}^{-2}} \right) \left(\frac{h}{100\,\mu\text{m}} \right)$$

where P is the gas pressure, j is the current density, and h is the anode–cathode distance. The prefactor is substantially smaller for hydrogen because of its low sputtering coefficient.

While different types of emitters can survive different amounts of damage, we can get a rough estimate of the required vacuum fairly easily. Assuming that the emitters will be degraded after 10 monolayers are sputtered away, the device pressure required to get the lifetime of 5000 h must be near 10^{-8} torr for a flat panel display and near 10^{-11} torr for a microwave amplifier. Both of these numbers are very clean vacuua, especially for a sealed system without active pumps. Noble gases and methane count toward this total, but are not significantly pumped by any getter material, so they are especially troublesome. They are removed only when they are ionized, accelerated, and buried in the gate or emitter. The system must be intrinsically very clean and designed not to evolve gases under electron and ion bombardment. It must be well-baked in order to obtain a commercially useful lifetime.

Ion bombardment can also damage or add permanent charge to insulators near the emitters. It has been suggested that ion bombardment can embed enough charge in unprotected insulators to cause breakdown and arcing between row and column lines (i.e., gates and cathodes) [5].

The energetic electron- and ion-driven chemistry inside a sealed operating display can also erode or corrode the emitter surface, which will degrade any special properties it may have. As a result, any emitter that depends on a specially prepared but reactive surface is bound to fail in a real display environment. The use of low work function materials is a good example, because most are reactive and exhibit a low work function only when they are clean and chemically intact. Even the relatively stable hydrogen-terminated (111) diamond surface will eventually lose its surface property due to ion impacts and hydrogen desorption. Diamond is an interesting material, because if it could be doped n-type, it should be a material with a negative electron affinity that would emit electrons at very low applied fields and without the steep exponential I–V characteristic that causes uniformity problems with all other practical field emitters. However, the negative electron affinity depends on the surface dipoles established by the hydrogen atoms, and it has been experimentally shown that ion bombardment would spoil the desirable surface property by removing the hydrogen atoms on the surface [34].

There are other mechanisms for device degradation, beyond sputtering and chemical damage to the emitters, although, like sputtering and chemical damage, most are problems caused by contamination. Once contaminants are attached to the emitters, electron-excited chemistry is possible, as are reactions between adsorbates. Certain adsorbates could react with the tip, displacing tip atoms and leading to corrosion. Other adsorbates might be attracted to the high-field region at the tip but might not desorb at any significant rate, so they might just pile up there, leading to changes even in the gross tip geometry. A runaway process is quite imaginable, where a spike of material (adsorbate or tip material) starts growing at the end of the emitter. As it grows, the current and electric field at the end increase, possibly making it grow faster. Any material with a dielectric constant greater than unity (which means essentially any material) experiences a force towards a high-field region due to basic electrostatics [35]. The large electrostatic forces under the large local fields necessary for field emission will surely lengthen such a material spike at the tip. Likewise, the electromigration process driven by the large current density near the tip will also aid in the accumulation of contaminants onto the tip. Electromigration is a process where the flux of electrons through a conductor can push along defects and grain boundaries; it is a thermally activated process where the normal diffusion of defects is given a down-stream bias by scattering with the moving electrons. While emitter materials are normally chosen to have low diffusion rates at room temperature (e.g., C, W, Si), the process can also apply to surface adsorbates, and to the tip itself at elevated temperatures.

As an aside, we believe that the research end of the field has spent far too much attention on analyzing data using the Fowler–Nordheim (F–N) equation [16]. The F–N theory will give field emission current densities for a clean and flat metal surface; however, it has been invariably applied to emitters that are never flat, often not metal, and rarely clean. Consequently, a lot of published results should be called into question. Along the same line, we would like to draw attention to work by Dean and Chalamala [36], where it was shown quite conclusively that emission from carbon nanotubes is dominated by surface adsorbates. Again, much work in the field (including some of the authors') has been studying adsorbate-covered surfaces, and while valid on its own terms, it is not entirely relevant to the clean surfaces that will be required for adequate lifetime in commercial devices.

Despite all these problems, there has been considerable progress. Most notably, displays produced at Candescent using metal FEAs have shown lifetime (the time to half-luminance or 50% degradation) of 25 000 h and reliability (the mean time to failure) of 36 000 h at 6 kV [29]. These are adequate for most display applications, e.g., laptop computers. Also, a collaboration between ISE Electronics and Mie University has reported device lifetimes exceeding 10 000 h with a carbon nanotube cathode [37].

2.3.1.5. Competing Technologies. The ultimate competition that FEDs and all other displays will face is an array of light emitting diodes (LEDs) or room temperature lasers. LEDs are now available in all three primary colors and with reasonable efficiencies; manufacturing cost is the current sticking point. The ideal efficiency (not yet achieved) of such a display is the best possible, with close to 100% of the incoming electrical energy converted to light, and there is little loss between the laser source and the viewer. The manufacturing problems, though, are hard: wafers of

LEDs are expensive because of the careful and precise growth procedures necessary to make sufficiently defect-free, III–V or II–VI semiconductor heterostructures. The mechanical spreading of LEDs across a large (e.g., notebook computer) display is also expensive: typical robotic pick-and-place operations cost of the order of $0.01 for each diode, which is far too expensive for the laying down of a million or more pixels. Likewise, testing and replacement of bad LEDs is nontrivial. However, there seem no physical laws preventing LED displays from being built, so we should assume that they will eventually be built, and built cheaply. Organic LEDs (OLEDs) and electroluminescent diodes (ELDs) are two other promising contenders. In particular, OLEDs are easier and cheaper to manufacture, and can be deposited on almost any substrate, including flexible ones such as plastics. However, they have not achieved the efficiency, lifetime and reliability needed for large scale commercial applications.

On a more practical level, liquid crystal displays are strong competition for smaller FEDs (14″ or less). LCDs have made tremendous manufacturing improvements, leading to rapidly dropping costs. Technically, the visual quality has also improved significantly in the last decade; viewing angles are now large enough so that the user's head position is unconstrained. FEDs will have to match LCD cost and image quality in order to be accepted.

For larger size displays, 36″ or larger with coarse pixels, plasma panels are the relevant competitor. They have a low cost per unit area, due to the thick film (no-lithography) production techniques. However, the technology is unlikely to scale to the fine pixel sizes needed for small and midsize displays. Driver electronics for displays is expensive, and the cost increases with the operating voltage. In an FED, that puts a premium on narrow cathode–gate spacings and electron emission at low electric fields. Because of this, FEDs are likely to have cheaper drive electronics than plasma panels, which require control voltage swings of approximately 100 V. Assuming FEDs can be built without lithography (see later), the cost competition between plasma panels and FEDs might hinge on the driver cost.

Finally, normal CRTs are the low cost competitor in all sizes, but they are heavy and bulky. Large CRTs (21″ and above) will not easily fit on office furniture, and as of 2001, we are starting to see some replacement of CRTs by more expensive AMLCDs, even for desktop applications.

2.3.1.6. Future Directions. Two features of FEDs come together to raise the possibility that they could be built cheaply without lithography. First, other than the emitters themselves, displays have large features. A reasonable pixel size for a directly viewed display is 200 μm, as set by the resolution of the human eye. There are no transistors, no busses and no windows, just an orthogonal array of gate and emitter strips. One could imagine using conventional printing processes to lay down or pattern materials. Printing has a negligible cost of a few cents per layer for a laptop-sized display, an order of magnitude cheaper than the per-layer cost of an AMLCD.

Second, it seems possible to make emitters without lithography. Certainly, carbon nanotube field emitters can be grown without lithographic patterning [38]. The gate electrode and layers to support the gate provide more of a challenge, as one needs to produce many closely spaced holes of relatively uniform size and density. As discussed above, statistical arguments make this possible: one does not have to get exactly the right number of emitters into a square array on a pixel. Techniques exist to create gate structures by using fine particles as masks [39,40] and individual ion damage tracks [41]. The latter technique has actually been used in making practical devices by Candescent.

Consequently, as these techniques are further developed, the entire device can become relatively simple to manufacture. We consider the subfield of zero-lithography displays to be interesting, under-explored, and quite possibly critical for making devices cheap enough to displace the entrenched technologies.

2.3.2. Microwave Power Amplifier Tubes

The modern electronic communication industry was born with the development of gridded vacuum tubes or triodes. These tubes made commercial broadcast radio and television possible. From the onset, the triodes suffered from many limitations, all related to the use of thermionic cathodes, which have to be heated above 800°C for electron emission to occur. In low power applications (<10 W), the power required to heat the cathodes can be greater than the power needed to operate the tubes. For high frequency (GHz) applications, the proximity of the hot cathode makes it difficult to stably position the control grid (gate) close enough to the cathode (25 μm or less) in order to get the transit time of the electrons short enough. These limitations led to the development of velocity modulated beam forming tubes such as klystrons and traveling wave tubes (TWTs), which possess better device characteristics but are an order of magnitude more costly to manufacture than the simple triodes. The overriding drawback of all vacuum tubes (other than cost) is their limited lifetime associated with the degradation of thermionic cathodes. The active material of the cathode (typically a low work function metal like barium) gradually evaporates. When it is gone, electron emission ceases, and the tube becomes useless.

The arrival of solid-state transistors and integrated circuits eliminated the use of vacuum tubes in all low power applications because of the elimination of heating power and solid-state devices' extremely long lifetime. However, semiconductor technology, even using new materials such as SiC and GaAs, does not easily solve the power amplification problem, particularly at GHz frequencies. To generate the 50–500 W needed in a wireless base station transmitter, solid-state amplifiers have to operate many transistors in parallel with complex microstrip combining circuits and bulky thermal management equipment. In contrast, for many high power and high frequency applications, vacuum tubes can be compact and efficient. They remain the amplifier of choice for radar, electronic warfare, and space-based communications.

The incorporation of cold cathodes in vacuum tubes promises to bring together the best features of both vacuum tubes (high power) and power transistors (long lifetime). Cold cathode tubes can be turned on instantaneously, without a tedious warm-up period. The grid can now be placed very close to the cathode (e.g., <10 μm) to enable low grid voltage (50–100 V) and high frequency (10 GHz or above) operation. In beam forming tubes, density modulation of electron beams by the grid through gated emission also becomes possible so that a long beam interaction section is no longer required, and the tubes can be shortened substantially.

Cold cathode tubes also share another attractive advantage over semiconductor devices: their high efficiency for linear amplification. In applications such as PCS/cellular base stations where one amplifier is shared between many narrow-band signals, it is imperative to operate an amplifier in the linear regime in order to avoid intermodulation of signals. Besides the elimination of heating power that contributes to the increased efficiency by a typical 5–10%, the main enhancement in klystrodes or TWTs is achieved by the incorporation of a collector or multiple collectors at the end of the electron beam's path that are biased at voltages lower than the accelerating anode voltage. As the bunched beam travels through the anode region where the rf power of the electron beam is coupled out, the electron beam (i.e., the dc current) is slowed down by the depressed potentials of the collectors. Most of the electrons then hit the collector with much reduced energies and create correspondingly less heat. Much of the kinetic beam energy can thus be fed back to the power supply, and operating efficiencies over 70% can be achieved at saturation powers and as high as 40% in the backed-off linear region. The amplifier tube can also be operated with power drawn on demand, as the depressed collectors can return as much as 95% of the beam energy back to the power supply when there is little rf power to disturb the electron velocities in the beam.

In contrast, an array of power transistors must be backed off from saturation to about 5% efficiency point to achieve adequately linear performance in a cell phone base station application. Additionally, constant dc biases have to be applied to the transistors when the amplifier is in operation, even during night time hours of very light loads. This constant energy drain further reduces the average efficiency of a solid-state amplifier system to about 3%. An efficient cold cathode amplifier system is attractive to the wireless communication industry because it can help to move the electronics from a building into a small box on the tower, close to the antennas. Highly efficient tubes are also especially valuable for space operation where available power is limited, and heat can only be dissipated by radiation.

Applications of field emitters to microwave tube amplifiers can be divided into two categories: triodes and beam forming tubes such as klystrons and TWTs. The main challenges in incorporating field emitters into these microwave tubes include: overcoming the initial transverse velocity produced at the FEAs and form a usable, narrow electron beam; producing stable, copious emission at high current densities in a tube environment; and developing a matching network to efficiently couple rf fields into the emitter/gate structure. We will delve into some specifics in the following by discussing the two categories separately.

2.3.2.1. Microwave Triodes.

A microwave triode is the vacuum equivalent of a field effect transistor: electrons are emitted from the cathode into vacuum, and a grid (gate) electrode varies the electric field between the emitter and the gate to control the emission current. Some of the electrons pass through apertures in the gate, accelerate towards the anode, and are collected to give a current through the device. The grid is traditionally biased negatively in a thermionic tube to minimize the grid current, although positive grid voltages can now also be used due to the development of shadow grid techniques on the cathode surface. In a cold cathode tube, the grid voltage is always positive in order to induce and control electron emission. If the device geometry is designed correctly, this switching action by the gate can yield an overall power gain.

However, conventional triodes (i.e., without shadow grids) do not exhibit power gain at frequencies above 4 GHz [1] because of the long transit time of the electrons between the cathode and the gate. At these higher frequencies, the electrons spend more than one-half cycle in the cathode–gate space, and the effect of the grid on the electrons is averaged away. The grid cannot be pushed closer to the cathode, which would reduce this transit time, largely because it needs to be hanging free in space and cannot be allowed to touch the hot cathode. Thermal expansion of the grid and cathode supports can easily move the grid by tens of microns under normal operating conditions. These deflections force the grid to be at least 25–50 μm from the hot cathode, which sets the transit time and, therefore, the maximum operating frequency.

In a tube with a field emitter cathode, it becomes possible to support the grid on pillars based on the cathode, as one does not need to worry about the barium from the hot cathode diffusing along the insulating pillars, because the cathode is neither heated nor barium-containing. These pillars are necessary for high performance designs, as the mechanical deflection from the electrostatic forces between the grid and cathode can become significant. Field emitter triodes can then have a much closer grid–cathode spacing, and thus a shorter transit time, which ceases to be the limit on the tube's maximum operating frequency. The new limit is determined by the ratio of transconductance to gate–cathode capacitance. It is reasonable to expect a cold cathode triode to operate at frequencies much higher than 5 GHz.

However, there are still serious technical difficulties remaining. To start with, the field emitter needs to operate reliably at high current densities. This is dictated both by application needs and device physics. For example, radar applications typically require current densities in the range of 2–10 A/cm^2 [42]. For a triode amplifier, the current needs to exceed 200 mA/cm^2 [43] in order to produce gain at frequencies all the way up to the transit time limit. If the current density is too low, the resistive losses in the grid can exceed the gain, rendering the device useless. The inter-electrode capacitances also become large at lower current densities, simply because the areas of the emitter and gate need to get bigger to maintain the same power levels, which leads to more losses in the grid. These current densities are large by field emitter standards, and we know of no devices that have achieved emitter lifetimes long enough to be commercially interesting (>5000 h) at adequate current densities for an amplifier.

The difficulty in obtaining high current densities with long lifetimes is made worse by the fact that field emission is intrinsically nonuniform. As discussed in detail earlier in Section 2.3.1.4.1 on cathode uniformity, the current from an array of emitters is often dominated by a few exceptionally good or hot tips, which can be either geometrically sharper or electrically more favorable (i.e., its work function is lowered by adsorbates). These "hot" emitters are overloaded and prone to destruction through overheating, diffusion, arcing or ion bombardment.

One of the most reliable schemes used to make the emission more uniform is to use a resistor to limit the current for each emitter. However, putting a pure resistance in series with each emitter makes it impossible to modulate the emission current at high frequencies; one needs a RC circuit where the capacitance is comparable to the gate–emitter capacitance [20] or perhaps a nonlinear resistor. NEC has recently introduced a vertical current limiter (VECTL) approach that uses a voltage controlled resistive element that provides arc protection but presents comparatively little resistance in normal high frequency operation [44–46]. The VECTL resistance is orders of magnitude lower than that of typical resistors used in displays. Under conditions of an arc, however, the bias across the VECTL increases, causing a constriction of the conductive channel that greatly increases the effective resistance. Because the VECTL structure presents a low resistance under normal operations, it may be compatible with modulating the emitters at GHz frequencies.

A second set of problems involves minimizing the emission current that hits the gate. Electrons that hit the gate reduce device efficiency and increase the amount of power required to drive the gate. Equally important, they heat the gate structure, which causes it to thermally expand, causing a variety of unwanted effects. This problem can be handled, in principle, by patterning the emitter to create a so-called shadow grid on the cathode so that the electrons aren't emitted from areas directly underneath the grid wires; thus avoiding emission from places that will bombard the gate. This technique, then, boils down to patterning the emitter and aligning the gate and emitter structures.

2.3.2.2. Microwave Beam Tubes.

Microwave beam tubes are useful at higher power levels and frequencies where triodes and semiconductor devices offer little competition. They also offer improved device performance such as higher efficiency and broad bandwidth. Normally, these tubes consist of a cathode, electrodes to accelerate and form a beam, an interactive structure such as a helix or cavity, magnets to focus and guide the beam, microwave signal input and output, and a collector to catch the electron beam. Electrons enter the interactive region during both the positive and negative phases of the microwave cycle as a narrow beam. Rather than directly turning the beam on and off, the interactive structure modulates the velocities of the electrons in the beam. The electrons entering during a positive phase are accelerated, while those entering during a negative phase are decelerated. The beam then propagates along the axis of the tube, and the faster elections catch up to the slower electrons in front of them. At the right point, the fast and slow electrons overlap, forming bunches along the beam. These bunches are essentially a series of pulses of electric current flowing by, and power can be extracted from the beam by coupling it to an output cavity. This mechanism effectively eliminates the transit time effects that limit the high frequency performance of a triode, but there are costs: the beam must have a very narrow velocity distribution, both in the longitudinal and transverse directions, and the tube structure is more complex and expensive to make.

Cold cathode tubes may be able to gain some advantages by operating in a klystrode mode [47], that is, applying a microwave signal to be amplified directly to the grid to create a current-density-modulated electron beam. This density modulated bunched electron beam, made possible by the short transit time between the emitter and grid when they are positioned very close (<10 μm), then accelerates through the tube and transfers its energy to an output cavity. The advantages of prebunching the beam at the grid are the high efficiency in the beam energy transfer and coupling due to reduced beam noise and a significant reduction in the required length of the interaction structure, resulting in smaller and lighter tubes. A shorter tube then relaxes the stringent requirement that the velocity dispersion of the electron beam must be small, which simplifies the emitter design. This is probably the most significant impact that the field-emitter technology can have on microwave power tubes, because it will potentially enable new classes of devices and the miniaturization of existing devices.

Over the years, there has been a considerable effort spent in building cold cathode beam tubes such as TWTs. Notable results came initially from a SRI and Teledyne collaboration in 1986 [48,49], and then more recently from NEC [44–46] when researchers there announced the successful operation of two 10 GHz TWT amplifiers incorporating Spindt type FEA cathodes. The demonstrated high currents (40 mA) and current densities (50 A/cm^2) as well as stability and longevity (5000 h at 1% duty cycle) represented a significant advance in FEA technology. However, density modulation was never attempted, and the device performance and lifetime still fall short of conventional TWTs.

One of the fundamental requirements for microwave beam tubes to work is that the electron beam must have a very narrow velocity distribution, both in the longitudinal and transverse directions. Transverse velocity dispersion broadens the beam, which leads to more current lost out of interception of electrons that are farther from the beam center and thus less efficient coupling of energy into and out of the beam. Intercepted electrons will also desorb atoms from the tube walls and spoil the vacuum. Longitudinal velocity dispersion smears out the bunching process, reducing the tube's gain and efficiency. Here, field emitters are at a substantial disadvantage. Thermionic emitters generate electron beams whose velocity dispersion is given by the temperature of the cathode. The transverse momentum from a hot cathode is roughly 1% of the forward momentum of a 1 keV beam. On the other hand, the transverse momentum from a field emitter array depends strongly on the design, but it can be as high as 50% of the forward momentum that the electrons have at the gate. This can be as much as 25% of the forward momentum of a 1 keV beam. With that much transverse momentum, it may be impossible even to focus the beam to a diameter smaller than the cathode, while substantial compression is typically achieved for thermionic beams. Reducing this transverse momentum is perhaps one of the most difficult technical challenges in this area.

An individual field emitter can have an extremely high brightness (current per unit area per unit solid angle) – as large as 10^8 A/(cm^2 sr). This huge number comes about because the emitter area can be very small, often only a few atoms across. Unfortunately, practical field emitters are limited to currents of only a few µA per tip, so many applications, especially microwave amplifiers, need more than one field emitter. That's when the trouble begins; once electron trajectories from multiple emitters begin to cross, one can no longer treat the electrons as coming from a compact source. Because electron lenses generally have strong aberrations, it rapidly becomes impractical to unmix the electron beams from the different emitters, and one must treat the source as if the electron emission were spread out across the whole macroscopic area of the emitter array, not just the microscopic points. This drops the effective brightness by a large factor: the square of the ratio of the tip spacing to the tip size (roughly a factor of a million), to a value that is close to the brightness that can be obtained from conventional thermionic emitters.

Once the electron beams mix and you lose information about which electron comes from which emitter, Liouville's theorem prevents you from ever increasing the brightness [50]. However, before they mix, one can build Einzel lenses onto individual clumps of emitters. These are typically small converging lenses built into the gate structure, and they function to reduce the perpendicular momentum of individual emitter sites before the electron beams merge and lose the information of the emitter's identity [51,52]. However, lens structures suitable for beam tubes are still far from practice. While the general concept is understood, no one has yet demonstrated sufficient control of the emitter growth, especially in combination with a complicated set of electrostatic lenses, to actually produce beams with a usefully small divergence.

2.3.2.3. Application Assessment.

These cold cathode microwave tubes are intended to be used for PCS/cellular base stations and/or mobile devices. However, the technological trend that might shut the door for this application is the expected reduction in cell size that will be required to handle the expected increases in traffic (on the assumption of constant bandwidth allocations). Smaller cell sizes mean lower power transmitters, and as the power decreases, semiconductor amplifiers become more practical. Also, field emitter reliability problems have to be completely solved before they can be taken seriously in satellite applications. Before that, there may be advantages for cold cathode tubes to be used in terrestrial applications that do not require long life, such as in missiles, decoys, jammers, etc. [42]. Such expendable applications do not have the same lifetime requirements for electron sources as space applications, but would benefit from the density modulated beams and the instant on/off capability of the cold cathode. However, very large current densities are needed in these devices.

It appears that among the major issues for microwave tube applications, reliability is probably the toughest to prove, but it will probably boil down to painstaking manufacturing practice. Much is still unclear about how one could control the beam spreading, and in the end it might still prove impossible. It is important to have a series impedance structure that can suppress arcs and ensure spatial current uniformity, but the structure must enable emission modulation at microwave frequencies while retaining the property of resistive or capacitive protection. The NEC approach may be promising, but a design that works at GHz frequencies in CW operation is still a substantial step beyond.

2.4. OUTLOOK

We are generally optimistic that field emitters will find their way to at least some applications. Probably they will succeed in one or more areas where other technologies are pushed to their limits, but we don't expect VME to sweep away semiconductor devices. Technology is advancing in the directions that will make field emitter devices possible: better cleanliness and more control of microscopic structures. However, there are competing technologies for all applications, and the ultimate commercial success of field emitters or VME may well be decided by the efforts of manufacturing engineers.

REFERENCES

1. J. A. Morton, *Bell Labs Res.* **27**, 166 (1949).
2. Z. W. Chen, Nanometric-scale electron beam lithography, in *Advances in Electronics and Electron Physics Vol. 83* (P. W. Hawkes, Ed.), Academic Press: New York, p. 108, 1992.
3. J. B. Pawley, LVSEM for high resolution topographic and density contrast imaging, in *Advances in Electronics and Electron Physics Vol. 83* (P. W. Hawkes, Ed.), Academic Press: New York, p. 203, 1992.
4. Avoiding a data crunch (feature article), *Scientific American*, May 2000.
5. A. A. Talin, K. A. Dean, and J. E. Jaskie, Solid State Electron. **45**, 963 (2001).
6. J. A. Castellano, *Handbook of Display Technology*, Academic Press: New York, p. 33, 1992.
7. R. F. Bell, G. P. Kochanski, and J. Thomson, Jr., Flat panel display apparatus and method of making same, US Patent 5,498,925 (1996).
8. C. J. Summers, in *Proc. 10th International Vacuum Microelectronics Conference*, Kyongju, Korea, p. 244, 1997.
9. G. P. Kochanski, C. A. Murray, M. L. Steigerwald, P. Wiltzius, and A. van Blaaderen, Display apparatus with coated phosphor and method of making same, US Patent 5,838,118 (1998).
10. R. V. Latham, *High Voltage Vacuum Insulation: The Physical Basis*, Academic Press: New York, p. 291, 1981.
11. R. Hawley, *Vacuum* **18**, 383 (1968).
12. R. V. Latham, Appendix—Flashover across solid insulators in a vacuum environment, in *High Voltage Vacuum Insulation: The Physical Basis*, Academic Press: New York, p. 229, 1981.
13. E. A. Chandross, S. Jin, G. P. Kochanski, J. Thomson, Jr., and W. Zhu, Method of making improved pillar structure for field emission devices, US Patent 5,704,820 (1998).
14. S. Jin, G. P. Kochanski, and W. Zhu, Multilayer pillar structure for improved field emission devices, US Patent 5,690,530 (1997).
15. N. Tizard-Gatel, A. Perrin, F. Levy, and J. F. Boronnat, in *Society for Information Display International Symposium Digest of Technical Papers*, Vol. 30, p. 1138, 1999.
16. R. H. Fowler and L. Nordheim, *Proc. R. Soc. London, Ser. A* **119**, 173 (1928).
17. P. R. Schwoebel and I. Brodie, *J. Vac. Sci. Technol. B* **13**, 1391 (1995).
18. A. Ghis, R. Meyer, P. Rambaud, F. Levy, and T. Leroux, *IEEE Trans. Electron Devices* **38**, 2320 (1991).
19. J. D. Levine, R. Meyer, R. Baptist, T. E. Felter, and A. A. Talin, *J. Vac. Sci. Technol. B* **13**, 474 (1995).
20. G. P. Kochanski, Flat panel field emission display apparatus, US Patent 5,283,500 (1994).
21. J. P. Barbour, W. W. Dolan, J. K. Trolan, E. E. Martin, and W. P. Dyke, *Phys. Rev.* **92**, 45 (1953).
22. W. A. Anderson, *J. Vac. Sci. Technol. B* **11**, 383 (1993).
23. T. Wolkenstein, *Electronic Processes on Semiconductor Surfaces during Chemisorption*, Plenum Publishing Corp.: New York, p. 35, 1991.
24. A. A. Holscher, *Adsorption Studies with the Field-Emission and Field-Ion Microscope*, Ph.D. Thesis, University of Leiden, p. 43 and vicinity, 1967.
25. Y. Wei, B. R. Chalamala, B. G. Smith, and C. W. Penn, *J. Vac. Sci. Technol. B* **17**, 233 (1999).

26. L. W. Swanson, *J. Vac. Sci. Technol.* **12**, 1228 (1975).

27. L. W. Swanson and L. C. Crouser, *J. Appl. Phys.* **40**, 4741 (1969).

28. L. W. Swanson and N. A. Martin, *J. Appl. Phys.* **46**, 2029 (1975).

29. G. Hopple and C. Curtin, *Inf. Display* **4&5**, 34 (2000).

30. I. Brodie, *Int. J. Electrostatics* **38**, 541 (1975).

31. M. L. Yu, B. W. Hussey, E. Kratschmer, T. H. P. Chang, and W. A. Mackie, *J. Vac. Sci. Technol. B* **13**, 2436 (1995).

32. C. F. Barnett, Atomic collision properties, in *A Physicist's Desk Reference* (Herbert L. Anderson, Ed.), American Institute of Physics: New York, p. 83, 1989.

33. D. P. Woodruff and T. A. Delchar, *Modern Techniques of Surface Science*, Cambridge University Press: New York, p. 248ff, 1988. Cambridge Solid State Science Series.

34. P. K. Baumann and R. J. Nemanich, *Surf. Sci.* **409**, 320 (1998).

35. W. R. Smythe, *Static and Dynamic Electricity*, Taylor and Francis: London, p. 18, Sec. 1.16–1.17, 1989.

36. K. A. Dean and B. R. Chalamala, *Appl. Phys. Lett.* **76**, 375 (2000).

37. Y. Saito, K. Hamaguchi, R. Mizushima, S. Uemura, T. Nagasako, J. Yotani, and T. Shimojo, *Appl. Surf. Sci.* **146**, 305 (1999).

38. C. Bower, W. Zhu, S. Jin, and O. Zhou, *Appl. Phys. Lett.* **77**, 830 (2000).

39. S. Jin, G. P. Kochanski, and W. Zhu, Flat panel display device, US Patent 5,808,401 (1998).

40. S. Jin, G. P. Kochanski, and J. Thomson, Jr., Spaced gate emission device and method for making same, US Patent 5,681,196 (1997).

41. T. S. Fahlen, in *Proc. 12th International Vacuum Microelectronics Conference*, Darmstadt, Germany, p. 56, 1999.

42. J. Dayton, M. A. Hollis, K. L. Jensen, R. A. Murphy, L. Parameswaren, and E. Wintucky, Assessment of the field emitter array TWT developed by NEC corporation, Internal NRL Report, 1997.

43. A. H. W. Beck, *Thermionic Valves*, Cambridge University Press: Cambridge, Ch. 9 and references within, 1953.

44. H. Makishima, H. Imura, M. Takahashi, H. Fukui, and A. Okamoto, in *Proc. 10th IVMC*, Kyongju, Korea, p. 194, 1997.

45. H. Takemura, Y. Yomihari, N. Furutake, F. Matsuno, M. Yoshiki, N. Takada, A. Okamoto, and S. Miyano, in *Tech. Dig. of IEDM*, p. 709, 1997.

46. H. Imura, S. Tsuida, M. Takahasi, A. Okamoto, H. Makishima, and S. Miyano, in *Tech. Digest of IEDM*, p. 721, 1997.

47. D. R. Hamilton, J. K. Knipp, and J. B. Kupler, *Klystrons and Microwave triodes*, Dover Publications: New York, 1966. MIT Radiation Laboratory series.

48. P. Lally and C. Spindt, in *Abstract for High Power Microwave Tube Conference*, Montery, CA, 1986.

49. P. M. Lally, Y. Goren and E. A. Nettesheim, *IEEE Trans. Electron Devices*, **36**, 2738 (1989).

50. L. D. Landau and E. M. Lifshitz, *Statistical Physics*, Part 1, 3rd ed. (Revised and expanded by E. M. Lifshitz and L. P. Pitaevskii), Pergamon Press: Oxford, England, p. 10, 1980.

51. J. E. Graebner, S. Jin, G. P. Kochanski, and W. Zhu, Microwave vacuum tube devices employing electron sources comprising activated ultrafine diamonds, US Patent 5,796,211 (1998).

52. T. S. Fahlen, Focusing properties of the thin CRT(tm) display, in *Society for Information Display International Symposium Dig. Tech. Papers*, Vol. 30, p. 830, 1999.

CHAPTER 3

Theory of Field Emission

KEVIN L. JENSEN

Naval Research Laboratory, 4555 Overlook Avenue SW, Washington, District of Columbia
20375-5347

"...we have our philosophical persons, to make modern and familiar things
supernatural and causeless."

W. Shakespeare, *All's Well That Ends Well*

3.1. INTRODUCTION

Vacuum microelectronics (VME) concerns itself with devices exploiting electron
ballistic transport within the vacuum after emission from microfabricated structures.
The small scale features of a typical VME cathode, perhaps the best known example
being the Spindt-type field emitter (a conical metallic tip protruding through a micron-
sized gate hole) [1,2], are required to generate the large fields needed to facilitate
electron emission. Other VME electron sources have a shorter but nevertheless distinguished pedigree, or are rapidly evolving, such as semiconductor field emitters,
carbon nanotubes, diamond and wide band gap semiconductors, and negative electron
affinity (NEA) materials, to name but a few. All seek to minimize the barrier to electron emission into vacuum by engineering the material properties (which typically
change the work function) or the physical geometry (which typically changes the field
enhancement) of the cathode. By comparison, thermionic cathodes are heated such
that a portion of the electron population has sufficient kinetic energy to surmount the
surface barrier.

The underlying connection between these electron sources lies in the barrier to
transport: its origin, dependence on material properties and applied field, and methods
to exploit its characteristics. The genesis of field emission theory lies in the early
work of Fowler and Nordheim [3] for one-dimensional (planar) emission, though the
form most commonly used in the burgeoning VME community is indebted to the
formulation of Murphy and Good [4]. More advanced treatments are found in (for
metals) Christov [5], Gadzuk and Plummer [6], and (for semiconductors) Stratton
[7], and Fursey [8]. Excellent books covering the theory of field emission from both
metals and semiconductors are found in Gomer [9] and Modinos [10]. Charbonnier has
given a clear description of the transition from field emission to thermionic emission
[11]. Most of tunneling theory in use is predicated on the Wentzel–Kramers–Brillouin
(WKB) approximation using a classical image charge potential; greater sophistication

Vacuum Microelectronics, Edited by Wei Zhu.
ISBN 0-471-32244-X. 2001 John Wiley & Sons, Inc.

has been undertaken by Cutler [12,13] and Mayer and Vigneron [14]. He, Cutler, Miskovsky, and others extended the one-dimensional (1-D) theory to account for surface curvature [15,16]. More recently, the usage of the 1-D theory to determine "emitting area" (capitalizing on the power of desktop computers to evaluate functions of elliptical integrals inherent in the analytical form of $J(F)$) has been definitively updated and extended by Forbes [17,18]. This brief and scattered recitation only hints at the richness of over five decades of history [19]: W. P. Dykes' prediction that cathodes based on field emission "should lead to a number of electron-beam devices, including oscilloscopes and other display tubes and even high-resolution television tubes" [20] is materializing.

Cataloging the copious theoretical literature* in VME would be superficial. While an effort has been made herein to sample from the sumptuous literature buffet, a different agenda is considered more useful. Experimental current–voltage ($I(V)$) relations are invariably rendered in "Fowler–Nordheim" (F–N) coordinates, in which $\ln[I(V)/V^2]$ is plotted as a function of $1/V$ for a multitude of field emission sources, be they a tungsten whisker, gated field emitter, carbon nanotube, or other exotic source. The ubiquitous linearity (more or less) of experimental data plotted in this manner indicates a commonality in the underlying theory of the emission mechanisms of many electron sources. The similarity is more intriguing when the experimental coordinates are naively derived from the 1-D, zero temperature, planar theory relating current density J to applied field F via the F–N relation

$$J(F) = a_{FN} F^2 \exp(-b_{FN}/F) \tag{1}$$

(where a_{FN} and b_{FN} are F–N parameters†) by arguing (erroneously) that current and voltage are proportional to current density and field, respectively. The geometry of a gated Spindt-type emitter could not be more different than that of the granular surface of a diamond film, and yet theoretically, there is a connection between the two, and their theoretical analysis shares more than just a passing similarity.

Richard Feynman once quipped that, though some understood relativity, no one understood quantum mechanics (the foundation of field emission theory). In symbolic deference to that sentiment, in the VME literature the F–N equation is more often than necessary conjured in an erroneous manner or without regard to its subtleties. The field emission theory presented herein endeavors to explore the physics behind Eq. (1) and its metamorphosis into the visually analogous, but nevertheless different $I(V)$ relation often substituted in its place. To avoid descent into a mathematical inferno, simple models chart the path from 1-D theory to the tools needed to analyze realistic field emission sources. In the published literature, such an effort is often (but not always) employed in the pursuit of models for emission from Spindt-type emitters and

---

*Proceedings of the *International Vacuum Microelectronics Conference* (IVMC) have been regularly published in the March/April issue of the *Journal of Vacuum Science and Technology B* since 1993. Proceedings of the first and second *International Vacuum Electron Sources Conferences* (1996 and 1998, respectively) have been published in *Applied Surface Science*. Two volumes on Materials Issues in Vacuum Microelectronics have been published as symposium proceedings by the Materials Research Society.

† The notation is fashioned after Modinos [10].

arrays. While the models discussed herein are widely applicable to electron sources in general, the scent of their FEA ancestry is never far away. When writing on decades of work in a fertile scientific area, presenting the logical evolution is unfortunately at the expense of the historical chronology and often bypasses deserving topics which do not fit comfortably in a tidy narrative.

3.2. ONE-DIMENSIONAL TUNNELING THEORY: METALS

3.2.1. Thermionic and Field Emission: Basic Concepts

3.2.1.1. Tunneling Probability. In preparation for the F–N equation, there is abundant utility in introducing the concepts and sharpening one's intuition by means of a model, which is spared the mathematical complexity but contains the physics. Consider the simplest barrier that allows for a distinction between thermionic emission and tunneling:

$$V(x) \equiv V_0\,\theta(x)\,\theta(x - L)$$

where $\theta(x)$ is the Heaviside step function (1 if the argument is greater than 0, and 0 otherwise), V_0 is the barrier height, and L is the width of the barrier. At this stage, they are merely parameters which characterize the potential; their physical significance will become apparent later on. Some notation is required. Let the mass of the incident electron be m and energy $E(k) = \hbar^2 k^2/2m$: we shall restrict our attention to cases in which the energy is parabolic in momentum. The maximum energy and momentum at $T = 0$ K are denoted by μ and k_F such that $\mu = E(k_F)$, where μ is the Fermi energy and k_F is the Fermi momentum. Introduce k_0 such that $E(k_0) = V_0$ and define $\kappa \equiv |k_0^2 - k^2|^{1/2}$. The symbols and values of the fundamental constants are given in Table 3.1.

Our treatment shall be a very concise rendition of a standard problem in quantum mechanics to illustrate methodology using a familiar example. For incident, reflected, and transmitted waves $\psi_{\text{inc}} = e^{ikx}$, $\psi_{\text{ref}} = r\,e^{-ikx}$, and $\psi_{\text{trans}} = t\,e^{ikx}$, respectively, continuity of the wave function and its first derivative at $x = 0$ and L implies two

TABLE 3.1. Fundamental Constants

Symbol	Definition	Formula	Value
m_0	Electron rest mass	–	510998.9 eV/c^2
$\hbar$	Planck's constant/2π	$h/2\pi$	0.6582119 eV fs
c	Speed of light	–	2997.925 Å/fs
q	Electron charge	–	1
ε_0	Permittivity of free space	$q^2/4\pi\alpha\hbar c$	5.526350×10^{-3} q^2/eV Å
α	Fine structure constant	–	$1/137.036$
a_0	Bohr radius	$\hbar/\alpha m_0 c$	0.5291772 Å
k_B	Boltzmann's constant	–	8.617341×10^5 eV/K

matrix equations for $k \leq k_0$

$$\begin{pmatrix} 1 & 1 \\ ik & -ik \end{pmatrix} \begin{pmatrix} 1 \\ r \end{pmatrix} = \begin{pmatrix} 1 & 1 \\ \kappa & -\kappa \end{pmatrix} \begin{pmatrix} a \\ b \end{pmatrix}$$

$$\begin{pmatrix} e^{\kappa L} & e^{-\kappa L} \\ \kappa e^{\kappa L} & -\kappa e^{-\kappa L} \end{pmatrix} \begin{pmatrix} a \\ b \end{pmatrix} = \begin{pmatrix} e^{ikL} & e^{-ikL} \\ ike^{ikL} & -ike^{-ikL} \end{pmatrix} \begin{pmatrix} t \\ 0 \end{pmatrix} \tag{2}$$

respectively, where a and b are the coefficients of the (exponential) wave function within the barrier. For $k > k_0$, κ is replaced by $i\kappa$ in Eq. (2). Standard matrix techniques show solutions for t and r are

$$r = -\frac{i(\kappa^2 + k^2)\sinh(\kappa L)}{2\kappa k \cosh(\kappa L) + i(\kappa^2 - k^2)\sinh(\kappa L)}$$

$$t = \frac{2\kappa k}{2\kappa k \cosh(\kappa L) + i(\kappa^2 - k^2)\sinh(\kappa L)} e^{-ikL} \tag{3}$$

The "transmission coefficient" is defined as the ratio between the incident and transmitted current, where the current is defined as $J(k) = (\hbar/2mi)(\psi^* \partial_x \psi - \psi \partial_x \psi^*)$ [21]. For incident plane waves with $k < k_0$, it may be shown that

$$T(k) = \frac{J_{\text{trans}}(k)}{J_{\text{inc}}(k)} = \frac{4\kappa^2 k^2}{4\kappa^2 k^2 + (\kappa^2 + k^2)^2 \sinh^2(\kappa L)} \tag{4}$$

Eqs. (3) and (4) contain many of the essential features that may be encountered when tunneling barriers characteristic of field emission are encountered here. They are as follows:

- κL is the product of the height of the barrier above $E(k)$ with the width of the barrier L, or the "area under the curve" defined by $\theta(k) \equiv \int_0^L (k_0^2 - k^2)^{1/2}\,dx$ (k_0 is, at present, independent of x).
- If $\exp(2\kappa L) = \exp(2\theta) \gg 1$ and $2\mu \approx V_0$ (as for metals), then $T(k) \approx C(k)\exp(-2\theta)$, where the coefficient $C(k) \approx (2k/k_0)^2$ as k tends to 0, but otherwise is of the order of unity near the maximum electron momentum k_F.
- If $\theta \geq 2$, $T(k) \approx [1 + (1/C(k))\exp(2\theta)]^{-1}$.* In other words, the exponential rise in the transmission coefficient tapers off (and does not exceed unity) before the apex of the potential is reached.

While a rectangular barrier admits a mathematical description decidedly simpler than the triangular barrier considered by Fowler and Nordheim, it nevertheless shares the essential features with regards to the behavior of $T(k)$.

---

*Compare to Eq. (1) of Christov [5] or Eq. (1.32) of Modinos [10], for which the coefficient of the exponential term is unity.

3.2.1.2. Current Density. The total current density J through the barrier is obtained by summing up the currents of all the individual electron wave functions with momentum $0 \leq k \leq k_F$. At finite temperatures, μ is no longer the energy of the most energetic electron, but rather the "chemical potential" defined such that the integral over the Fermi Distribution Function $f_{FD}(E)$ over all momentum space at any temperature reproduces the electron density [22]. The electrons' energy is characterized by a Fermi distribution function $f_{FD}(E)$: $f_{FD}(E < \mu) = 1$ at $T = 0$ K, and is zero otherwise. The three-dimensional (3-D) nature of k-space has so far been ignored. Considering it, $E(k) = \sum_i (\hbar k_i)^2/(2m)$, where the sum is over the momentum components along the x, y, and z axes. The current density through the barrier is given by the integral of the product of the electron charge, velocity, distribution function and tunneling probability, or

$$J(\beta,L) = \frac{q}{2\pi} \int_0^\infty \frac{\hbar k_x}{m} f(k_x) T(k_x) \, dk_x \tag{5}$$

where the electron velocity into the barrier is given by $\hbar k_x/m$. If we let $E_x \equiv (\hbar k_x)^2/2m$, then it is clear that the integrand of Eq. (5) is proportional to the energy distribution in E_x of the transmitted electrons. The integral of Eq. (5) is often represented as $(q/2\pi\hbar) \int f(E_x) T(E_x) \, dE_x$, a form particularly convenient for deriving the F–N equation here.

The distribution function $f(k_x) \equiv f(k)$ is obtained from the Fermi distribution by (in cylindrical coordinates $dk_y \, dk_z = 2\pi k' dk'$)

$$f(k) = \frac{g}{(2\pi)^2} \left(\int_0^\infty \frac{2\pi k' dk'}{e^{\beta[E'+E-\mu]} + 1} \right) = \frac{m}{\pi\beta\hbar^2} \ln\left(e^{\beta(\mu-E(k))} + 1 \right) \tag{6}$$

where $\beta = 1/k_B T$, and k' is identified with the parallel momentum to the surface (in cylindrical coordinates), and $g = 2$ accounts for electron spin. In Eq. (6) and hereafter, k and E without a subscript are understood to be k_x and E_x, respectively. At $T = 0$ K, Eq. (6) reduces to $f(k) = (k_F^2 - k^2)/(2\pi)$ for $k \leq k_F$. The current integrand is sharply peaked at $k \approx k_F$. κ may be Taylor expanded about k_F to second order in the exponential approximation to $T(k)$. Recalling the discussion following Eq. (4), the leading term in the total current density through the barrier for large $\kappa(k_F)L$ is then

$$J_{rect}(L) = \frac{4\mu\phi^2}{(\mu + \phi)^2 \pi^2 \hbar L^2} \exp\left(-2\frac{L}{\hbar}\sqrt{2m\phi} \right) \tag{7}$$

where the term corresponding to the lower limit of integration in Eq. (5) is negligible. The parameter ϕ representing the height of the barrier above μ, i.e., $\phi = V_0 - \mu$, has been introduced, and the subscript "rect" denotes a rectangular barrier. Eq. (7) contains enough physics to understand why Eq. (1) is plausible for more general potentials. In field emission, the length scale L is related to the applied field $F(\text{eV/Å}) = qE_{\text{vac}}$ (where E_{vac} is the electric field) by $FL \approx \phi$. With this replacement, Eq. (7) has the same parametric dependence upon applied field F that Eq. (1) exhibits, and identifies b_{FN} as proportional to $\phi^{3/2}$ and a_{FN} inversely proportional to ϕ. More complex potentials exhibit similar F and ϕ behavior: the major difference is an alteration of the coefficients implicit in a_{FN} and b_{FN}. The conclusions for the simple barrier remain valid in general:

- Analytical current density formulae result from a Taylor expansion (conveniently chosen about the Fermi momentum $k \approx k_{\text{F}}$ (or $E(k) \approx \mu$) in $T(k)$;
- Because $\beta\mu \gg 1$ (e.g., at $T = 300$ K and $\mu = 4.5$ eV, $\beta\mu \approx 174$), only those electrons with $E(k) \approx \mu$ contribute to J, and J is only weakly dependent on temperature;
- An adequate approximation to $T(k)$ is the "area under the curve" approach familiar from the lowest order term in the WKB approximation to the wave function, as the coefficient $C(k)$ is of the order of unity near $k = k_{\text{F}}$;
- Eq. (7) degrades if $\beta\mu$ or $\kappa(k)L$ are not large, or if significant emission occurs near the barrier maximum (the thermal tail of $f(k)$ is not negligible)—as occurs for thermionic emission.

The second bullet deserves particular notice because — contrary to standard descriptions of the WKB approach — the potential is *not* a slowly varying function of position for the square barrier considered here over a length scale characteristic of $1/k_{\text{F}}$. When the barrier is wide ($k_{\text{F}}L \gg 1$), then Eq. (4) may be approximated by $T(k) = 0$ for $E(k) < V_0$ and unity otherwise. At finite temperatures for $E(k) > \mu$, and using the approximation $\ln(1 + \varepsilon) \approx \varepsilon$ for small ε, $f(k)$ assumes the Maxwell–Boltzmann statistics form, and (using $(\hbar^2 k/m)\mathrm{d}k = \mathrm{d}E$) the current density becomes

$$J_{\text{RLD}}(T) = \frac{q}{2\pi\hbar} \int_{\mu+\phi}^{\infty} \frac{m}{\pi\beta\hbar^2}\, e^{\beta[\mu-E]}\, \mathrm{d}E = \frac{qm}{2\beta^2\pi^2\hbar^3}\, e^{-\beta\phi} \tag{8}$$

Eq. (8) is the well-known *Richardson–Laue–Dushman (RLD) equation*. When there is no applied field, ϕ is identified as the work function Φ, which varies from material to material, and ranges from 2 to 6 eV [23].

Eqs. (7) and (8) contain the essential features of the tunneling and thermionic emission phenomena, but at a minimum of mathematical complexity (realistic potentials alter that). Furthermore, they represent opposite limits of the current integrand in Eq. (5) using the transmission coefficient of Eq. (4) and the supply function of Eq. (6). As a result, this simple model is well suited to chart the transition from thermionic emission to field emission. Figure 3.1 shows (as a function of energy) the supply function for $T = 500$, 1000, and 2000 K, and transmission coefficient for $L = 5$, 10, and 20 Å, where $\mu = \phi = 2$ eV. Oscillations that are visible for the wide barrier in $T(k)$ are due to the abrupt nature of the potential which introduces sinusoidal behavior — potentials which vary smoothly strongly suppress such oscillations. The

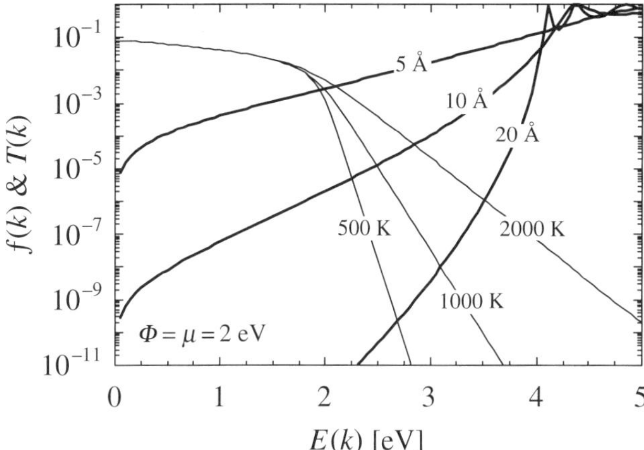

FIGURE 3.1. Transmission coefficient and supply function for a rectangular barrier for various widths (5, 10, and 20 Å) and temperatures (500, 1000, and 2000 K). The values of the chemical potential and work function were made small so as to highlight the field, thermal-field (T-F), and thermionic regions.

"knee" of the supply function marks the approximate location of the chemical potential: it is seen that each $\ln(T(E(k)))$ is approximately linear in this region. Consequently approximating $\ln(T(E))$ by a linear function is appropriate as long as the temperature is low enough that the decline in the supply function sufficiently dominates the rise in the transmission coefficient as energy increases. As shown in Fig. 3.2, the product of the supply function and the transmission coefficient is then typically peaked for low temperatures/narrow barriers near $E \approx \mu$ for "field" ($L = 10$ Å, $T = 500$ K) conditions and for high temperatures/wide barriers near $E = \mu + \phi$ for "thermal" ($L = 20$ Å, $T = 2000$ K) conditions. When the barrier is sufficiently narrow and the temperature sufficiently high, then a thermal-field (T-F) region arises where both processes are comparable. In Fig. 3.2, the oscillations which appear for high $E(k)$ are an artifact of the reflection probability associated with a square barrier — simple triangular barriers would not have such oscillations. The energy spread, or full width at half maximum (FWHM), is determined by the rapidity with which $T(E)$ rises and $f(k(E))$ falls, and therefore, is seen to depend strongly on conditions.

3.2.2. The Fowler–Nordheim Equation

3.2.2.1. *The Triangular Barrier Approximation.* The triangular potential barrier considered by Fowler and Nordheim is $V(x) = V_0 - Fx$ for $x \geq 0$ and 0 otherwise [24].* While the derivation, which follows, differs from theirs, it is in

---

* It was this form of the potential considered by Fowler and Nordheim [24]; inclusion of the image charge appeared in their subsequent work. The present Airy function approach deviates from that of Fowler and Nordheim in order to enable generalizations below.

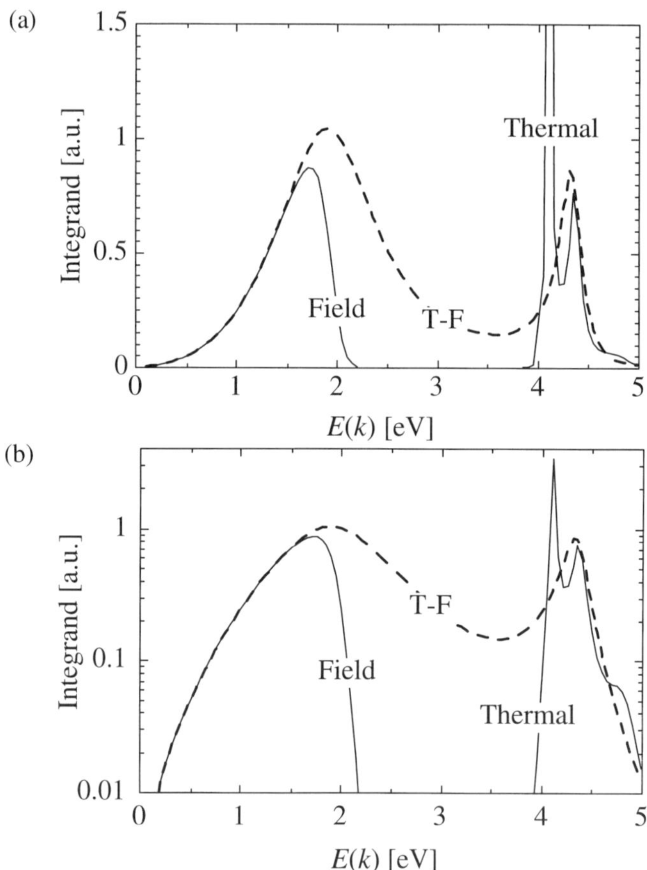

FIGURE 3.2. (a) Product of the transmission coefficient and supply function as shown in Fig. 1 versus energy. The curves represent the energy distribution of the emitted electrons. (b) Same as part(a) but for the log of the product of $T(E)f(k(E))$.

anticipation of later sections. Schrödinger's equation for $x > 0$ may be cast in the form of Airy's differential equation $(\partial_\omega^2 - c^2\omega)\psi(z) = 0$, where $z = (k_0^2 - k^2 \pm fx)/f^{2/3}$, $f = 2m|F|/\hbar^2$, $\omega = |z|$ and $c^2 = z/\omega$ [25]. At present, restrict attention to the $(-f)$ case. Introduce $Zi(c,\omega)$, where

$$Zi(c,\omega) = -\frac{1}{4}(c-1)(c^2+2c+3)Ai(c^2\omega)$$

$$+\frac{1}{4}(c+1)(3c^2-2c+1)Bi(c^2\omega) \tag{9}$$

$c = \pm 1$ represents exponential behavior, and $c = \pm i$ represents sinusoidal behavior. The equations for $Zi'(c,\omega)$ are the same with the replacements $Ai \Rightarrow Ai'$ and $Bi \Rightarrow Bi'$, for which the primes indicate differentiation with respect to argument.

For example, $Zi(1, \omega) = Bi(\omega)$ and $Zi(i, \omega) = Ai(-\omega) - iBi(-\omega)$. Airy functions are related to Bessel functions of order $1/3$, and are often encountered in physics problems dealing with propagation: Zi is, therefore, related to the Hankel function. The transmitted wave function becomes

$$\psi(x) = t(k)\frac{Zi(c(x), \omega(x))}{Zi(c_0, \omega_0)} + r(k)\frac{Zi(-c(x), \omega(x))}{Zi(c_0, \omega_0)} \tag{10}$$

where subscript 0 on c and ω indicate evaluation at $x = 0$, and for field emission boundary conditions, $r(k) = 0$ (no electrons incident from the right, or vacuum, side) but is included here for completeness. The form of Eq. (10) ensures that the asymptotic limit for vanishing field gives rise to propagating or exponentially decaying solutions ($t(k)\exp(i\kappa x)$ or $t(k)\exp(-\kappa x)$, respectively, where $\kappa = |k_0^2 - k^2|^{1/2}$), depending on whether the incident particle energy is above or below V_0, and further, that the coefficients $r(k)$ and $t(k)$ remain (numerically) manageable — that is, calculating their magnitude does not require (or exceed) double precision limits in, for example, Fortran. $T(k)$ becomes

$$T(k) = \frac{f^{1/3}|t|^2}{\pi k |Zi(1, \omega_0)|^2} \tag{11}$$

As before, matching the wave function and first derivative at $x = 0$ for $k > k_0$ gives

$$\begin{pmatrix} 1 & 1 \\ ik & -ik \end{pmatrix}\begin{pmatrix} 1 \\ r \end{pmatrix} = \begin{pmatrix} 1 & 1 \\ -\dfrac{f^{1/3}Zi'(i, \omega_0)}{Zi(i, \omega_0)} & -\dfrac{f^{1/3}Zi'(-i, \omega_0)}{Zi(-i, \omega_0)} \end{pmatrix}\begin{pmatrix} t' \\ r' \end{pmatrix} \tag{12}$$

where t' and r' are the coefficients for $x = 0^+$. Unlike the case $k > k_0$, under-the-barrier transport requires one more matching if the Zi functions (instead of the Airy functions themselves) are used. For finite f, the wave function emerges from the barrier, so that ($\omega = 0$ because $E(k) = V(x)$)

$$\begin{pmatrix} \dfrac{Zi(1, 0)}{Zi(1, \omega_0)} & \dfrac{Zi(-1, 0)}{Zi(-1, \omega_0)} \\ -\dfrac{f^{1/3}Zi'(1, 0)}{Zi(1, \omega_0)} & -\dfrac{f^{1/3}Zi'(-1, 0)}{Zi(-1, \omega_0)} \end{pmatrix}\begin{pmatrix} t' \\ r' \end{pmatrix} = \begin{pmatrix} \dfrac{Zi(i, 0)}{Zi(1, \omega_0)} & \dfrac{Zi(-i, 0)}{Zi(-1, \omega_0)} \\ -\dfrac{f^{1/3}Zi'(i, 0)}{Zi(1, \omega_0)} & -\dfrac{f^{1/3}Zi'(-i, 0)}{Zi(-1, \omega_0)} \end{pmatrix}\begin{pmatrix} t \\ 0 \end{pmatrix} \tag{13a}$$

After isolating the coefficient vector and employing the Wronskian of Airy functions, Eq. (13a) is equivalent to

$$\begin{pmatrix} t' \\ r' \end{pmatrix} = \begin{pmatrix} -i & \dfrac{iZi(1, \omega_0)}{Zi(-1, \omega_0)} \\ \dfrac{Zi(-1, \omega_0)}{Zi(1, \omega_0)} & 1 \end{pmatrix}\begin{pmatrix} t \\ 0 \end{pmatrix} \tag{13b}$$

Eq. (13b) is required to relate left-hand-side (LHS) coefficients to right-hand-side (RHS) coefficients in Eq. (12) when $E(k) - V(x)$ changes sign, that is, when a transition from below the potential to above occurs as when the wave function emerges from the barrier; an analogous equation holds when transitioning from above the potential to below. Solving Eqs. (12) and (13) for t when $k < k_0$ results in

$$t = \frac{2k\,Zi(1,\omega_0)}{f^{1/3}\,(Zi'(1,\omega_0) + i\,Zi'(-1,\omega_0)) - ik\,(Zi(1,\omega_0) + i\,Zi(-1,\omega_0))} \tag{14}$$

Asymptotically ($\omega \gg 1$; recall that c can assume any of the four values ± 1, $\pm i$)

$$Zi(c,\omega) \approx \frac{1}{8\sqrt{\pi}\,\omega^{1/4}}([c+3]\,[c^2+1] + 2\sqrt{2}[c-1][c^2-1])\exp\left(\frac{2}{3}c\omega^{3/2}\right)$$

$$Zi'(c,\omega) \approx c^3\sqrt{\omega}\,Zi(c,\omega) \tag{15}$$

so that Eq. (11) becomes (to leading order in the denominator)

$$T(k) \approx \frac{16\kappa k}{4\,k_0^2\exp\left(\dfrac{4}{3f}\kappa^3\right) + 8\,\kappa k + k_0^2\exp\left(-\dfrac{4}{3f}\kappa^3\right)} \tag{16}$$

Notice the similarity to Eq. (4). To obtain $J(F)$, the same approximations as for the square barrier are performed: neglect terms other than the dominant exponential term in the denominator, Taylor expand κ to 1st order in $(\mu - E)$ in the exponent (but not in the co-efficient of the exponent), and integrate with the zero-temperature $f(k)$. After some re-arranging, it may be shown that

$$J(F) \approx \frac{q}{4\pi^2\hbar}\frac{\sqrt{\mu}}{[\mu+\phi]\sqrt{\phi}}F^2\exp\left(-\frac{4}{3\hbar F}\sqrt{2m\phi^3}\right) \tag{17}$$

Eq. (17) is the oft-cited *F–N equation* and corresponds to Eq. (28) of Fowler and Nordheim's seminal article of 1928 [24]. Note that

- $\mu + \phi$ is the height of the *triangular barrier*. When $\mu \approx \phi$, then $(\mu/\phi)^{1/2}/(\mu + \phi) \approx 1/(2\phi)$;
- Eq. (17) has been derived for $T = 0$ K;
- the argument of the exponential term in Eq. (16), analogous to the square barrier, can be obtained for $\theta(k) \equiv \int_0^L (k_0(x) - k^2)^{1/2}\,dx$ as before, except that $k_0(x)$ is position dependent, and the upper limit of integration is defined by $k_0(L) = k$;
- the approximation $\beta\mu \gg 1$ has been used; if this condition is violated (e.g., when the electron density is small, as for semiconductors), then neglected terms corresponding to the lower limit of integration in Eq. (5) must be retained;
- image charge corrections have not been included.

3.2.2.2. The Chemical Potential.   Thus far, μ has been treated as an input para-meter, but it is obtained from the electron (number) density ρ through the Fermi–Dirac distribution according to [26]

$$\rho = 4\frac{M_{\mathrm{c}}}{\sqrt{\pi}}\left(\frac{m}{2\pi\beta\hbar^2}\right)^{3/2}\int\limits_0^\infty \frac{\sqrt{y}\,\mathrm{d}y}{1+e^{y-\beta\mu}} \equiv N_c\frac{2}{\sqrt{\pi}}F_{1/2}(\beta\mu) \tag{18}$$

where M_{c} is the number of equivalent minima in the conduction band (e.g., 1 for metals, 6 for silicon, etc.), $F_{1/2}(x)$ is the Fermi–Dirac integral, and m is the effective mass of the electron (equal to the rest mass $m_0 \approx 0.511$ MeV/c^2 for metals). Given the high electron density for metals, band bending under the influence of high applied fields is negligible (but not for semiconductors), in which case μ is, to a good approximation, equal to its bulk value. Eq. (18) holds for any temperature: at 0 K, $\mu = \mu_0$ is the "Fermi energy E_{F}" [27,28]. If $M_{\mathrm{c}} = 1$ and $T = 0$ K, then $\rho = k_{\mathrm{F}}^3/(3\pi^2)$, as known from statistical mechanics [22]. Tables of electron concentrations ρ and Fermi energies E_{F} may be found in the literature (e.g., Table 1 in Chapter 5 of Ref. [28]). For large values of $\beta\mu_0$, asymptotic expansions which allow iterative solution may be employed, such as [22,25,29]

$$\frac{2}{3}(\beta\mu_0)^{3/2} = \int\limits_0^\infty \mathrm{d}x\,\frac{\sqrt{x}}{1+e^{x-\beta\mu}} \approx \frac{2}{3}(\beta\mu)^{3/2}\left(1+\frac{1}{8}\left(\frac{\pi}{\beta\mu}\right)^2 + \frac{7}{640}\left(\frac{\pi}{\beta\mu}\right)^4\right)$$

$$\tag{19a}$$

which, upon inversion, yields

$$\mu(T) = \mu_0\left[1 - \frac{1}{12}\left(\frac{\pi}{\beta\mu_0}\right)^2 - \frac{1}{80}\left(\frac{\pi}{\beta\mu_0}\right)^4 + \cdots\right] \tag{19b}$$

3.2.2.3. Image Charge Potential. A charged particle outside the surface of a metal induces a charge distribution on the surface to screen the charged particle's electric field inside the bulk. The surface charge acts as though a charged particle of equal and opposite sign is located equi-distant from the surface as the original charge. If the charge is an electron, then the force the electron feels due to the induced "image charge" is equal to $-q^2/4\pi\varepsilon_0(2x)^2$, where x is the distance from the surface of the metal to the external electron. Extracting the electron to infinity requires work to be done such that

$$V_{\mathrm{image}}(x) = \int\limits_x^\infty -\frac{q^2}{16\pi\varepsilon_0 x'^2}\,\mathrm{d}x' = -\frac{q^2}{16\pi\varepsilon_0 x} \tag{20}$$

The conclusions of this classical description are retained in form when quantum mechanical considerations are taken into account [30]: the image charge contribution affects the barrier near the surface and the form of Eq. (20) holds to within a few angstroms of the metal/vacuum interface. The ubiquitous constant $q^2/16\pi\varepsilon_0 \approx 3.6$ eV Å shall be designated Q. Asymptotically for $F = 0$, the difference between the barrier height and the chemical potential is the work function, so $\phi = \Phi$. The image charge potential for field and thermionic emission then takes the form

$$V(x) = \begin{cases} \mu + \Phi - Fx - \dfrac{Q}{x} & x \geq x_{\min} \\[2mm] 0 & x < x_{\min} \end{cases} \tag{21}$$

where $V(x_{\min}) = 0$. The zeros of $V(x) - E(k)$ occur at the locations

$$x_{\pm} = \frac{h \pm \sqrt{h^2 - 4QF}}{2F} \tag{22}$$

where the $(\pm)$ subscript is associated with the larger and smaller root, respectively, and h is given by $h(E) = \mu + \phi - E(k)$. The triangular barrier is recovered in the limit $Q = 0$. The difference $x_{+} - x_{-}$ shall be designated L_0.

3.2.2.4. Approximate Solutions to T(k).

If the wave function is represented as $\psi = \exp(i S(x)/\hbar)$ and the phase $S(x)$ is expanded in powers of $\hbar$ such that $S(x) \approx S_0(x) + \hbar S_1(x) + \cdots$, then Schrödinger's equation can be written as

$$(\partial_x S_0)^2 + 2m(V(x) - E) + \hbar \left(2\partial_x S_0 \partial_x S_1 - i\partial_x^2 S_0\right) + \cdots = 0 \tag{23}$$

As before, introduce $\kappa(x) \equiv (k_0(x)^2 - k^2)^{1/2}$, and restrict attention to tunneling, where $k_0(x) > k$. Equating equal powers of $\hbar$ gives [21]

$$S_0(x) = \int^{x} \hbar\kappa(x')\, dx'; \qquad S_1(x) = \frac{1}{2}i\, \ln(\kappa(x)) \tag{24}$$

$S_0/\hbar$ is thus identified as the "area under the curve" expression encountered previously. In the pursuit of an analytical total current density formula for the image charge potential, it is convenient to ignore the $S_1(x)$ term. The "area under the curve" formulation of the transmission coefficient, introduced in the discussions following Eqs. (4) and (17) and established by Eq. (24) for "smoothly" varying potentials (those in which $|\hbar\partial_x S_1(x)/\partial_x S_0(x)| \ll 1$), takes the form

$$\theta(E) = \frac{1}{\hbar} \int_{x_-}^{x_+} \sqrt{2m\left(h(E) - Fx - \frac{Q}{x}\right)}\, dx \tag{25}$$

where $\theta = S_0(x)/\hbar$, and $h(E) = \mu + \phi - E(k)$ is the barrier height. Recall that $T(k) \approx \exp(-2\theta)$, Eq. (25) can be recast in a form appropriate for evaluation by Gaussian quadrature. Define $x = x_- + L_0 \sin^2 s$, then

$$\theta(E) = 2\frac{\sqrt{2\,Fm}\,L_0^2}{\hbar} \int_0^{\pi/2} \frac{\cos^2(s)\sin^2(s)}{\sqrt{x_- + L_0\sin^2(s)}}\, ds \equiv 2\frac{\sqrt{2\,Fm}\,L_0^{3/2}}{\hbar} R\left(\frac{x_-}{L_0}\right) \tag{26}$$

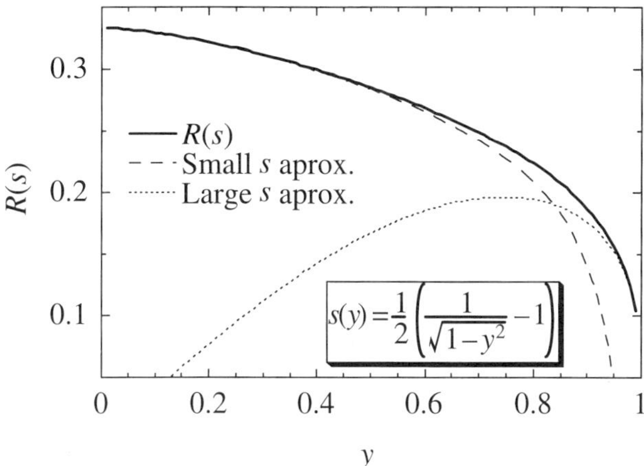

FIGURE 3.3. A comparison of the function $R(s)$ used in the *WKB* evaluation of the image charge potential with its asymptotic expansions.

where the introduced function $R(s)$, as shown in Fig. 3.3, is defined by [31]

$$R(s) = \int_0^{\pi/2} \frac{\cos^2(x)\sin^2(x)}{\sqrt{s+\sin^2(x)}}\, dx$$

$$\approx \frac{\sqrt{s+1}}{4224}\left(\pi\left[336\ln\left(\frac{s}{s+1}\right) - 151\right]s + 1408\right) \quad (s \to 0) \quad (27)$$

$$\approx \frac{\pi\sqrt{s}}{4(4s+1)} \quad (s \to \infty)$$

Taylor expanding $\theta(E)$ about μ such that

$$\theta(E) \approx \theta(\mu) + (\mu - E)(\partial_E\theta)|_{E=\mu} \equiv \frac{b_{FN}}{2F} + \frac{1}{2}c_{FN}(\mu - E) \quad (28)$$

and inserting $T(k) = \exp(-2\theta)$ into Eq. (5) at finite temperatures results in

$$J(T,F) = \frac{m}{2\pi^2\beta\hbar^3}\exp\left(-\frac{b_{FN}}{F}\right)\int_0^{\infty}\ln\left(1 + e^{\beta(\mu-E)}\right)\exp[-c_{FN}(\mu - E)]\, dE$$

$$= \frac{m}{2\pi^2\hbar^3 c_{FN}^2}\exp\left(-\frac{b_{FN}}{F}\right)\left[\frac{\pi c_{FN}}{\beta\sin\left(\dfrac{\pi c_{FN}}{\beta}\right)}\right] \quad (29a)$$

Eq. (29a) has been called the "thermal-field emission" equation [10]. Use has been made of the exact integral (applicable when $\exp(-\beta\mu) \approx 0$, i.e., the lower limit of the integral in Eq. (29a) can be neglected):

$$\int_0^\infty x^{\mu-1}\,\ln(1+x)\,\mathrm{d}x = \frac{\pi}{\mu\,\sin(\mu\pi)} \tag{29b}$$

where $0 < \mathrm{Real}(\mu) < 1$ [32]. The term in square brackets in Eq. (29a) is close to unity for field emission at room temperature (e.g., for $F = 0.448$ eV/Å at $T = 300$ K and molybdenum parameters, it is 1.028), and so is often neglected — as in Eq. (1) — but if fields become low or temperatures high, a positive convexity is introduced on a F–N plot. Eq. (29) is related to the elliptical integral functions $v(y)$ and $t(y)$ commonly used through the identification

$$a_{\mathrm{FN}}(F) = \frac{1}{16\,\pi^2\hbar\Phi\,t(y)^2}$$

$$b_{\mathrm{FN}}(F) = \frac{4}{3\,\hbar}\,\sqrt{2m\Phi^3}v(y) \tag{30}$$

$$c_{\mathrm{FN}}(F) = \frac{2}{\hbar F}\sqrt{2m\Phi}\,t(y)$$

where $y^2 = 4QF/\Phi^2$. The $v(y)$ and $t(y)$ have been tabulated by Forbes [17,18]. In terms of the $R(s)$ function, they are given by

$$v(y) = 3(1 - y^2)^{3/4}\,R\left(\frac{1}{2}\frac{1 - \sqrt{1 - y^2}}{\sqrt{1 - y^2}}\right)$$

$$t(y) = \left(1 - \frac{2}{3}y\frac{\partial}{\partial y}\right)v(y) \tag{31}$$

All the coefficients in Eq. (30) are either explicitly or (through the definition of y) implicitly dependent on applied field: linearity of Eq. (29) on a F–N plot is not *a priori* guaranteed, though it may be made so by linearizing $v(y)$ in F and approximating $t(y)$ by a constant. For $T = 0$ K, a widely used approximation due to Spindt, Brodie, and co-workers [2] is $t(y) \approx \sqrt{1.1}$ and $v(y) \approx v_0 - v_1 y^2$, where they chose $v_0 = 0.95$, and $v_1 = 1$. A slightly better approximation based on Eq. (26) and the demand that $v_1 \equiv 1$ gives $t(y) = \sqrt{1.1164}$ and $v_0 = 0.93685$ [25]. Quadratic $v(y)$ approximations are reasonable for intermediate values of y. They are useful in making estimates of effective emission areas (below) and work functions from plots of experimental data. However, as Forbes has shown, such estimates depend on where the tangent lines to the theoretical F–N plots are taken, and they degrade near the values $y = 0$ or 1, and so some care is called for [17,18]. If a quadratic $v(y)$ is used, then the exponential term containing v_1 can be absorbed into a_{FN} and then b_{FN} redefined to not include v_1. The resulting $J(F)$, being identically linear on an F–N plot, is often used for estimating Φ_{exp} based on the slope and emission area based on the intercept.

TABLE 3.2. Material Parameters for Si and Mo at $T = 300$ K

Symbol	Description	Silicon	Molybdenum
M_c	Conduction band eqv. minima	6	1
Φ	Experimental work function	$\chi - \mu$	4.41 eV
χ	Electron affinity	4.05 eV	$\mu + \Phi$
ρ_0	Bulk electron density	$\approx 10^{-6}$ Å^{-3}	0.06463752 Å^{-3}
μ_0	Bulk chemical potential	-0.08611525 eV	5.873 eV
$Q(K_s)$	$\alpha\hbar c(K_s - 1)/4(K_s + 1)$	3.041785 eVÅ	3.599911 eVÅ
m	Electron effective mass	0.3282800 m_0	m_0
K_s	Dielectric constant	11.9	$\gg 1$
N_c	Density coefficient	2.831983×10^{-5} Å^{-3}	2.509416×10^{-5} Å^{-3}

The approximation in Eq. (29) is predicated on $\beta\mu \gg 1$, which is manifestly true for metals. For semiconductors, the lower limit cannot be casually neglected. As shown by Stratton [7] and elsewhere,* including the lower limit results in the replacement

$$\frac{c_{FN}\pi/\beta}{\sin(c_{FN}\pi/\beta)} \Rightarrow \frac{c_{FN}\pi/\beta}{\sin(c_{FN}\pi/\beta)} - (1 + c_{FN}\mu)\, e^{-c_{FN}\mu} \qquad (32)$$

where terms comparable to the next correction, $\delta = c_{FN}^2 \exp[-\mu(\beta + c_{FN})]/[\beta(c_{FN} + \beta)]$, and smaller have been discarded (i.e., for Mo and Si parameters at $T = 300$ K and $F \approx 0.5$ eV/Å, $\delta \approx 0(10^{-33})$ and $0(10^{-12})$, respectively; the parameters are given in Table 3.2).

The zero-temperature approximation to Eq. (29) is evidently obtained by approximating $\zeta / \sin(\zeta) \approx 1$, where $\zeta = \pi c_{FN}/\beta$. The opposite limit, where thermionic emission dominates, is less evident. There, fields are typically of such a size that the factor L_0 in Eq. (26) renders θ large unless the energy of the electron is close to the barrier maximum. From the image charge potential in Eq. (21), the barrier maximum V_{max} occurs at $x = (Q/F)^{1/2}$ and is equal to $V_{max} = \mu + \phi$, where $\phi = \Phi - (4QF)^{1/2}$. $T(E(k))$ then vanishes for $E \leq V_{max}$ and ϕ is replaced by $\Phi - (4QF)^{1/2}$ in the exponential term in Eq. (8).

3.2.3. Beyond the Fowler–Nordheim Equation

3.2.3.1. Many-Body Effects: Barrier Origin and the Image Charge. The
image charge approximation given in Eq. (21) is classical in origin, but is nevertheless a good approximation for $x > 3$ Å [33–35] (depending on the metal surface). Closer to the surface, the quantum mechanical nature of the electron degrades the

---

* The corresponding equation C1 in Ref. [25] contains a typographical error compared to that in Ref. [31], which has been corrected here.

approximation. Further, the status of the barrier (and hence, work function) has so far been consigned to an input parameter based on experimental data. A proper account of the electron wave nature and density affects both approximations. The environment in which the electrons propagate as plane waves has so far been assumed to be a uniform background potential and a step function of height $\mu + \Phi$ at the metal/vacuum interface. The origin of the step function is the change of electron density from metallic values ($\approx 10^{22}$ electrons/cm^3) to a negligible fraction of that in vacuum, plus a dipole introduced by electron wave function penetration into the barrier, plus the effects of the ionic cores of the metal atom lattice constituting the crystal in which the electrons propagate. Mathematically, these three terms are captured, respectively, in the equation for the barrier height[†]

$$V_0 = -\frac{\partial}{\partial \rho} \left[\rho \, \varepsilon_{\mathrm{xc}}(\rho) \right] + \Delta\phi + \varepsilon_{\mathrm{ion}} \tag{33}$$

where ρ is the electron density encountered in Eq. (18), and $\varepsilon_{\mathrm{xc}}$ is the exchange-correlation energy per particle [27] accounting for many-body effects: because of the interaction of the electrons, there is a reduction in probability of electrons being near each other in addition to the Pauli exclusion principle, and both account for the many-body interaction [36]. The first term in Eq. (33) shall be referred to as V_{xc}. The (bulk) electron density is often parameterized by r_s such that one electron is contained in a sphere of radius $r_\mathrm{s}a_0$ (a_0 is the Bohr radius 0.529Å, so that r_s is dimensionless), or $\rho = 1/[(4\pi/3)\,(r_\mathrm{s}a_0)^3]$: for example, a density of 10^{22} electrons/cm^3 corresponds to $r_\mathrm{s} \approx 5.44$, whereas a density of 10^{18} electrons/cm^3 would correspond to $r_\mathrm{s} \approx 117$.

The replacement of the ionic lattice constituting the crystal which forms the metal by a uniform positive background ("jellium" model) is predicated on the fact that the electron concentration does not build up in the vicinity of the ionic cores: the conduction electrons are not bound to the cores by the same barrier which confines the core electrons, and so conduction electrons spend only a small portion of their time there [35]. Nevertheless, the stability of simple metals in comparison to free atoms is because of the decrease in the ground state energy due to ionic cores, thereby increasing the binding energy of the electrons [28]. A satisfactory treatment involves the use of pseudo-potentials [37] to model the contribution of the ion lattice, as done by Lang and Kohn [34], but that is beyond the scope of the present treatment. An albeit crude estimation of the ionic core term $\varepsilon_{\mathrm{ion}}$, based on the Wigner–Seitz model, proceeds by demanding that the derivative of the electron wave function vanish at the boundary for a spherically symmetric potential which vanishes for $r \geq r_\mathrm{s}a_0$,[*]

[†] Apart from the usage of ρ instead of n for density, the notation shall follow Ref. [25]. A spurious density coefficient in the first term of Eq. (3) in that reference has been corrected in Eq. (33).

[*] The approach here is based on the approximate treatment in Ref. [25], and based on Haas and Thomas [23], but far more accurate approaches are in the literature: see, for example Ref. 38 and references therein.

for which

$$\varepsilon_{\text{ion}} = \frac{6}{5}\frac{\alpha_{\text{fs}}\hbar c}{r_s a_0}\left(1 - 5\left(\frac{a_i}{r_s a_0}\right)^2\right)$$
(34)

where a_i is of the same order as the ionic radius of the metal atom — but is nevertheless different. Ignorance of ε_{ion} has been shifted to ignorance of a_i, but once the value of a_i has been found by other means (e.g., forcing V_0 in Eq. (33) to be equal to $\mu + \Phi_{\text{exp}}$, where the "exp" denotes "experimental"), investigations may proceed for different temperatures and field configurations. Even so, the conceptual motivation for introducing Eq. (34) — namely, the conduction band characteristics are dictated by s–p electron behavior and not complicated by d electron behavior — is adequate only for simple bulk metals and not true in general. Side-stepping these complications is warranted only because the present interest is to develop generalizations to the F–N equation, which nominally account for many-body effects.

The exchange-correlation potential requires greater attention. The total energy of an electron gas E divided by the number of electrons N is denoted as ε and is composed of three parts: a kinetic term ε_{ke}, an exchange term ε_{ex} (which accounts for the same spin electrons' tendency not to occupy the same space), and a correlation term ε_{cor} (which is simply the difference between the first two terms and the "true" energy); ε_{cor} is called the "stupidity" energy by Feynman. For a uniform electron gas, the kinetic energy and density terms are

$$\rho = \frac{g}{(2\pi)^3}\int f_{\text{FD}}(E)\,\mathrm{d}^3k = \frac{g}{(2\pi)^3}\int_0^{k_F} 4\pi k^2\,\mathrm{d}k$$

$$\varepsilon_{\text{ke}} = \frac{g}{(2\pi)^3}\int E(k)\,f_{\text{FD}}(E)\,\mathrm{d}^3k = \frac{g}{(2\pi)^3}\int_0^{k_F}\frac{\hbar^2 k^2}{2m}\,4\pi k^2\,\mathrm{d}k$$
(35)

where $g = 2$ is the number of spins, and $\mathrm{d}^3k = 4\pi k^2\,\mathrm{d}k$. In Eq. (35), $f_{\text{FD}}(E)$ is the Fermi–Dirac distribution in the zero-temperature limit, $f_{\text{FD}}(E)|_{T=0\,\text{K}} = \theta(\mu - E) = \theta(k_F - k)$, where θ is the Heaviside step function. Thus

$$\varepsilon_{\text{ke}} = \frac{3}{5}\mu\rho = \frac{3}{5}\left(\frac{9\pi}{4}\right)^{2/3}\frac{1}{r_s^2}Ry$$
(36)

where $Ry \equiv \alpha\hbar c/(2a_0) = 13.6045$ eV (Rydberg energy), and $a_0 = 0.529\,\text{Å}$ (Bohr radius).

Evaluation of the potential energy requires greater detail. In the second quantization formulation, two creation operators fill up the holes created by two destruction operators [39] — that is, electron $\boldsymbol{k}$ scatters to state $\boldsymbol{p} - \boldsymbol{q}$ and electron $\boldsymbol{p}$ scatters to state $k + q$ (the exchange process, q being the momentum associated with the coulomb interaction) — thereby giving rise to the first order energy shift. Evaluating their contribution amounts to finding (for unit volume)

$$E_{ex} = \frac{g}{(2\pi)^3} \frac{\alpha \hbar c}{2} \int f_{FD}(E)\, d^3k \int \frac{1}{q^2} \theta\left(k_F - |k + q|\right) d^3q \qquad (37)$$

where the $1/q^2$ arises from the coulomb potential. Performing the k-integration and taking the ratio with the number of particles N to get ε_{ex} results in

$$\varepsilon_{ex} = \frac{\alpha \hbar c}{16\pi k_F^3} \int_0^{2k_F} (q + 4k_F)(q - 2k_F)^2\, dq = \frac{3}{2\pi}\left(\frac{9\pi}{2g}\right)^{1/3} \frac{1}{r_s} Ry \qquad (38)$$

The correlation energy ε_{cor} is composed of the summation of a large number of Feynman diagrams and is beyond the scope of the present treatment, but is discussed at length in both Feynman [27] and Fetter and Walecka [39]. Intermediate values of r_s — as nature would have it — typifies the electron density of metals characteristic of field emitters, for which a convenient asymptotic formula must be generated. A connection formula between the high density (low r_s) expansions and the low density limit, where the electrons become localized, or "crystalize" into a "Wigner Solid" (body-centered cubic lattice), suffices. A popular one due to Wigner* is

$$\varepsilon_{cor}|_{Wigner} \approx -\left(\frac{0.88}{r_s + 7.8}\right) Ry \qquad (39)$$

A form which has the correct asymptotic limits as well as the proper minimum is [25]

$$\varepsilon_{cor} \approx -\left(\frac{130.78}{r_s^6 + 4388.4} + \frac{0.87553}{r_s + 3.0016\sqrt{r_s} + 5.6518}\right) Ry \qquad (40)$$

where the first term in parentheses is designed to achieve the minimum value of $\varepsilon_{cor}(r_s)$ as found by Ishihara and Montroll [40].

The dipole term $\Delta\phi$ is obtained from Poisson's equation once the electron density as a function of position is known. Consider again the rectangular barrier (the triangular potential case proceeds analogously, but the manipulations of the Airy function-based $r(k)$ and $t(k)$ are more involved than the hyperbolic trigonometric functions associated with the rectangular barrier). In terms of the normalized wave

---

*Units used in the literature vary widely and are not often shown explicitly. Here, r_s is a dimensionless parameter and the Rydberg ($Ry \approx 13.6\,eV$) is the unit of energy, following Feynman. Lang and Kohn used two Rydbergs for energy and a Bohr radius $a_0 \approx 0.529$ Å for length, thereby making (for example) the 0.88 in Eq. (39) become 0.44 in their Eq. (2.15a) in Ref. [34]. Other conventions abound.

function $\psi_k(x)$, the quantum mechanical density $\rho(x)$ is given by

$$\rho(x) = \frac{1}{2\pi} \int_0^\infty f(k) |\psi_k(x)|^2 \, dk \tag{41}$$

where $f(k)$ is the supply function encountered in Eq. (6). Eqs. (18) and (41) for $\psi_k(x) = e^{ikx}$ (no reflection) are equivalent. Writing $r(k) = R(k)\exp(-2i\varphi)$ and using Eq. (3) yields

$$R(k) = \frac{(\kappa^2 + k^2)\sinh(L\kappa)}{\sqrt{(\kappa^2 + k^2)^2 \ \sinh^2(L\kappa) + 4\,\kappa^2 k^2}} \tag{42}$$

for the magnitude and

$$\varphi(k) = \frac{1}{2}\left(\arctan\left[2\frac{\kappa k \cosh(L\kappa)}{(k^2 - \kappa^2)\ \sinh(L\kappa)}\right] + \pi\right) \tag{43}$$

for the phase. With $\psi_k(x) = [\exp(ikx) + R\exp(-i(2\varphi + kx))]/\sqrt{2}$ in Eq. (41), the evaluation of $\rho(x)$ for $x \leq 0$ is possible, which, given the involved k-dependence entailed in Eqs. (42) and (43), must be performed numerically. For $T = 0$ K and high and/or wide barriers, an analytical expression can be obtained which is identical to the triangular potential case for high barriers and/or low fields:

$$\lim_{\beta, L, V_0 \to \infty} \rho(x) = \frac{g \, k_{\mathrm{F}}^3}{6\,\pi^2}\left[1 + 3\frac{\cos(\zeta)}{\zeta^2} - 3\frac{\sin(\zeta)}{\zeta^3}\right] \tag{44}$$

where the coefficient is the bulk density, $\zeta \equiv 2k_{\mathrm{F}}(x - x_0)$, and $x_0 = 1/k_0$. $\zeta(x)$, therefore, depends on both work function (due to k_{F}) and barrier height (due to x_0). The trigonometric functions in Eq. (44) are responsible for the "Friedel oscillations" in the electron density near the interface, visible in Fig. 3.4 for varying chemical potential and work function. Changes in the chemical potential have greater impact on the profile due to the wavelength associated with the most energetic electrons. Eq. (44) holds as the limiting case for both triangular and rectangular potentials. This is significant because of the following:

- For metals, the thermal Fermi–Dirac distribution is close to the $T = 0$ K limit (recall the weak temperature dependence of Eq. (29a) where $\beta = 1/k_B T$), and so $\rho(T, x) \approx \rho(0\ \mathrm{K}, x)$.
- To leading order, the density profile is dependent on barrier height, not shape, as in Fig. 3.4(b).
- To leading order, changes in the potential barrier height result in a translation of the density profile by $x_0 = \hbar/(2m V_0)^{1/2}$ where $V_0 = \mu + \Phi$; that is, the curves of Fig. 3.4(b) would overlap if plotted versus $(x + x_0)$ instead of x.

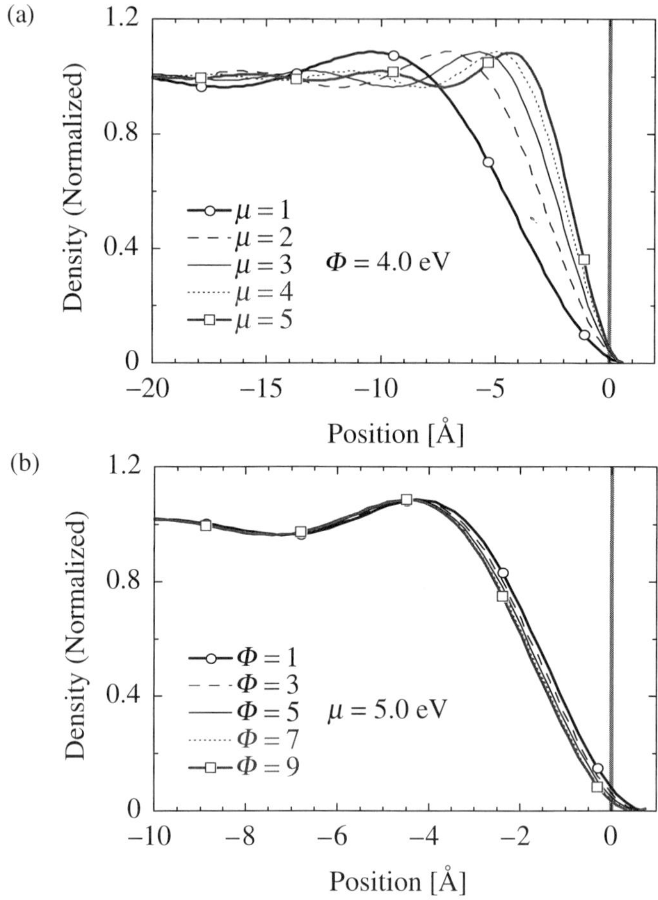

FIGURE 3.4. (a) Electron density as a function of position for various chemical potentials for a work function of 4.0 eV and a rectangular barrier profile. Undulations in the density are associated with Friedel oscillations. (b) Same as part (a) but for various work functions with a chemical potential of 5.0 eV.

- To preserve global charge neutrality, the origin of the background positive charge x_i (typically negative) is different from the electron density, and depends on μ and V_0 (the wave function has been calculated for a potential of an assumed shape, not one self-consistently generated using Poisson's equation)

The spilling out of electrons into vacuum leads to a dipole-induced potential difference $\Delta\phi$ at the interface. A crude approximation of $\Delta\phi$, suggested by the variational approach of Smith [41], is obtained by using a hyperbolic tangent approximation to the density

$$\rho_{\text{aprx}}(x) = \frac{1}{2}\left(\frac{g k_{\text{F}}^3}{6\pi^2}\right)\{1 - \tanh\left[\lambda k_{\text{F}}(x - x_i)\right]\} \tag{45}$$

x_i is chosen such that $\zeta_0 \equiv \zeta(x = x_i)$ and $\rho(x_i)$ is half of its bulk value due to the (artificial) symmetry of Eq. (45) about the ion origin. From Eq. (44), $\zeta_0 = -2.4983$, and will be approximated by $-5/2$ below. λ is obtained by setting $\partial_x \rho = \partial_x \rho_{\mathrm{aprx}}$ at $x = x_i$, for which $\lambda = 1.2483$ and will be approximated by $5/4$ below. The crude hyperbolic approximation of Eq. (45) allows for the analytical integration of Poisson's equation to give

$$\Delta\phi \approx \frac{\alpha\pi\hbar c g k_{\mathrm{F}}}{36\lambda^2}; \qquad x_i \approx -\frac{5}{4k_{\mathrm{F}}} + x_0 \tag{46a}$$

Eq. (46a) captures the basic behavior: as the barrier height changes, the magnitude of the dipole contribution and the location of the background positive charge origin shift to accommodate. The crudeness of the hyperbolic model, appealing though it is, overestimates the dipole contribution. If Eq. (44) for the density is used, a better approximation is obtained:* the insistance that an explicit integration over the ion and electron densities give zero net charge (plus a judicious neglect of smaller order terms) gives the following approximations

$$\Delta\phi \approx \frac{a_0}{\pi x_0} \left((4 - \pi k_{\mathrm{F}} x_0) k_{\mathrm{F}} x_0 - (1 - k_{\mathrm{F}}^2 x_0^2) \ln\left[\frac{1 + k_{\mathrm{F}} x_0}{1 - k_{\mathrm{F}} x_0}\right] \right) Ry$$

$$x_i \approx \frac{84 x_0}{168 - 36 k_{\mathrm{F}}^2 x_0^2 - 5 k_{\mathrm{F}}^4 x_0^4} \ln\left(\frac{17}{25} k_{\mathrm{F}}^2 x_0^2\right) \tag{46b}$$

The evaluation of the barrier height V_0 must be iteratively obtained due to the dependence of x_0 on V_0 (and x_i on x_0), and must proceed in conjunction with determining the value of a_i in Eq. (34), as $\varepsilon_{\mathrm{ion}}$ is sensitive to a_i's value, and it is not *a priori* the ion radius.

Sodium is a good test case. Using the parameters $r_{\mathrm{s}} = 3.930$ and $\Phi_{\mathrm{exp}} = 2.28$ eV, an iterative evaluation performed with a_i as an adjustable parameter until the predicted barrier height Eq. (33) matches $\mu + \Phi_{\mathrm{exp}}$ for $F = 0$ gives $a_i = 0.9882$ Å (compared to the actual ion radius of 0.9504 Å), $V_{\mathrm{xc}} = 5.532$ eV, $\Delta\phi = 1.064$ eV, $\mu = 3.244$ eV, $\varepsilon_{\mathrm{ion}} = 1.071$ eV,[†] x_0 and x_i are 0.830 Å and -0.441 Å, respectively. Better results are obtained if the finite temperature–finite barrier height formalism is retained, Poisson's equation is calculated numerically, Gaussian quadrature integration is used to calculate the electron density using numerically evaluated transmission coefficients [25], and x_i is chosen separately so as to insure global charge neutrality. The calculation is non-trivial, given the Friedel oscillations which compromise the analytical approximations and the fact that $|r(k)|$ need no longer be close to unity. The results of such a calculation for sodium at $T = 300$ K give $a_i = 0.8944$Å, $\Delta\phi = 0.6158$ eV,

---

* The expression for $\Delta\phi$ combines Eqs. (14) and (15) in Ref. [25] and corrects numerical coefficients. The equation for x_i is based on an extension of the (easier) $x > 0$ integral of Eq. (12) in Ref. [25]. It is worth repeating that high barrier and low temperature are assumed.

† Compare to Table IV of Lang in Ref. [35], which used the pseudo-potential analysis.

$\mu = 3.247$ eV, $\varepsilon_{\mathrm{ion}} = 0.6210$ eV, and $x_i = -0.427$ Å, respectively, which gives a sense of the validity of the analytical model behind Eq. (46).

For another example, consider molybdenum, the widely used material in Spindt-type field emitter arrays. For $r_s = 2.922$, $T = 300$ K, $\Phi_{\mathrm{exp}} = 4.41$ eV, and $a_i = 0.7596$ Å (compared to an ionic radius of 0.9300 Å), it is found that $\Delta\phi = 0.8485$ eV, $\mu = 5.5873$ eV, $\varepsilon_{\mathrm{ion}} = 0.6210$ eV, $x_0 = 0.5917$Å, and $x_i = -0.3175$Å, respectively.

These models are heuristic. Better methods to estimate the work function, from Bardeen [33] to the density functional formalism of Lang and Kohn [34,35] and beyond [42,43*], yield the surface energy and a variety of related effects with accuracy. As done in the density functional approach developed by Lang and Kohn, the exchange-correlation term is used as a potential in Schrödinger's equation. For the present, all that is required is to show that the effective barrier height has a component due to a change in electron density from bulk values to negligible (vacuum) levels and another component due to the degree with which electron penetration occurs into the barrier. Regarding the latter, penetration will be enhanced if the barrier height decreases or is thinner (i.e., the applied field is increased), or if a greater portion of the electrons can be induced to higher energy (i.e., the temperature is raised), both affecting the apparent barrier height. Therefore, the "work function" in the F–N equation will be temperature and field dependent.

3.2.3.2. Temperature and Field Dependent Barrier Height.

The image charge potential requires modification to accommodate many-body effects. The primary effect of barrier height is to shift the origin of the electron density ($\zeta(x)$ in Eq. (44)) by an amount x_0. The image charge potential will depend on the origin of the electron density, so the term Q/x in Eq. (21) should be replaced by $Q/(x + x_0)$, a conclusion which can be justified through more rigorous theory [44–46]. Here, $x = 0$ is identified as the *electron* coordinate system origin, and so there will be differences in sign compared to the literature. The potential will resemble $V(x) = \mu + \Phi(T, F) - Fx - Q/(x + x_0)$.

Because x_0 depends, to the first order, on barrier height, not on shape (as for the sufficiently abrupt potentials of thermal and field emission), approximating the image charge potential by a triangular barrier potential of the same height is adequate to obtain the electron density. From that density, Poisson's equation may be numerically solved to obtain the dipole and exchange-correlation potential for arbitrary field (once a_i is determined from the $F = 0$ case), and the process iterated until the predicted barrier height is convergent with the assumed barrier height [25]. Due to the relation of the dipole term to the penetration of electrons into the barrier, the apparent height of the barrier will increase with temperature and field.

There is no *a priori* reason to expect the ratio of the increase in barrier height $\Phi(F) - \Phi_0$ (where $\Phi_0 = \Phi(F = 0)$) and the product Fx_0 to be linear: in fact, $\Phi(F) - \Phi_0 \approx (Fx_0)^{1-\delta}$. In practice, δ is numerically found to be small so that pursuing its evaluation is unnecessarily fastideous (e.g., for molybdenum and cesium

---

* An interesting alternative method is due to [43].

parameters with $Fx_0 \leq 0.11$, $\delta_{\mathrm{Mo}} \approx 0.15$ and $\delta_{\mathrm{Cs}} \approx 0.12$, respectively). The simple approximation $\delta \approx 0$ is, therefore, adopted, and the effective potential is

$$V(x) = \mu + \Phi_0(T) - F(x - x_0) - Q/(x + x_0) \tag{47}$$

where $\mu + \Phi_0$ is the asymptotic barrier height in the absence of an applied field. In the same way, the simplest approximation for the temperature dependence of $\Phi(T)$ is to mimic empirical fits* and let $\Phi_0(T) = \Phi_0(0) + \alpha T$, where the constant α follows the notation of Haas and Thomas (and is not the fine structure constant). In the range of temperatures typical of thermionic cathodes ($T \approx 1000 - 1400$ K), α is approximately constant and of the order of $k_{\mathrm{B}} \approx 8.617 \times 10^{-5}$ eV/K. Because thermionic cathodes operate with an extraction grid held at kilovolt potentials hundreds of microns from the surface, a field (weak by field emission standards) is usually present which lowers the effective barrier height by approximately $(4QF)^{1/2}$ (as per Eq. (21)). For example, simulation suggests that for Cs at $F \approx 100$ V/μm and 40 V/μm, α is approximately $0.14k_{\mathrm{B}}$ and $0.12k_{\mathrm{B}}$, respectively, numbers comparable to experimental values for other materials (e.g., $5.8k_{\mathrm{B}}$ for Ba, $0.29k_{\mathrm{B}}$ for Mo, and $3.0k_{\mathrm{B}}$ for Si). Because the emitter surface is not atomically flat due to adsorbates, field enhancement occurs and makes the local field F appear larger than the macroscopic field $F_0 \approx V_{\mathrm{g}}/D$, where V_{g} is the grid (or anode) potential and D is the distance from the surface to it. Evaluations of α are, therefore, pedagogical rather than predictive, but they are in the ball park.

Finally, consider the effect of shifting the origin of the background positive charge by x_i. From classical electrostatics, a sheet of charge generates a field of magnitude $\sigma/2\varepsilon_0$, where σ is the surface charge density [47]. In the present case, there is global charge neutrality, so the extent that this field acts is finite. In an *ad hoc* manner, the field from a sheet of charge $\sigma = \rho x_i$ acting over a length scale x_i gives rise to a potential difference $\rho x_i^2/2\varepsilon_0 = 8Qk_{\mathrm{F}}^3 x_i^2/(3\pi)$. Incorporating this effect into the potential of Eq. (47) gives rise to the *analytical image charge potential* for approximating the effects of dipole and exchange-correlation potentials in the classical image charge model:

$$V_{\mathrm{analytic}}(x > 0) = \mu(T) + \Phi_0(T) + \frac{8}{3\pi} Qk_{\mathrm{F}}^3 x_i^2 - F(x - x_0) - \frac{Q}{(x + x_0)} \tag{48}$$

For $x \leq 0$, a reasonable approximation is to assume that $V(x)$ for $x < 0$ decays in a manner compatible with Eq. (45) (recall that electron density and $V(x)$ are related via Poisson's equation), but that obscures the ripples in $V(x)$ (much reduced compared to the rectangular barrier) due to the Friedel oscillations in the density. That is not our concern, however. Rather, Eq. (48) allows for a transparent assessment of the various effects on the F–N equation. By shifting to a coordinate system $y = x + x_0$, the classical image charge potential is recovered as long as an "effective" work function

---

* See the tables of work function in Haas and Thomas [23].

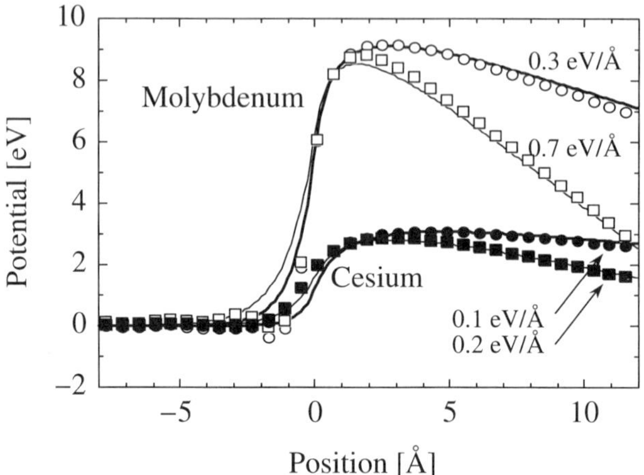

FIGURE 3.5. A comparison of the analytical image charge model with the simultaneous numerical solution of Schrödinger's equation and Poisson's equation for Mo and Cs parameters at various fields. Open/closed circles and squares: numerical results from the Schrödinger/Poisson approach for Mo/Cs, respectively. Lines: curves from the analytical image charge model.

is defined [48]

$$\Phi_{\text{effective}}(T) = \Phi_0(T) + \frac{8}{3\pi} Q k_{\text{F}}^3 x_i^2 + 2F x_0 \tag{49}$$

Clearly, $\Phi_{\text{effective}}$ is to be identified with the experimental work function, and Φ_0 is found such that $\Phi = \Phi_{\text{exp}}$ at $F = 0$ eV/Å. Use of Eqs. (48) and (49) mimics a self-consistently evaluated exchange-correlation potential (simultaneous iterative solution of Poisson's and Schrödinger's equations) and the dipole term rather well, as shown in Fig. 3.5 for both high (Mo) and low (Cs) values of the work function.

In contrast to Eq. (49), usage of the F–N equation in the literature assumes that Φ is a constant value. What effects arise by neglecting the other factors in Eq. (49)? If the barrier height is not allowed to vary, the changes due to F and T will be attributed to the field enhancement factor $\beta_g = F/V$, where V is the potential of the anode or extraction grid/gate. The slope of the F–N equation is then proportional to $\Phi^{3/2}/\beta_g$. An error $\delta\Phi$ in work function estimates will generate an error of $\delta\beta_g \approx (3\beta_g \delta\Phi)/(2\Phi)$ in field enhancement factor estimates (and visa versa). Consider, as an example, Mo values such that when the modifications indicated in Eq. (49) are neglected, then $\delta\Phi/\Phi \approx 0.15$ for typical fields, giving $\delta\beta_g/\beta \approx 0.23$.

A fair portion of $\delta\Phi$ is traceable to the image charge term. The sanctity of the classical image charge approximation is so widely embraced that suggesting Q/x needs modification requires support. Ancona performed a density gradient analysis of field emission from metals at high fields and concluded [49] "... the original

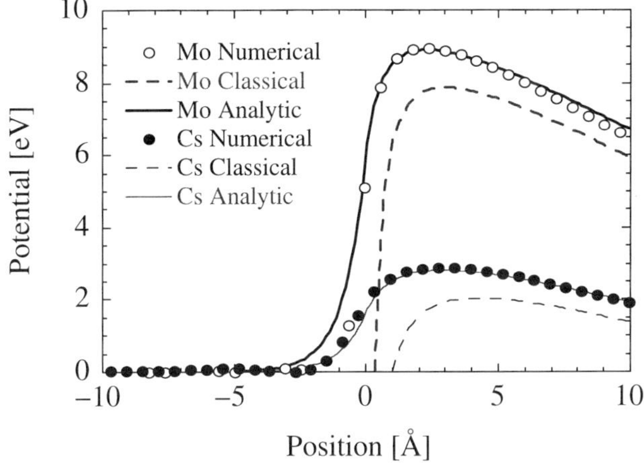

FIGURE 3.6. A comparison of the analytical image charge model with the simultaneous numerical solution of Schrödinger's equation and Poisson's equation as well as the classical image charge model for Mo ($\Phi = 4.41$ eV, $F = 0.4$ eV/Å) and Cs ($\Phi = 2.14$ eV, $F = 0.2$ eV/Å)

Fowler–Nordheim approach with a simple triangular barrier gives much more accurate results than when it is 'corrected' for the image force." In effect, the electron density profile at high fields is such that the barrier is not being lowered by the $\sqrt{4QF}$ factor predicted by the classical image charge potential, but by a lesser factor. That finding is, in fact, compatible with the analytical image charge potential, which likewise predicts that the barrier will not lower to the extent predicted by classical image charge theory, as shown in Fig. 3.6, where the numerical solution is compared with the analytical and classical image charge models.

In terms of current density, the slope and intercept of $J(F)$ on a F–N plot (i.e., $1/F$ vs. $\ln(J(F)/F^2)$) is given in Fig. 3.7 for fields between 0.15 and 1.0 eV/Å. At high fields, the analytical image charge model deviates towards the triangular barrier (no image) approximation. At lower fields, $J(F)$ approaches the classical image charge case, indicating that the barrier more closely matches that encountered in thermionic emission, where an applied field does reduce the apparent barrier height by the amount $\sqrt{4QF}$.

As Ancona argued, the tunneling barrier is dynamically related to the behavior of the electron density near the interface, due to the nature of the exchange-correlation and dipole potentials. Assuming a static equilibrium-based potential, such as the classical image charge approximation, is provisional. Nevertheless, the classical image charge F–N equation is dogma in the field emission literature: to ignore it is at one's own peril. Bowing to orthodoxy in what follows, Φ will simply be taken as the experimental work function (unless otherwise stated) with the caveat that the usage of Eq. (49) in Eq. (29) allows for an approximate recovery of the correct tunneling theory.

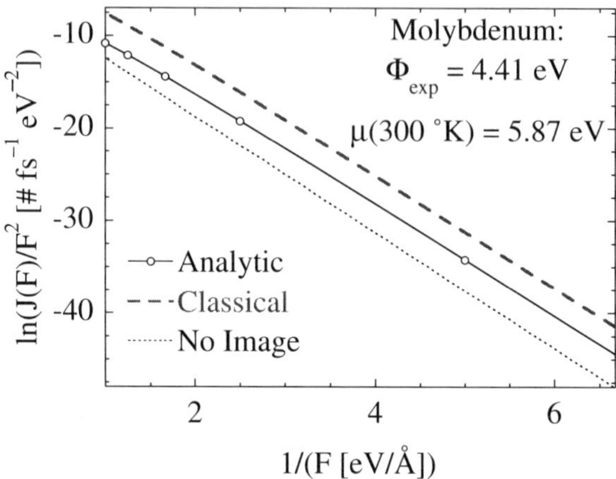

FIGURE 3.7. A comparison of the current density as a function of applied field in F–N coordinates ($1/F$ vs. $\ln(J(F)/F^2)$) for the triangular barrier, classical image charge, and analytical image charge potential barriers.

3.2.3.3. *Small Chemical Potential.*

Pivotal to the derivation of the F–N equation was the expansion of $\theta(E)$ about $E \approx \mu$ in Eq. (25). This is adequate for metals because $\theta(E)$ is approximately linear near $E \approx \mu$, and the tunneling probability of the low energy electrons is completely negligible. However, when the electron density is small, as for semiconductors, $T(k)$ has begun its inexorable slide towards unity near $k \approx k_F$, and k_F can be comparable to k_0 when the incident electron energy is near the barrier height. Consequently, μ is no longer a good parameter to expand E about. What is to be done? The blunt answer is to integrate Eq. (5) numerically using $T(k)$ obtained through Eq. (25) directly. But an analytical solution is nevertheless obtainable as long as Eq. (16) can be approximated by

$$T(k) \approx (2k/k_0)^2 \, \exp(-2\theta(E(k))) \tag{50}$$

where k_0 is defined by the barrier height $\mu + \phi$ (e.g., $\phi \approx \Phi - \sqrt{4QF}$ in the classical image charge approximation). Ignore for the moment effective mass variations and M_c from Eq. (18), but they are easily included. The motivation is that while the tunneling barrier is not triangular, it is nevertheless sufficiently triangular-like to expropriate the triangular barrier's momentum-dependent coefficient. Let the value of $E(k)$ which maximizes $(\hbar k/m)T(k)f(k)$ be denoted E_0, and introduce the slope terms, A and B

$$A \equiv 2\,\theta(E_0); \qquad B \equiv -\frac{\partial}{\partial E}\,(2\theta(E))\,|_{E=E_0} \tag{51}$$

where the notation evokes Eqs. (1) and (28). For metals, $E_0 \approx \mu$, but for semiconductors at low field, that can be a poor choice. Retaining the quadratic coefficient of

Eq. (50) in the current integral in Eq. (29a) gives, to leading order [31]

$$J(F) = \frac{4}{15}\mu^{5/2}\,\Phi\left(2,\frac{7}{2},-B\mu\right) + 2\beta^{-2}\,\mu^{1/2}\left(\frac{\pi^2}{12} - e^{-\beta\mu}\right)$$

$$\Phi(a,b,x) = \frac{\Gamma(c)x^{1-c}}{\Gamma(c-a)\Gamma(a)}\int_0^x \frac{t^{a-1}e^t}{(x-t)^{1+a-c}}\,dt \tag{52}$$

where $\Phi(a,b,x)$ is the *degenerate hypergeometric function* $_1F_1(a,b,x)$.* When graphed on a F–N plot compared to Eq. (29a), it appears that the straight line plot has only been shifted by a small amount, but that belies the difference; the presence of the $(2k/k_0)^2$ coefficient alters the energy distribution so that the contribution from the low energy electrons to the tunneling current is overestimated ("area under the curve" approximations to WKB will not drive the transmission coefficient to zero as the momentum vanishes, but rather to a finite nonzero value).

3.2.3.4. *Quadratic Potentials.*

With the exception of the image charge component, the tunneling barriers considered have so far been linear. Often, the strength of the fields causes such a rapid reduction in the potential that this approximation is adequate over the short distances characteristic of tunneling. Nevertheless, instances arise where the approximation requires modification. Cutler and Nagy [12] considered modifications to the field emission potential due to quantum effects that add a term proportional to x^{-2}. The depletion or accumulation of electrons or holes near the Schottky barrier at the interface of two materials [50,51] can be simply modeled by a quadratic $V(x) = \mu - \Delta + (\phi + \Delta)(W - x)^2/W^2$, where ϕ is the height of the Shottky barrier, Δ is the energy difference between the conduction band and the vacuum level, and $W = [K_s(\phi + \Delta)/(2\alpha\pi\hbar cN_D)]^{1/2}$ is the width of the depletion layer (N_D is the doping density) [52]. A novel and interesting theoretical analysis combining the concepts of internal field emission through the Schottky barrier in a diamond thin film with the Nottingham cooling effect has been performed recently by Miskovsky and Cutler [53]. The field enhancement associated with the sharp geometry of modern field emitter sources [54,55] add effects which may be approximated by the addition of a term γx^2 to the image charge potential.

Recall the treatment of Eq. (25), and the parameter $L_0 = x_+ - x_-$, where $x_\pm$ are zeros of the potential defined in Eq. (22). Analogously, when the potential acquires a quadratic term γx^2, it has three roots

$$x_j = -2\Re\cos\left(\xi + \frac{\pi}{3}(4j + 1)\right) + \frac{F}{3\gamma} \tag{53}$$

---

* See Section 9.21 of Ref. [32].

where $j = 0$, 1, or 2, and the symbols $\mathfrak{R}$ and ξ are given by

$$\mathfrak{R} = \frac{1}{3\gamma}\sqrt{F^2 - 3\gamma h}$$

$$\xi = -\frac{1}{3}\arccos\left(\frac{2F^3 - 9Fh\gamma + 27Q\gamma^2}{2(F^2 - 3h\gamma)^{3/2}}\right)$$

(54)

The roots x_0 and x_1 are analogous to the $x_\pm$ roots (they converge as γ vanishes), but x_2 amounts to mathematical convenience rather than physical significance. Note that x_0 is *not* the $x_0 = 1/k_0$ parameter related to the barrier height. Define $L = x_1 - x_0$. $\theta(E)$ becomes

$$\theta(E) = \frac{2\,L^2}{\hbar}\sqrt{2\,m\,\gamma}\int_0^{\pi/2}\sqrt{(x_2 - x_0) - L\,\sin^2(s)}\,\frac{(\cos(s)\,\sin(s))^2}{\sqrt{x_0 + L\,\sin^2(s)}}\,ds$$

(55)

If γ is small, then $x_2 - x_0 \gg L$ so that, to leading order

$$\theta(E) \approx \frac{2}{\hbar}\sqrt{2m\gamma(x_2 - x_0)L^3}\left(R\left(\frac{x_0}{L}\right) - \frac{L}{15(x_2 - x_0)} - \frac{L^2}{105(x_2 - x_0)^2}\right)$$

(56)

$\theta(E)$ can be linearized in $(\mu - E)$ as in Eq. (28), yielding

$$\theta(E) \approx [c_0 - c_1(\mu - E)]\theta_0(E)$$

(57)

where θ_0 is $\theta(E)$ with $\gamma = 0$, and c_0 and c_1 are to be determined from Eq. (55) or (56). The form of Eq. (29) may be retained as long as the definitions in Eq. (30) are modified according to

$$v(y) \rightarrow c_0 v(y)$$

$$t(y) \rightarrow c_0 t(y) - \frac{2}{3}\Phi c_1 v(y)$$

(58)

As shown by Forbes et al. [55], c_0 and c_1 may be approximated by

$$c_0 \approx 1.2472 - 0.25409F$$

$$c_1 \approx -0.03\ \text{eV}^{-1}$$

(59)

for Mo parameters, where F [eV/Å] is the local field in the vicinity of the tunneling barrier, not an asymptotic or macroscopic field. The performance of Eq. (29) with Eqs. (58) and (59) compares well to exact treatments for metallic parameters.

Increasing the factor b_{FN} by approximately 25%, as suggested by Eqs. (30), (58) and (59) can profoundly affect the estimates of current density: $J(F)$ can be reduced by orders of magnitude depending on the model of the potential and the magnitude of the local field. Physically, as the field decreases, the width of the barrier $L(E)$ increases dramatically. Conversely, as the field increases, the quadratic term becomes increasingly irrelevant and $L(E) \approx L_0(E)$: the tunneling barrier more closely resembles the standard image charge potential to the tunneling particle. The overall effect is to introduce negative convexity on a F–N plot of $J(F)$ as the current density decreases rapidly at small fields.

3.2.4. General Potentials: The Airy Function Method

A general potential $V(x)$ may be broken up into N linear segments (not necessarily of equal length) such that

$$V(x_n \leq x < x_{n+1}) \equiv V_n + s_n F_n(x - x_n) \tag{60}$$

where $s_n = \pm 1$, $F_n \geq 0$, and n is an index ranging from 0 to N. There is no requirement that $V(x_n - \delta) = V(x_n + \delta)$ for infinitesimal δ. In fact, Airy function methods have been applied to resonant tunneling diodes with considerable success [29,56–60]. The introduction of the s sign parameter is necessary to keep track of the proper sign for derivatives of the wave function. Appropriating the Zi functions encountered in the treatment of the triangular barrier, the wave function for $x_n \leq x < x_{n+1}$ is represented as

$$\begin{pmatrix} \psi(x) \\ \partial_x \psi(x) \end{pmatrix} = \begin{pmatrix} \dfrac{Zi(c, \omega)}{Zi(c_0, \omega_0)} & \dfrac{Zi(-c, \omega)}{Zi(-c_0, \omega_0)} \\ \dfrac{s f^{1/3} Zi'(c, \omega)}{Zi(c_0, \omega_0)} & \dfrac{s f^{1/3} Zi'(-c, \omega)}{Zi(-c_0, \omega_0)} \end{pmatrix} \begin{pmatrix} t_n \\ r_n \end{pmatrix} \tag{61}$$

where the generalized $c(x)$ and $\omega(x)$ functions are now defined by

$$c(x) = \begin{cases} i & k_0^2 + s f(x - x_n) - k^2 < 0 \\ 1 & k_0^2 + s f(x - x_n) - k^2 \geq 0 \end{cases} \tag{62}$$

$$\omega(x) = f^{-2/3} |v + s f(x - x_n) - k^2|$$

The n-subscript which should appear on s, f, and k_0 (and by extension, on c and ω) has been suppressed for notational simplicity. c_0 and ω_0 are evaluated at $x - x_n = 0$. By definition, the following relations hold

$$(k_0)_n \equiv \frac{1}{\hbar} \sqrt{2m V(x_n)}; \qquad f_n \equiv \frac{2m}{\hbar^2} |V(x_{n+1}) - V(x_n)| \tag{63}$$

If $c(x)$ should change from i to 1, or visa versa, as x ranges from x_n to x_{n+1}, then a transitional matrix relating the coefficients for transport over/under the barrier has to be inserted to match the coefficients for transport under/over the barrier, as in Eq. (13).

The transition matrices are, for under to over (exponential $c = 1$ to sinusoidal $c \equiv i$: $s = -1$)

$$\begin{pmatrix} t'_n \\ r'_n \end{pmatrix} = \begin{pmatrix} -i & \dfrac{i\,Zi(1,\omega)}{Zi(-1,\omega)} \\ \dfrac{Zi(-1,\omega)}{Zi(1,\omega)} & 1 \end{pmatrix} \begin{pmatrix} t_n \\ r_n \end{pmatrix} \tag{64a}$$

where use has been made of $\omega_n(x) = 0$. Primes on r and t denote coefficients to the left of the transition point. For over to under (sinusoidal $c = i$ to exponential $c = 1$: $s = +1$)

$$\begin{pmatrix} t'_n \\ r'_n \end{pmatrix} = \frac{1}{2} \begin{pmatrix} i & \dfrac{Zi(i,\omega)}{Zi(-i,\omega)} \\ -\dfrac{i\,Zi(-i,\omega)}{Zi(i,\omega)} & 1 \end{pmatrix} \begin{pmatrix} t_n \\ r_n \end{pmatrix} \tag{64b}$$

These transition matrices are not required when $V(x)$ changes discontinuously at x_n.

Matching of the wave function and its first derivative to the left and right of each x_n results in a string of matrix multiplications prescribed by Eq. (61). Whenever the energy transitions from above to below — or visa versa — the transition matrices of Eq. (64) must be inserted. The boundary conditions are $t_0 = 1$ (incident particles from the "left") and $r_{N+1} = 0$ (no incident particles from the "right"). The current to the right of x_N is then given by (analogous to Eq. (11))

$$T(k) = \frac{f_N^{1/3}|t_N(k)|^2}{\pi\,k|Zi\,(c(x_N),\,\omega(x_N))\,|^2} \tag{65}$$

where $s_N = -1$ or current would not flow. The results for the triangular barrier are recovered in the limit $N = 2$.

For the field emission problem, such a method is preferable to a plane wave approach, where the potential is decomposed into many flat segments, and incident and reflected plane waves are matched. This is because a surprisingly small number of trapezoidal segments (e.g., $N \approx 12$ for the typical metallic field emission potential, slightly more for depletion layer potentials) are needed for reasonable accuracy. At large x, the potential is approximately linear. By comparison, numerous rectangular segments are required to account for even one trapezoidal region of any appreciable extent. The price paid is the cost of numerically evaluating the Zi functions. Power series expansions based on those for the Airy functions are available for small argument. For large argument, the asymptotic form, of which Eq. (15) is the limiting case, is

$$Zi(c,\omega) = Fi(c,\omega)\omega^{-1/4}\,\exp\left(\frac{2}{3}c\omega^{3/2}\right) \tag{66a}$$

Polynomial expansions exist for the $Fi(c, \omega)$ function [25], but the following Padé approximations are useful if ω is sufficiently large:

$$Fi(c, \omega) = \frac{1}{8\sqrt{\pi}}[(c + 3)(c^2 + 1) + 2\sqrt{2}(c - 1)(c^2 - 1)]\left(\frac{96c\omega^{3/2} - 67}{96c\omega^{3/2} - 77}\right) \quad (66b)$$

It is clear from Eq. (66) why the Zi functions are useful for the tunneling problem. The exponential growth and decay of Airy functions is captured in the exponential part of Eq. (66a): the remaining terms vary smoothly. By taking ratios of Zi functions in Eq. (64) these exponential terms are eliminated, rendering the coefficients t_n and r_n easily represented in single-bit precision on a desktop computer even when the field vanishes. Consider a potential variation, for which a typical matrix element resembles

$$\frac{Zi(1, \omega)}{Zi(1, \omega_0)} \approx \sqrt{\frac{\kappa_0}{\kappa}} \exp\left(\frac{2}{3f}(\kappa(x)^3 - \kappa(0)^3)\right) \quad (67a)$$

where $\omega(x) = \kappa(x)/f^{2/3}$ and the notation follows that used in treating the triangular barrier. For small f, the exponential becomes

$$\frac{Zi(1, \omega)}{Zi(1, \omega_0)} = \left(\frac{\kappa_0^2}{\kappa_0^2 + sfx}\right)^{1/4} \exp\left(s\kappa_0 x + \frac{f}{4\kappa_0}x^2\right) \quad (67b)$$

where, for convenience, x designates $(x - x_n)$. Analogous results hold for $c = i$. Thus, the wave function in Eq. (61) reproduces the familiar exponential behavior from the rectangular barrier encountered in Eq. (2) as the field vanishes, even though the underlying Airy functions explode or vanish in magnitude.

The Airy function approach may appear somewhat draconian, given the finesse and success of the WKB approach. Nevertheless, occasions do arise where its use is necessitated. Consider the surface band structure for surface layers on silicon (Figure 1 of Johnston and Miller [72]), or the potential barriers associated with deep-level defects in external fields for charged and neutral impurities within the tunneling region [51,62–64]. A coulomb charge embedded in the oxide of a silicon emitter will introduce a potential profile shown in Fig. 3.8(a). The "area-under-the-curve" WKB formulation will miss the resonant peak captured by the Airy function method, as shown in Fig. 3.8(b). While the enhanced tunneling current presumably occurs only in the vacinity of the charge, it is orders of magnitude greater than the tunneling current in the absence of a charge, and therefore, can significantly contribute to the overall current [64]. Methods to account for resonance effects necessarily transcend the simple F–N approximation in Eq. (1).

3.2.5. Emission from Semiconductors

For a chemical potential of 5 eV typical of metals at room temperature, the electron density is 5.08×10^{23} cm^{-3}. For a slab of electrons 6 Å deep, this amounts to an equivalent surface density $\sigma \approx 0.3$ Å^{-2}. On the other hand, using the relation

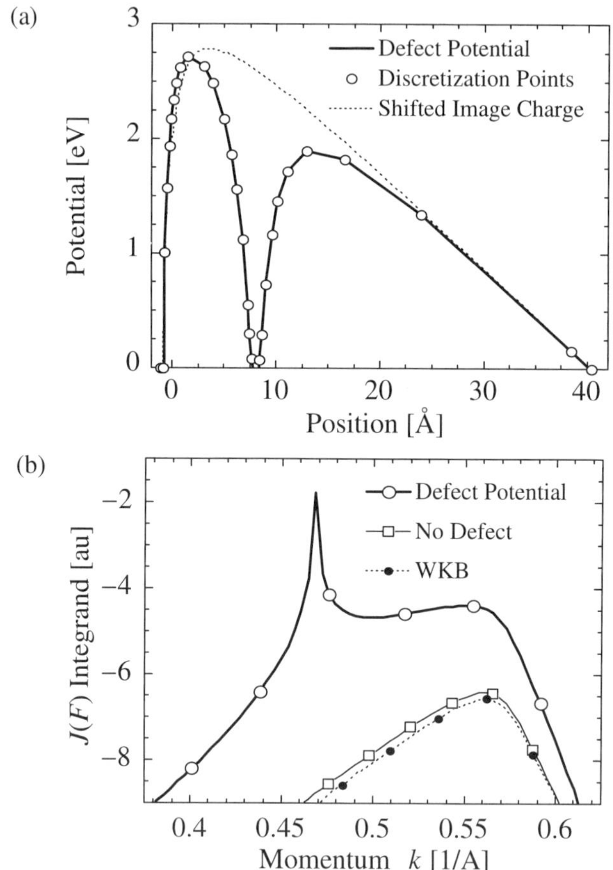

FIGURE 3.8. (a) Model of the inclusion of a coulomb potential inside an oxide for silicon parameters. Straight lines between the open circles constitute the piecewise linear potential barrier used in the Airy function analysis shown in part (b). A comparison of the transmission coefficient as calculated using the Airy function method with and without a coulomb potential with the potential calculated by the WKB approximation.

$F = \sigma/\varepsilon_0$, the surface density needed to shield an applied field of 0.5 eV/Å is only 0.0028 Å^{-2}. A minor increase in electron density at the surface of the metal is, therefore, adequate to shield the interior from even strong electric fields. In contrast, for a fairly highly doped semiconductor such that the electron density is 10^{18} cm^{-3} ($\mu \approx -0.086$ eV for $M_c = 6$ and $m \approx 0.3283\ m_0$), the equivalent surface density for the aforementioned model slab is 6×10^{-6} cm^{-2} and is far less than that needed to shield an external field. Even though accumulation at the surface increases the electron density by two to three orders of magnitude over bulk values, penetration effects will extend far into the semiconductor in comparison to metals.

The effects of a small chemical potential on the F–N equation have already been addressed in Eqs. (32) and (52), but there are other effects to be considered. Band bending will tend to alter the barrier height in a field-dependent manner. Emission may occur from the valence band, surface states, and defects within the band gap, and will depend on the type of doping (n- or p-type) used [65]. The presence of defects within the band gap can contribute to the current, especially for low electric fields [66, 67]. Hole current must be accounted for. The effective mass in a semiconductor is not, in general, equal to the electron rest mass in vacuum. The emitted current from wide band gap semiconductors can be strongly influenced or dictated by transport through the Schottky barrier characterizing the back contact, and transport through the wide band gap material has numerous complications of its own [61,68–71]. Surface layers or adatoms [72], and for analogous reasons, layered semiconductor structures which can give rise to a resonant tunneling effect [73], complicate matters by introducing modifications to either the potential barrier or the supply function. Band bending at the surface is complicated by the fact that a triangular well-like potential has discrete energy states (given approximately by zeros of the Airy function) rather than a continuum of states [74]. Oxides or adsorbates may exist on the surface which may have charged inclusions [64] that preclude a simple Fowler–Nordheim-like relationship, and which may contribute to the large fluctuations observed in field emission from silicon tips [75].

The complexity of these issues precludes their adequate treatment in any abbreviated analysis. Stratton has dealt with some of these issues in his seminal article in a manner unlikely to be improved upon [7]. Here, the effects of band bending, where emission is, for sake of argument, presumed to come primarily from the conduction band (high fields and doping, and/or low temperatures), is considered. A simplified account of the *zero emitted current approximation* (ZECA), treated in greater detail by Modinos, will be used to give an indication of changes required for semiconductors.

3.2.5.1. Boltzmann Transport Equation and ZECA.

Electron transport through semiconductors is described by the Boltzmann Transport Equation (BTE), which, in one dimension (and no magnetic fields) becomes [76]

$$-\partial_t f(x, k; t) = \frac{\hbar k}{m}(\partial_x f) - \frac{1}{\hbar}(\partial_x V)(\partial_k f) - \partial_c f|_{col} \tag{68}$$

The last "collision" term governs scattering events and may be represented by the relaxation time approximation [77]. The scattering time τ is related to the distance electrons travel between scattering events, and for typical silicon parameters, it is of the order of several hundred femtoseconds [78]. The mean free path $l = \tau v_{th}$, where v_{th} is the thermal velocity $(8k_B T/\pi m^*)^{1/2} \approx 2$ Å/fs, and is thus several hundred angstroms at room temperature for silicon parameters. The distribution function $f(x, k; t)$ relaxes to a thermal Fermi–Dirac distribution encountered in Eq. (6), except that in the presence of a varying potential $V(x)$, the chemical potential μ must be replaced by the electrochemical potential $\mu(x)$ [79]. Under time-independent equilibrium conditions, the collision term drops out. Using the thermal Fermi–Dirac distribution function from Eq. (6), Eq. (68) then shows that $\mu(x) = \mu_0 + \phi(x)$, where μ_0 is the bulk value and V is replaced by ϕ as part of a notation change in preparation for using Poisson's equation.

The redistribution of electrons accounted for by $\phi(x)$ dictates the extent of band bending under an applied field F_{vac}.[†] At the surface, $\phi(0^-) = \phi_{\text{s}}$, which will be assumed to be measured with respect to the conduction band edge E_{c}. By virtue of being subject to band bending, the electron affinity χ of a semiconductor is specified rather than work function. ϕ_{s} may be numerically calculated from Poisson's equation once $F_{\text{s}} = F(0^-) = F_{\text{vac}}/K_{\text{s}}$ is specified, where F_{vac} is the externally applied field. Poisson's equation becomes

$$\partial_\phi F^2 = \frac{\rho - \rho_0}{K_{\text{s}}\varepsilon_0}$$

$$= \frac{N_{\text{c}}}{K_{\text{s}}\varepsilon_0} \frac{2}{\sqrt{\pi}}(F_{1/2}[\beta(\mu_0 + \phi)] - F_{1/2}(\beta\mu_0)) \tag{69a}$$

giving

$$F_{\text{vac}}(\phi_{\text{s}}) = \left(\frac{4N_{\text{c}}}{K_{\text{s}}\varepsilon_0\sqrt{\pi}\beta} \left[\int_0^\infty \sqrt{x}\, \ln\left(\frac{e^{\beta\mu - x} + 1}{e^{\beta\mu_0 - x} + 1} \right) dx - \beta\phi_{\text{s}} F_{1/2}(\beta\mu_0) \right] \right)^{1/2} \tag{69b}$$

Two limits, however, are instructive. When the field F is small, band bending is not strong, and the electrochemical potential can be negative even at the surface. If $\beta\mu$ is negative and its magnitude sufficiently large, $F_{1/2}(\beta\mu) \approx (\pi/4)^{1/2}\, e^{\beta\mu}$ so that

$$F_{\text{vac}}(\phi_{\text{s}}) \approx \left(\frac{2N_{\text{c}}K_{\text{s}}}{\beta\varepsilon_0} \exp(\beta\mu_0)\, [\exp(\beta\phi_{\text{s}}) - \beta\phi_{\text{s}} - 1] \right)^{1/2} \tag{70a}$$

a form known from semiconductor device analysis (though F instead of F_{vac} is typically specified). In the opposite limit $\beta\phi_s \gg 1$, which is more typical of field emission conditions, a convenient formula is given by (where $\beta\mu \gg 1$) [80]

$$F_{\text{vac}}(\phi_{\text{s}}) \approx \left(\frac{2\pi^2 N_{\text{c}}K_{\text{s}}}{3\beta\varepsilon_0} \left(\frac{\beta\mu}{\pi} \right)^{1/2} \left[\frac{8}{5}\left(\frac{\beta\mu}{\pi} \right)^2 + 1 \right] \right)^{1/2} \tag{70b}$$

A comparison of the exact and approximate forms are shown in Fig. 3.9, from which it may be concluded that for typical field emission fields, Eq. (70b) suffices. The ZECA approximation for field emission from semiconductors, in its simplest form and for $\beta\mu$ sufficiently large, is then given by Eqs. (29) and (32) with the modifications:

- The chemical potential μ replaced with $\mu_{\text{s}} = \mu_0 + \phi_{\text{s}}$
- The surface potential term ϕ_{s} is determined from Eq. (70) and F_{vac}

[†] Regarding units: herein, potentials and fields are multiplied by electron charge so that units are in electron volts and eV/Å, respectively. This will alter the customary sign notation in Poisson's equation and render $\rho(x)$ a number density rather than a charge density.

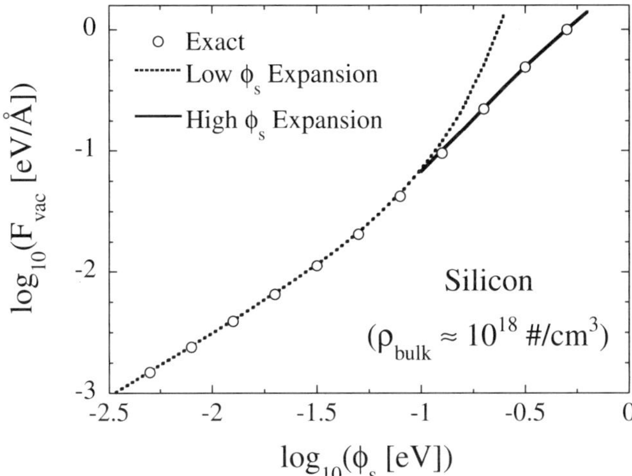

FIGURE 3.9. Relationship between the vacuum field and the electrochemical potential parameter ϕ_s in the *ZECA* for silicon parameters (see Table 3.2). For fields typical of field emission phenomena ($F \geq 0.01$ eV/Å), the large asymptotic approximation is adequate.

- The "work function" Φ is set equal to $\chi - \mu_s$
- The image charge term Q replaced throughout the F–N equation with the dielectric modification [47]

$$Q(K_s) = \frac{K_s - 1}{K_s + 1} \frac{\alpha \hbar c}{4} \tag{71}$$

3.2.5.2. The Wigner Distribution Function Approach.

The simplified ZECA assumption regarding the electron density profile is not sustainable near the surface (or interface), as it suggests that the electron density increases until the origin. That approximation runs afoul of several issues. The equilibrium distribution is maintained by scattering events. By way of example, consider silicon: at room temperature, an electrochemical potential at the surface of 0.2 eV implies an external field of $\approx$0.4 eV/Å (for a bulk doping of 10^{18} cm^{-3}), so that the characteristic length scale over which the internal potential decreases is roughly $(K_s\phi_s)/F_{\text{vac}} \approx 8$ Å, a good deal shorter than the electron's mean free path. Assuming that the distribution is thermal near the origin is, therefore, problematic. Even for a completely thermal distribution (vanishing relaxation time), the wave nature of the electron and the behavior of Eq. (44) show that $\rho(x)$ begins to decline in magnitude — rather than increase — at a distance approximately π/k_F prior to the barrier. What does this entail for the electrochemical potential?

The BTE does not include tunneling effects, and so it will be unable to anticipate the dipole term due to electron penetration of the barrier. Tunneling phenomena are intimately intertwined with the quantum mechanical nature of the incident electrons. Thus, the density implicit in Eq. (18) is not the same as the density of Eq. (41): they become equivalent far into the bulk, where the out of phase oscillations of the wave function for various momenta average out and the bulk density is recovered, but near the origin and into the tunneling potential, the density is decreasing and decaying because the wave functions for all k approximately vanish at the origin, a type of behavior not incorporated into the simplified ZECA approach. A modification of the BTE is, therefore, required to introduce the quantum effects.

Wigner suggested a quantum distribution function $f(x, k)$ based on the Fourier transform of a generalization of Eq. (41) for the electron density matrix $\rho(x, y)$

$$f(x, k) = \frac{1}{\pi} \int_{-\infty}^{\infty} e^{-2iky} \, \mathrm{d}y \int_{-\infty}^{\infty} \mathrm{d}k' \, f(k') \psi_{k'}^*(x + y) \, \psi_{k'}(x - y) \tag{72}$$

To be specific, the Wigner Distribution Function (WDF) $f(x, k)$ is not a probability distribution function as it takes on negative values, but it reduces to the Boltzman distribution in the classical limit [81]. However, the WDF phase space description [82] is remarkably useful: it can be used to calculate averages, suggests a particle trajectory interpretation analogous to the BTE, and has proven itself remarkably adept for modeling the high frequency behavior of tunneling structures (in particular, but not limited to, resonant tunneling structures) [78,83–87]. The time evolution of $f(x, k; t)$ is obtained from that for the density matrix, and is

$$\partial_t f(x, k; t) = -\frac{\hbar k}{m} \partial_x f(x, k; t) + \int_{-\infty}^{\infty} V(x, k-k') f(x, k'; t) \, \mathrm{d}k'$$

$$V(x, k-k') = \frac{2}{i\pi\hbar} \int_{-\infty}^{\infty} (V(x + y) - V(x - y)) \, e^{2iy(k-k')} \, \mathrm{d}y \tag{73}$$

The behavior of the potential $V(x)$ now figures prominently in determining when quantum effects arise: when the potential is smoothly varying, only the first order term in a Taylor expansion of $V(x + y) - V(x - y)$ needs to be retained in Eq. (73), for which the integral containing $V(x, k - k')$ becomes proportional to $\partial_x V(x) \partial_k f(x, k)$, and the BTE results. When the next order term, proportional to $\partial_x^3 V(x) \partial_x^3 f(x, k)$, is no longer negligible — such as near the abrupt potentials at the surface/interface characteristic of field emission and tunneling (and for such abrupt potentials, all of the higher order gradients must be retained) — quantum effects are introduced, thereby breaking the ZECA approximation relating the electrochemical potential to the solution to Poisson's equation.

The numerical evaluation of Eq. (73) entails the solution of a rather substantial matrix equation which, as recently as the mid-1990s, required supercomputer resources to model dynamic behavior, though the algorithms* can now be implemented on faster desktop computers. When solved in parallel with Poisson's equation, a self-consistent equilibrium $f(x, k; t)$ can be dynamically obtained by letting the system evolve from

---

*A low memory version that avoids the largely empty off-diagonal block matrices associated with the differential operator is found in Ref. [29].

an initial state, where the relaxation is introduced by a collision term analogous to that used in the BTE. From the resulting distribution function, density and current are calculated from the first two "moments" of $f(x, k; t)$ by (compare to Eq. (41))

$$\rho(x; t) = \frac{1}{2\pi} \int_{-\infty}^{\infty} f(x, k; t)\, dk$$

$$j(x; t) = \frac{1}{2\pi} \int_{-\infty}^{\infty} \frac{\hbar k}{m} f(x, k; t)\, dk$$

$$(74)$$

Because $f(x, k; t)$ is time dependent, both the density and current are likewise time dependent, and so the Wigner function method may be used to model dynamic response, for example, after a sudden change in external field [31]. After convergence to steady state conditions, the self-consistent potential and density profiles are analogous to those given by ZECA, but with the important modification that the density $\rho(x)$ near $x \approx 0$ is more spread out, the value of ϕ_s is larger than ZECA suggests, and $\rho(x)$ is smaller than predicted by Eq. (18) with $\mu = \mu_0 + \phi_s$.

Given the importance of barrier height in the F–N equation, such differences matter. However, the effort required to numerically solve Eq. (73) for semiconductors is a disincentive to do so in comparison to the relative simplicity of field emission from metals. In addition, while simulating the dynamic evolution of the electron density, temperature effects, and scattering quite well, the Wigner approach nevertheless is ill-equipped (in contrast to the Airy function approach) to provide estimates of tunneling current unless the applied field is sufficiently high, as finite difference errors will overshadow the current density when the field is small. In contrast, the ZECA methodology breaks down at high fields (it overestimates band bending and hence μ) for which the WDF is ideal. ZECA is nevertheless useful at low fields for understanding field emission from systems where band bending and accumulation occurs, and the electron density is comparatively low. Still, Schrödinger/Poisson approaches are useful for modeling transport [80,88,89], nonequilibrium dynamics [90], Nottingham heating/cooling from metallic emitters [91], and internal field emission in a diamond thin film [53].

3.3. EMISSION FROM MULTI-DIMENSIONAL STRUCTURES

Much of the discussion below regarding the estimation of total current from 3-D structures assumes that the structure under investigation resembles a sharpened whisker or, if gated, the conical Spindt-like field emitter tip. Other electron sources with differing geometries exist. Nevertheless, field emission from sharpened structures enjoys an august history. Notable work in the 1950s and 1960s was performed at Linfield Research Institute in pursuit of microwave devices [92,93]. In the 1960s, Shoulders conceived of developing microfabricated field emission structures [94], and by the 1970s, these structures were successfully fabricated and analyzed by Spindt, Brodie, and colleagues [1,2]. By the 1980s, field emitter array technology was rapidly developing, and the first IVMC was inaugurated in the late 1980s by H.F. Gray and C.A. Spindt in Williamsburg, Virginia. Compared to its far younger brethren, Spindt-type field emitters have enjoyed considerable attention early on, and the history of modeling 3-D field emitter structures reflects that. The techniques surveyed herein, therefore, tilt towards conical (hyperbolic) and ellipsoidal structures, but nevertheless have wider applications.

The simplest approach to evaluate total field emission current from a structure is to integrate the 1-D current density as a function of a varying local electric field over a 3-D surface, as done in the boundary element methods considered here or finite element methods [95–98]. Otherwise the 3-D nature of the emitter surface [99] is computationally daunting.

3.3.1. Field Enhancement and Emission Area

Consider a conducting sphere of radius a_s held at potential V_0. It is an elementary problem in classical electrostatics to show that (in polar coordinates) the potential experienced by a unit point charge at (r, θ) is given by

$$V(r, \theta) = V_0 - F_0 r \, \cos(\theta) \left[1 - \left(\frac{a_s}{r} \right)^3 \right] - \frac{2Qa_s}{\left(r^2 - a_s^2 \right)} \tag{75}$$

where F_0 is the macroscopic or asymptotic field and the z-axis is the axis of rotational symmetry. Imagine slicing the sphere in half and placing it on a conducting plane. As long as the radius of the hemisphere is sufficiently large, the image charge contribution does not need to be augmented by additional image charge terms, and Eq. (75) remains satisfactory to model field emission from a hemispherical bump or boss: with the sphere grounded, V_0 becomes the energy difference between the vacuum level and the conduction band. Fields are largest at the apex, so on-axis ($\theta = 0$) behavior is considered. If the coordinates are recast as $z - a_s \equiv x = 0$, then for values of x smaller than the sphere radius, a Taylor expansion of Eq. (75) gives

$$V(x) = \left(V_0 + \frac{Q}{2a_s} \right) - \left(3F_0 + \frac{Q}{4a_s^2} \right) x - \frac{Q}{x} + \left(\frac{3F_0}{a} + \frac{Q}{8a_s^3} \right) x^2 \tag{76}$$

Independent of radius, the effect of the hemisphere is to increase the field at the apex by a factor of 3, which is the "field enhancement" associated with a sphere. A quadratic term is also introduced to the potential from the spherical geometry. For example, for typical parameters of $V_0 = 10$ eV, $F_0 = 0.1$ eV/Å, and $a_s \approx 50$ Å, Eq. (76) predicts a barrier height of 10.036 eV, an apex field of 0.30036 eV/Å, and a quadratic coefficient of 6.0036×10^{-3} eV/Å^2. Evidently, unless the radius is small, the Q-dependent terms (with the exception of Q/x) are often negligible, and shall be neglected below.

The off-axis field on the sphere is given by $F(\theta) = 3F_0 \cos(\theta)$. The current density, being exponentially sensitive to the applied field, drops precipitously as a function of θ.

Using the quadratic approximations to $v(y)$ and $t(y)$ in Eq. (30)

$$\frac{J(F(\theta))}{J(F_{\text{tip}})} \approx \cos^2(\theta) \, \exp\left(-\frac{4v_0}{3\hbar F_{\text{tip}}} \sqrt{2m\Phi^3} \frac{(1-\cos(\theta))}{\cos(\theta)}\right) \tag{77}$$

where $v_0 \approx 0.93685$, and $F_{\text{tip}} \equiv F(0)$. The rapidity with which the current density falls off is, therefore, seen to be apex-field dependent — the smaller the field, the more rapid the decline. Two observations: first, for typical Mo parameters at $F_{\text{tip}} \approx 0.5$ eV/Å, the current density has decreased by a factor of 2 by $\theta \approx \pi/10\,(18°)$ and a factor of 10 by $\theta \approx \pi/6\,(30°)$ in comparison to the apex value; second, because of the F_{tip} dependence of Eq. (77), the emission area A cannot be constant [100], as often assumed (e.g., setting $I(V) = AJ(F)$), but is dependent on the apex field. For a hemisphere

$$I_{\text{hemi}}(V_{\text{g}}) = 2\pi a_{\text{s}}^2 J(F_{\text{tip}}) \int_0^1 x^2 \, \exp\left(-\frac{b_{\text{FN}}^0}{F_{\text{tip}}} \frac{(1-x)}{x}\right) \, \mathrm{d}x \tag{78}$$

where b_{FN}^0 is b_{FN} with $v(y)$ replaced by v_0, and the relationship $F_{\text{tip}} = \beta_{\text{g}} V_{\text{g}}$ is implicitly assumed: if the asymptotic field is due to an anode a large distance D away and held at a potential V_{g}, then $\beta_{\text{g}} \approx 3/D$. The area factor b_{area}, defined by the coefficient of $J(F_{\text{tip}})$, is an integral over the surface, and is explicitly a function of the apex field. This is but the simplest definition: when inferring an emission area from experimental data, subtleties in the treatment of $v(y)$ arise, as well as in the definition of "emission area" itself, and has been extensively discussed by Forbes [17,18,55]. The integral in Eq. (78) can be represented by $I_{\text{tip}}(V) = b_{\text{area}} J(F_{\text{tip}})$. To a good approximation [101]

$$b_{\text{area}}(F_{\text{tip}}) \approx 2\pi a_{\text{s}}^2 \left(\frac{F_{\text{tip}}}{b_{\text{FN}}^0 + 4F_{\text{tip}}}\right) \left[1 + \frac{4F_{\text{tip}}^2}{\left(b_{\text{FN}}^0 + 5F_{\text{tip}}\right)^2}\right] \tag{79}$$

For metals, $F_{\text{tip}}/b_{\text{FN}}^0 \leq 0.15$ for $F_{\text{tip}} < 0.9$ eV/Å. Eq. (79) appears to suggest that $<30\%$ of the hemisphere emits at a current density of $J(F_{\text{tip}})$, which is an erroneous interpretation: the current density decreases away from the apex, so that more of the surface emits current than simply b_{area}. The "characteristic emission area" [55] (the area which encloses 99.9% of the emitted current) is approximately a factor of $e^1 = 2.72$ larger than indicated by Eq. (79), and therefore, the "characteristic current density" is $e^{-1}J(F_{\text{tip}})$.

Eq. (79) and the expression for the field enhancement factor β_{g} appear to be of limited utility given the idealized model considered. β_{g} will undoubtedly change as progressively more conical or needle-like structures are considered. After considerably greater mathematical excursions associated with hyperbolic/ellipsoidal geometries, the same characteristic dependence is still found: b_{area} will be, to leading order, linearly dependent upon the apex field. Therefore, the inference $I = AJ$ (and from it the equation $I(V_{\text{g}}) = AV_{\text{g}}^2 \exp(-B/V_{\text{g}})$ widely used to model experimental data) is faulty and will require investigation to see in what manner the inference should be made.

3.3.2. Hyperbolic and Ellipsoidal Structures

A method for solving boundary value problems is to approximate the conducting surfaces by "planes" in an orthogonal coordinate system [102]. Poisson's equation is then separable, and the field and potential easily solved [103,104]. The prolate spheroidal coordinate system is, therefore, ideal for modeling structures which can be approximated by hyperbolic or ellipsoidal surfaces of revolution, and has often been used for field emission problems [15,105–107]. Recently, Forbes and Jensen [108] applied the methodology to emission from ellipsoids retaining the notation of Morse and Feshbach; the treatment here synopsizes that account, but will use the notation of earlier treatments [52,101].

3.3.2.1. Two-Dimensional Wedge. For generalized coordinates (q_1, q_2, q_3), the gradient and the Laplacian are given by

$$\vec{\nabla} = \sum_{i=1}^{3} \hat{i} h_i^{-1} \partial_i$$

$$\nabla^2 = (h_1 h_2 h_3)^{-1} \sum_{i,j,k=1}^{3} \frac{1}{2}(1 + \varepsilon_{ijk})\varepsilon_{ijk} \partial i \left(\frac{h_j h_k}{h_i} \partial i \right)$$

(80)

where ε_{ijk} is the Levi–Civita symbol, and $(q_1, q_2, q_3) = (x, y, z)$ in Cartesian coordinates. The scale factors h_i are given by

$$h_i = \sqrt{(\partial_i x)^2 + (\partial_i y)^2 + (\partial_i z)^2}$$

(81)

In two-dimensional (2-D) hyperbolic coordinates (α, β, γ), the equation for a wedge takes the form:

$$x = a_h \sinh(\alpha) \sin(\beta)$$

$$y = a_h \cosh(\alpha) \cos(\beta)$$

(82)

$$z = \gamma$$

where a_h is the distance from the origin to the focus of the hyperbola defined by constant β. Solving $\nabla^2 \psi = 0$ under the boundary conditions, $\psi(\beta = \beta_0) = 0$ ("wedge") and $\psi(\beta = \pi/2) = V_0$ ("anode"), results in

$$\psi(\beta) = V_0 \frac{(\beta - \beta_0)}{\left(\frac{\pi}{2} - \beta_0\right)}$$

$$\vec{F}(\alpha, \beta) = (-\hat{\beta}) \frac{V_0}{a_h \left(\frac{\pi}{2} - \beta_0\right) \sqrt{\sinh^2(\alpha) + \sin^2(\beta)}}$$

(83)

If the 2-D radius of curvature at the apex of the wedge is defined by $\partial_x^2 y|_{x=0} = 1/a_s$, and the distance from the wedge to the anode as z_0, then $\tan^2 \beta_0 = a_s/z_0$. In the limit that β_0 is small, the field at the apex is $F_{tip} \approx 2V_0/\pi \sqrt{a_s z_0}$.

3.3.2.2. Rotationally Symmetric (Conical/Ellipsoidal) Geometry. The relation between the cylindrical and hyperbolic coordinates is

$$\rho = \frac{a_h}{2} \sinh(\alpha) \sin(\beta)$$
$$z = \frac{a_h}{2} \cosh(\alpha) \cos(\beta) \tag{84}$$

And further, $x = \rho \cos(\delta)$ and $y = \rho \sin(\delta)$. Let $\psi = X(\alpha)Y(\beta)Z(\delta)$. Laplace's equation becomes

$$[\partial_\alpha (\sinh(\alpha)\partial_\alpha) - n(n+1)\sinh(\alpha)] X(\alpha) = 0$$
$$[\partial_\beta (\sin(\beta)\partial_\beta) + n(n+1)\sin(\beta)]Y(\beta) = 0 \tag{85a}$$

Solutions exist in the form:

$$X_n(\alpha) = a P_n(\cos(i\alpha)) + a' Q_n(\cos(i\alpha))$$
$$Y_n(\beta) = b P_n(\cos(\beta)) + b' Q_n(\cos(\beta)) \tag{86b}$$

where $P_n(x)$ and $Q_n(x)$ are Legendre polynomials of the first and second kind, and the a's and b's are constants. Whereas ellipsoids would make use of the X_n solutions, hyperboloids use the Y_n solutions. Hyperbolic emitters are defined by setting $\beta = \beta_0$. The hyperbolic surface is at zero potential and the anode (characterized by $\beta = 0$) at V_0, for which $n = 0$. A diode geometry is presupposed — no gate structure has been introduced as yet. The potential and field are

$$\psi(\beta) = V_0 \left[1 - \frac{Q_0(\cos(\beta))}{Q_0(\cos(\beta_0))} \right]$$
$$\vec{F}(\alpha, \beta) = (-\hat{\beta}) \frac{V_0}{Q_0(\cos(\beta_0))} \frac{1}{a_h \sin(\beta)\sqrt{\sinh^2(\alpha) + \sin^2(\beta)}} \tag{86}$$

where ψ denotes potential and $Q_0(\xi) = (1/2) \ln[(1+\xi)/(1-\xi)]$ for $\xi \leq 1$. In terms of the apex field ($\alpha = 0$), the field along the surface of the hyperboloid is given by

$$F(\alpha, \beta_0)|_{hyperbola} = F_{tip} \frac{\sin \beta_0}{\sqrt{\sin^2 \beta_0 + \sinh^2 \alpha}} \tag{87}$$

Eq. (87) is a general result. It shall now be related to the parameters typical of emitters. Let the radius of curvature at the apex (as before) be denoted by a_s and the distance between emitter apex and the anode by D, then the apex radius is

$$a_s \equiv -\left(\partial_\rho^2 z\right)^{-1}\Big|_{\rho=0} = -\frac{a_h}{2}\frac{\sin^2(\beta_0)}{\cos(\beta_0)} \tag{88}$$

Likewise the cone angle is defined by

$$\cos(\beta_0) = \sqrt{\frac{D}{D + a_s}} \tag{89}$$

The tip, or apex, field is then

$$F_{tip} = \frac{2\cos(\beta_0)}{a_s\cos(\beta_0)\ln\left(\dfrac{1+\cos(\beta_0)}{1-\cos(\beta_0)}\right)} V_0 \tag{90a}$$

In the limit that $D \gg a_s$, as is often the case, Eq. (90a) reduces to

$$F_{tip} \approx \frac{2}{a_s\ln\left(\dfrac{4D}{a_s}\right)} V_0 \tag{90b}$$

as shown by Martin et al. [105].

Eq. (89) is an awkward equation. It suggests that the tip-to-anode separation bears some relation to the cone angle, and experimentally that is manifestly not true. The mathematics indicates only that *if* the physical system approximately matches that coordinate system, *then* the field at the apex is approximately given by Eq. (90), *not* that Eq. (90) holds for an arbitrary experimental arrangement. It is nevertheless reasonable to expect that a diode described by the hyperbolic model will qualitatively follow the behavior of Eq. (90), e.g., the apex field and anode voltage share an inverse relationship to tip radius and a logarithmic dependence on the tip-to-anode separation.

With F_{tip} and a method of calculating the apex current density $J(F_{tip})$, the total current from a hyperbolic diode may be evaluated by finding the analog of Eq. (79)

$$b_{area}(F_{tip}) = \frac{1}{J_{FN}(F_{tip})}\int_\Omega J_{FN}(F)\,d\Omega$$

$$= \int_0^\infty \left(\frac{F(\rho)}{F_{tip}}\right)^2 \exp\left(b_{FN}^0\left(\frac{F(\rho)-F_{tip}}{F(\rho)F_{tip}}\right)\right)\sqrt{1+(\partial_{\rho}z)^2}\,2\pi\rho\,d\rho \tag{91a}$$

The integral in Eq. (91a) is the total current from an emitter tip. Performing the integration over the surface of the hyperbola [109]

$$b_{\text{area}}(F_{\text{tip}})|_{\text{hyp.diode}} \approx 2\pi a_{\text{s}}^2 \left(\frac{F_{\text{tip}} \cos^2(\beta_0)}{b_{\text{FN}}^0 + F_{\text{tip}} \sin^2(\beta_0)} \right) \tag{91b}$$

and considering the same parameters as before (i.e., $F_{\text{tip}}/b_{\text{FN}}^0 \leq 0.15$) and $\beta_0 \approx 18°$, then the area factor constitutes $\leq 26.8\%$ of the πa_{s}^2 area.

For the ellipsoid [55]

$$\begin{aligned}
F_{\text{tip}}|_{\text{ellipsoid}} &= \frac{F_0}{\sinh^2(\alpha_0) Q_1(\cosh(\alpha_0))} \\
&= \frac{1}{2} \frac{(2R^2 - 3)}{\ln(2R) - 1} F_0
\end{aligned} \tag{92a}$$

where α_0 defines the surface of the ellipsoid, F_0 is the asymptotic field (produced by a distant anode), $Q_1(\xi) = \{(\xi/2) \ln[(\xi + 1)/(\xi - 1)] - 1\}$ is a Legendre polynomial of the second kind, and $\xi > 1$. The second line in Eq. (92a) is in the large R limit, where R is the ratio between the height (major axis radius) and the width (minor axis diameter), and the height of the ellipsoid is given by $R^2 a_{\text{s}}$. In the opposite limit, $F_{\text{tip}} \approx 3F_0$, as for the hemisphere. Analogous to Eq. (91a), the current density is integrated over the ellipsoid, but the upper limit in the ρ integration is $W = Ra_{\text{s}}$, so that

$$b_{\text{area}}(F_{\text{tip}})|_{\text{ellipsoid}} \approx 2\pi a_{\text{s}}^2 \left(\frac{F_{\text{tip}}}{b_{\text{FN}}^0 + F_{\text{tip}}} \right) \tag{92b}$$

As may have been expected, because of the greater curvature of the ellipsoid in comparison to the hyperboloid, the area factor is slightly larger for a given apex field.

3.3.3. Gated Geometries and Triodes

The simplest analytically solvable triode geometry, designated elsewhere as the "Saturn model" [110,111], is composed of a sphere representing the emitter, a charged ring of radius $a_{\text{g}} = r_{\text{g}} \sin(\alpha)$ suspended above the sphere at a distance $z_{\text{g}} = r_{\text{g}} \cos(\alpha)$ from the sphere's center representing the gate, and finally, a background field parallel to the axis of symmetry and characterized by an asymptotic value of F_{a} accounting for the anode. The ring potential can be expanded in terms of Legendre polynomials $P_l(\cos \theta)$ and is [112]

$$\begin{aligned}
V(r, \theta) &= -F_{\text{a}} r \cos(\theta) + \frac{Q_{\text{g}}}{r_{\text{g}}} \sum_{l=0}^{\infty} \left(\frac{r}{r_{\text{g}}} \right)^l P_l(\cos \alpha) P_l(\cos \theta) \\
&\quad + \sum_{l=0}^{\infty} A_l r^{-(l+1)} P_l(\cos \theta)
\end{aligned} \tag{93}$$

where the terms correspond to the anode, gate, and sphere potentials, respectively. The gate term arises from an expansion of the complete elliptical integral $K(p)$, which arises from the integral over the charged ring. The A_l is determined by the requirement that on the sphere's surface, $V(a_s, \theta) = 0$. Likewise, Q_g is determined by specifying the gate potential at some position near the ring. Doing so, the field along the surface is given by

$$F(a_s, \theta) = -3F_a \cos(\theta) - F_g \sum_{l=0}^{\infty} (2l + 1) \left(\frac{a_s}{r_g}\right)^l P_l(\cos\alpha) P_l(\cos\theta) \tag{94}$$

where $F_g = Q_g/(r_g a_s)$. It is seen that Eq. (94) decomposes into an anode part and a gate part. Consequently, the field-enhancement factor approximation $F_{tip} = \beta_g V_g$, where V_g is gate potential, is not valid unless the anode contribution is negligible. If the tip of the sphere lies approximately within the ring plane and the anode contribution is in fact negligible, then the apex field is approximately given by

$$F_{tip}|_{saturn} \approx \frac{\pi}{a_s \ln\left(\dfrac{8a_g}{t}\right)} V_g \tag{95}$$

where t is a parameter dependent upon gate (and possibly tip) parameters. Eq. (95) is useful for qualitative purposes; quantitatively, compared to numerical simulation using boundary element techniques (discussed here), it gives poor estimates of the field-enhancement factor. Nevertheless, it provides an indication of how the apex field may vary with gate radius. A "hybrid" approximation between the hyperbolic emitter and the Saturn model combines the dependencies and is

$$F_{tip}|_{hybrid} \approx \frac{\pi}{a_s \ln\left(\dfrac{ka_g}{a_s}\right)} V(z_0) \tag{96}$$

where $z_0 = a_s \cot^2 \beta_0$ and specifies the apex, $V(z)$ is the potential on axis, and k is to be determined. Eq. (96) is constructed explicitly to develop an analytical model of a triode.

It is convenient to redefine the z-axis so that the apex of the emitter is at $z = 0$, or

$$z \equiv z_0 - \frac{1}{2}a_h \cos(\beta)\cosh(\alpha) \tag{97}$$

In experiments, the gate is not a ring of charge, and a sheet of charge representing the remainder of the gate [113] must be added. The potential on axis becomes

$$V(z) = V_g + \frac{F_0}{2}\left[z + \sqrt{z^2 + a_g^2} + \frac{1}{2}\frac{a_g^2}{\sqrt{z^2 + a_g^2}} - \frac{z_g^2}{z + a_c}\right] \tag{98}$$

where $F_0 = (V_a - V_g)/D$ is the field between anode and gate, away from the gate hole. z_g and a_c are determined from boundary conditions that $V(0) = 0$ and $\partial_z V(0) = F_{tip}$, and they are approximately given by $z_g = V_g(2/F_0 F_{tip})^{1/2}$ and V_g/F_{tip}, respectively. Consequently, Eq. (98) may be approximated by

$$V(z) = \frac{F_{tip}z}{V_g + F_{tip}z}\left(1 + \frac{F_0 z}{V_g}\right)V_g \tag{99}$$

where the *ad hoc* term in parentheses insures proper behavior as z goes to D. Using Eq. (99) in Eq. (96) allows F_{tip} to be determined:

$$F_{tip} \approx \left(\frac{\pi}{\ln\left(k\dfrac{a_g}{a_s}\right)} - \tan^2(\beta_0)\right)\frac{V_g}{a_s} \tag{100a}$$

where β_0 is the cone angle. The factor k must be determined by other means. Using boundary element simulations, it can be shown [114] that over a range of tip and gate radii k may be approximated by

$$k \approx \frac{1}{54}\left(86 + \frac{a_g}{a_s}\right)\cot(\beta_0) \tag{100b}$$

The relationship in Eq. (100a) is only one possible analytical form (k being determined by a fit to boundary element simulations). It degrades if β_0 is too small or the tip and gate parameters too eccentric (a_g/a_s is too large), at which point other approximations come into vogue. Another approximation that does not have the small β_0 anomaly is given by

$$k \approx \frac{16a_g}{25a_s} + \left[11 - \frac{7a_g}{5a_s}\tan(\beta_0)\right] \tag{100c}$$

Depending on the geometric parameters, Eqs. (100b) and (100c) have ranges of validity. For Spindt-type gated field emitters, Eq. (100) provides a successful interpretation of experimental data, from emission distribution to space charge effects and tip/array behavior and to a prediction of device performance [115,116] and tip morphology due to ion bombardment [117]. The parametric approximation to k in Eq. (100b) causes tip currents calculated by the hybrid model to be within a factor of 2 from those calculated by boundary element methods. That demerit is offset by the inherent advantages of an easily calculated analytical model of tip current.

3.3.4. Nanoprotrusions

Fitting data using the analytical model above gives values of the "effective" tip radius a_s as (typically) 30–70 Å. The use of "effective" requires emphasis: on the basis of scanning tunneling microscope (STM) or transmission electron microscope (TEM) images, Spindt-type emitters have tip radii of the order of several hundred angstroms [118]. Those images show protrusions and local undulations, which belie the approximations inherent in the analytical model. TEM images of $a_s \approx 50$ Å tips fabricated from molybdenum [119] and silicon [120,121] likewise show undulations and bumps. A hyperbolic or ellipsoidal model is, therefore, a substantial idealization of an actual emitter structure. Several researchers [122,123] point out that actual emission sites may be significantly smaller than the apparent SEM/TEM tip radius: the current from the tip may be due to local emission sites (grain boundaries, clusters of atoms, or nanoprotrusions on the emitter). Given the nature of the nanoprotrusions and the tips upon which they occur, a subset of the emitters from the entire array are responsible for the bulk of the current [124–126]. Consequently, the calculated "effective" tip radii correlates with the arguments of Purcell et al. [123], Charbonnier [122], Fursey et al. [54], and others that the emitter contains nanoprotrusions whose dimensions are of the order of 20–30 Å. In fact, a given emitter may contain a varying number of such nanoprotrusions that may be created by ion impact, removed by sputtering processes, and which may further migrate to the apex under the strong gradients along the tip. By taking the tip radius a_s as an "effective" radius of the emitter, these nanoprotrusions, or "bumps on tips" [127], can be accommodated.

An approximation of the increase in the field from a bump on tip is given by imagining a hemispherical "nanoprotrusion" on a larger hemispherical tip. Recall that for a background field F_0, the field at the apex of the hemisphere is $3F_0$. The hemispherical "nanoprotrusion" will contribute another factor of three, so that overall, the ratio of the nanoprotrusion apex field with the background field will be of the order of $3(3F_0) = 9F_0$. More realistically, a crude model of a nanoprotrusion is obtained by the potential composed of a flat plane, a boss, and a charge suspended just above the boss in cylindrical coordinates and ignoring the image charge term

$$
V(\rho, z) = V_0 - F_0 z \left[1 - \left(\frac{a}{\sqrt{z^2 + \rho^2}} \right)^3 \right] + \frac{q}{\sqrt{((z-a)^2 + \lambda \rho^2)}} \tag{101}
$$

where q and λ are chosen to obtain a nanoprotrusion of a desired size, e.g., $q \approx 50$ eV/Å and $\lambda \approx 12$ approximate a nanoprotrusion 13.24 Å high and 14.2 Å width at half-height $a \approx 250$ Å and $F_0 \approx 0.1$ eV/Å, for which $F_{tip} \approx 5.56 F_0$, smaller than the crude hemisphere-on-hemisphere model because of the substantially larger "base" associated with Eq. (101) — the equipotential surfaces resemble more of a cusp than a hemisphere.

3.3.5. Boundary Element Methods

Progress beyond the analytical models requires the use of numerical techniques to account for finite gate size and thickness, closely spaced anodes, oxide layers, and other geometrical/material specifications. Modeling the geometry of an emitter structure may invoke the axial or translational symmetry of the emitter to utilize an elegant 2-D boundary element approach [128–130]. Consider as an example a rotationally symmetric hyperbola of revolution, which mimics several of the "vertical emitter" structures. It consists of an anode, a gate with hole, and a base plane with an emitter. The apex of the emitter is characterized by length scales orders of magnitude smaller than the gate, and requires a nonuniform grid of the computational domain, making finite-difference formulations numerically costly, more so because emission occurs from a small region.

The boundary element method can bypass issues of fine grid spacing. The potential is obtained by integrating over the surface charge densities σ on the emitter, gate, and anode:

$$\phi(\vec{r}) = \frac{1}{4\pi\varepsilon_0} \int_{\Omega} \frac{\sigma(\vec{r}')}{|\vec{r} - \vec{r}'|}\, d\Omega \tag{102}$$

where $d\Omega$ is the differential surface element (circular ribbons for rotational symmetry) with a constant surface charge density. Rendering Eq. (102) in discrete form in cylindrical coordinates (ρ, z) gives a matrix equation such that

$$\vec{\phi} = \overline{\overline{M}} \cdot \vec{\sigma}$$

$$[\overline{\overline{M}}]_{i,j} = \frac{1}{\pi\varepsilon_0}\sqrt{1 + s_j^2} \int_{\rho_j}^{\rho_{j+1}} \rho' \frac{K(p)}{\gamma}\, d\rho' \tag{103}$$

where the arguments of the complete elliptical integral $K(p)$ are given by

$$p = \frac{1}{\gamma}\sqrt{4\rho_{i+1/2}\rho'}; \qquad \gamma = \sqrt{(\rho_{i+1/2} + \rho')^2 + (z_{i+1/2} - z')^2} \tag{104}$$

where the subscript "$i + 1/2$" denotes the middle of a ribbon of width ε_i (not to be confused with permittivity of free space ε_0) and the coordinates are given by

$$\rho_{j+1} = \rho_j + \varepsilon_j; \qquad s_j = \left(\frac{z_{j+1} - z_j}{\rho_{j+1} - \rho_j}\right) \tag{105}$$

$$z' = z_{i+1/2} + s_i(\rho' - \rho_{i+1/2})$$

The elliptical integral $K(p)$ is defined by

$$K(p) = \int_{0}^{\pi/2} \frac{d\theta}{\sqrt{1 - p^2\sin^2\theta}} \tag{106}$$

When $i = j$, $K(p)$ contains a logarithmic singularity of the form $(1/2)\ln[16/(1 - p^2)]$ as p approaches 1. The singular portion can be analytically integrated. Dropping the $(i + 1/2)$ subscript on ε, ρ, and s

$$\int_{\rho_i}^{\rho_{i+1}} \frac{\rho'}{2\gamma} \ln\left(\frac{16}{1 - p^2}\right) d\rho'$$

$$\approx \frac{\varepsilon}{(24\rho)^2}\left([96\rho^2 - (s^2 + 2)\varepsilon^2]\left[3\ln\left(\frac{16\rho}{\sqrt{(s^2 + 1)\varepsilon}}\right) + 1\right] + 192\rho^2\right) \quad (107)$$

Eq. (107) is a good approximation for large ρ, but for small ρ neglected terms are of the same order of magnitude. In practice, the singular part of the self-interaction term can be calculated by Eq. (107) and a Gaussian quadrature routine used to evaluate the remainder whether ρ is small or not.

The surface charge densities σ_i are found by inverting the matrix Eq. (103), from which the field at the surface $F_i = \sigma_i/\varepsilon_0$ is obtained. The fields give rise to a current density per ribbon from the F–N equation. The total current can be found by summing over the current contributions from each ribbon:

$$I(V_g) = 2\pi \sum_{i=1}^{N_{\text{tip}}} \sqrt{1 + s_i^2}\,\rho_{i+1/2}\varepsilon_i\, J(F_i) \quad (108)$$

where only the ribbons constituting the emitter surface are considered, and the factors inside the summation are associated with the area of the ribbon.

Away from the conducting surfaces, each ribbon may be approximated by a ring of charge located at $\rho_{i+1/2}$, with a total charge Q_i (which implicitly contains factors $1/4\pi\varepsilon_0$). The potential at any point (ρ, z) can then be calculated by the contribution from N ribbons defining the emitter, gate, and anode via

$$\phi(\rho, z) = \sum_{i=1}^{N} Q_i \frac{2}{\pi} \frac{K(p_i)}{\gamma_i} \quad (109)$$

Using Eq. (109), an estimate of the equipotential lines may be constructed, or the fields needed for, e.g., a particle trajectory simulation, may be generated.

The boundary element method can be readily implemented on a personal computer and can be used for configurations that are analytically intractable, or to provide the value of k in the hybrid model of Eq. (96) (the basis for the fits of Eq. (100)). Rotational symmetry is not a prerequisite, only a convenience to reduce the number of boundary elements required: the method can be generalized to small areas instead of ribbons. A further capability is that many device simulations require estimates of capacitance between the various components. Summing up subsets of Q_i to form a unit (e.g., Q_g for the gate) held at the same potential (e.g., ϕ_g for the gate), the capacitances between the components are given by (b, g, t, and a refer to base, gate, tip, and anode for a gated Spindt-type emitter) [130]

$$
\begin{bmatrix} Q_t \\ Q_b \\ Q_g \\ Q_a \end{bmatrix} = \begin{bmatrix} C_{tt} & C_{tb} & C_{tg} & C_{ta} \\ C_{bt} & C_{bb} & C_{bg} & C_{ba} \\ C_{gt} & C_{gb} & C_{gg} & C_{ga} \\ C_{at} & C_{ab} & C_{ag} & C_{aa} \end{bmatrix} \begin{bmatrix} \phi_t \\ \phi_b \\ \phi_g \\ \phi_a \end{bmatrix} \tag{110}
$$

where $C_{ij} = C_{ji}$. By setting the various surfaces to zero potential and performing the simulation, the various C_{ij} may be extracted.

3.3.6. The Statistical Hyperbolic/Ellipsoidal Model

Few devices require a single emitter in isolation. From displays and rf amplifiers to numerous other applications, an array of tips must act in concert to provide the current required. Large numbers are in general required, and where there are large numbers, there is statistical variation. Measuring 1 µA from a single tip or 10 µA from 10 tips does not imply that 1 A can be drawn from 10^6 such tips in an array.

The obvious variation arises from the small changes in effective tip radius and work function that can occur as a natural consequence of any number of fabrication techniques and natural processes. As seen from the equations for F_{tip} (see Eq. (90)), apex radius affects emitter performance to a greater degree than the other geometrical factors. Further, practical "vacuums" are anything but that, as a variety of contaminating species exist, even in high vacuum conditions, that can stick on the emitter surface and change the work function characteristics: oxides often form; nanoprotrusions are generated and removed by sputtering; and charged inclusions can occur. Consequently, variation is to be expected. Given a distribution of emitters with various tip radii and effective work functions, the total current is simply

$$
\begin{aligned}
I_{array}(V_g) &= \sum_{j=1}^{N_{tip}} I_{tip}(V_g, a_i, \Phi_i) \\
&= N_{tip} \Sigma(V_g) I_{tip}(V_g, a_s, \Phi)
\end{aligned} \tag{111}
$$

where the statistical factor Σ is defined by this equation, and I_{tip} is the tip current of the sharpest and lowest work function emitter — for sake of argument, it is assumed that these conditions arise on the same emitter, but that need not be true. Given the exponential dependencies inherent in field emission, it is clear that a fraction (sometimes a small fraction) of the emitters will dominate the total current. The need for the exact distribution is, therefore, mitigated. For a pedagogical description, a simple distribution suffices: let the tip radii be linearly distributed according to $a(s) = a(s)(1 + s)$, and the work function similarly by $\Phi(s) = \Phi + s\Delta\Phi$. Those emitters for which s is small dominate, suggesting that $\ln(I_{tip})$ be Taylor expanded in s.

Analytical formulae exist [116], but given the variety of parameters which can occur, such formulae are less satisfactory than simply finding the derivatives numerically using an Euler scheme, in which $\partial f(x) \approx [f(x + \delta) - f(x)]/\delta$, due to the complex dependencies involved. Let $b_X = -\partial_s \ln(I_{\text{tip}}(X))|_{s=0}$, where X is either $a(s)$ or $\Phi(s)$, from which

$$\frac{I_{\text{array}}(V_{\text{g}})}{N_{\text{tip}} I_{\text{tip}}(V_{\text{g}})} \approx \left(\frac{1}{\Delta s} \int_0^{\Delta s} \exp(-b_a s)\, ds\right) \left(\frac{1}{\Delta \Phi} \int_0^{\Delta \Phi} \exp(-b_\Phi s')\, ds'\right) \tag{112a}$$

The integrals are straightforward and give

$$\Sigma(V_{\text{g}}, \Delta s, \Delta \Phi) = \left(\frac{1 - \exp(-b_a \Delta s)}{b_a \Delta s}\right) \left(\frac{1 - \exp(-b_\Phi \Delta \Phi)}{b_\Phi \Delta \Phi}\right) \tag{112b}$$

The approximation of a linear distribution of tip radii and work function results in

$$I_{\text{array}}(V_{\text{g}}) \approx N_{\text{tip}} \Sigma(V_{\text{g}}, \Delta s, \Delta \Phi) b_{\text{area}}(V_{\text{g}}) J(F_{\text{tip}}(V_{\text{g}})) \tag{113}$$

Eq. (113) is the essence of the "statistical hyperbolic model (SHM)." The area and field-enhancement factors change for an elliptical structure, but the general form of the equation holds. As an example, consider Mo parameters at 300 K: $\Phi_{\text{exp}} \approx 4.41$ eV, $\mu = 5.87$ eV, $a_{\text{s}} \approx 40\,\text{Å}$, $a_{\text{g}} \approx 0.2\,\mu\text{m}$, $\Delta \Phi \approx 0.3$ eV, $\Delta s \approx 0.3$, $\beta_0 \approx 18°$, $N_{\text{tips}} = 10^8$ (if the tip-to-tip spacing is 1 μm, then the emission area is 1 cm^2), and the current range is from $I(V_{\text{min}}) = 1\,\mu\text{A}$ to $I(V_{\text{max}}) = 10$ A. Then simulation suggests that $V_{\text{max}} \approx 56$ V, $V_{\text{min}} \approx 25$ V, $B_{\text{FN}} \approx 655$ V (where B_{FN} is the negative slope of the approximately straight line of $\ln(I(V)/V_{\text{g}}^2)$ vs. $1/V_{\text{g}}$. Note that A_{FN} and B_{FN} are intercept and slope parameters of $\ln(I/V^2)$ vs. $1/V$, and are analogous (but not equivalent) to a_{FN} and b_{FN} derived from $\ln(J/F^2)$ vs. $1/F$). For these parameters, $\Sigma(V_{\text{g}})$ is very close to linear and given by $-0.0602 + 0.00516 V_{\text{g}}$, and therefore, ranges from 7 to 23% (likewise, $b_{\text{area}}[\text{Å}] \approx -30.5\,\text{Å} + 16.7 V_{\text{g}}$ [V]). Is this reasonable? While the exact metallic parameters are open to some ambiguity, two features are encouraging: first, the predicted B_{FN} is in the range of reported for SRI emitters with that geometry; second, the percent of working tips given by Σ is comparable to the numbers reported by Constancias and Baptist [125], who reported an analogous linear dependence on voltage (because the geometric parameters and current regime are different, an exact correspondence will not be obtained).

3.3.7. Emission from Spindt-Type Field Emitters

Eq. (113) allows for the unique determination of the "effective" tip radius and the distribution factors Δs and $\Delta \Phi$ from experimental data. B_{FN} is only weakly dependent on Δs, so it may be used to determine a_{s} (e.g., by Newtonian iteration), after which A_{FN} may be used to set bounds on the values of Δs and $\Delta \Phi$: because there are three unknowns (a_s, Δs, $\Delta\Phi$) to be determined from two experimental parameters (B_{FN} and A_{FN}), the values of Δs and $\Delta\Phi$ must be estimated by recourse to other means, such as TEM/SEM photographs and the work function associated with adsorbates.

The estimation of the effective radius is contingent upon the assumption of the work function Φ value for the surface barrier, though Φ varies from one crystallographic plane to another [23], making the usage of an average value commonplace. For $I(V)$ data found in the literature, the calculated minimum tip radius is typically 20–80Å. Work function variation is presumably of the order of 0.5–1 eV, because adsorbate work functions are often of the order of 5–6 eV* — a criteria that reduces one of the three unknowns (a_s, Δs, $\Delta\Phi$). A refined estimate can be made by considering the evaluation of (V) for a single tip whose performance degrades over time due to adsorbates. Calculated tip radius variation is then approximately 50%, which is "reasonable" in the sense that larger apex radii would start to feature bumps and nanoprotrusions themselves. The results of the fitting of experimental data with the SHM supports the dual assertions: (i) the emission sites are nanometer scale protrusions, and (ii) a subset (in cases, a *small* subset) of emitters is responsible for most of the current.

Experimental techniques exist to reduce the range of the uncertainty parameters Δs and $\Delta\Phi$. Methods to combat the changes induced by adsorbates and variations in radii are well known [2,93,132–134]: standard bake-out procedures, running an emitter in the presence of H_2 or other gases [135,136], or electron bombardment of anodes and collectors. Such "conditioning" or "seasoning" improves emission, presumably by reducing $\Delta\Phi$. Another technique, known as "field forming," has been shown by SRI to produce a dramatic improvement in F–N characteristics of Mo Spindt-type FEAs (a decrease in B_{FN} and increase in A_{FN}), presumably by reducing the variation Δs in the effective tip radius. In Figure 11 of Spindt et al. [137], the effects of conditioning were demonstrated to give a factor of 2 reduction in B_{FN} and a factor of 40 improvement in A_{FN}. Fitting of the experimental data, as in Fig. 3.10, provides a remarkably good account of the changes, even though some of the parameters (such as cone angle) used are provisional. For the "SRI-Before," the parameters are $T = 300$ K, $\Phi = 4.41$ eV, $\mu = 5.873$ eV (corresponding to an electron density of 6.4634×10^{22} cm^{-3}), $\beta_0 = 13.2°$, and $a_g = 0.325$ μm for an array containing $N_{tips} \approx 10\,000$. $\Delta\Phi$ was assumed to be 1 eV, in keeping with the general range of adsorbates. A tip radius of 30.3 Å and radii distribution factor $\Delta s \approx 45.0$ reproduce the "Before" experimental data well. After field forming and treatment, the radius and distribution factors change. Those values giving a reasonable fit to the data indicate the tip effective radius has been reduced to 14.6 Å and the distribution terms to $\Delta\Phi \approx 0.2$ eV and $\Delta s = 0.2$. The SHM predicts, through statistical factors, that initially only a small number of tips ($<0.1\%$) were participating, yet after conditioning, $\Sigma \approx 20$–40%.

Field forming itself is a complex interplay between surface tension forces, which tend to blunt the emitter tip, and electrostatic forces, which tend to sharpen it.

---

* In earlier versions, $\Delta\Phi$ was not included so that Δs was presumed to range from on the order of 0.1 to on the order of 1000.

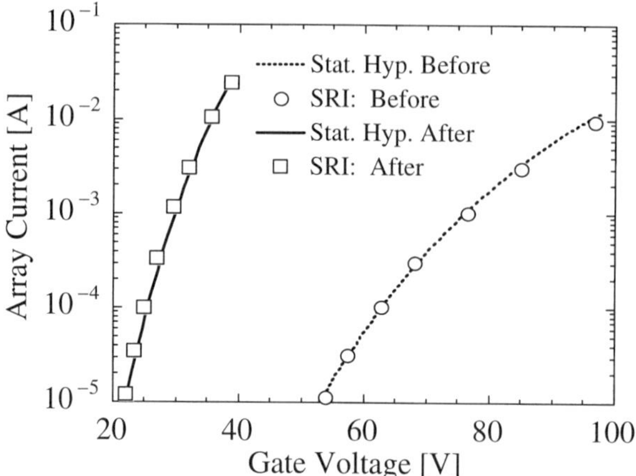

FIGURE 3.10. The effects of field forming and conditioning on the emission current from a Mo-FEA fabricated by SRI (experimental data courtesy of C. A. Spindt).

Electrostatic forces will tend to dominate when the field exceeds

$$F \approx \sqrt{\frac{16\pi\gamma}{r}} \tag{114}$$

where the expression for F is in electrostatic units, r is the radius of the emitter and $\gamma \approx 2100$ dyne/cm for tungsten [138]. An analysis of molybdenum field emitters by Spindt and Brodie has shown that, at the fields present at the apex, field forming does occur and can be used to "season" the tips [139]. The ability for material to migrate is temperature-dependent, so the application of a high field at elevated temperatures correlates with the evolution of sharper structures. The magnitude of the field is such that nanoprotrusions, if present, can migrate as well [140]. While the conditioning study of Spindt et al. [137] was not directed to investigate the role of nanoprotrusions, later studies by Mackie et al. [141] investigated the possible influence of nanoprotrusions directly. That study showed coatings such as zirconium carbide, which have been shown to have a high sputter threshold and greater chemical inertness, appear to impede nanoprotrusion generation and migration under high fields. The nanoprotrusions were then shown to have exceptionally small emission areas, which supports the present theoretical analysis.

3.3.8. Single Tip Emission Distribution Experiment

Are there grounds for the plausibility of the SHM? That is, are there reasons to believe that an idealized mathematical model has any semblance to bumpy, noisy, and inscrutable emitters of any configuration? $I(V)$ data, such as those considered previously, are supportive but not conclusive, as the integration of any number of current densities over a like number of surface structures will give tip current predictions of the right magnitude as long as an adjustable parameter such as a_s is available. Rather, it must be ascertained if the distribution of emitted electrons from a real structure is at all in accordance with theory predictions for an idealized one.

The emittance of an electron source is a measure of the spread of the electron beam, and may be taken as the weighted average of the product of the lateral displacement of the beam from the symmetry axis with the angle of the electron trajectory. A conformal mapping analysis to determine the intrinsic emittance of a field emitter for a 2-D wedge geometry was performed by Liu and Lau [142]. The intrinsic emittance for a wedge was found to be comparable to a conventional (thermionic) cathode. Further, it was argued that the 2-D results overestimate the 3-D case.

At approximately the same time, an experiment to determine the distribution of emitted electrons from a single tip was performed by Phillips et al. [143] at the Naval Research Laboratory using singly addressable emitters fabricated by SRI. The experiment itself was crucial in that a characterization of the emittance of high-brightness electron sources is imperative for the design of devices which make use of them, in particular for electron guns to be used in traveling wave tubes (TWT's), where tight tolerances are imposed [144]. The experimental measurements were performed in an UHV environment (3×10^{-10} torr) to minimize damage to the field emitter from ion bombardment. A novel and intricate design incorporating a laser interferometer and Burleigh's inchworm^{TM} allowed for the manipulation and recording of the absolute position of a microfabricated electron detector with nanoscale precision. The inchworm was a linear induction motor actuated by the piezoelectric effect capable of being moved in steps of 3–4 nm. Laser interferometry allowed for measurement of the detector position from outside the UHV with a precision of 79 nm.

Current was collected by a Faraday cup anode of radius $\Delta_0 = 50 \, \mu\text{m}$. The experimental distribution was found for two values of gate-to-anode separation $D_{\text{exp}} \approx 700 \pm 100 \, \mu\text{m}$ and $975 \pm 100 \, \mu\text{m}$, for which the anode current collected by the Faraday cup on axis was measured to be $I_a = 15.406 \, \text{nA}$ and $12.269 \, \text{nA}$, respectively. The depth of the Faraday cup was 1 μm, and the back plate was biased 5 V higher than the front, which was held at 125 V. The current emitted by the conical single tip Spindt-like emitter was measured to be $I_{\text{tip}} = 0.29 \, \mu\text{A}$ at a gate voltage $V_g = 64 \, \text{V}$. These values constitute the extent of the specified empirical parameters used to construct the theoretical model.

The theoretical analysis proceeds as follows. The distribution factors Δs and $\Delta \Phi$ are zero. Therefore, $I_{\text{tip}}(V_g) = b_{\text{area}}(V_g) J(F_{\text{tip}}(V_g))$. The work function was presumed to be that of molybdenum, for which $\Phi = 4.41 \, \text{eV}$. The tip radius a_s was adjusted until the theoretical value of I_{tip} is equal to the experimental value, from which it is determined that $a_s = 34.83 \, \overset{\circ}{\text{A}}$. The asymptotic field was assumed to be $F_0 = (V_a - V_g)/D$, where V_g and V_a are the gate and anode potentials, respectively. As in the derivation of the analytical tip current, $z_g \approx V_g \sqrt{2/(F_0 F_{\text{tip}})}$ is the radius of an equivalent hemispherical boss which approximates the single emitter structure (see Eq. (98)), and which serves as a convenient boundary between descriptions of electron dynamics influenced predominantly by the emitter tip versus dynamics influenced predominantly by the gate and anode geometry.

In polar coordinates $(r = a_s, \theta)$, the distribution of emission along the surface of the emitter apex is given by $f(\theta) = J(\theta)/J(0) \approx \exp[-(1 + b_{FN}^0/(2F_{tip}))$ $(\sin^2 \theta / \cos^2 \beta_0)]$. Consider electron trajectories starting from the emitter tip and ending at the surface of the boss (hemisphere of radius z_g). Let the polar coordinates on the boss be $(r = z_g, \theta_g)$ ("g" denotes gate). If electron motion was radial, then θ would equal θ_g, but that approximation was found experimentally to be inadequate [109]: rather, (and not surprisingly) electrons follow the strong field lines. Following field lines and noting that $F(\theta_g) = 3F_0 \cos(\theta_g)$ on the boss, the distribution $f(\theta_g)$ over the boss is shown to be

$$f(\theta_g) = \exp\left(-2\chi \sin^2 \theta_g\right)$$

$$\chi = \frac{1}{2}\left(1 + \frac{b_{FN}^0}{2F_{tip}} + \frac{1}{2}\sin^2 \beta_0\right) \tag{115}$$

For the additional experimental parameters $a_g = 0.6\ \mu$m and $\beta_c = 15°$, it is found that $b_{FN}^0 = 5.927$ eV/Å, $F_{tip} = 0.587$ eV/Å and $\chi = 3.041$. The root-mean-square (rms) value of θ_g and the transverse energy E_t for electrons emerging from the boss are then

$$\langle\theta_g^2\rangle = \sqrt{\frac{1}{24}\left(15 - \frac{16I_1(\chi) - I_2(\chi)}{I_0(\chi)}\right)}$$

$$E_t = \frac{1}{2}\left(1 - \frac{I_1(\chi)}{I_0(\chi)}\right)V_g \tag{116}$$

where $I_n(z)$ is a Bessel function of order n. Using the asymptotic forms of the Bessel functions, $\theta_{rms} \approx \sqrt{(4\chi + 2)}/(4\chi)$ and $E_t \approx (4\chi + 1)V_g/(4\chi)^2$.

For the experimental single emitter parameters, the rms spread angle θ_{rms} is, from numerical integration, Eq. (116), and the asymptotic expression, 18.75°; 18.33°, and 17.72°, respectively. Analogously, from Eq. (116) and the asymptotic expression, $E_t = 5.98$ and 5.69 eV, respectively, for a gate voltage of 64 V. For electron beams for power tube applications, a significant E_t is undesirable; focusing grids have been proposed as a means of reducing its effect [145].

At the anode, particle conservation from the boss to the anode plane is invoked to show that in cylindrical coordinates (ρ, z)

$$J_{anode}(\rho) = \left(\frac{a_s}{\rho_0}\cos^2 \beta_c\right)^2 J_{FN}(F_{tip})\exp(-(\rho/\sigma)^2)$$

$$\sigma = \sqrt{\frac{2}{4\chi - 1}}\,\rho_0 \tag{117}$$

where $\rho_0 = v_g\Delta t + z_g$, v_g is the electron velocity at the boss, i.e., $\frac{1}{2}mv_g^2 = V(z_g) \approx$

V_g, and Δt is the time of flight from boss to anode, $\Delta t = [(2D - z_g)\sqrt{V_g} - z_g\sqrt{V_a}]/v_g(\sqrt{V_g} + \sqrt{V_a})$; in the limit that z_g is vanishingly small, the influence of the boss diminishes and Δt approaches the *space charge free* expression [146]. Electrons on a parallel path to the anode will be deflected towards the Faraday cup because of the higher back plate bias. In order to determine that portion of the incident beam that is captured by the Faraday cup, a particle simulation is required. Based on such a simulation [109], the effective capture radius is $\Delta = 62.69$ and 63.07 µm for $D_{\text{theory}} = 766.4$ and 856.1 µm, respectively. The effective anode distance D_{theory} is obtained by demanding that the total current (Eq. (117) multiplied by a differential area element and integrated over all ρ) at the anode equals the experimentally measured tip current. Both theoretical values lie near the range of experimental uncertainty.

Integrating Eq. (117) over the circular Faraday cup hole with radius Δ and approximating the result by a Gaussian distribution $\Delta I(\rho)$ gives [147]*

$$\Delta I(\rho) = b_\Delta J_{\text{FN}}(F_{\text{tip}}) \exp[-(\rho/\Sigma)^2]$$

$$b_\Delta = \pi \Delta^2 \left(\frac{a_s \sigma}{\rho_0 \Sigma} \cos^2 \beta_c \right)^2 \tag{118}$$

$$\Sigma = \sigma \left/ \sqrt{1 - \frac{1}{2}\left(\frac{\Delta}{\sigma}\right)^2} \right.$$

A comparison to the experimental data is shown in Fig. 3.11. The remarkable agreement indicates that the SHM can qualitatively predict the behavior of field emitter structures — for instance, to compare differing arrays evaluated in different test stations or subject to conditioning.†

3.3.9. Experimental Considerations

Simple models and their generalization have so far served to identify the current–voltage relations of emitters in terms of fundamental geometrical and material parameters. With few exceptions, parameters such as emission site radius, effective work function, and field are not readily measurable, and their approximate values must be determined by other means. The well-known fluctuations in current at constant voltage over time associated with field emission — attributed to changes due to sputtering and the migration and generation of nanoprotrusions, local emission sites, and even embedded charge — is so ubiquitous that bringing attention to it is mundane. In the case of field emission from diamond films, the exact mechanisms of emission are still the subject of active investigation [51,148–151].

---

* Small differences in the SHM are due to the usage of an analytical FEA model, the application of an iterative approach to determine Δ and D, and the inclusion of $(\rho/\Sigma)^2$ in $\Delta I(\rho)$ in Eq. (118).

† Such analyses have been discussed in Ref. [115], though no factor for work function variation was included, making the Δs factor for the tip radius variation larger than otherwise.

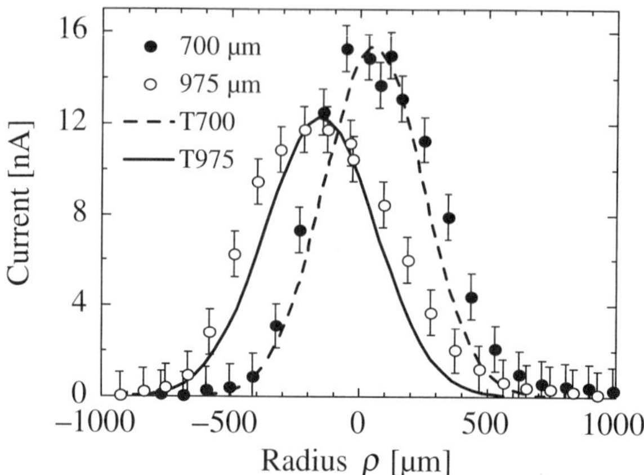

FIGURE 3.11. Comparison of the emission distribution from a single Mo field emitter fabricated by SRI with the predictions of the SHM (experimental data courtesy of P. M. Phillips).

The not-so-infrequent reporting of model-dependent parameters as though they were experimentally measured ones results in dubious comparisons. Three examples suffice. First, macroscopic fields (anode-to-cathode voltage difference divided by separation and measured in V/µm) reported for emission from films (e.g., diamond) are sometimes compared to the much larger but microscopic and local fields (extraction grid voltage multiplied by a field-enhancement factor and measured in eV/Å) associated with sharpened structures (e.g., FEAs). The comparison of a macroscopic to a microscopic field has no meaning in the absence of a statement about the field enhancement at the emission site of the former. Second, the anode–cathode system is rarely parallel plate, so that quotes of "field" being anode voltage over anode-to-gate separation $F_0 = (V_a/D)$ are misleading. A spherical anode above a flat cathode produces a field at the cathode surface approximately equal to $(V_a/D)[a(2D + a)/(D + a)^2]$, where a is the anode ball radius — e.g., if $D = 500\,\mu$m and $a = 200\,\mu$m, then the field at the cathode is 30% of (V_a/D), showing that statements of field without reference to anode geometry are disputable. Third, estimates of Φ equal to a (small) fraction of an electron volt are reported under the assumption that the field-enhancement factor is unity. All that can be ascertained from the slope on a F–N plot (assuming $b_{FN}/F \approx B_{FN}/V_g$) is the ratio $\Phi^{3/2}/\beta_g$ (but see Eq. (126) for a modification). Specification beyond that requires techniques such as the field emission retarding potential (FERP) method [152] or the comparison of field emission to photoemission distributions [68], for which reasonable work function/barrier height parameters and large beta factors are obtained.

Gate and anode voltages and emitter-to-anode separations can be experimentally measured. Field enhancement factors, effective work functions at the emission site, and local fields are typically inferred. The two most common parameters to characterize emission are A_{FN} and B_{FN}. They are often determined by a linear fit to data represented on a F–N plot, though the transformation to F–N coordinates alters the

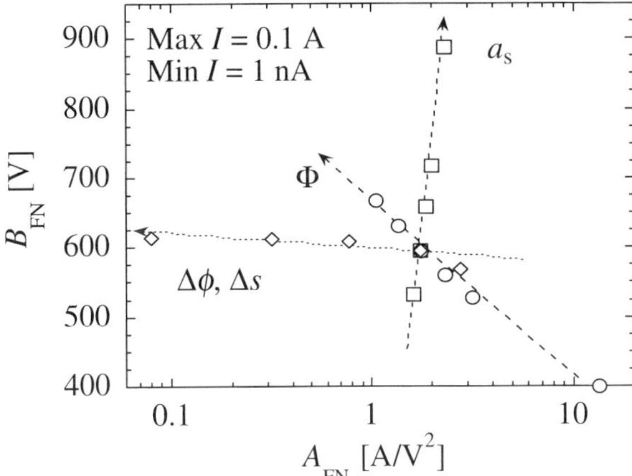

FIGURE 3.12. Changes in the F–N A_{FN} and B_{FN} parameters as a function of the effective tip radius, work function, and statistical distribution parameters $\Delta\Phi$ and Δs. The F–N fit used a Legendre polynomial analysis with slopes and intercepts evaluated at midpoint for current with the maximum and minimum values held fixed at 0.1 A and 1.0 nA, respectively. The tip radii were taken to be 30, 35, 40, 45, and 60 Å. The work functions were taken to be 3.2, 4.0, 4.2, 4.4, 4.6, and 4.8 eV. The statistical model parameters likewise were taken to be $\Delta\Phi$ (eV) $= \Delta s = 0.1, 0.3, 0.6, 1$ and 2. The arrows indicate direction of increasing parameters. The intersection point corresponds to $a_s = 35$ Å, $\Phi = 4.4$ eV, and $\Delta\Phi = \Delta s = 0.3$ (eV).

error bars associated with the data, especially for low voltages. On occasion the data themselves are such that calling them "linear" is charitable. Moreover, small changes in the parameters result in a scattering of the A_{FN} and B_{FN} parameters. To mimic that scatter, consider the results of the (metallic) SHM of an FEA when its underlying parameters (tip radius, work function, and distribution values) are altered. The arrows in Fig. 3.12 show the direction in which the F–N parameters change when a change in tip radius, work function, or statistical parameter is made. Given the dynamic nature of the emission surface, a scatter of the F–N parameters is perhaps not surprising.

The F–N fit parameters are nevertheless the tools of the trade. A discussion of experimental factors that affect these parameters constitutes the next topic. The discussion is rooted for convenience in the language of FEAs (tips, gates, arrays, etc.), but with some flexibility in interpretation, it can be extended to accommodate a variety of cold cathode structures.

3.3.10. Fowler–Nordheim Parameters

When discussing experimental $I(V)$ data, simple treatments of the F–N equation often "derive" the total current vs. extraction potential, $I(V)$, from the current density vs. local field relation $J(F)$ by the replacements $F = \beta V$ and $I = AJ$. That prescription is unsatisfactory. While the effects of a distant anode may be negligible so that $F_{tip} \approx \beta_g V_g$, certainly the area factor is not independent of the applied voltages. The geometrical models indicate why: because of the sensitivity of $J(F)$ on field and the variation of the field across the emitter surface, as the field declines, the area over which significant emission occurs declines as well; as in Eqs. (91) and (92), emission area is approximately proportional to extraction potential.

Often, $I(V)$ relations are quoted for *arrays,* not single emitters. $J(F)$ is for a particular *emission site.* The effects of a statistical distribution are seen to introduce yet another factor of extraction voltage to the coefficient of the $I(V)$ exponential. At best, then, the current–voltage relation for an array of emitters should be $I(V) = AV^{n+2}\exp(-B/V)$, where $n \approx 2$ (one factor from Σ, one factor from b_{area}). Here, A and B are related to, but not equal to, A_{FN} and B_{FN}. The discussion so far is general: the same arguments can be applied to emissive materials where (i) the emission sites have differing geometrical and material properties, and (ii) there are a large number of them. What implications does this have for the F–N fit of experimental data?

Answering that question requires investigating what constitutes a best fit of the data. A function $y(x)$, continuous in a region $x_0 - \delta \leq x \leq x_0 + \delta$, can be expanded in terms of Legendre polynomials (here, x_0 is the midpoint, not an image charge parameter). Let $y_m(x)$ be that polynomial of degree m, which minimizes the L_2 norm of $p_m(x) - y(x)$, where $p_m(x)$ are all polynomials of degree m or less [153]. Then

$$y_m(x) = \sum_{k=0}^{m} C_k P_k(z)$$

$$C_k = \left(k + \frac{1}{2}\right) \int_{-1}^{1} y(x) P_k(z)\, dz \tag{119}$$

where $z = (x - x_0)/\delta$. Legendre polynomials satisfy $-1 \leq P_k(z) \leq 1$, and

$$(k + 1)P_{k+1}(z) = (2k + 1)z P_k(z) - k P_{k-1}(z) \tag{120}$$

with $P_0(z) = 1$ and $P_1(z) = z$. The coefficient of linear correlation R gives a measure of how well $y(x)$ is approximated by $y_m(x)$:

$$\sigma_{fg}^2 = \int_{x_0-\delta}^{x_0+\delta} (f - \langle f \rangle)(g - \langle g \rangle)\, dx$$

$$\langle f \rangle = \frac{1}{2\delta} \int_{x_0-\delta}^{x_0+\delta} f\, dx \tag{121}$$

and $R = \sigma_{xy}/(\sigma_{xx}\,\sigma_{yy})^{1/2}$. The linearity of $y(x)$ is indicated by how close R^2 is to unity.

For the SHM, taking the log of $I(V)$ results in $y(x) = -n \ln(x/x_0) + \ln(A) - Bx$, where $n \approx 2$ and $x = 1/V$. Introduce the functions

$$S = \ln\left(\frac{x_0 + \delta}{x_0 - \delta}\right); \qquad U = \ln\left(1 - \left[\frac{\delta}{x_0}\right]^2\right) \tag{122}$$

then the first few C_k are

$$C_0 = \ln(A) - Bx_0 + \frac{n}{2\delta}[-x_0 S + (2 - U)\delta]$$

$$C_1 = -B\delta + \frac{3n}{4\delta^2}\left[(x_0^2 - \delta^2)S - 2x_0\delta\right] \tag{123}$$

$$C_2 = \frac{5n}{12\delta^3}\left[-3x_0(x_0^2 - \delta^2)S + 2(3x_0^2 - 2\delta^2)\delta\right]$$

As another example, when data has weak convexity, then $y(x) \approx \ln(A) - Bx + Dx^2$. Alternatively, an array of emitters with a Gaussian distribution in B values gives $D = \sigma^2/2$, where σ is the standard deviation [154]. In this case, all C_k for $k > 2$ vanish, and

$$C_0 = \ln(A) - Bx_0 + \tfrac{1}{3}D(\delta^2 + 3x_0^2)$$

$$C_1 = (-B + 2Dx_0)\delta \tag{124}$$

$$C_2 = \tfrac{2}{3}D\delta^2$$

Regardless of how the C_n are obtained, for F–N $I(V)$, the *least squares linear fit* uses the first two terms of the Legendre expansion. For experimental data analysis, $I(V)$ is defined as $A_{FN} V^2 \exp(-B_{FN}/V)$ so that

$$A_{FN} = \exp\left(C_0 - \frac{x_0}{\delta}C_1\right); \qquad B_{FN} = -\frac{C_1}{\delta} \tag{125}$$

Eq. (125) is the least squares approximation to the F–N parameters whether or not higher order terms in the Legendre polynomial expansion are retained. As such, curvature effects due to space charge for high voltages or noise for low voltages are in part mitigated. If higher terms were retained, A_{FN} and B_{FN} would no longer be constant but would depend on $x = 1/V$ and therefore, become local intercept and slope parameters (see discussion in Ref. [113]). In terms of data, x_0 represents the midpoint of $1/V$, and δ governs the range over which $1/V$ varies. Because a F–N plot of SHM $I(V)$ data is concave, Eqs. (124) and (125) then state that the B_{FN} parameter will depend on x_0 and δ: linear fits to the high voltage regime will give (not surprisingly) different F–N slopes than fits to the low voltage regime. Specifically, $\ln\{I(V)/V^2\} \approx \ln(A) - B/V + n\ln(x_0 V)$. Circumstances are set to insure that A and A_{FN} have the same units. The midpoint and range parameters are $x_0 = (V_{\max}^{-1} + V_{\min}^{-1})/2$ and $\delta = (V_{\min}^{-1} - V_{\max}^{-1})/2$) so that the predicted F–N parameters are (to leading order in δ/x_0)

$$A_{\mathrm{FN}} \approx A \exp\left\{n\left[1 + \frac{2}{3}\left(\frac{\delta}{x_0}\right)^2\right]\right\}$$

$$B_{\mathrm{FN}} \approx B + \frac{n}{x_0}\left[1 + \frac{1}{2}\left(\frac{\delta}{x_0}\right)^2\right]$$

(126)

Consequently the value of the slope depends on where the midpoint of the voltage range is, and then on the extent of the voltage range (which may seem counter to experimental intuition). The effect is seen in the light variation of B_{FN} in Fig. 3.12 when the statistical parameters change, as the voltage range for the sharpest, lowest work function emitters changes to keep the maximum and minimum current values fixed. For example using generic parameters (maximum and minimum voltage of 120 and 80 V, respectively) and $n = 2$, Eq. (126) suggests A_{FN} will be 7.8 times larger than A and B_{FN} will be 196 V larger than B. These findings are very close to a linear fit of output data from an SHM simulation. In summary, when experimental F–N parameters are used to generate a theoretical model, the effects of the voltage dependence of the area factor and statistical distribution of emission sites must be accounted for.

3.3.11. Transconductance

The exponential dependence of the apex current density on the local apex field indicates that small changes in apex field result in substantial changes in current per tip. This general feature of cold cathodes, namely, the current is exponentially related to the inverse of the field, dictates that small changes in modulation potential result in almost complete turn-off of the electron beam. By comparison, space charge limited thermionic emitters — where $I(V)$ is proportional to $V^{3/2}$ — require the voltage to decrease substantially to turn off the emission current. "Transconductance" refers to the rate at which current changes with voltage, or $g_m = \partial I/\partial V_g$. As drive power is related to gate modulation amplitude, devices from flat panel displays to rf applications seek to improve efficiency by increasing transconductance, or put differently, by reducing the gate potential modulation amplitude required for effective beam turn-off. In practice, this amounts to a demand to reduce the F–N B_{FN} value. Why is this so?

As suggested in Eq. (126), many disparate factors may be folded into — or more correctly, hidden in — the linear fit to $I(V)$ data. Assume that this has been done for two different emitters so that over a range of desired currents and voltages, each array is characterized by an A_{FN} and B_{FN}. From the definition of transconductance, it follows

$$g_m = \frac{(B_{\mathrm{FN}} + 2V)}{V^2} I(V)$$

(127)

For arrays of *comparable* geometry, emission area, and operating conditions, the voltage required for a given current level scales linearly (roughly) with B_{FN}: for example, FEA ring cathodes of similar layout but with different B_{FN} values fabricated by two separate organizations (SRI and MIT) produced similar current levels for gate voltages near 10–12% of their respective B_{FN} values [155]. For a given current and peak voltage to B_{FN} ratio (V_{pk}/B_{FN}), it is found that $(g_m)_1/(g_m)_2 \approx (B_{FN})_2/(B_{FN})_1$, or in other words, to increase the transconductance for a given current, the B_{FN} value must be reduced. By way of comparison, for a thermionic cathode, $g_m = 3I(V)/(2V)$.

3.3.12. Current Ratios and Modulation Frequency

For several devices which utilize field emission cathodes, three parameters individually or in combination bear on the utility of the cathode to the application. These parameters are beam turn-off conditions, average to peak current ratio, and pulse repetition frequency (alternatively, frequency of gate modulation). In analyzing these parameters, the cold cathode in question is assumed to be characterized by F–N A_{FN} and B_{FN} parameters. Using notation established in the literature for emission gating of the electron beam, the peak, minimum, and average quantities (e.g., current and voltage) are identified by the subscripts "pk," "min," and "ave." When the voltage is sinusoidally modulated, the amplitude of modulation is designated by the subscript "rf."

The minimum to maximum current ratio $\zeta = I_{min}/I_{pk}$ may be used to identify what minimum voltage is required to ensure that the beam is turned off. It is approximated by

$$\frac{V_{min}}{V_{pk}} \approx \frac{(B_{FN} + 2V_{pk})}{(B_{FN} + 2V_{pk}) - V_{pk}\ln(\zeta)} \tag{128}$$

By comparison, the same ratio for thermionic emitters, operated in space charge limited mode, is given by $\zeta^{2/3}$, a consequence of $I_{therm}(V) = PV^{3/2}$ (Child's Law) [146]. For example, if $\zeta = 0.01$ and $V_{pk} \approx 0.1 B_{FN}$, then V_{min} is 63% of V_{pk} —whereas for a thermionic cathode, V_{min} would be 1% of V_{pk}. Thus, a cold cathode will "turn off" for voltages that are a sizable fraction of the peak voltage, which has profound implications for devices that seek to minimize the drive power (as in displays or amplifiers) or which require complete beam turn-off (as in radar) [156].

If the extraction grid or gate potential is sinusoidally modulated such that $V(t) = V_{pk} - V_{rf}(1 - \cos(\omega t))$, then the average to peak ratio of the emitted current is given by [25]

$$\frac{I_{ave}}{I_{pk}} = \frac{\frac{1}{2\pi}\int_0^{2\pi} I_{FN}(V(\omega t))\, \mathrm{d}(\omega t)}{I_{pk}}$$

$$= e^{-z}I_0(z) - \frac{(B_{FN} + V_{pk})V_{pk}}{(B_{FN} + V_{pk})^2}\{[2zI_0(z) - (2z+1)I_1(z)]ze^{-z}\} \tag{129}$$

where $z = (V_{rf}/V_{pk}^2)(B_{FN} + 2V_{pk})$ and $I_n(z)$ is a hyperbolic Bessel function of order n. When V_{rf} is sufficiently small, the second term is small. Asymptotically, $I_0(z) \approx e^z/(2\pi z)^{1/2}$, so that, to leading order, $I_{ave}/I_{pk} \approx 1/\sqrt{2\pi z}$. For example, consider the idealized but nevertheless typical-valued case where $40V_{rf} = 10V_{pk} = B_{FN}$ ($z = 3$), Eq. (129) gives $I_{ave}/I_{pk} = 0.22462$, of which 0.24300 is due to the first term. Comparing again to thermionic emitters, the average to peak ratio for the parameters considered above ($4V_{rf} = V_{pk}$) is 0.66312, another indication of the inherently large transconductance associated with field emission. Even for full modulation ($2V_{rf} = V_{pk}$) of a thermionic cathode, the average to peak ratio is 0.42441. The ability to completely suppress the beam with voltage swings that are a fraction of the average voltage makes field emission cathodes suitable candidates for Class B or C amplifiers [131,157,158].

3.3.13. Protective Resistance

The benefits of including a protective resistor in an array are several [124]. First, it homogenizes the field emission across the array (increase in uniformity). Second, it protects the array from catastrophic arc damage by limiting the current due to an abnormal event which would otherwise melt the gate and short out the array. Finally, it has been shown that the robustness of the array is improved, in that it can be turned on without requiring bake-out after exposure to air [159].

In the event of an arc, enough stored energy (from the base–gate capacitor) is released to the gate and, if not stymied, causes it to melt and splatter, thereby destroying the site, or (if enough energy is released) possibly the array. Resistive protection, by providing a voltage drop in proportion to the emitted current, thereby suppresses current runaway and limits damage. A simple model of such resistance is considered here, though the complexity of various approaches has achieved some sophistication [160].

How does the presence of resistance affects the F–N characterization of the emitter? For low voltages (and therefore currents), the voltage drop $I(V)R$ is irrelevant, but in the range when current is significant

$$I(V) = A_{FN}(V - I(V)R)^2 \exp\left(-\frac{B_{FN}}{V - I(V)R}\right) \tag{130}$$

For solutions to exist, $V > I(V)R$, so that the current cannot exceed some maximum value. If $I(V)R/V$ is small, then

$$I_R(V) \approx \left[1 + \left(\frac{B_{FN}}{V} + 2\right)\frac{I_0(V)R}{V}\right]^{-1} I_0(V) \tag{131}$$

where the "0" subscript denotes Eq. (130) with $R = 0\ \Omega$.

The terms R and I in Eq. (131) are array parameters. Often, the resistance per tip is specified. The connection between the array resistance and the tip resistance is naively

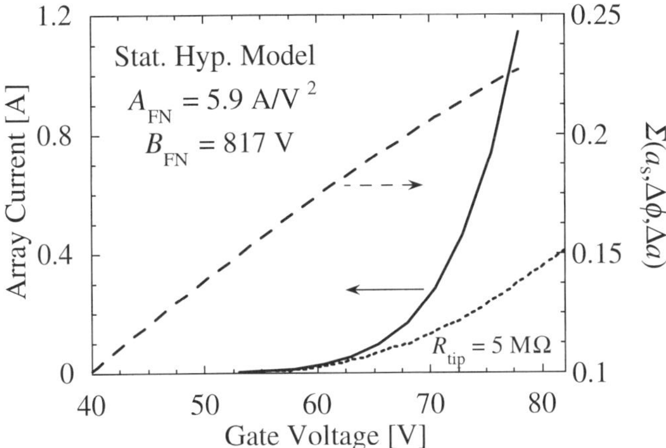

FIGURE 3.13. Array current vs. voltage for Mo-like parameters: $\Phi = 4.41$ eV, $T = 300$ K, $\rho(\mu) = 5.0 \times 10^{22}$ cm^{-2}, $a_s = 40$ Å, $a_g = 0.5$ μm, $\beta_g = 15°$, and $N_{tips} = 10^6$. Effects of a 5×10^6 Ω resistor are also shown.

ascertained by considering the voltage drop $I_{array}(V)R_{array} = N_{tips}I_{tip}(V)R_{array} \equiv I_{tip}R_{tip}$. However, if there is a distribution of tip radii or work functions — and no array is perfectly uniform — then it is the resistance of the dominant emitters that is germane, and not the average over the array. For example, in the case of the SHM, the resistance for the dominant emitters is closer to $R_{tip} \approx N_{tips}\Sigma(\Delta s, \Delta\Phi)R_{array}$; the effects of such a resistance is shown in Fig. 3.13.

A resistor may also be connected to the gate to protect the array in the event of an arc by providing negative feedback [136]. The current that appears on the RHS of Eq. (130) is then replaced with the intercepted, or gate, current. Under normal operation, a small fraction ΔI_{array} is intercepted, so that $I_0(V)$ is replaced with ΔI_{array} in Eq. (131), where ΔI is often of the order of 1% or so (in the presence of space charge, it can become significantly greater).

3.3.14. Space Charge

Space charge, or the presence of electrons between cathode and anode, can serve to reduce the extraction field used to draw emitted charge away. The effect is well known for thermionic cathodes, and is used to smooth out or reduce the noise of emitted current in a space charge limited (SCL) cathode [146]. The thin gates typical of VME cold cathode sources, being microfabricated, are often ill equipped to handle large fractions of the emitted current. For example, failure of an FEA is often preceded by a rise in gate current: current striking the gate is argued to liberate adsorbates which ionize, impact the emitter, generate nanoprotrusions which may direct further current to the gate, and so lead to a runaway condition [122].

The maximum one dimensional current that may be drawn from cathode to anode is known as "Child's Law." In its most common form, which assumes that there is no initial velocity of the electron at the cathode, it is [114]

$$I_{\text{Child}}(V_{\text{anode}}) = \frac{qmAv_a^3}{18\pi\alpha\hbar cD^2} \tag{132}$$

where $v_a = (2qV_{\text{anode}}/m)^{1/2}$ is the electron velocity at the anode, D is the cathode to anode separation, and A is the cathode area (such that $I = JA$). When the emitted current from a cold cathode approaches the maximum current specified by Child's Law, then the gate interception increases.

Other factors come into play such that the 1-D Child's limit can be exceeded. By the time an emitted electron has moved approximately a gate distance away from the field emitter tip, it has achieved a velocity $v_{\text{g}} \approx (2qV_{\text{g}}/m)^{1/2}$. When the emitted current has an initial velocity, current in excess of Eq. (132) can be drawn, and may be shown to be [161,162]

$$I_{\max}(V_{\text{gate}}, V_{\text{anode}})|_{1\text{D}} = I_{\text{Child}}(V_{\text{gate}})\left[1 + \frac{v_a}{v_{\text{g}}}\right]^3 \tag{133}$$

where the approximation $2a_{\text{g}} \ll D$ was implicitly used. Eq. (133) is still 1-D: if emission is considered to be from a strip of width W (as opposed to an area $A \gg D$), then as shown by Luginsland et al. [163], another factor is introduced such that

$$I_{\max}(W, V_{\text{gate}}, V_{\text{anode}})|_{2D} = I_{\text{Child}}(V_{\text{gate}})\left[1 + \frac{v_a}{v_{\text{g}}}\right]^3\left(1 + \frac{0.3145}{W/D} - \left(\frac{0.02}{W/D}\right)^2\right) \tag{134}$$

The maximum current will likewise be larger for emission from a small square area, given the ability of the pencil beam to expand in yet another dimension: for a beam of radius r_{b} with no initial velocity [164]

$$I_{\max}(r_{\text{b}}, V_{\text{anode}})|_{3D} = I_{\text{Child}}(V_{\text{anode}})\left[1 + \left(\frac{D}{2r_{\text{b}}}\right)^2\right] \tag{135}$$

It is clear that a finite sized pencil beam with an initial velocity can transport a fair amount of current above the 1-D Child's law limit when beam expansion is allowed along axes perpendicular to the transport axis. Nevertheless, when space charge effects are encountered, the reflection of current back to the gate causes the gate current to rise from 1 to over 5% rapidly [114].

The interception of current by the gate causes a F–N plot of anode current as a function of gate voltage to exhibit a plateau effect (which would be in addition to any effect due to a voltage drop caused by the presence of a gate resistor). The effect of such a plateau, if not excluded from experimental data, is to cause the fitted B_{FN} value to be underestimated. As many devices that require low voltage operation desire low B_{FN}, not discounting space charge effects can cause better emitter performance to be inferred than warranted. Consequently, total current (gate plus anode) and the effect of any resistive voltage drop induced by protective resistances must be accounted for in specifying the F–N parameters.

3.3.15. Transit Time

The amount of time an electron spends under the influence of the emitter-to-gate's electric field is referred to as the "transit time." The transit time, therefore, has direct bearing on the switching speed or pulse repetition frequency limits of an electron source [116]. For thermionic emitters with a cathode-to-grid separation of $z_g = 250$ μm (the limit of the technology for microwave power tubes) and an extraction field of 0.264 eV/μm, the transit time τ equals $\sqrt{(2mz_g/F_0)} \approx 104$ ps, corresponding to a cutoff frequency of $f = 1/(2\pi\tau) \approx 1.53$ GHz. Since transit time scales with the square root of the cathode-to-grid separation, decreasing that distance by a factor of 2 will only raise the cut-off frequency by a factor of 1.4. Electron sources for which the extraction potential is provided by gates that are in line with, or close to, the emitter site present a different limit. For FEAs transit time may be calculated from the potential in Eq. (99) by neglecting $F_0 z / V_g$ in comparison to unity and integrating the inverse velocity from 0 to $z_g = V_g\sqrt{(2/F_0 F_{tip})}$

$$\tau = \frac{\sqrt{(m/2)V_g}}{F_{tip}}[\sqrt{s_g(s_g+1)} + \ln(\sqrt{s_g+1} + \sqrt{s_g})] \tag{136}$$

where $s_g = \sqrt{(F_{tip}/2F_0)}$. In the limit of large s_g, $\tau \approx [mV_g/(4F_0 F_{tip})]^{1/2}$. Using parameters from the single tip emission distribution experiment but with $V_g = 54.5$ V (so that $F_{tip} = 0.5$ eV/Å), and $F_0 = 2$ eV/μm (higher F_0 is required to avoid space-charge effects), the transit time and f are 0.0955 ps and 1667 GHz, respectively, three orders of magnitude larger than the thermionic cathode case.

3.4. CONCLUSION

An abbreviated account of 75 years of field emission theory can only give a cursory flavor of the richness of the efforts to describe emission from the cornucopia of electron emissive materials. To make the didactic transition from 1-D to 3-D theory, reliance on a particular geometry (hyperbolic cone) was often invoked, which tended to circumscribe the theoretical issues treated herein, and which may have concealed the fact that electron emission physics is as complex and diverse as the variety of sources discussed in the literature. While the present treatment confined its focus to the F–N equation's derivation, application, and deficiencies, theoretical issues nevertheless abound which have not been addressed, e.g., Monte Carlo simulations [165] of electron transport through bulk material, quantum mechanical treatments of the atomic nature of the emission site, and the nature of the Schottky contact between differing materials (e.g., carbon nanotube and metal), to name but a few. In practice, the rich physics and engineering possibilities of cold cathodes, beyond being simply as a thermionic cathode replacement, has only been partially exploited. Because of their unique characteristics and capabilities, cold cathode insertion into a variety of devices is in fact being realized beyond flat panel displays and rf amplifiers. As we have seen, a qualitative account of their behavior is often obtainable with the simple models described herein.

ACKNOWLEDGMENTS

Colleagues too copious to enumerate, through interaction, discussion, at times patient guidance, and permission to use data, deserve substantial gratitude. In particular, the following individuals have been inordinately helpful (alphabetically): M. Ancona, S. Bandy, I. Brodie, F. A. Buot, F. Charbonnier, R. Forbes, A. K. Ganguly, B. Gilchrist, M. Green, M. Hollis, M. A. Kodis, Y. Y. Lau, W. Mackie, C. M. Marrese, R. A. Murphy, D. Palmer, L. Parameswaran, R. K. Parker, P. M. Phillips, A. K. Ragagopal, J. L. Shaw, A. Shih, C. A. Spindt, D. Temple, D. Whaley, and E. G. Zaidman. Support from the *Naval Research Laboratory* and the *Office of Naval Research* is most gratefully acknowledged.

REFERENCES

1. C. A. Spindt, *J. Appl. Phys.* **39**, 3504 (1968).

2. C. A. Spindt, I. Brodie, L. Humphery, and E. R. Westerberg, *J. Appl. Phys.* **47**, 5248 (1976).

3. R. H. Fowler and L. W. Nordheim, *Proc. R. Soc. (London)* **A119**, 173 (1928).

4. E. L. Murphy and R. H. Good, *Phys. Rev.* **102**, 1464 (1956).

5. S. G. Christov, *Phys. Stat. Sol.* **17**, 11 (1966).

6. J. W. Gadzuk and E. W. Plummer, *Rev. Mod. Phys.* **45**, 487 (1973).

7. R. Stratton, *Phys. Rev.* **135**, A794 (1964).

8. G. N. Fursey, *Appl. Surf. Sci.* **94**, 44 (1996).

9. R. Gomer, *Field Emission and Field Ionization*, American Institute of Physics: New York, 1993; American Vacuum Society Classics.

10. A. Modinos, *Field, Thermionic, and Secondary Electron Emission Spectroscopy*, Plenum: New York, 1984.

11. F. M. Charbonnier, *Appl. Surf. Sci.* **94/95**, 26 (1996).

12. P. H. Cutler and D. Nagy, *Surf. Sci.* **3**, 71 (1964).

13. H. Q. Nguyen, P. H. Cutler, T. E. Feuchtwang, N. Miskovsky, and A. A. Lucas, *Surf. Sci.* **160**, 331 (1985).

14. A. Mayer and J. P. Vigneron, *J. Phys. Condens. Matter* **10**, 869 (1998).

15. J. He, P. H. Cutler, N. M. Miskovsky, T. E. Feuchtwang, T. E. Sullivan, and M. Chung, *Surf. Sci.* **241**, 348 (1991).

16. P. H. Cutler, J. He, J. Miller, N. M. Miskovsky, B. Weiss, and T. E. Sullivan, *Prog. Surf. Sci.* **42**, 169 (1993).

17. R. G. Forbes, *J. Vac. Sci. Technol. B* **17**, 526 (1999).

18. R. G. Forbes, *J. Vac. Sci. Technol. B* **17**, 534 (1999).

19. Proceedings of the 42nd International Field Emission Symposium, *Appl. Surf. Sci.* **94/95** (1996); see in particular, the papers of Forbes, Charbonnier, and Fursey.

20. W. P. Dyke, *Scientific American*, 108, Jan. 1964.

21. L. Schiff, *Quantum Mechanics*, 3rd ed., McGraw-Hill: New York, 1968; R. Shankar, *Quantum Mechanics*, Plenum Press: New York, 1980.

22. R. Kubo, H. Ichimura, T. Usui, and N. Hashitsume, *Statistical Mechanics*, North-Holland: New York, p. 241, 1965. See in particular Chapter 4.

23. G. A. Haas and R. E. Thomas, Thermionic emission and work function, in *Techniques of Metals Research, Vol. VI Part 1* (R. F. Bunshah, Ed.), Wiley: New York, 1972. The appendices contain a comprehensive list of work function values for the elements.

24. R. H. Fowler and L. Nordheim, *Proc. Roy. Soc. Lond.* **A 119**, 173 (1928).

25. K. L. Jensen, *J. App. Phys.* **85**, 2667 (1999).

26. S. M. Sze, *Physics of Semiconductor Devices*, 2nd ed., Wiley: New York, 1981.

27. R. P. Feynman, *Statistical Mechanics: A Set of Lectures,* W. A. Benjamin Inc.: Reading, MA, 1972.

28. C. Kittel, *Introduction to Solid State Physics*, 7th ed., Wiley: New York, 1996.

29. K. L. Jensen and A. K. Ganguly, *J. Appl. Phys.* **73**, 4409 (1993).

30. M. K. Harbola and V. Sahni, *Phys. Rev. B* **39**, 10437 (1989).

31. K. L. Jensen, *J. Vac. Sci. Technol. B* **13**, 516 (1995).

32. I. S. Gradshteyn and I. M. Ryzhik, *Table of Integrals, Series, and Products*, Academic Press: New York, 1980. See Eq. (10) of Section 4.293.

33. J. Bardeen, *Phys. Rev.* **58**, 727 (1940).

34. N. D. Lang and W. Kohn, *Phys. Rev. B* **1**, 4555 (1970); *ibid.*, **7**, 3541 (1973).

35. N. D. Lang, The density-functional formalism and the electronic structure of metal surfaces, in *Solid State Physics, Vol. 28*, (H. Ehrenreich, F. Seitz, and D. Turnbull, Eds.), Academic Press: New York, p. 225, 1973.

36. J. Bardeen, *Phys. Rev.* **49**, 653 (1936).

37. A. Kiejna, *Phys. Rev. B* **47**, 7361 (1993) and references therein.

38. A. Kiejna and K. F. Wojciechowski, *Metal Surface Electron Physics*, Pergamon: Oxford, 1996.

39. A. L. Fetter and J. D. Walecka, *Quantum Theory of Many-Particle Systems*, McGraw-Hill: New York, 1971. See in particular Sections 3 and 12.

40. A. Ishihara and E. W. Montroll, *Proc. Nat. Acd. Sci.* **68**, 3111 (1971).

41. W. Jones and N. H. March, *Theoretical Solid State Physics, Vol. II: Non-equilibrium and Disorder*, Dover: New York, Sec.10.11.1, 1985.

42. A. Kiejna and K. F. Wojciechowski, *Metal Surface Electron Physics*, Pergamon: Oxford, 1996.

43. I. Brodie, *Phys. Rev.* **51**, 13660 (1995).

44. J. A. Appelbaum and D. R. Hamann, *Phys. Rev. B* **6**, 1122 (1972).

45. A. G. Eguiluz and W. Hanke, *Phys. Rev. B* **14**, 10433 (1989).

46. L. G. Il'chenko and Y. V. Kryuchenko, *J. Vac. Sci. Tech. B* **13**, 566 (1995).

47. L. Eyges, *The Classical Electromagnetic Field*, Dover: New York, 1972.

48. K. L. Jensen, *J. Appl. Phys.* **85**, 4455 (2000).

49. M. G. Ancona, *Phys. Rev. B* **46**, 4874 (1992)

50. W. Mönsch, *Semiconductor Surfaces and Interfaces*, 2nd ed., Springer-Verlag: Berlin, Ch. 2, 1995.

51. A. T. Sowers, B. L. Ward, S. L. English, and R. J. Nemanich, *J. Appl. Phys.* **86**, 3973 (1999).

52. K. L. Jensen, J. E. Yater, E. G. Zaidman, M. A. Kodis, and A. Shih, *J. Vac. Sci. Tech. B* **16**, 2038 (1998).

53. N. M. Miskovsky and P. H. Cutler, *Appl. Phys. Lett.* **75**, 2147 (1999).

54. G. N. Fursey and D. V. Glazanov, *J. Vac. Sci. Technol. B* **16**, 910 (1998).

55. R. Forbes, Private communication.

56. H. Inaba, K. Kurosawa, and M. Okuda, *Jpn. J. Appl. Phys.* **28**, 2201 (1989); E. N. Glytsis, T. K. Gaylord, and K. F. Brennan, *J. Appl. Phys.* **66**, 6158 (1989); K. F. Brennen and C. J. Summers, *J. Appl. Phys.* **61**, 614 (1987); C. M. Tan, J. Xu, and S. Zukotynski, *J. Appl. Phys.* **67**, 3011 (1990).

57. H. Inaba, K. Kurosawa, and M. Okuda, *Jpn. J. Appl. Phys.* **28**, 2201 (1989).

58. E. N. Glytsis, T. K. Gaylord, and K. F. Brennan, *J. Appl. Phys.* **66**, 6158 (1989).

59. K. F. Brennen and C. J. Summers, *J. Appl. Phys.* **61**, 614 (1987).

60. C. M. Tan, J. Xu, and S. Zukotynski, *J. Appl. Phys.* **67**, 3011 (1990).

61. J. Robertson, *J. Vac. Sci. Technol. B* **17**, 659 (1999).

62. W. Mönch, *Semiconductor Surfaces and Interfaces*, 2nd ed., Springer: New York, p. 86, 1995.

63. S. D. Ganichev, E. Ziemann, W. Prettl, I. N. Yassievich, A. A. Istratov, and E. R. Weber, *Phys. Rev. B* **61**, 10361 (2000).

64. K. L. Jensen and J. L. Shaw, *Mater. Res. Soc. Symp. Proc.* **621**, R3.3 (2000).

65. Z.-H. Huang, P. H. Cutler, N. M. Miskovsky, and T. E. Sullivan, *J. Vac. Sci. Technol. B* **13**, 526 (1995).

66. W. Zhu, G. P. Kochanski, S. Jin, and L. Seibles, *J. Appl. Phys.* **78**, 2707 (1995).

67. W. Zhu, G. P. Kochanski, S. Jin, and L. Seibles, *J. Vac. Sci. Technol. B* **14**, 2011 (1996).

68. O. Gröning, O. M. Küttel, P. Gröning, and L. Schlapbach, *J. Vac. Sci. Technol. B* **17**, 1970 (1999).

69. M. S. Chung, B.-G. Yoon, J. M. Park, P. H. Cutler, and N. M. Miskovsky, *J. Vac. Sci. Technol. B* **16**, 906 (1998).

70. P. Lerner, N. M. Miskovsky, and P. H. Cutler, *J. Vac. Sci. Technol. B* **16**, 900 (1998).

71. S. R. P. Silva, G. A. J. Amaratunga, and K. Okano, *J. Vac. Sci. Technol. B* **17**, 557 (1999).

72. R. Johnston and A. J. Miller, *Surf. Sci.* **266**, 155 (1992).

73. V. G. Litovchenko, A. A. Evtukh, Yu. M. Litvin, N. M. Goncharuk, and V. E. Chayka, *J. Vac. Sci. Technol. B* **17**, 655 (1999).

74. V. G. Litovchenko and Yu. V. Kryuchenko, *J. Vac. Sci. Technol. B* **11**, 362 (1993).

75. J. L. Shaw, Private communication.

76. O. Madelung, *Introduction to Solid State Theory*, 2nd ed., Springer-Verlag: New York, 1978. See Section 4.2 "The Boltzmann Equation."

77. J. M. Ziman, *Principles of the Theory of Solids*, 2nd ed., Cambridge University Press: Cambridge, 1972.

78. K. L. Jensen, *J. Vac. Sci. Technol. B* **13**, 505 (1995).

79. J. S. Blakemore, *Semiconductor Statistics*, Dover: New York, 1987.

80. K. L. Jensen, Y. Y. Lau, and D. S. McGregor, *Appl. Phys. Lett.*, **77**, 585 (2000).

81. L. E. Reichl, Probability distributions in dynamical systems, in *A Modern Course in Statistical Physics*, U. of Texas Press, Austin, TX, Ch. 7, Sec. E–G, 1980.

82. Y. S. Kim and M. E. Noz, *Phase Space Picture of Quantum Mechanics*, World Scientific: New Jersey, 1991.

83. J. R. Barker, D. W. Lowe, and S. Murry, in *The Physics of Submicron Structures* (H. L. Grubin, K. Hess, G. J. Iafrate, and D. K. Ferry, Eds.), Plenum: New York, p. 277, 1984.

84. W. Frensley, *Phys. Rev. B* **36**, 1570 (1987); *ibid, Solid State Electronics* **31**, 739 (1988).

85. N. C. Kluksdahl, A. M. Kriman, C. Ringhofer, and D. K. Ferry, *Solid State Electronics* **31**, 743 (1988).

86. F. A. Buot and K. L. Jensen, *Phys. Rev. B* **42**, 9429 (1990).

87. K. L. Jensen and A. K. Ganguly, *J. Vac. Sci. Technol. B* **11**, 371 (1993).

88. F. Rana, S. Tiwari, and D. A. Buchanan, *Appl. Phys. Lett.* **69**, 1104 (1996).

89. R. Waters and B. Van Zeghbroeck, *Appl. Phys. Lett.* **75**, 2410 (1999).

90. S. A. Barengolts, M. Yu. Kreindel, and E. A. Litvinov, *Surf. Sci.* **206**, 126 (1992).

91. N. M. Miskovsky, S. H. Park, J. He, and P. H. Cutler, *J. Vac. Sci. Technol. B* **11**, 366 (1993).

92. W. P. Dyke, J. K. Trolan, W. W. Dolan, and G. Barnes, *J. Appl. Phys.* **24**, 570 (1953).

93. F. M. Charbonnier, J. P. Barbour, L. F. Garrett, and W. P. Dyke, *Proc. IEEE*, 991 (1963) and references therein.

94. K. R. Shoulders, *Adv. Comput.* **2**, 135 (1961).

95. E. G. Zaidman, *IEEE Trans. Electron Devices* **40**, 1009 (1993).

96. D. W. Jenkins, *IEEE Trans. Electron Devices* **40**, 666 (1993).

97. W. D. Kesling and C. E. Hunt, *J. Vac. Sci. Technol. B* **11**, 518 (1993).

98. D. Nicolaescu and V. Avramescu, *J. Vac. Sci. Technol. B* **12**, 749 (1994).

99. A. A. Lucas, H. Morawitz, G. R. Henry, J.-P. Vigneron, Ph. Lambin, P. H. Cutler, and T. E. Fuchtwang, *Phys. Rev. B* **37**, 10708 (1988).

100. D. Nicolaescu, *J. Vac. Sci. Tech. B* **11**, 392 (1993).

101. K. L. Jensen and E. G. Zaidman, *J. Vac. Sci. Technol. B* **12**, 776 (1994); see K. L. Jensen and E. G. Zaidman, *J. Vac. Sci. Technol. B* **14**, 1873 (1996) for the correct α factor.

102. J. D. Jackson, *Classical Electrodynamics*, 2nd ed., Wiley: New York, Ch. 3, 1975.

103. G. Arfken, *Mathematical Methods for Physicists*, 3rd ed., Academic Press: Orlando, FL, 1985.

104. M. Morse and H. Feshbach, *Methods of Theoretical Physics II*, McGraw-Hill: New York, 1953.

105. E. E. Martin, F. M. Charbonnier, W. W. Dolan, W. P. Dyke, H. W. Pitman, and J. K. Trolan, Research on field emission cathodes, *Technical Report 59–20*, Wright Air Development Division, Jan. 1960.

106. P. H. Cutler, J. He, N. M. Miskovsky, T. E. Sullivan, and B. Weiss, *J. Vac. Sci. Technol. B* **11**, 387 (1993).

107. D. A. Kirkpatrick, A. Mankofsky, and K. T. Tsang, *Appl. Phys. Lett.* **60**, 2065 (1992).

108. R. Forbes and K. L. Jensen, Unpublished.

109. K. L. Jensen, P. Mukhopadhyay-Phillips, E. G. Zaidman, K. Nguyen, M. A. Kodis, L. Malsawma, and C. Hor, *Appl. Surf. Sci.* **111**, 204 (1997); *ibid, Appl. Phys. Lett.* **68**, 2807 (1996).

110. K. L. Jensen, E. G. Zaidman, M. A. Kodis, B. Goplen, and D. N. Smithe, *J. Vac. Sci. Technol. B* **14**, 1942 (1996); *ibid*, 1947 (1996).

111. D. Nicolaescu, *J. Vac. Sci. Technol. B* **14**, 1930 (1996).

112. K. L. Jensen, E. G. Zaidman, M. A. Kodis, B. Goplen, and D. N. Smithe, Theory and simulation of a gated field emitter: Analysis and incorporation into macroscopic device characterization, *NRL Technical Memorandum*, NRL/FR/6840-95-9782, 1995.

113. K. L. Jensen, M. A. Kodis, R. A. Murphy, and E. G. Zaidman, *J. Appl. Phys.* **82**, 845 (1997).

114. K. L. Jensen, *J. Appl. Phys.* **83**, 7982 (1998).

115. K. L. Jensen, *Recent Res. Dev. Vacuum Sci. & Tech.* **1**, 45 (1999).

116. K. L. Jensen, *Phys. of Plasmas* **6**, 2241 (1999).

117. C. M. Marrese, J. E. Polk, K. L. Jensen, A. D. Gallimore, C. A. Spindt, R. L. Fink, and W. D. Palmer, "Chapter 11: Performance of Field Emission Cathodes in Xenon Electric Propulsion System Environments," Micropropulsion for Small Spacecraft (Progress in Astronautics and Aeronautics Vol. 187), M. M. Micci, A. D. Ketsdever (eds.), American Institute of Aeronautics and Astronautics, Reston, VA, 2000.

118. I. Brodie and P. R. Schwoebel, *Proc. IEEE* **82**, 1006 (1994).

119. W. D. Goodhue, P. M. Nitishin, C. T. Harris, C. O. Bozler, D. D. Rathman, G. D. Johnson, and M. A. Hollis, *J. Vac. Sci. Technol. B* **12**, 693 (1994).

120. D. Palmer, H. F. Gray, J. Mancusi, D. Temple, C. Ball, J. L. Shaw, and G. E. McGuire, *J. Vac. Sci. Technol. B* **13**, 576 (1995).

121. J. Itoh, Y. Nazuka, S. Kanemaru, T. Inoue, H. Yokoyama, and K. Shimizu, *J. Vac. Sci. Technol. B* **14**, 2105 (1996).

122. F. Charbonnier, *J. Vac. Sci. Technol. B* **16**, 880 (1998).

123. S. T. Purcell, V. T. Binh, and R. Baptist, *J. Vac. Sci. Technol. B* **15**, 1666 (1997).

124. J. D. Levine, *J. Vac. Sci. Technol. B* **14**, 2008 (1996).

125. C. Constancias and R. Baptist, *J. Vac. Sci. Technol. B* **16**, 841 (1998).

126. N. A. Cade, R. Johnston, A. J. Miller, and C. Patel, *J. Vac. Sci. Technol. B* **13**, 549 (1995).

127. R. L. Hartman, W. A. Mackie, and P. R. Davis, *J. Vac. Sci. Technol. B* **12**, 778 (1994).

128. A. Renau, F. H. Read, and J. N. H. Brunt, *J. Phys. E: Sci. Instrum.* **15**, 347 (1982).

129. R. L. Hartman, W. A. Mackie, and P. R. Davis, *J. Vac. Sci. Technol. B* **14**, 1952 (1996).

130. E. G. Zaidman, K. L. Jensen, and M. A. Kodis, *J. Vac. Sci. Technol. B* **14**, 1994 (1996).

131. See Section II of F. M. Charbonnier, J. P. Barbour, L. F. Garrett, and W. P. Dyke, *Proc. IEEE*, **51**, 991 (1963).

132. P. R. Schwoebel and C. A. Spindt, *Appl. Phys. Lett.* **63**, 33 (1993).

133. P. R. Schwoebel and C. A. Spindt, *J. Vac. Sci. Technol. B* **12**, 2414 (1994).

134. C. A. Spindt, C. E. Holland, P. R. Schwoebel, and I. Brodie, *J. Vac. Sci. Technol. B* **14**, 1986 (1996).

135. P. R. Schwoebel and C. A. Spindt, *Appl. Phys. Lett.* **63**, 33 (1993).

136. D. Temple, W. D. Palmer, L. N. Yadon, J. E. Mancusi, D. Vellenga, and G. E. McGuire, *J. Vac. Sci. Technol. B* **16**, 1980 (1998).

137. C. A. Spindt, C. E. Holland, A. Rosengreen, and I. Brodie, *IEEE Trans. Elec. Dev.* **38**, 2355 (1991). See in particular Fig. 11.

138. J. P. Barbour, F. M. Charbonnier, W. W. Dolan, W. P. Dyke, E. E. Martin, and J. K. Trolan, *Phys. Rev.* **117**, 1452 (1960); P. C. Bettler, F. M. Charbonnier, *Phys. Rev.* **119**, 85 (1960).

139. C. A. Spindt and I. Brodie, Molybdenum field-emitter arrays, in *Tech. Dig. of the IEEE Intl. Electron Devices Meeting*, 12.1.1, 1996.

140. F. Charbonnier, *J. Vac. Sci. Technol. B* **16**, 880 (1998).

141. W. A. Mackie, T. Xie, and P. R. Davis, *J. Vac. Sci. Technol. B* **17**, 613 (1999).

142. Y. Liu and Y. Y. Lau, *J. Vac. Sci. Tech. B* **14**, 2126 (1996).

143. P. M. Phillips, C. Hor, L. Malsawma, K. L. Jensen, and E. G. Zaidman, *Rev. Sci. Instruments* **67**, 2387 (1996).

144. D. R. Whaley, B. Gannon, C. R. Smith, C. M. Armstrong, and C. A. Spindt, *IEEE Trans. Plasma Sci.*, **28**, 727 (2000).

145. C. M. Tang, T. A. Swyden, and K. A. Thomason et al., *J. Vac. Sci. Technol. B* **14**, 3455 (1996).

146. C. K. Birdsall and W. B. Bridges, *Electron Dynamics of Diode Regions*, Academic Press: New York, p. 36, 1966.

147. K. L. Jensen, *Recent Res. Devel. Sci. & Tech.* **1**, 45 (1999).

148. C. Bandis and B. Pate, *Phys. Rev. B* **52**, 12056 (1995).

149. R. Schlesser, M. T. McClure, B. L. McCarson, and Z. Sitar, *J. Appl. Phys.* **82**, 5763 (1997).

150. J. Robertson, *Mat. Res. Soc. Symp. Proc.* **621**, R 1.1 (2000).

151. R. Forbes, *Mat. Res. Soc. Symp. Proc.* **621**, R 3.1 (2000).

152. W. A. Mackie, J. E. Plumlee, and A. E. Bell, *J. Vac. Sci. Technol. B* **14**, 2041 (1996).

153. D. M. Young and R. T. Gregory, *A Survey Of Numerical Mathematics, Vol. 1*, Dover: New York, p. 318, 1973.

154. J. D. Levine, *Le Vide les Couches Minces*, Supplement au N271, Mars–Avril 1994 (extended abstracts of the 7th International Vacuum Microelectronics Conference, Grenoble, France, 1994).

155. K. L. Jensen, R. H. Abrams, and R. K. Parker, *J. Vac. Sci. Technol. B* **16**, 749 (1998).

156. K. L. Jensen, *Physics of Plasmas* **6**, 2241 (1999).

157. M. A. Kodis, K. L. Jensen, E. G. Zaidman, B. Goplen, and D. N. Smithe, *J. Vac. Sci. Tech. B* **14**, 1990 (1996).

158. M. A. Kodis, K. L. Jensen, E. G. Zaidman, B. Goplen, and D. N. Smithe, *IEEE Trans. Plasma Sci.* **24**, 970 (1996).

159. J. D. Levine, R. Meyer, R. Baptist, T. E. Fetler, and A. A. Talin, *J. Vac. Sci. Technol. B* **13**, 474 (1995).

160. L. Parameswaran and R. A. Murphy, *Mat. Res. Soc. Symp. Proc.* **621**, R 4.6 (2000).

161. J. Luginsland, S. McGee, and Y. Y. Lau, *IEEE Trans. Plasma Sci.*, **26**, 901 (1998).

162. B. E. Gilchrist, K. L. Jensen, A. D. Gallimore, and J. G. Severns, *Mat. Res. Soc. Symp. Proc.* **621**, R 4.8 (2000).

163. J. W. Luginsland, Y. Y. Lau, and R. M. Gilgenbach, *Phys. Rev. Lett.*, **77**, p. 4668, 1996.

164. S. Humphries, *Charged Particle Beams*, John Wiley and Sons, Inc.: New York, 1990.

165. P. H. Cutler, Z.-H. Huang, n. M. Miskovsky, P. D'Ambrosio, and M. Chung, *J. Vac. Sci. Technol. B* **14**, 2020 (1996).

CHAPTER 4

Spindt Field Emitter Arrays

CHARLES A. (CAPP) SPINDT, IVOR BRODIE, CHRISTOPHER E. HOLLAND
AND PAUL R. SCHWOEBEL

SRI International, 333 Ravenswood Avenue, Menlo Park, California 94025-3493

4.1. INTRODUCTION

The objective of this chapter is to introduce the Spindt cathode and review the fabrication technology, operating characteristics, and applications of this microfabricated, gated, field emission, cold cathode. In doing this, we will not attempt to cover many of the basic physical issues involved in depth, but rather will refer to the ample body of literature covering the fundamentals of field emission as appropriate. Our aim is to acquaint the readers with Spindt-type microfabricated field emitters and the similarities and differences between these microfabricated field emitters and classic etched-wire field emitters.

We begin in Section 4.2 with a brief history of the development of the Spindt cathode and the resulting vacuum microelectronics (VME) technological field. Section 4.3 is devoted to the fabrication technology, and the many options and variables that one can employ to tailor the emitter arrays for given applications. These include various lithography technologies, materials compatibility issues, management of thin film growth dynamics during depositions, materials considerations for specific applications, and ongoing developments with this new technology. In Section 4.4 emitter array performance is discussed. A review of modeling tools available for emitter analysis is followed by "intrinsic" emitter array performance under the best conditions that we can produce. We then discuss means for dealing with the "real" operating environments that may be encountered in practical applications. As a part of this, long-term and short-term stability (lifetime and noise) is also discussed. In addition, we will explore the use of the Fowler-Nordheim (F-N) equation [1] for evaluating and designing emitter arrays for specific applications, using the well-known F-N a and b parameters and cathode capacitance obtained from experimental results. The section concludes with a summary of overall performance at the time of this writing. Some less common applications that may not be covered elsewhere are discussed in Section 4.5. This section will be brief and specific, as the major applications

Vacuum Microelectronics, Edited by Wei Zhu.
ISBN 0-471-32244-X. 2001 John Wiley & Sons, Inc.

are covered in depth in Chapters 7 and 8. Finally, we give an overall summary in Section 4.6, along with acknowledgments.

The reader is reminded that much of the work discussed here is ongoing, and some of the results and conclusions presented may be tentative, and thus represent only our best assessment of the situation at the time of this writing. With that in mind, we will always try to make it clear when we are indulging in speculation.

4.2. A BRIEF HISTORY OF THE SPINDT CATHODE

Spindt cathodes consist of microfabricated metal field emitter cones formed on a conducting electrode (base) by thin film deposition processes. Each cone has its own concentric aperture in an accelerating grid electrode (or gate) which is insulated from the base electrode by a thin dielectric layer, as shown in Fig. 4.1. Typically, the cone height and thickness of the dielectric layer are about 1 μm, the tip radius about 200 Å, the aperture diameter about 0.5 μm, and the tip-to-tip spacing from 1 to 5 μm. However, all the dimensions can be controlled over a wide range. In particular, the aperture diameter can be as small as 0.1 μm and the tip radius down to about 50 Å. The array functions as a field emission source of electrons when a positive potential is applied to the gate relative to the tips. No heat is needed.

The basic ideas that led to the development of microfabricated field emitter array (FEA) came out of a very ambitious notion conceived by Ken Shoulders and Dudley Buck at MIT in the 1950s [2]. They proposed to employ thin film deposition and micromachining techniques to fabricate integrated vacuum field-effect devices to 1000Å sizes (i.e. vacuum integrated circuits based on field emission cold cathodes). To be seen in proper perspective it must be remembered that this was done before there were solid-state integrated circuits. Their vision has proven to be remarkably clear, but at that time was 40 years beyond technology's reach.

Shoulders brought the basic concepts that he and Buck developed to the Stanford Research Institute (also known as "SRI" at that time, and now officially as SRI International) in 1958, and with US Department of Defense support, he initiated a research

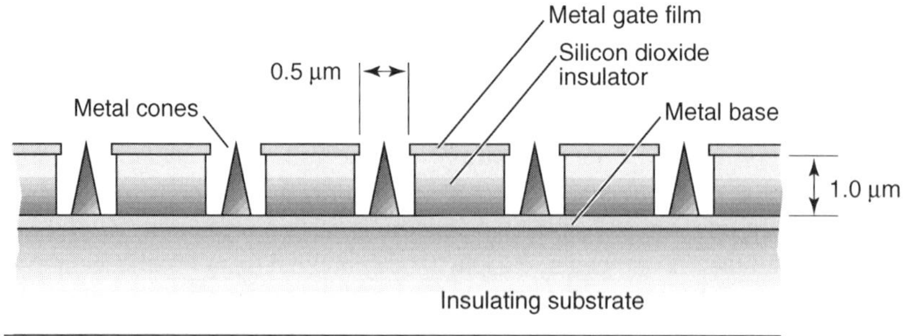

FIGURE 4.1. The Spindt microfabricated field emitter array.

program to develop microfabricated vacuum integrated circuits [3]. As a part of these activities he proposed a thin display tube based on matrix-addressed arrays of micro-fabricated field emitters — the field emission display, now commonly known as an FED [4]. This was the genesis of the VME technology that is presently being researched in essentially every industrial country in the world. As a part of Shoulder's program, Capp Spindt developed a process for microfabricating arrays of miniature metal field emitter cones in micron-size cavities with an integrated extraction electrode (gate). These structures were shown to produce detectable electron emission to an external anode with 20 V applied between the gate and the cones, and several microamperes of emission with 100 V applied between the cones and the gate. These results were first reported at the 1966 IEEE Conference on Tube Techniques [5], and in the open literature in 1968 [6]. Shoulders left SRI in 1968 to pursue other interests, and the emitter array development work continued at a low level, with internal funding by SRI until 1973. In 1973, support for development of a cold cathode for microwave tubes based on this microfabricated FEA technology was received from the NASA Lewis Research Center under the direction of Dr. Ralph Forman. At the same time Ivor Brodie, a well-known vacuum-tube expert, joined SRI as Director of the Applied Physics Laboratory, where the work was being done. One of Brodie's first acts was to initiate publication of the significant advances that had been made in the technology up to that time [7]. These two events, along with the concurrent development of large-area silicon emitter arrays at Westinghouse by Thomas and coworkers [8], were critical in taking the technology to the point that others began to take notice. In particular Richard Greene and Henry Gray at the Naval Research Laboratory (NRL), where *thermionic* vacuum integrated circuits developed at SRI by Geppert [9] were being evaluated, became interested in the technology and actively promoted it within the government as well as initiating NRL in-house work. In 1985 the French group at LETI, led by Robert Meyer, announced preliminary results with their efforts to produce a flat panel, field emission display (FED) based on the Spindt cathode [10]. The enormous commercial potential for such a product was immediately recognized by many throughout the world, and activity in the field (which by now was known as Vacuum Microelectronics) accelerated virtually overnight. SRI had also been working on a FED panel, and in 1987 Chris Holland, who was leading the panel fabrication effort at SRI, reported the first three-color FED panel [11]. At this time it became clear that many teams throughout the world were independently researching VME technology, but most were working in isolation and even innocent ignorance of others that were active in the field. Recognizing this, Richard Greene at NRL and Ivor Brodie at SRI International assigned their colleagues, Henry Gray at NRL and Capp Spindt at SRI, the task of organizing the first International Vacuum Microelectronics Conference (IVMC) which was held in Williamsburg, Virginia, in 1988. The IVMC has been attracting between 200 and 300 researchers annually from all over the world ever since, and has had a major impact on the establishment of VME as a recognized emerging technological field with applications ranging from flat panel displays to spacecraft charge management systems and microwave tube amplifiers.

4.3. FABRICATION TECHNOLOGY

In this section we will discuss the fundamentals of the Spindt-cathode fabrication process. As with most processes, the basic procedure is quite straightforward, but there are many subtleties that can impact the results that one obtains. We will begin by describing the evolution of the basic process, and then discuss the parameters of importance and some of the variations of the process that can be employed to suit specific needs and applications.

4.3.1. The Original Spindt-Cathode Fabrication Process

The underlying theory for field emission is expertly covered by Jensen in Chapter 3, and will not be repeated here. However, a brief discussion of the benefits to be had by miniaturizing field emitter structures is presented to orient the reader, and show how microfabrication technology allows us to take advantage of the unique characteristics of field emission.

The basic appeal of the field electron emission phenomena is that it provides very efficient, cold, electron emission; extremely high emission current density; and the potential for very high total currents with emitter arrays. These attributes have not been widely utilized because of the difficulties with reliability, lifetime, and emission instabilities associated with classic etched-wire field emitters. These problems exist because of the high voltages required for emission with classic emitters (generally 1 kV or greater), and associated sputtering effects which have an impact on the emitter-tip surface condition and thereby the effective work function and local field enhancement factors.

F-N theory for emission from metals under intense electric fields predicts emission current I for a given applied voltage V as follows:

$$I = aV^2\, e^{-b/V} \tag{1}$$

where

$$a \cong 1.5 \times 10^{-6} \frac{A}{\phi}\, e^{10.4/\phi^{1/2}} \beta^2$$

$$b \cong 6.44 \times 10^7 \phi^{3/2}/\beta$$

given A is the emitting area in cm^2, ϕ the cathode work function in eV, and β the geometric factor in cm^{-1} that determines the electric field E at the cathode such that $E = \beta V$.

Casual inspection of these relationships indicates that, in order to obtain high currents at low voltages, one desires to have a as low as possible ϕ, and as high as possible A and β. The effective work function ϕ is a property of the emitting surface, taking into account the emitter-tip material plus any impurities and adsorbates on the surface at the vacuum/emitter interface. This is a dynamic surface due to the nature of the field emission process. Thus, one should think in terms of a dynamic effective work function that takes into account the operating environment and changes as the operating environment changes. As a practical matter, when dealing with cold emitters, the effective work function is a factor that is difficult to control and is generally accepted as is for the given situation (typically 4–5 eV for commonly used materials).

On the other hand, the dominant factors determining the β factor are the emitter-tip radius r and the spacing R between the emitter tip and the field-forming anode or gate. Following a simple concentric sphere model to illustrate the situation, Brodie and Schwoebel [12] show that

$$\beta = \frac{R}{kr(R - r)} \approx \frac{1}{kr} \quad \text{for } r \ll R \tag{2}$$

where $1 < k < 5$ accounts for the effect of the emitter shank [13].

Applying dimensions that would be normal in classic emitter work, such as $R = 1$ mm and $r = 2000$ Å, we see that $\beta = 5.00 \times 10^4/k$ cm^{-1}. Recalling that the field at the tip is $F = \beta V$ and that the onset of field emission is generally accepted to occur at fields of the order of 10^7 V/cm, we see that an applied voltage of the order of 1000 V is required to produce emission.

On the other hand, using dimensions to be expected for microfabricated emitters, i.e. $r = 250$ Å and $R = 5000$ Å, we obtain a β of $4.17 \times 10^5/k$ cm^{-1}. From this we see that an applied voltage of the order of 100 V will produce the required field with a structure having the microfabricated dimensions. Numerous more rigorous treatments all leading to the same conclusions can be found in the literature. See for example the work of Marcus et al. [14], Zaidman et al. [15], Temple et al. [16], Jensen et al. [17], and Kang et al. [18].

The above relationships illustrate the intuitively satisfying notion that reducing the spacing between the electrodes and minimizing the emitter-tip radius are means to increase the field at the emitter tip for a given applied voltage. Furthermore, a close-spaced integrated gate electrode will shield the emitter tip from sputtering as a result of ions formed between the gate and any higher voltage accelerating electrode in an electron-beam forming system. That is, any ions formed more than a few microns away from the emitting tip would not be focused back onto the tip because of electrostatic shielding by the gate electrode, and those formed very close to the tip would cause minimal damage because of low-voltage differentials in that region, despite the very high fields.

In addition, the application of microfabrication technology to the forming of emitter arrays allows one to take advantage of using arrays of emitter tips operated in parallel to produce large total currents, as well as high current densities averaged over the area occupied by the array. The integrated gate electrode prevents mutual shielding between tightly packed tips, which was a difficulty with etched-wire emitters. Finally, applying microfabrication technology to the forming of FEAs with nanometer-scale dimensions allows for parallel processing of up to billions of emitter tips per square centimeter on a substrate. These can be used as a large-area, matrix-addressable, emitter arrays such as in the display application, or be diced into thousands of individual emitter arrays for economic, high-volume fabrication of cathodes for applications such as microwave tubes.

To summarize, classic etched-wire field emission work has demonstrated that there is great potential for developing a generally useful and improved cathode, based on the field emission phenomena. The work done in this field by pioneers such as Dyke and Dolan and coworkers, who reported current densities exceeding 10^8 A/cm^2 and peak pulse currents exceeding 6 A from a single tip [19], made it clear that field emission cathodes could improve the performance of many devices if a means eliminate the difficulties associated with the high-voltage requirements and resultant stability and lifetime issues could be found. This was the motivation behind the development of the Spindt-cathode process, which was originally reported in 1966 as a method for fabricating an encapsulated micron-size vacuum diode [5].

The original rectifying diode fabrication process is shown in Fig. 4.2. A sapphire substrate was coated with a thin film of evaporated molybdenum, followed by an evaporated aluminum oxide insulating layer, and then a second molybdenum layer to form a metal–oxide–metal structure on the sapphire. This thin film sandwich structure was then sprayed with finely dispersed 0.5-μm diameter polystyrene balls and coated with an evaporated film of aluminum as shown in Fig. 4.2(a). The polystyrene balls were then washed away, exposing the top molybdenum layer where the balls had shadowed the aluminum. A chemical etch of the molybdenum using the aluminum as a resist mask was then performed to form holes in the molybdenum film, followed by an etch of the aluminum oxide using the molybdenum as a resist mask, thereby producing the structure shown in Fig. 4.2(b). The substrate was then mounted on a rotatable shaft in a vacuum deposition system equipped with two electron-beam heated molybdenum evaporators. One evaporator was positioned so that the evaporated molecular beam arrived at the substrate at a grazing angle such that a significant portion of the deposited material condensed on the upper inner edges of the holes. This material caused the etched holes to close as the substrate was rotated about an axis perpendicular to the substrate surface. The second molybdenum evaporator was positioned so that it deposited material perpendicular to the substrate surface and into the bottom of the cavities etched in the top molybdenum and aluminum oxide layers. The relative rates of deposition between the grazing-angle source and the source normal to the surface of the substrate controlled the shape taken on by the molybdenum deposited into the bottom of the cavities. When properly done a sharp cone was formed in the bottom of each cavity, with the tip of the cone approximately in the plane of the top molybdenum film as the cavity was sealed off by the combination of deposition from the two evaporators as shown in Fig. 4.2(c).

Diode rectification was obtained as a result of field emission from the sharp tips when an ac voltage was applied to the structure. When the top film was biased to a sufficient positive potential (~45 V) relative to the tips, current flowed in the forward direction as a result of field emission from the sharp tips. Conversely, in the reverse polarity the top film had no such field enhancing features and there was no current flow [5,6].

Armed with this result, it was quickly realized that, if the top film was open instead of closed, the electrons could be emitted out into a vacuum space. The device

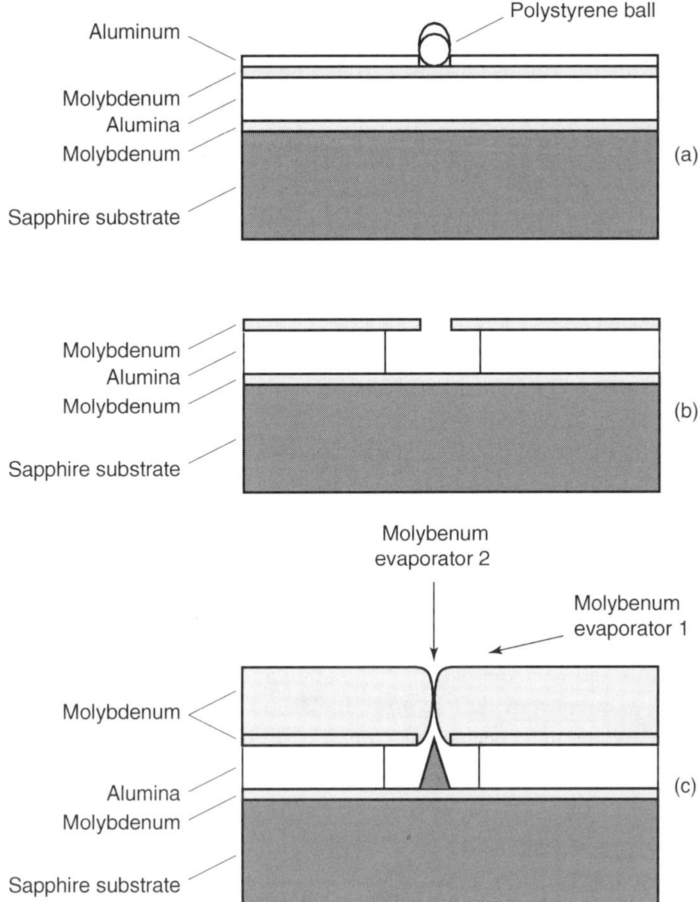

FIGURE 4.2. Fabrication of the original microfabricated, field emission, rectifying diode. (a) Patterning holes with polystyrene balls as shadow masks; (b) etched holes in the metal–oxide–metal structure; (c) emitter-cone formation and encapsulation by physical vapor deposition.

could then be used as an array of field emitting cold cathodes. An array of field emitters was something that had been recognized to have great potential by workers using one-dimensional (1-D) linear arrays of classic etched-wire field emitters at the Linfield Research Institute to achieve high pulse current densities for X-ray production [19]. However, this etched-wire approach had proven difficult to produce in two-dimensional (2-D) arrays because of the problems with uniformity and mutual electrostatic shielding between the emitter tips.

Three modifications to the fabrication process described earlier were made to produce a microfabricated FEA cathode. The first replaced the original grazing-angle molybdenum deposition with a selectively etchable material that would serve as a sacrificial lift-off layer as well as helping to close the gate aperture so that a cone would be formed in the cavity. After the cone had been formed, all of the material deposited on the gate could be removed by dissolving the lift-off layer, thereby exposing the cone. The first material chosen for the sacrificial layer was aluminum, as it had been used as the masking layer for etching the cavities, and therefore it was known that it could be easily removed chemically without damaging the molybdenum gate and cones.

A second modification followed shortly after the first results were obtained. It is the nature of evaporated metal films to form nucleation sites on the substrate that merge into a continuous film as the deposition progresses [20]. This can be of no particular consequence when the molecular beam of evaporant is directed normal to the surface of the substrate, but when the deposition is from a grazing angle relative to the substrate surface, the nucleation sites shadow the substrate and build up protruding peaks in the direction of the deposition source. This effect produces a very rough film, even though the substrate is rotated during the deposition to improve uniformity and coat all of the top inner edge of the cavity in the gate film. For the initial feasibility studies this was acceptable, but it soon became evident that the uniformity of shape from cone to cone in an array suffered from this rough closure of the gate aperture during the cone-formation process.

The remedy turned out to be relatively simple. Evaporated oxides tend to condense on the substrate without forming agglomerated nucleation sites, and therefore produce a smooth film even when deposited at a grazing angle to the substrate surface. With this in mind, aluminum oxide was used in place of the aluminum for the sacrificial lift-off layer, and did indeed produce a much smoother closure of the edge of the gate cavity as shown in Fig. 4.3.

(a)   (b)

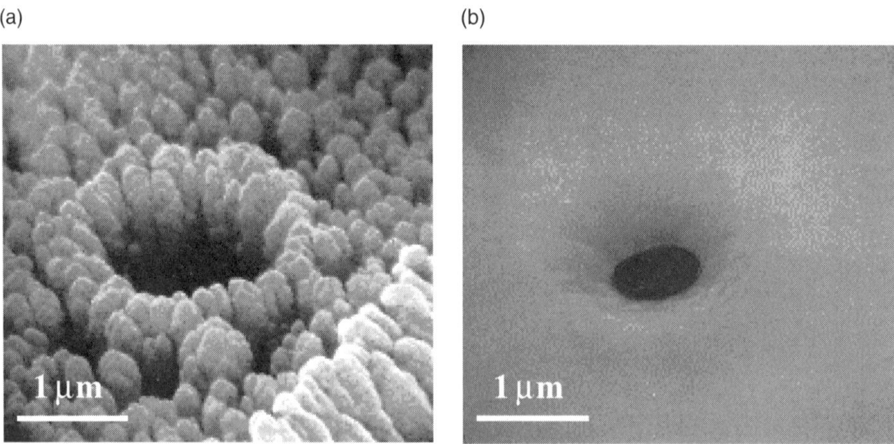

FIGURE 4.3. Scanning electron micrographs showing (a) a very rough aluminum lift-off layer deposited at a grazing angle while rotating the substrate and (b) a relatively smooth lift-off layer of aluminum oxide deposited at a grazing angle while rotating the substrate.

However, the insulating layer was also evaporated aluminum oxide and therefore was also dissolved by the lift-off etchant (phosphoric acid at 120°C). This difficulty was overcome by a third modification to the diode process, which was to bake the structure prior to the cone-deposition process at 1000°C in vacuum. This procedure hardened the aluminum oxide dielectric layer so that phosphoric acid (at 120°C) had virtually no effect on it, whereas the unbaked evaporated aluminum oxide lift-off layer etched readily in the same phosphoric acid. With these modifications the process worked reasonably well, and emission of a few microamperes to an external anode was achieved with 60 V applied between the tips (base) and gate [5].

Aside from providing a smoother, more uniform lift-off layer, aluminum oxide turned out to have the additional benefit of providing a means of stress management in the deposited films. In general, it is known that most metals tend to deposit with internal tensile stress, and dielectrics with internal compressive stress. In certain cases there are means to manage the internal stress in depositing films by manipulating deposition conditions such as temperature, deposition rate, and residual gas pressure during the process. However, as a practical matter these factors are difficult to incorporate reliably into a deposition process. The general topic of internal stress in deposited films is important, but quite involved, not completely understood, and beyond the scope of our discussion here. For a complete summary of film stress the reader is referred to [21].

For our purposes it is sufficient to know that molybdenum tends to deposit with internal tensile stress while aluminum oxide deposits with internal compressive stress, and the two, when codeposited, tend to cancel each other. Thus, continuing to deposit the aluminum oxide while the molybdenum is forming the emitter cone tends to neutralize the tensile stress in the molybdenum and allows much thicker films — and thus taller cones — to be deposited without the risk of breaking the sample up. Continuing with the aluminum oxide deposition throughout the cone deposition process also provides for an additional control parameter for the shape of the emitter cone that is formed by the process. This is done by adjusting the relative rates of deposition between the grazing angle aluminum oxide lift-off/hole-closure material and the molybdenum cone-formation material which contributes relatively little (but significantly) to the rate of the gate-hole closure.

It is important to note, however, that the contribution of the aluminum oxide material to the rate of hole closure is limited to the early stages of the cone-deposition process. The reason for this is illustrated in Fig. 4.4. By the nature of the film growth process, the top inner-edge of the gate hole develops an increasingly large radius of curvature as the deposition process proceeds. Then, at some point, the minimum inner radius of the gate aperture will be "over the horizon" with respect to a line of sight to the aluminum oxide evaporator. When this occurs, the aluminum oxide no longer contributes to the closure of the hole, but only to stress relief in the molybdenum film on the top surface of the gate film. Then, because the contribution of the molybdenum deposition to the rate of hole closure is much less than that of the aluminum oxide, the hole will — from that point on in the process — close at a slower rate. The result is that the cone forms with a smaller half-angle and a larger aspect ratio, resulting

(a)

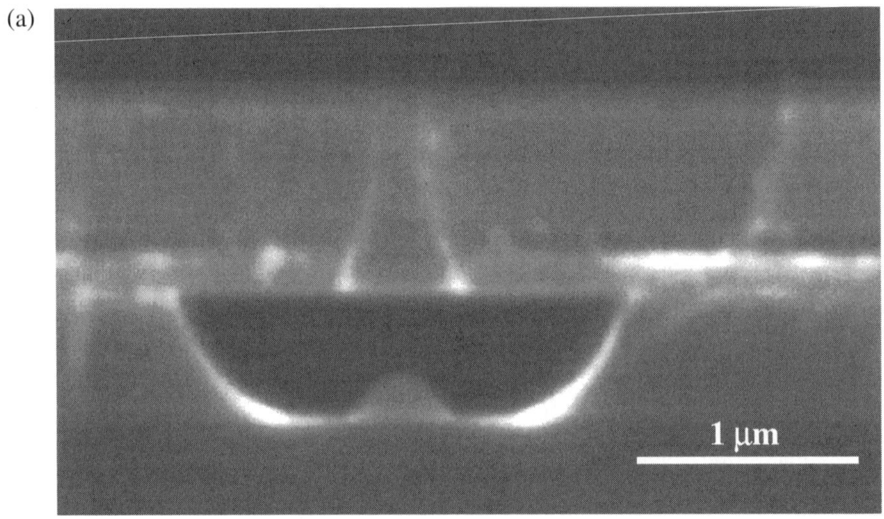

(b)

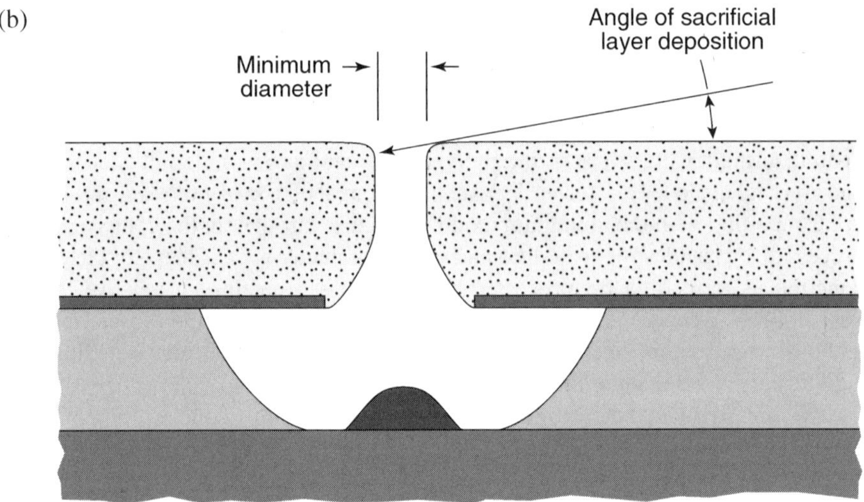

FIGURE 4.4. A cross section of a scanning electron micrograph and sketch of the sacrificial lift-off layer, showing a lack of hole closure past a point where the minimum diameter is shadowed by the edge of the hole. The last 0.5 μm of the aluminum oxide lift-off material was deposited with the molybdenum cone material evaporator turned off in order to illustrate the effect.

in a pleasing "inverted golf-tee" shape. There are several more subtle factors that can enter into the picture and have an influence (sometimes dramatic) on the shape of the emitter cone that is produced by this basic process, and these will be discussed in detail in the next section.

4.3.2. Improvements and Variations to the Original Process

The basic process described in the previous section can be applied to any flat, smooth, vacuum-compatible substrate, by using a chemical vapor deposited (CVD), evaporated, spin-on glass (SOG), or sputtered oxide dielectric layer. However, as the technology progressed, and a variety of potential applications emerged, it became obvious that silicon substrates are much more convenient than sapphire or glass in situations where a common emitter base electrode is acceptable. The original reason was because of the high-quality oxide that can be thermally grown on silicon. However, shortly after beginning to work with silicon substrates, it became clear that there are several additional advantages to be had by using silicon substrates. Some of these are as follows:

1. A number of the microfabrication tools that have been developed for the semiconductor industry can be used in processing silicon wafers (e.g., optical steppers, electron-beam lithography, chemical vapor deposition, reactive ion etchers, chemical–mechanical polishing, etc.).
2. A wide range of silicon resistivities is available, and, in some instances, very highly resistive silicon can be used, thereby effectively incorporating an integrated emission-buffering resistor with each tip.
3. Devices can be integrated with the tips on a silicon wafer to control emission.
4. Silicon is probably the flattest, smoothest, vacuum-compatible substrate available at low cost.

For these reasons, common-base emitter arrays are now most often fabricated on silicon substrates, as illustrated in Fig. 4.5, with the following process steps:

1. Obtain n-type silicon wafers. The conductivity range can be anywhere between 0.01 and 2000 Ω cm, depending on the application one has in mind for the emitter array.
2. Oxidize the wafers using standard thermal oxidation techniques to the desired thickness. Again the optimum thickness will depend on the intended application and normally ranges from 0.5 to 2.0 μm. Factors involved are emitter packing density, emission modulation frequency (if any), and intended gate-aperture size.
3. Coat the oxide layer with a uniform layer of gate metal. The material and thickness (typically $\sim$0.2 μm) are design issues that again depend on the gate-aperture size, cone material to be used, intended application, and compatibility with the chemistry involved in the lift-off process. Historically molybdenum

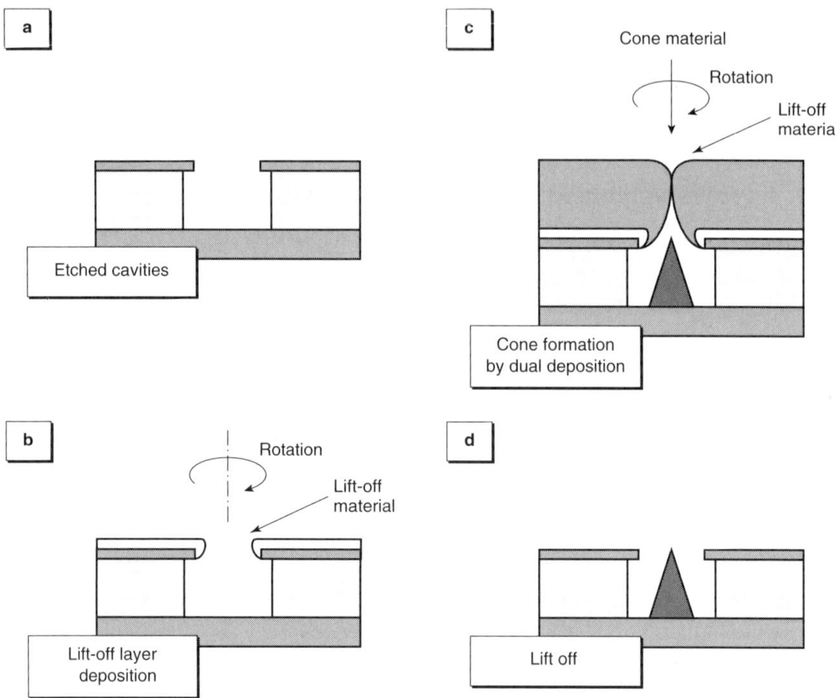

FIGURE 4.5. The original Spindt-cathode fabrication process. (a) Cavities are etched in the top two layers of a metal/dielectric/metal (or silicon) stack using resist patterning and wet or dry etching. (b) A sacrificial lift-off layer is deposited onto the top layer and inner walls of the upper portion of the cavity by grazing-angle deposition while rotating the substrate on an axis perpendicular to the surface. (c) Cones are formed in the cavities by depositing metal perpendicular to the substrate surface. The grazing-angle deposition can be used continuously or intermittently during this step to help control the hole-closure rate and to manage stresses in the deposited layer as desired. Alternatively, it can be terminated during the cone deposition if the conditions permit. (d) Lift-off is done by a wet etch of the lift-off material with a solvent that attacks only the lift-off material.

has been used most commonly; however, niobium, chromium, platinum, and nickel as well as others have also been utilized.

4. Coat the gate material with a photo- or electron-sensitive resist and pattern an array of holes in the resist with the lithography tool of choice. Photolithography has been the most commonly used method; however, electron-beam lithography, focused ion-beam lithography, and particle tracking have also been employed. This is an important step, as hole size and uniformity are critical parameters in the operation of an emitter array.

5. Transfer the hole pattern in the resist to the gate film by developing the resist and etching the gate material through the patterned resist. Wet or dry chemistry can be used, but a polishing etch is strongly recommended. The reason for this

is to obtain a smooth surface on the inner edge of the gate aperture so that a smooth, uniform gate-aperture closure — and the best possible cones — will be obtained during the subsequent cone-formation procedure.

6. At this point one has several options on how to etch cavities in the oxide down to the silicon substrate base. The easiest is to simply use a buffered hydrofluoric acid (BOE) wet etch which stops at the silicon base. However, this is an isotropic etch, and may not be appropriate if the hole pattern is densely packed. This is because the isotropic etch will undercut the gate film and break into the neighboring cavities if the hole pitch is equivalent to the oxide thickness. If this happens, the gate film may no longer have adequate support to withstand the rigors of the subsequent cone-formation processes. Under these conditions, an anisotropic dry reactive ion etch (RIE) must be used. In this case, selectivity in the RIE etch rates between the materials in the structure must be considered. In particular, the selectivity in etch rates between the oxide and the silicon at the bottom of the cavity is an issue with regard to obtaining uniformity in cavity depth across the array. This is best managed by stopping just short of a complete etch of the oxide with RIE and finishing up with a brief isotropic hydrofluoric acid wet etch. The isotropic wet etch stops at the silicon, but also undercuts the edge of the gate film. However, if done properly, the etch is not of sufficient duration to undercut the gate enough to break thorough the sidewalls to the neighboring cavities. This slight undercut has the additional feature of moving the triple point between the gate metal, oxide, and vacuum away from the field-stress-raising bottom-inner edge of the gate film — thereby reducing the risk of electrical flash-over damage to the structure during operation. The risk of electrical flash-over can be further reduced by undercutting the gate film with a brief BOE isotropic etch prior to the anisotropic RIE of the oxide.

7. Mount the silicon substrate in a vacuum deposition system and rotate the substrate about an axis perpendicular to its surface. A lift-off layer (e.g., aluminum oxide) is deposited by thermal evaporation onto the rotating substrate at an angle of about 30° from the substrate surface. In this way the diameter of the holes in the gate film is decreased to the desired size by film buildup on the upper-inner edges of the holes, as shown in Fig. 4.5(b). The desired size is determined by the thickness of the gate film and oxide layer, and relies on the dynamics of the following emitter-cone growth process. This will vary with the operating environment in each deposition system, and must be determined experimentally. It is important to note that, because deposition is essentially "line-of-sight," no lift-off material reaches the bottom of the cavity.

8. Deposit the cone material through the partially closed holes and into the bottom of the cavities by evaporation from a source positioned to deposit material perpendicular to the substrate surface. As the diameter of the cavities continues to decrease because of the grazing angle deposition, the material depositing in the bottom of the cavities forms a cone defined by the shadow masking effect of the closing hole in the gate film. A point is formed on the cone when the hole completely closes, as shown in Fig. 4.5(c). Considerable control of the cone height, angle, and tip radius is obtained by choice of starting gate-hole size,

gate thickness, thickness of the oxide layer, and deposition parameters, and the cones are reproduced exactly from cavity to cavity if the gate holes are of the same size. Factors to be considered in the choice of the emitter-cone material are discussed in Section 4.4.4.

9. Dissolve the lift-off layer (with phosphoric acid at 120°C if aluminum oxide is used), thereby releasing the cone material deposited on the top surface of the gate film during the cone-formation process. The cones remain in place because vacuum evaporation is line-of-sight, and no lift-off material reaches the bottom of the cavities.

A rigorous rinse and cleaning process, followed by vacuum bake-out and storage in vacuum until the emitters are to be used, completes the process. The details of the cleaning process are dependent on the materials used for the gate and cones. We have found that hydrogen firing (at 600°C) as a last clean-up step prior to vacuum storage is beneficial unless materials that readily form hydrides have been used in the structure. Figure 4.6 shows a scanning electron micrograph of a typical Spindt-cathode array fabricated on a silicon substrate by this process.

Portion of 50,000-tip array with
a packing density of 10^8 tips/cm^2

(a)

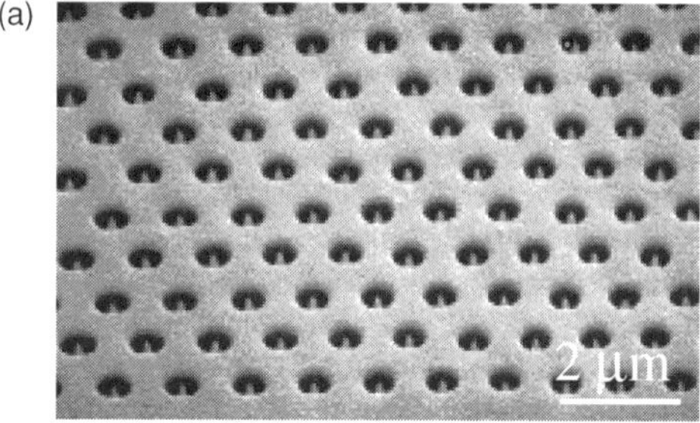

Cross section of a tip

(b)

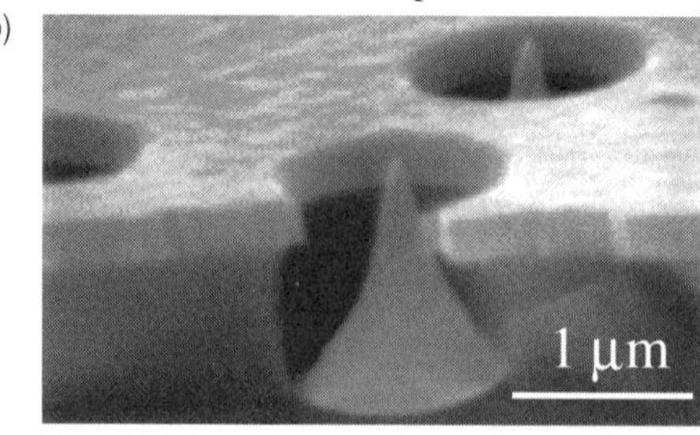

FIGURE 4.6. Scanning electron micrographs of Spindt cathodes: (a) a portion of a 50 000-tip array with a tip packing density of 10^8 tips/cm^2 and (b) a cross-sectional view of a single molybdenum Spindt emitter [34].

4.3.2.1. *Lithography for Patterning Gate Apertures.* The size, shape, and uniformity of the gate apertures are critical parameters that must be carefully controlled in order to obtain the desired results. The original technique of using 0.5-µm diameter polystyrene balls as shadow masks worked well for developing the basic process and demonstrating emission. However, it is clear that random arrays of this kind do not lend themselves to the fabrication of routinely reproducible arrays of emitters. One of the researchers in our laboratory at SRI, Gene Westerberg, was developing a multiple-electron-beam lithography system at the time this work was evolving, and it was immediately obvious that it was a "natural" as a tool to pattern gate apertures for the emitter arrays [22]. The system is shown schematically in Fig. 4.7. A fine-mesh screen was illuminated by an electron beam defined by a small "point-source" aperture positioned relatively far away from the screen (e.g., $Z_o = 1$ m).

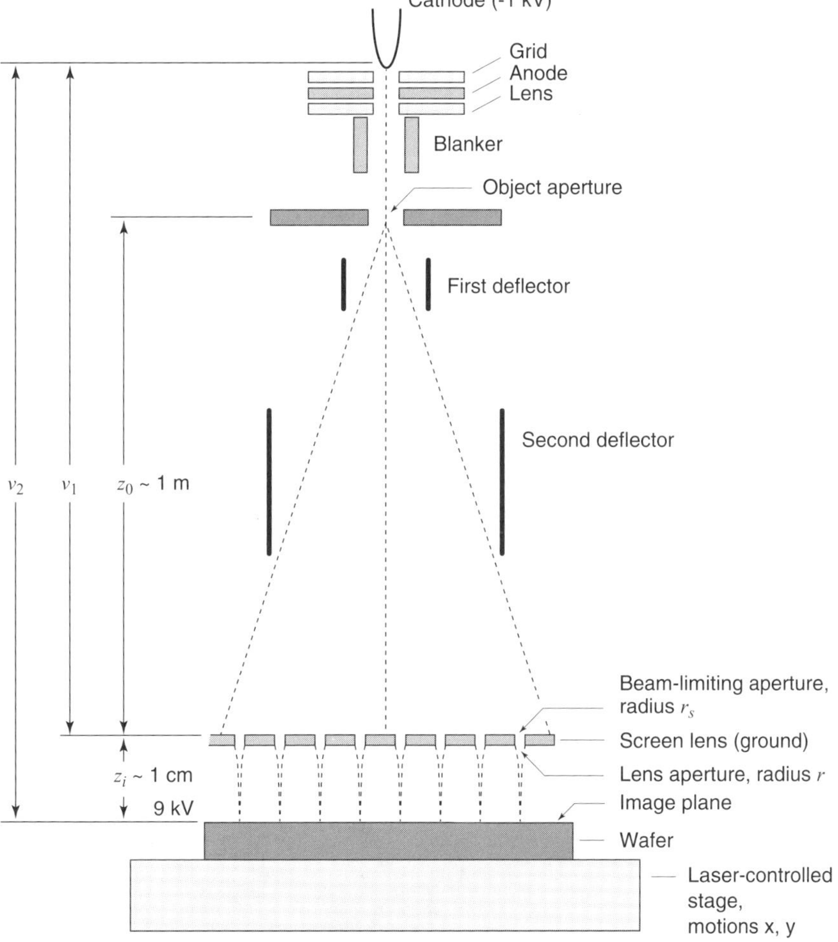

FIGURE 4.7. A schematic drawing of the SRI screen-lens multiple-electron-beam lithography system. Screen lens at ground, cathode at -1 kV, and wafer at $+9$ kV [22].

Then a target substrate coated with an electron-sensitive resist is positioned under the screen at a distance of about 1 cm. A voltage of several kilovolts applied between the target and screen creates an electrostatic field that causes each of the openings in the screen to perform as a tiny electrostatic lens and project a demagnified image of the object aperture onto the target. The demagnification factor (M) is set by the geometry of the system and is simply twice the distance between the target and screen divided by the distance between the screen and object aperture ($M = Z_i/2Z_o$). For example, the demagnification factor for the dimensions shown in Fig. 4.7 ($Z_o =$ 1 m and $Z_i = 1$ cm) is 5×10^{-3}. That is, size of the image on the target is 5×10^{-3} times that of the object aperture. Furthermore, there is an image of the object aperture formed by every opening in the screen. In the screen-lens system built at SRI, a 2000-mesh screen (2000 lines/in.) was used for the screen lens, and a submicron diameter beam was formed by each screen opening. Thus an orthogonal pattern of submicron spots on 0.5-mil or 12.5-μm centers was printed in the electron-sensitive resist by each exposure. For higher image packing densities a step-and-repeat process was used. For example, Fig. 4.8 shows groups of 4×4 emitter tips on 1.25-μm centers with a tip-to-tip spacing within each 4×4 group of 2.5 μm. This was done with 16 step-and-repeat exposures in about 15 s. This property of parallel exposures is the primary feature of this lithography tool. An entire wafer can be covered with a pattern of spots, as shown in Fig. 4.8, in the time that it takes to do one pixel of 4×4 emitter tips, i.e. about 15 s. In 25 step-and-repeat exposures a uniform pattern of tips on 2.5-μm centers could be printed, covering an entire 6-in. wafer in less than a minute of writing time. This unique, one-of-a-kind, laboratory-built experimental tool was used at SRI to pattern gate apertures from the early 1970s to the mid-1990s when submicron commercial optical-stepper lithography became readily available.

Two other techniques have been used for patterning sub-micron gate apertures over large areas. The first is a particle-tracking method whereby high-energy atomic-size particles are used to bombard a resist-coated substrate. The particles leave a damaged track through the resist that is subsequently developed to form a randomly arranged pattern of very tiny holes ($\leq 1000\,\text{Å}$). The gate apertures are then etched in the exposed

FIGURE 4.8. A scanning electron micrograph of a portion of an emitter array patterned with the SRI screen-lens electron-beam lithography system.

material on the target. Candescent Technology Corporation has used this method to produce very small gate-aperture emitters [23], and this is the reason that Candescent has been able to operate its display panels with peak cathode driving voltages of about 10–15 V. These are the lowest operating voltages reported at the time of this writing, and it is due entirely to the small dimensions they are able to achieve over large areas with the particle-tracking method.

Another small gate-aperture patterning method is laser interferometry used by MIT's Lincoln Laboratory to fabricate very high packing density emitter arrays. This method involves splitting a laser beam and using the two beams to form a standing-wave interference pattern on a photoresist-coated substrate, thereby exposing the photoresist with lines and spaces determined by the wavelength and incident angle of the beams. The substrate is then rotated $90°$, and the exposure repeated to generate a screen-like pattern in the resist. Holes with packing densities of up to 10^9 holes/cm^2 over very large areas have been patterned by this method [24–27]. The VME team at MIT's Lincoln Laboratory fabricated Spindt-type molybdenum emitter arrays using this lithography technique, and was able to achieve emission current densities of up to 2400 A/cm^2 averaged over the area occupied by the array of emitter tips [25]. Figure 4.9 is a scanning electron micrograph of a portion of such an array.

These more exotic methods for obtaining sub-micron gate apertures are very good for achieving low operating voltages and high current densities, but they are not readily available to most workers. However, there are other "tricks" developed by clever workers to achieve this end without having to employ exotic tools. One of

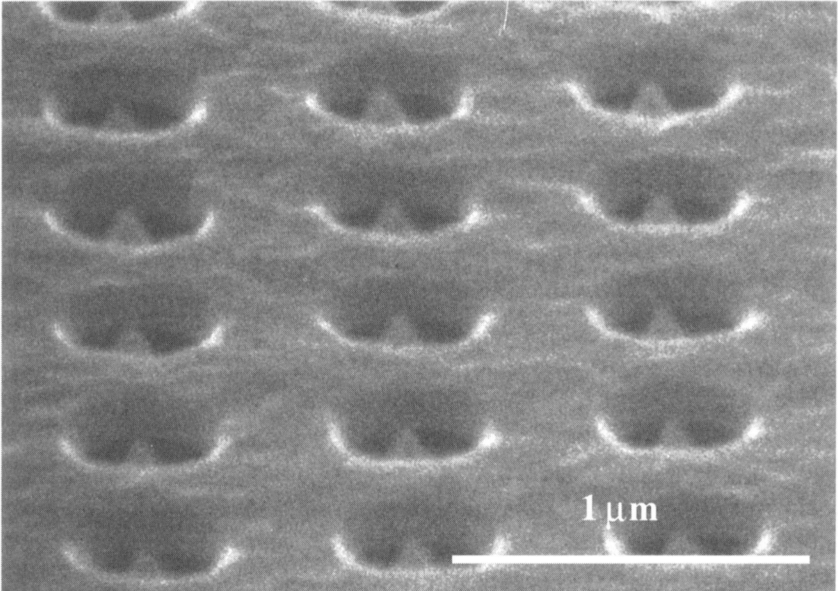

FIGURE 4.9. A scanning electron micrograph of a hole array patterned with laser interferometry and tips formed at MIT's Lincoln Laboratory. The tip packing density is about 10^9 tips/cm^2. (Courtesy of MIT's Lincoln Laboratory.)

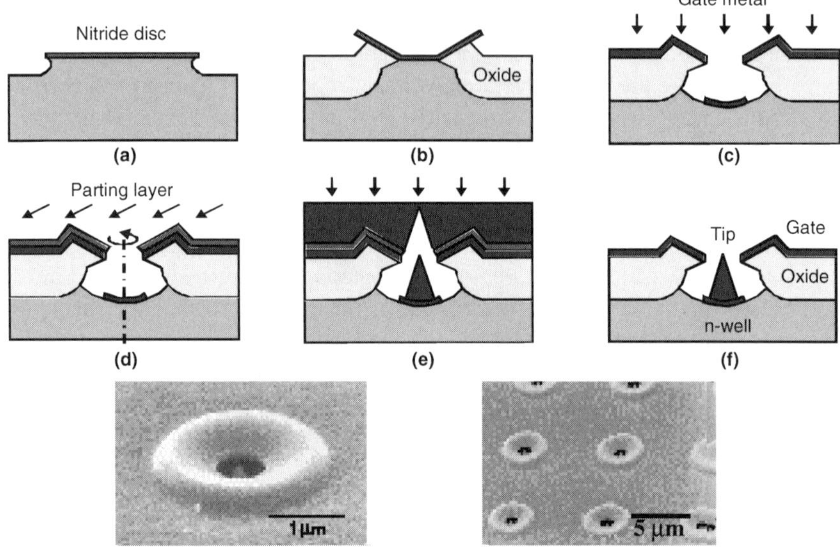

FIGURE 4.10. A method of forming submicron gate apertures with micronscale lithography using the LOCOS process after Oh (1998) (Courtesy of Prof. Jong Duk Lee, Seoul National University). (a) A silicon nitride is patterned into ~2-μm diameter disks with conventional photolithography. The silicon is then etched isotropically to facilitate oxidation under the disk as shown. (b) The exposed silicon is then thermally oxidized to the desired gate insulation layer thickness. (c) The nitride is removed and the exposed silicon etched isotropically to form a gate cavity. Molybdenum is then evaporated perpendicularly to form a gate electrode. The gate-aperture diameter is now ~0.5 μm. (d) A sacrificial lift-off layer of aluminum is deposited. (e) Molybdenum is deposited perpendicular to the surface to form an emitter cone in the cavity. (f) The aluminum lift-off layer is removed by selective etching that completes the process. The two SEM micrographs at the bottom show a portion of an array of sub-micron gate diameter emitters spaced on 8-μm centers. The close-up view at the left shows the gate diameter to be 0.6 μm [28].

the more ingenious is one developed at Seoul National University in Prof. Jong Duk Lee's group, employing the well-known local oxidation of silicon or LOCOS process [28]. The process can be used to produce submicron diameter gate apertures, using conventional micron size lithography technology. The procedure is to pattern micron size nitride disks on a silicon substrate, and then grow a thermal oxide on the silicon, using standard furnace oxidation. The oxide intrudes under the nitride disk in the commonly called "bird's beak" fashion, thereby creating a oxide layer with holes in it under each nitride disk, as shown in Fig. 4.10. The nitride is then removed, and a cavity etched in the exposed silicon. A gate metal is deposited as shown, and a cone is then grown in the cavity by the standard Spindt process. Very nice, uniform, sub-micron gate-diameter emitter arrays have been fabricated with this technique.

4.3.2.2. Sacrificial Lift-Off Layer Issues. Management of the lift-off layer is a critical issue from the viewpoint of emitter-cone uniformity over an array and

repeatability from array to array. The best possible gate-aperture lithography is of no value if the lift-off layer causes nonuniform, rough closure of the gate aperture during the cone-formation process. As mentioned in the description of the original cone-formation process, the texture with which a film forms is dependent on several parameters such as the angle of incidence of the molecular beam, the substrate temperature, and the nature of the evaporant material itself. For example, aluminum oxide can form a very smooth layer even at very grazing deposition angles to the substrate and at elevated temperature. Conversely, aluminum can produce a very rough surface even when deposited normal to the substrate surface if the substrate temperature is not carefully controlled, and will always produce a rough film when deposited at a grazing angle to the substrate surface. These effects are a consequence of the nucleation and growth dynamics of thin films (see, e.g., [20]).

Another important consideration regarding the lift-off layer is the issue of the selective etching required to achieve a successful lift-off. This is a very important factor in the emitter-cone fabrication process. When using molybdenum for the gate film and cones, hot phosphoric acid removes the aluminum oxide lift-off layer cleanly without attacking the gate or cones. However, if the lift-off etch is prolonged there can be etching of the silicon around the base of the cone at the bottom of the cavity. If not managed well, this attack of the silicon can actually undercut the cone and release it from the silicon base. This is not normally an issue, but if, for example, a multiple-cone deposition process (discussed in the next section) is used to achieve a high aspect ratio cone (large height-to-base diameter ratio), the resulting repeated exposure to phosphoric acid can result in release of the cone. If this becomes a problem, there are alternate material combinations, such as water-soluble materials (e.g., sodium chloride), and lift-off techniques for the cone-formation process that can be used.

One of the simplest means for avoiding having the silicon etched away from under the cone is to use a substrate such as glass or sapphire coated with molybdenum as the base material. Then there is nothing but molybdenum and chemically vapor deposited silicon dioxide in the cavity, and these are not attacked by hot phosphoric acid. These alternate substrates also make it possible to use potassium hydroxide which gives a very fast etch of the aluminum oxide lift-off layer, but cannot be used except very briefly when working with silicon substrates other than heavily doped p^{++} silicon.

The use of alternate substrate materials still restricts the choice of gate and cone materials to those that are not etched by phosphoric acid or potassium hydroxide. There are a number of materials that are better choices than molybdenum for special applications, such as operation in relatively hostile environments of the kind that might be found in some satellite orbits that contain relatively high partial pressures of atomic oxygen, for example. In addition, other materials may naturally form higher aspect-ratio cones, or more uniform shapes because of the nature of the nucleation and growth process of films. The issue comes down to the etchants required to remove the lift-off layer with regard to etch selectivity between the other materials in the structure and the lift-off layer.

A final issue is the potential for electrochemical etching of metals of different work functions (or ionization potentials) in the cathode structure. For example, we found that molybdenum cones were attacked when gold contact pads were incorporated into the cathode structure [29].

Ideally one would prefer to have a lift-off layer that would dissolve readily in common solvents, as most materials that would be considered as candidates for cone formation would not be attacked by such materials. However, the lift-off layer must be compatible with the vacuum deposition procedures required for the cone-formation process, and this limits the choices. Happily, such materials do exist. For example, sodium chloride, which dissolves readily in water, has been used in our laboratory for years as a lift-off layer in a volcano field-ionization source fabrication process [30].

4.3.2.3. *Management of Cone-Formation Parameters During Deposition.*

There are a host of issues that come into play when considering the cone-formation process and the cone material. These include the system geometry with regard to placement of the grazing angle lift-off material evaporator and the cone-formation evaporator, the effective source sizes of the two evaporators, the source-to-substrate distance, the deposition rate, and relative deposition rates between the two evaporators, the substrate temperature, the pressure in the deposition chamber, and the materials used for the lift-off layer and the cone formation. A few of the basic issues regarding the emitter-cone material are as follows:

1. It must have a high enough melting point and low enough vapor pressure to withstand vacuum bakeout temperatures of up to at least 400°C.

2. It is desirable to have a low work function, combined with low vapor pressure and good chemical stability (unfortunately, these properties tend to be somewhat mutually exclusive).

3. The material should be tolerant of at least short-term exposure to the atmosphere.

4. It must strongly adhere to the base material in the cathode structure.

5. It must have appropriate chemical solvency properties with respect to the requirements for etching of the lift-off layer used in the cone-formation process.

6. It is desirable to have appropriate physical properties with regard to diffusion activation energies and thermal/field forming as they relate to cone aspect-ratio effects as discussed later.

7. A material that forms high aspect-ratio emitter cones at manageable substrate temperatures is preferable.

8. In most cases, the material must be available in a pure form and suitable for vacuum evaporation. An exception would be when using the collimated sputtering technique as described by van Veen et al. [31], in which case sintered powders and alloys as well as pure elements can be used.

The two primary concerns in choosing an emitter-cone material are the nature of the nucleation and growth dynamics of the material as it deposits in thin film form, and the material's behavior as a field emitter in the operating environment. The first is basic for all applications. If the material does not form a relatively smooth film so that uniform cones are obtained, or grows with excessive stresses during deposition so that it breaks up when deposited in the thickness required to form emitter cones, it is not suitable for this application. The material's behavior as an emitter depends on its work function, conductivity, thermal stability, and compatibility with the vacuum environment in which it is operated.

4.3.2.3.1. The Effects of Substrate Temperature During the Cone-Formation Process.

The emitter-cone formation process is very straightforward in principle, but fraught with subtleties and pitfalls in the details of the procedure. A very important parameter in the cone-formation process is the substrate temperature during the cone deposition. There are two reasons for this. The first is that, when using the standard aluminum oxide lift-off layer, the aluminum oxide becomes increasingly difficult to etch as the substrate temperature increases during deposition. In fact, the first cathode structures were made with an aluminum oxide dielectric layer, which was rendered completely resistant to phosphoric acid by baking to $1000°C$ [6]. This troubling issue must be balanced with the increased tensile-stress build-up at lower substrate temperatures, exhibited by most metals as they deposit onto a substrate [21]. If the temperature is not high enough, the films will break up because of tensile stress before the required deposition has been completed. On the other hand, if the temperature is too high, the aluminum oxide lift-off layer will be difficult or impossible to remove without damaging other parts of the structure.

A second more subtle effect is the rate of closure of the gate aperture as a result of the arrival of the cone-formation material after the grazing-angle deposition has "gone over the horizon," as shown in Fig. 4.4. The tiny radius of curvature of the top edge of material that is causing the gate-hole diameter to decrease during the cone-formation process comes into play as a factor in the diffusion rate of material arriving at that surface. This is because of an increase in the effective surface tension at that site due to the curvature relative to the surface tension on the nearby flat surface of the gate film. Thus, if the temperature exceeds a critical value relative to the diffusion-activation energy of the cone material depositing on the surface, diffusion away from the relatively high-energy edge of the gate aperture during the cone deposition will reduce the rate of closure of the gate aperture, thereby increasing the aspect ratio of the cone that is formed within the cavity. The effect is analogous to the build-up or blunting of emitter tips by diffusion as described by Herring [32] and discussed in Section 4.4.3. When depositing molybdenum emitter cones onto a substrate the effect can increase the aspect ratio of the cones by a factor of 2 or more when the substrate temperature is approximately $500°C$ or higher [33]. Figure 4.11 is a scanning electron micrograph showing a very high aspect ratio cone made in this way. With materials having lower diffusion-activation temperatures it is possible to prevent the gate aperture from closing at all and produce a vertical post rather than a cone. Figure 4.12 is an example of a nickel "post" rather than cone deposited in a cavity because of the excessive substrate temperature during the deposition process. The same temperature-dependent effect has also been reported by van Veen and coworkers when depositing cones by sputtering through a collimator while heating the substrate [31]. Clearly this phenomenon can be used to form high aspect ratio cones in deep cavities for robust, low-capacitance emitter arrays. However, it must be carefully managed with regard to compatibility with the materials used and the lift-off process.

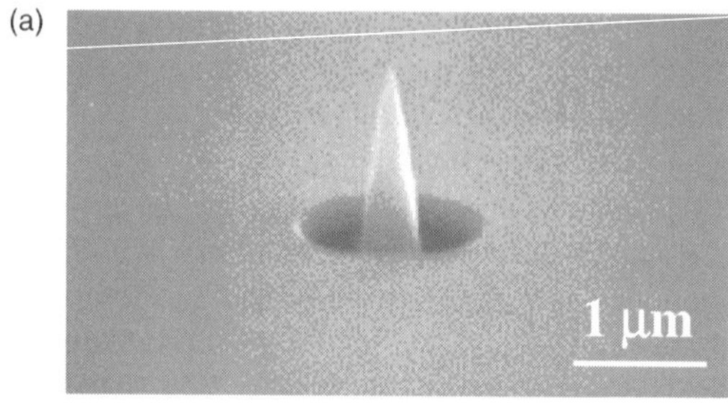
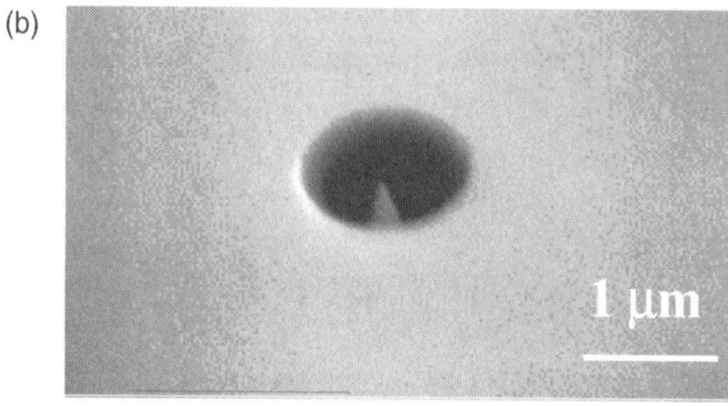

FIGURE 4.11. Examples of molybdenum cones deposited at (a) ∼500°C, and (b) ∼200°C. The gate-aperture diameter is 1.4 μm and the oxide thickness is 1.35 μm in both cases [33].

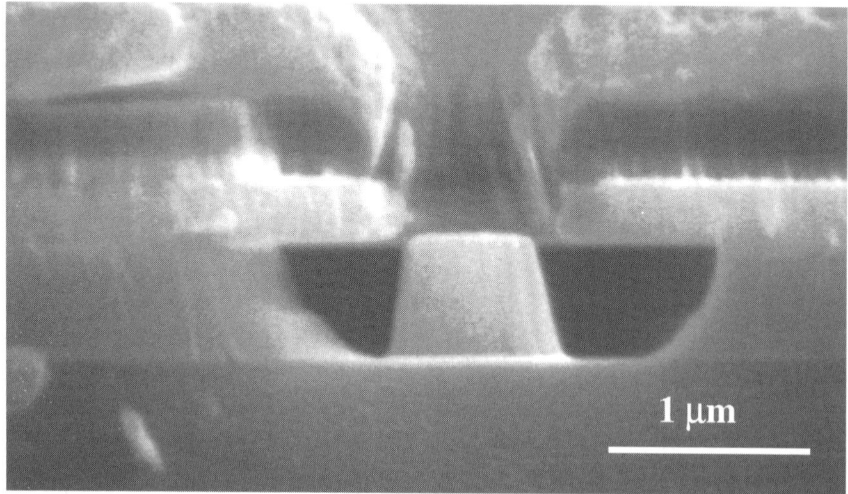

FIGURE 4.12. Nickel "posts" formed to illustrate that elevated substrate temperature can result in essentially no closure of the gate aperture during the cone-deposition process using certain materials.

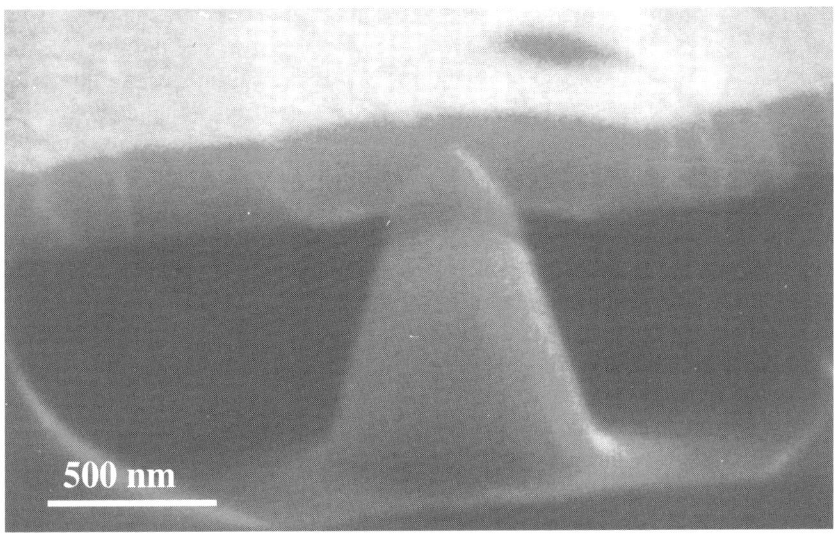

FIGURE 4.13. A cone deposited using titanium at 10^{-6} torr for most of the deposition, followed by deposition in 10^{-4} torr of nitrogen to form reactively deposited titanium nitride at the tip. Scatter due to the increased pressure increases the rate of closure of the gate aperture, thereby forming a much shallower cone half-angle near that tip.

4.3.2.3.2. System Pressure During the Cone-Formation Process. In general, a cone with a sharp tip and small half-angle is desirable. Along with the substrate temperature, materials selection, deposition rates, and system geometry, the system pressure is an important factor in the shape of cone that is achieved. If the mean free-path is not sufficiently long with respect to the distance the molecular beam must travel to reach the substrate, scatter due to collisions will result in a relatively nonparallel "broad source" molecular beam arriving at the substrate. This will cause rapid closure of the gate film, and a large half-angle, low aspect ratio cone. This effect is illustrated very clearly in Fig. 4.13, which shows a cross section of a titanium cone with a titanium nitride tip. The cone was formed by depositing most of the titanium cone under low (10^{-6} torr) pressure conditions, and then near the end of the process increasing the pressure to high in the 10^{-4} torr range with nitrogen so that the remainder of the cone was formed of reactively deposited titanium nitride. The scanning electron micrograph in Fig. 4.13 shows a distinct change in the half-angle of the cone near the apex due to the different nature of the deposited material at that point (i.e., titanium nitride compared to titanium). If the pressure in the system had been in the 10^{-4} range throughout the deposition run, the cone would have been very short and squat, and completely unsatisfactory. However, the relatively large half-angle at the apex is acceptable if it comes with a reasonably sharp tip.

4.3.2.3.3. Emitter-Cone Shape and Uniformity. In most applications emitter-tip uniformity in an array is important, and in many it is critical. When discussing field emission from microfabricated polycrystalline emitters, there are two levels of uniformity that must be considered. The first is the "macro-scale" uniformity, which refers to the gate-hole diameter, overall tip radius, cone half-angle, and tip position relative to the gate aperture. The other is the "micro" issue of what the atomic-scale nature of the emitter surface is with regard to grain orientation, grain boundaries, impurity concentration, and coverage with adsorbates. Neither is simple matter, and both must be addressed if uniformity is an issue in a given application.

Macro-scale uniformity begins with the formation of the gate aperture. This is both a lithography (patterning) and an etching issue. In our experience to date, optical-stepper lithography has produced the most consistently uniform gate-aperture patterns, with electron-beam lithography a close second. It is assumed, of course, that the developing, rinsing, and descumming of the exposed resist are always done properly. The etching of the gate pattern must also be done carefully in order to preserve uniformity. Generally, rapid etches tend to promote undercutting of the resist which is undesirable, and etches that evolve gas can be particularly troublesome with regard to etch uniformity due to resist lifting and bubbles adhering to the surface and thereby masking local areas. We have found that an electrolytic polish etch can produce uniform gate apertures with very smooth sidewalls. As mentioned earlier, the surface texture is important because of the following process steps involving deposition of the lift-off layer onto the inner edges of the gate apertures while rotating the substrate as shown in Fig. 4.5. Clearly, if the inner edge is a rough surface, as is often produced by chemical etching, this grazing-angle lift-off layer deposition will accentuate the roughness and promote irregular closure of the gate aperture during the cone-formation process. This is obviously detrimental to the form of the cone that is grown within the cavity as well as to the uniformity of the emitter-cone shapes from cavity to cavity.

Given that uniform, smooth gate apertures are formed, the next process in the cathode fabrication sequence is to deposit the sacrificial lift-off layer at a grazing angle to the substrate surface while rotating the substrate in the manner shown in Fig. 4.5. If acceptable uniformity is to be achieved, this grazing angle deposition must form a smooth film on the inner edges of the gate aperture and top surface of the gate film. This can be achieved by using an oxide for the lift-off layer. As discussed earlier, aluminum oxide (and indeed, most oxides) form a very smooth layer even when deposited at extreme grazing angles to the substrate surface, whereas aluminum and most other metals tend to produce a relatively rough film when deposited at grazing angles onto even very smooth substrate surfaces. These variations are due to differences in the nucleation and growth dynamics of these particular materials. Another aforementioned important feature — that we repeat for emphasis — is that oxides generally tend to deposit with internal compressive stresses, and thus reduce the effective tensile stress that is inherent in most deposited metals [21].

The next step is to deposit the emitter cone into the cavity. Historically this has been a molybdenum deposition normal to the substrate surface while continuing with the rotation and grazing-angle deposition of an alumina lift-off layer. It is not strictly necessary to continue with the lift-off layer deposition throughout the cone-formation deposition with all cone materials, but there are four general reasons why it is desirable:

1. Maintaining the grazing-angle deposition during the cone-formation deposition gives the operator one more parameter of control over the shape of the metal cone that is being formed. This is by controlling the relative rates of deposition of the lift-off material and the cone-formation material to achieve the desired gate-aperture closure rate and thereby the cone shape and height.

2. The aforementioned film-stress management.

3. As also previously discussed, metal cone material deposited by itself will tend to form a rough closure of the inner surface of the gate aperture because of the grazing-angle deposition effect. Mixing the metal oxide into the deposit reduces this effect.

4. The composite metal/metal oxide lift-off film is easier to remove in the oxide lift-off etchant than is a pure metal film.

With careful attention to all the issues discussed earlier, it is still problematic as to whether uniformity from emitter-cone to emitter-cone will be achieved. Indeed, it is likely that each individual cone will not be precisely in the form of a perfect cone because of the dynamics of the film nucleation and growth process and the polycrystalline nature of the cone that is formed by most materials used in this process. Figure 4.14 is a very high magnification transmission electron micrograph of a molybdenum emitter tip grown by the process described previously. The form is not

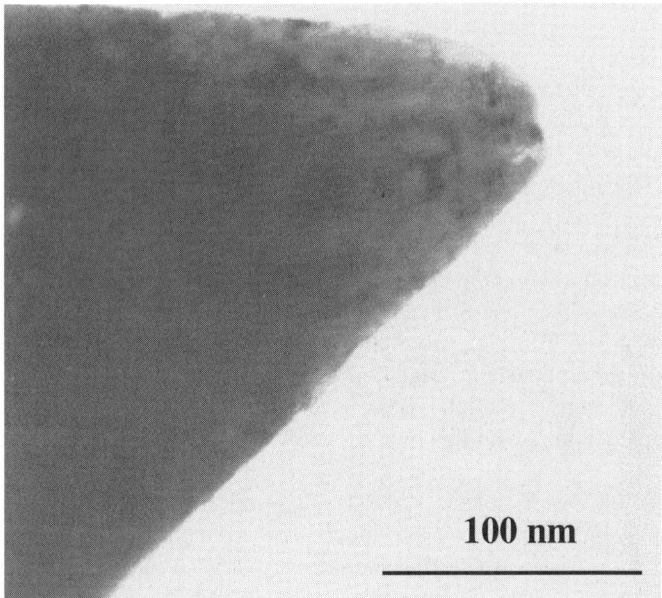

FIGURE 4.14. A transmission electron micrograph of a relatively large-radius, SRI microfabricated, molybdenum emitter tip, showing the polycrystalline nature of the material. (Courtesy of John Macauley, Bell Labs.)

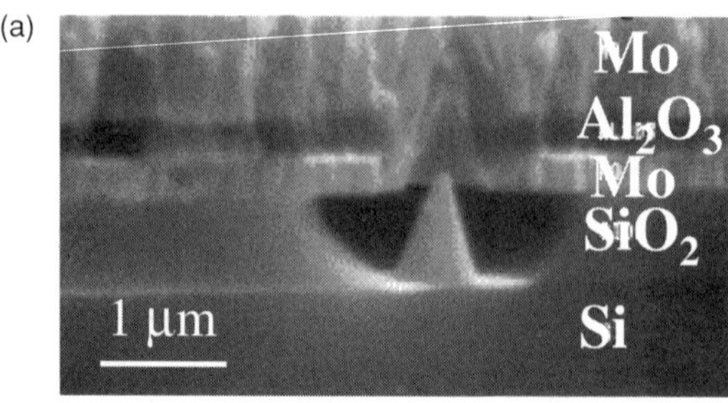

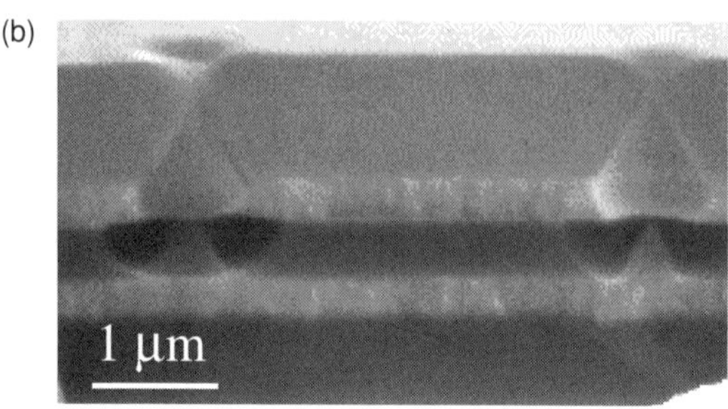

FIGURE 4.15. SEM micrographs showing cross sections of cones deposited with (a) molybdenum, exhibiting a relatively grainy hole closure and (b) chrome silicide, showing a very smooth closure, demonstrating the effects of materials related film growth dynamics on texture and uniformity.

all that bad, but close examination shows slight irregularities at the very tip — right where all the emission activity is. Clearly, not all the tips in an array would have this same form. As a practical matter, at some point the smoothness that can be achieved at the tip of the emitter cone will reach a materials-dependent irreducible minimum that relates to the inherent nucleation and growth properties of the cone material.

With this in mind, experiments were conducted using alternate cone-formation materials that were chosen with a view to film growth properties. Figure 4.15 shows a scanning electron micrograph cross section of a portion of an array formed using a chrome silicide as the cone-formation material along with a molybdenum emitter array for comparison. It is clear from the micrograph that the silicide forms a much smoother, more amorphous-appearing film than the molybdenum on the macro-scale; however, to date high-resolution TEMs have not yet been made of the details of the emitter tips obtained with this ongoing work. This brings us to the micro-scale uniformity issue.

Micro-scale uniformity relates to the details of micro-irregularities at the tip, as shown in Fig. 4.14, and on down in scale to atomic dimensions. In addition to the atomic-scale characteristics of the tip material, one must also consider that the emitter-tip surface is normally covered with adsorbates. It is a routine practice for surface scientists doing classic field electron-emission microscopy (FEEM) with etched tungsten wire emitter tips to remove adsorbates, particularly oxygen, which is very strongly adsorbed, by flash-heating the tip to at least 1500°C prior to initiating emission. This heating also serves to form the tip [13]. Clearly this is not possible with microfabricated emitter arrays of the kind we are discussing here. Therefore, we must be aware that the nature of the emitter surface that we are dealing with is dynamic and that emission performance relies on properties of the materials (the cone and adsorbates) that form the surface layer at the solid/vacuum interface on the emitter tip. As a practical matter, this issue can be the dominant reason for emission nonuniformity that one observes from emitter tip to emitter tip and, for that matter, over the surface of a given single emitter tip. This issue will be discussed further in Section 4.4. Having said that, we note that in spite of these very real micro-scale uniformity issues, the statistics of large numbers can salvage the situation when working with emitter arrays. This effect has been demonstrated experimentally and is also discussed in Section 4.4. In addition, the FED community has been successful in achieving very good emission uniformity over large areas by using integrated emission-buffering resistors as well as averaging the emission over a large number of emitters.

4.3.2.3.4. Process Variations and Alternate Materials for Specific Applications. The FED panel application has dominated the interest in the development of microfabricated FEAs, and most FED groups have the basic emitter array fabrication and operation well in hand. It is a help to them that the emission levels required for FED are quite modest with regard to the emission capabilities of the emitters. They are normally working in the nanoampere per tip range with display panels, and it has been well established by Spindt et al. [34] that 10–100 μA/tip is readily achievable. More recently, short pulses in the mA range have been achieved (see Fig. 4.37 later in Section 4.4.8.1).

Several other applications are being investigated, albeit at a relatively much lower level of effort, and some of these require much more of the emitter arrays than the FED application from both an emission level and operating environment point of view. These include ionizers for mass spectrometry [35,36], vacuum ion gauges [37,38], microwave tubes [33,39–42], and more recently, spacecraft charge management [43]. These applications present cathode designers with a variety of specific requirements relating to transconductance, capacitance, and environmental issues. These aspects of VME will be discussed in depth in Chapters 7 and 8 on applications; however, for the purpose of completeness in our coverage of Spindt-cathode fabrication technology, we will next discuss some materials issues with regard to specific applications and operating environments.

4.3.3. Ongoing Developments and Materials Issues

Applications where the environment cannot be carefully controlled present a materials compatibility issue. Molybdenum has performed well as the cone material for many workers in a variety of applications where it has been possible to maintain a reasonable degree of control over the environment. However, in applications such as ionizers for mass spectrometry, vacuum gauges, and spacecraft, the commonly used molybdenum tip may not do well with regard to resisting corrosion from the environment. Integrated resistors help with survivability under conditions where emission bursts because of the dynamics of adsorption/desorption in a relatively high-pressure environment, but if there are high partial pressures of corrosive materials such as oxygen and water vapor, the molybdenum will not hold up. However, there are other materials that may very well manage to operate satisfactorily in such environments. Mackie at the Linfield Research Institute has shown that the transition metal carbides are remarkably robust field emitters, and has demonstrated, along with similar results obtained at SRI, that coatings of carbides of the transition metals on molybdenum emitter arrays can improve the emission characteristics significantly [34,44,45]. It is also well known that iridium performs well under the conditions of field ionization with oxygen [46], and it can be inferred from this that an iridium emitter would perform well in an atomic-oxygen rich environment, as is found in some satellite orbits. Indeed, iridium has been shown to be resistant to oxygen at 2000 K [47]. The development of alternate lift-off layer processes has opened the way for investigations of many other cone materials using the Spindt-cathode fabrication technology, and these investigations are ongoing at this time.

4.3.3.1. Emitter-Cone Aspect Ratio and Cathode Capacitance.

The issue of the capacitance C between the base and gate of the cathode is of importance for two reasons. First, the energy W stored in a structure under an applied operating voltage V is $W = CV^2/2$. It is desirable to keep the stored energy to a minimum so that there is the least possible energy available to couple into any transient event that has the potential to be damaging. The issue of stored energy is also a reason why it is desirable to have the lowest possible operating voltage, as the stored energy is proportional to the voltage squared. Second, when the emission is being modulated (as in microwave applications) or switched (as in display) the power required to drive the device is proportional to the capacitance squared, and inversely proportional to the transconductance squared [33]. Thus, as a general goal for almost all applications, one wishes to keep the operating voltage and the capacitance as low as possible.

The most direct way to reduce the operating voltage is to make the smallest possible gate apertures, and fabricate the emitter cones to be at least level with the top of the gate film and with the sharpest possible tips. Unfortunately, the size of the cone scales with the gate diameter by the nature of the fabrication process. For example, reducing the gate diameter by a factor of 2 requires reducing the thickness t of the dielectric layer by a factor of 2, thereby doubling the capacitance. Of course, cutting the gate-aperture diameters by a factor of 2 may also allow one to reduce the tip pitch (spacing between tips) by a factor of 2, thereby reducing the area A needed for a given number of emitters by a factor of 4. Then one would realize a net reduction in capacitance by a factor of 2, as capacitance is proportional to A/t. In addition, the smaller dimensions increase the field-forming β factor, thereby reducing the operating voltage for a given emission current. However, the issue to be kept in mind with the size reduction (aside from limits imposed by the lithography tools that are commonly available) is that the cone-formation process scales. Thus if the gate aperture is reduced, the oxide layer must be reduced in the same ratio. At an oxide thickness of about 0.5 μm or less,

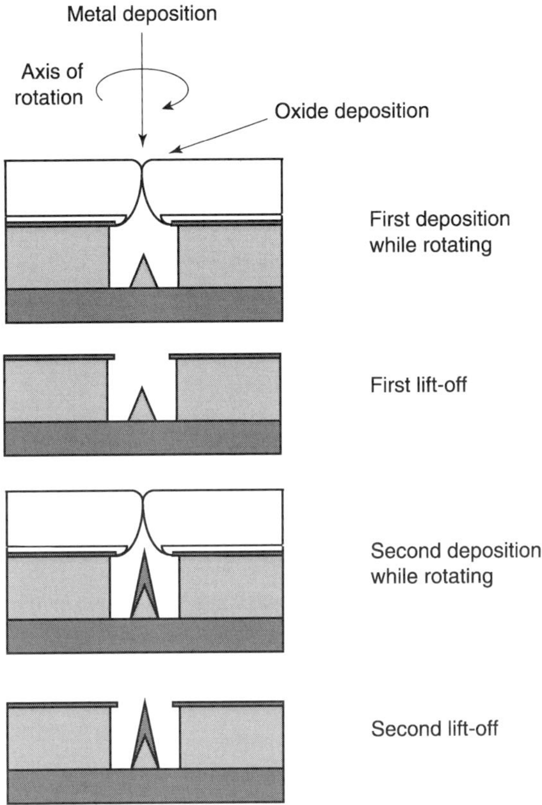

FIGURE 4.16. Forming high aspect ratio cones by the multiple cone-deposition method [29].

defect density in the interlayer dielectric tends to become an issue with regard to step coverage and pinhole effects.

The solution is to fabricate high aspect ratio emitter cones so that the oxide thickness can be preserved as the gate-aperture diameter is decreased. One of two methods or a combination of the two methods can do this. The first is the high aspect ratio cone growth method achieved by heating the substrate during the cone growing process, as described earlier and shown in Fig. 4.11. The second is to do a double or multiple emitter-cone deposition process [48], as shown in Fig. 4.16. Using this technique, it is possible to double the aspect ratio of the emitter cone with a double deposition, or even quadruple the ratio with multiple depositions. For molybdenum, which tends to produce a cone with an aspect ratio of one under normal substrate temperatures, this means that up to 2 μm of oxide can be used with gate apertures about 0.5 μm in diameter. Figure 4.17 is a scanning electron micrograph of molybdenum emitter cones formed with four depositions in a 2 μm thick oxide and gate apertures of 0.58 μm. Using other cone materials that tend to naturally produce higher aspect ratio cones should produce the same effect with fewer depositions.

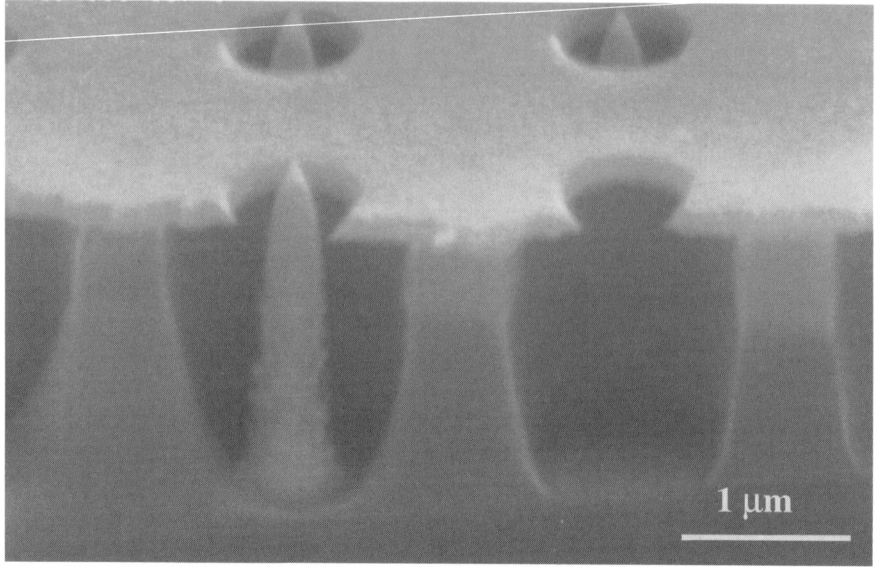

FIGURE 4.17. A scanning electron micrograph of high aspect ratio molybdenum cones made with the multiple cone-deposition process. The cones are 2 μm tall with gate apertures 0.6 μm in diameter.

Indeed, it is quite reasonable to speculate that single depositions of very high aspect ratio cones will be possible in the light of preliminary evidence in hand, showing nickel "cones" growing essentially straight up in the form of "posts" when deposited at elevated substrate temperatures (Fig. 4.12). In principle one could build posts to the desired height, and then cool the substrate so that the gate aperture closes as the deposition progresses, thereby forming a cone at the end of the post. Finally, the tip radius is perhaps the most important parameter with regard to operating voltage, as can be seen from the simple analysis and references given in Section 4.3.1, showing $\beta = R/kr(R - r)$, where r is the tip radius, R the tip-to-counter electrode spacing, and $1 < k < 5$ accounts for the influence of the emitter shank [12]. Manipulating the tip radius when using the Spindt-cathode fabrication process is complicated and difficult as many factors are involved. Deposition rate, effective source size, cone material, substrate temperature, gate-aperture size, distance between the deposition source and substrate, system pressure, etc., can all have an impact on the cone tip radius. However, if all else is essentially constant, as it normally is when using the same system for all cone depositions, smaller gate apertures generally produce smaller tip radii. This is simply a result of having the masking edge of the gate aperture closer to the forming cone, thereby reducing the penumbra effect in the arriving molecular beam. As this ongoing work continues the full potential of these various methods for producing low-capacitance, low-voltage emitter arrays will be explored in depth, as capacitance and drive voltage are critical in several important applications.

4.3.4. Wedges and Edges as Emitter Surfaces

A discussion of fabrication technology would be incomplete without a discussion of other emitter geometries that have been considered and can be fabricated with variations of the basic Spindt-cathode formation process described earlier or simple planar processing techniques. Wedges, as in for example [34]; "horizontal" edge emitters, as in [49,50]; "vertical" edge emitters, as in [51,52]; and "volcanoes," as in [53], have all been investigated at one time or another. The primary appeal of wedges as emitters is that, compared to tips, the configuration can increase the ratio of the surface area that is under high enough electric field stress to be liable to field-emit electrons relative to the total area occupied by the emitter array. The downside is that, because of the geometry, the voltage required to achieve sufficient field for emission is higher for edges than it is for tips by roughly a factor of 2 [34]. In addition, the actual emission is not uniformly distributed along the emitter edge, but rather occurs at several localized sites along the edge, just as there are several active emitting sites on an emitter tip rather than uniform emission over the whole surface of the tip. Thus, the performance of an edge emitter relies on the density of emission sites distributed along the edge. While the notion that there could be many more emitter sites per unit area occupied by an array of edges or wedges than there would be in an array of emitter tips is intuitively satisfying, there have been no data produced to verify this idea. Indeed, data have been published showing over 2000 A/cm^2 averaged over the total area occupied by the array [25,33]; to our knowledge at this time, there have been no reports of any other form of cathode remotely approaching this value.

4.3.5. Volcanoes

Another form of edge emitter is a structure developed at SRI as an improvement to a "multi-point" field-ionization source for mass spectrometry, which has been dubbed as the microvolcano ionizer for obvious reasons [53] (Figs. 4.18 and 4.19). The device does not strictly fall in the category of Spindt cathodes, but is very closely related, and the fabrication technology should be of general interest to VME practitioners. Furthermore, although it was developed specifically as a field ionizer, it can also function as a field-electron emitter — albeit demonstrably not as well as an emitter cone does. Structurally, the microvolcano is simply a Spindt cathode with a hole through the center of the cone and all the way through the substrate. The idea is to mount the device at the entry to a mass analyzer, such as a quadrupole mass spectrometer, and introduce the sample gas to be analyzed to the spectrometer through the hole in the emitter cone. When a sufficient negative voltage is applied to the gate relative to the volcano structure, molecules passing near the volcano rim are field ionized and accelerated into the mass analyzer. The features of this approach are a relatively low voltage required for field ionization (100 V rather than 10 000 V) and a relatively high ionization efficiency compared to classic field ionization as developed primarily by Beckey's group in Bonn [54]. The method used to fabricate a cone with a hole all the way through it is shown in Fig. 4.18, and described in detail elsewhere [30,53]. Briefly, a metal-coated 1.5-μm thick, 100-μm square silicon dioxide membrane was

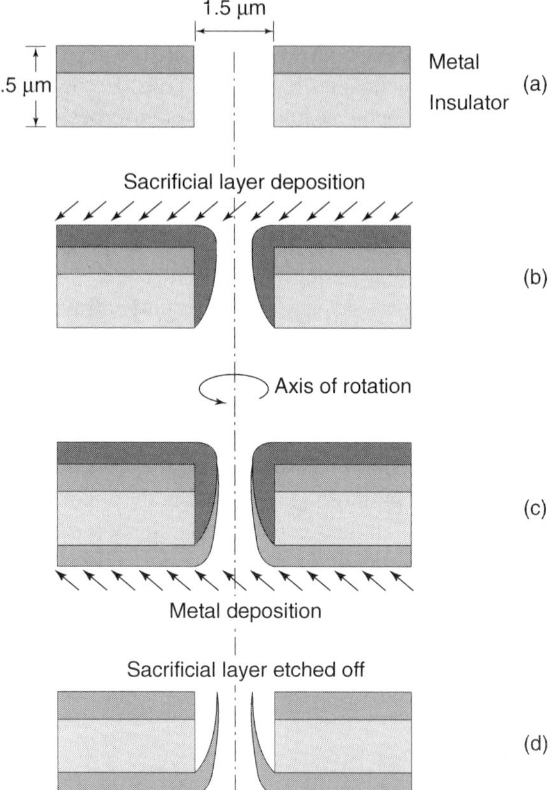

FIGURE 4.18. The process flow for fabricating microvolcano field ionizers. A hole, or array of holes, is patterned in a metal/insulator membrane. The metal side of the membrane is then coated with a sacrificial material by physical vapor deposition while rotating the substrate. The insulator side is then coated with metal in the same fashion. Etching away the sacrificial material completes the process [53].

formed on a thermally oxidized $\langle 100 \rangle$ silicon substrate. An array of 1.5-µm diameter holes on 10-µm centers was etched in the membrane using electron-beam lithography and anisotropic etching to produce a structure in which each hole is as shown in Fig. 4.18(a). The holes were then partially closed by angle evaporation of a selectively etchable sacrificial material onto the membrane from the metal counterelectrode side of the structure as shown in Fig. 4.18(b). The microvolcano-forming material was then angle evaporated from the opposite side of the membrane, partially closing the hole from that side with a metal that is insoluble in etchants used for removing the sacrificial layer, as shown in Fig. 4.18(c). Finally, the sacrificial layer was removed by wet etching, leaving hollow cones centered in the cavities, as shown in Figs. 4.18(d) and 4.19.

(a)
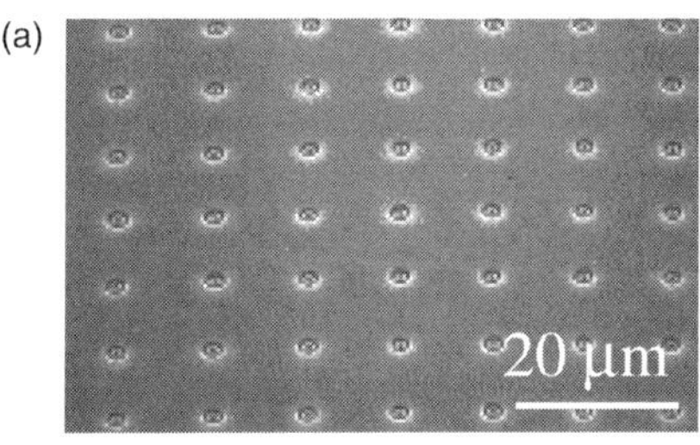

(b)
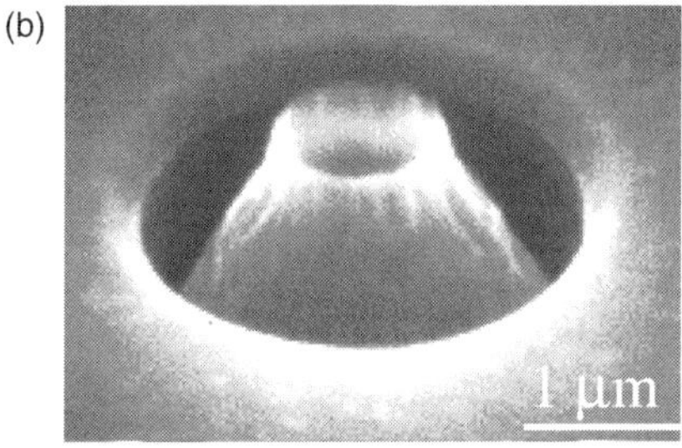

FIGURE 4.19. Scanning electron micrographs of the SRI microvolcano field-ionization source [53].

Tests with the microvolcano structure confirmed that the device produced useful field-ion current for mass spectrometry at less than 100 V [53], as compared to 10 kV for conventional field ionizers [54]. Figure 4.20 shows ion-count data produced by a 10×10 array of microvolcanoes, as a function of the voltage applied between the volcano and the integrated counterelectrode with a sample pressure of 0.2 torr of toluene vapor on the inlet side of the structure. The output current was measured with a microchannel plate multiplier and counting circuit as is commonly done with field-ion mass spectrometry. The background vacuum was 6×10^{-7} torr. Maintaining a high vacuum in the chamber was made possible by the low gas conductance of the submicron-diameter orifices of the microvolcanoes. The noise level was about 100 counts/s, and the background signal due to a combination of organic vapors in the chamber and residual sample gas in the volcano structure was about 500 counts/s without the toluene sample. The threshold of detectable field ionization of the toluene was about 48 V, and the count rate increased exponentially to about 2000 counts/s at 70 V, where a saturation effect appeared because of the limited dynamic range of the detector used in making this measurement.

Mass spectra were obtained using the microvolcano with a standard quadrupole mass spectrometer. A variety of simple hydrocarbon species were used in the first tests, including toluene, benzene, n-propylbenzene, n-butylbenzene, acetone, and butane. In each case the microvolcano produced spectra that were characteristic of field ionization, with greater than 98% of the observed signal attributable to M^+ ions of the sample species. Figure 4.21 is the mass spectrum produced by the microvolcano field

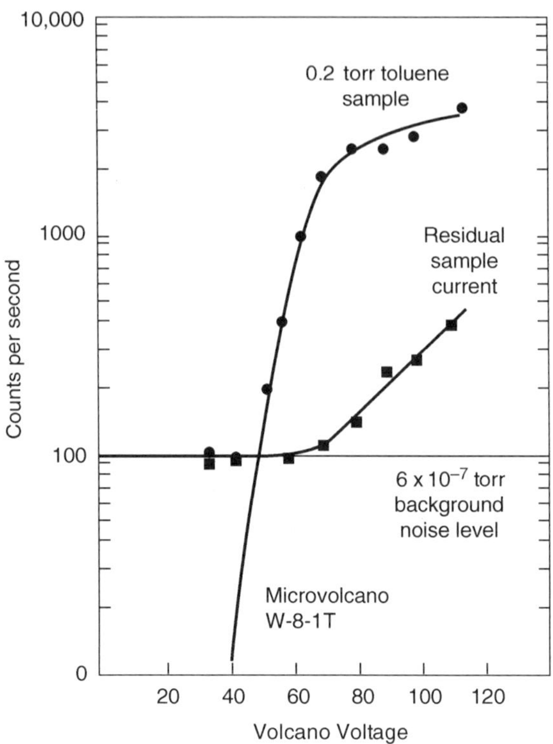

FIGURE 4.20. Voltage–current characteristics of a 100-tip microvolcano array (100-μm square area) with 0.2 torr of toluene in the inlet [53].

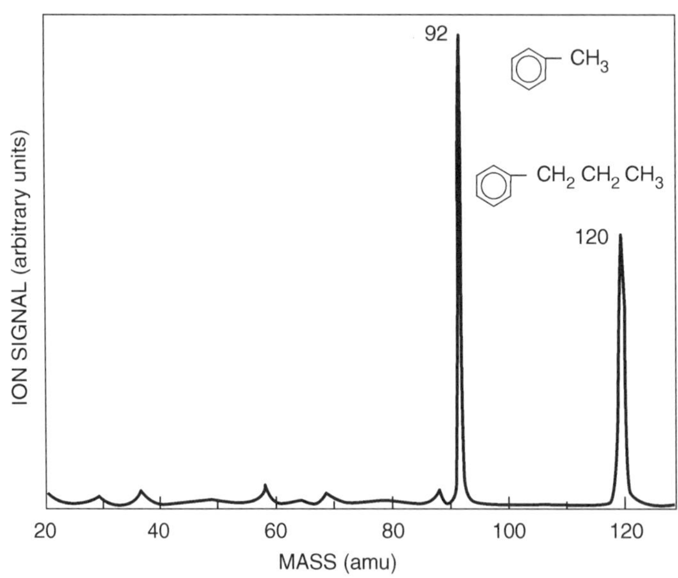

FIGURE 4.21. Mass spectra of a toluene–*n*-propylbenzene (2:1) mixture taken with a microvolcano field ionizer and a quadrupole mass analyzer [53].

138

ionizer and a quadrupole mass spectrometer analyzer from a mixture of toluene and
n-propylbenzene, showing the clear identification of the parent ions, and essentially
no clutter due to fragmentation. This nonfragmenting feature enabled Aberth and
coworkers at the Linus Pauling Institute of Science and Medicine to detect mass
peaks related to sarcoma cancer in mice by mass spectrometric analysis of urine
samples using an SRI volcano field ionizer [55].

The microfabricated field-ion source was conceived and developed for field-ion
mass spectrometry of hydrocarbon and biological samples. However, there are other
possible applications, the most interesting of which is the idea of adapting the technology to liquid-metal field-ion sources. Mitterauer made small liquid-metal field-ion
sources for space applications (e.g., propulsion and spacecraft surface-charge management) and has proposed adapting the microfabricated field ionizer to this use [56].
In addition, focused ion beams are used in microelectronic fabrication, and there may
be potential for improved performance with low-voltage microfabricated sources.
Finally, there is an ongoing need for a compact, portable detector capable of sampling the environment and detecting materials that are risky for health or security. For
such applications, a microvolcano ionizer coupled with an ion-mobility analyzer has
many potential advantages.

4.3.6. Fabrication Summary

As discussed previously, variations in basic Spindt-cathode fabrication process can be
used to form other emitter geometries such as wedges and volcanoes. The results have
shown that the volcano configuration can be very useful as a field ionizer, but like the
wedge, has not yet produced electron emission performance that compares favorably
with the emitter tip array. To date, the most fruitful advances in fabrication technology
from a performance point of view have been improvements in the basic emitter array
geometry with regard to size and packing density. Not surprisingly, reductions in gate-
aperture diameter and tip radius, and increases in emitter-tip packing density, have
consistently produced larger emission current for a given applied voltage between the
base and gate electrodes.

Emitter material is another issue that has been explored with regard to improving
stability and emission for a given applied voltage. So-called diamond-like-carbon
(DLC) films have been of particular interest to many workers. A great deal of effort
has gone into exploring DLC, and a wide range of results have been reported. Some
of the most interesting data to date have been obtained with thin coatings of DLC
on otherwise standard molybdenum Spindt emitters. For example, Jung et al. [57]
reported a 30% decrease in required voltage as well as improved stability with a
20-nm coating of nitrogen-doped DLC on molybdenum emitter arrays. Carbon and
other alternate emitter materials will be covered in depth in Chapter 6, and thus this is
only mentioned here as an acknowledgement that materials as well as geometry can
be an important parameter in emitter array performance.

In general, for almost all applications one wishes to obtain the lowest possible
operating voltage, largest current density, best emission uniformity, best emission

stability, and maximum reliability possible. All of these suggest using the smallest possible gate diameter, highest practical packing density, and largest possible number of emitters operating in parallel. The packing density and gate-aperture diameters are linked as a practical matter. There is some minimum spacing between emitter tips, below which an arc failure of one will propagate to its nearest neighbor, triggering a chain-reaction failure of the emitter array. A second issue is the scaling of the emitter cavity and cone structures. Generally, the cones form with approximately a 1:1 base diameter-to-cone height aspect ratio. Thus a 0.15-μm diameter gate aperture would normally require a 0.15-μm thick insulating oxide layer. Such a thin oxide layer produces a relatively large capacitance and introduces a high risk with regard to area defect density for most practical cathode sizes. Nevertheless, small-area arrays of Spindt-type cathodes with these dimensions and a tip-to-tip spacing of 0.32 μm, or approximately 10^9 tips/cm^2, have been successfully fabricated by MIT's Lincoln Laboratory for a microwave application using laser interferometric lithography. The team at Lincoln Lab has reported current densities exceeding 2000 A/cm^2 with small areas of these arrays [25]. In addition, Candescent is routinely producing field emission display panels with 0.15 μm-diameter gate apertures and an average packing density of 4 holes/μm^2 (4 $\times$ 10^8 tips/cm^2) using the ion-tracking process [23]. Candescent reports that this high packing density combined with the very small feature sizes allows them to operate at very low cathode-drive voltages of about 10–15 V. Both of these hole-patterning techniques have the advantage of being applicable to relatively large areas. In addition, both could benefit from using the techniques described in Section 4.3.3.1 for fabricating high aspect ratio emitter cones so that a relatively thick dielectric layer can be used, thereby reducing the capacitance and defect-density issues associated with large areas.

For now, these ~0.1-μm-diameter gate dimensions seem to be approaching the limits for the practical gate-aperture size using the standard Spindt-cathode fabrication technology because of the nature of the sacrificial layer used in the process and the minimum thickness of the layer that will provide for reasonably reliable lift-off. Smaller gate apertures will require methods that do not use a sacrificial lift-off layer.

Finally, the issue of high aspect ratio cones is very important in conducting some basic studies of microfabricated emitter-tip properties. Specifically, the source of the well-known flicker noise associated with the emission could be effectively studied by field electron emission microscopy (FEEM) and field-ion microscopy (FIM). However, high-resolution FIM requires a much larger, at least by a factor of 5, electric field on the emitter tip than does electron field emission. Thus, FIM would require an increase (by about a factor of 5) in the voltage normally applied across the base to gate insulator, and with most cathode configurations this would lead to a voltage breakdown before field ionization was achieved. However, by using the multiple cone-deposition process, or the high temperature cone-deposition process to form high aspect ratio cones, it may well be possible to achieve field-ionization and field-ion imaging with single microfabricated emitter tips. This will be explored further in the next section.

4.4. PERFORMANCE

In this section we will discuss the basic emission properties of the as-fabricated Spindt cathode, followed by processing procedures for best performance, the factors that can impact performance during operation, ongoing investigations into emitter behavior, and prospects for continued improvements in performance and reliability.

We will begin with a discussion of the modeling tools available for aiding our understanding of emitter performance, followed by actual performance of the Spindt cathode with molybdenum emitter tips, carefully processed and operating under good ultra-high vacuum conditions. In other words, the characteristics of the cathodes are under "ideal" laboratory conditions. We will then look at the issues involved with "real" operating environments and the impact of these operating conditions on emitter performance. Following this, we will explore the options available to make the "real" environment cathode-friendly, and/or what we can do with the cathodes to make them more resistant to a less than friendly environment. Depending on the application, one or the other option may be the better choice. For example, there is nothing we can do about the environment we may encounter in outer space. However, we are finding that modern vacuum technology may well enable us to create a more favorable vacuum environment in a given vacuum tube if thoughtful and careful tube processing practice is followed.

Stability, or "$1/f^n$" emission noise, as expressed in the mean square current fluctuation i^n [58]: $\langle i^n \rangle \approx A(I^\alpha/f^n)\Delta f$ $1 < n < 2$, $\alpha \approx 1$, where f is the frequency and I the average current, is more or less of a problem, depending on the particular application that one is dealing with. We will discuss ongoing studies of the basic source of emission noise in Spindt cathodes, along with processes that have been shown to reduce emitter noise significantly.

Lifetime can be thought of as long-term stability, and to be meaningful, it must be defined in the context of a particular application. For example, the voltage required to maintain a given emission will vary over wide ranges with an emitter array used as an electron source for a vacuum gauge or a mass spectrometer ionizer because of adsorption and desorption of gases on the emitter tips under varying pressure conditions. This is not a major problem with these applications as the voltage required for a given emission is not normally an issue, and in addition, the emission variations are generally reversible as the environment changes (Figs. 4.28 and 4.29).

On the other hand, in a flat panel display, the end of life has been defined by at least one FED group [59] as the point at which the peak luminance for given applied voltages falls to 50% of the original value. In principle, a roughly 10% increase in the voltage applied between the base and gate of the emitter array would produce a sufficient increase in emission current to compensate for such a loss, and this is easily done, for example, in a vacuum-gauge application. However, in the display application, the voltage available from the driver circuitry is limited.

Reliability is related to lifetime, but is thought of as concerning a sudden end to life due to an event rather than a long-term decay to a point of inadequate performance for a given application. Because the emitter array is a microstructure and is often operated in an environment containing relatively large-scale, high-voltage electrodes, reliability can be an issue due to damage resulting from a brief arc or "tic" that would otherwise be a normal event of no consequence. There is little that can be done to the emitter array itself to enable it to withstand a direct arc from a high-voltage electrode. Thus the challenge is for the tube engineers to provide corona shields and to produce surfaces in the tube that have little tendency to arc. At the time of this writing, this issue has received little attention from the vacuum-tube community, and one can only speculate as to how successful attempts at arc suppression in vacuum tubes will be with regard to emitter array reliability.

4.4.1. Modeling Electron Trajectories from the Emitter Array

The objective of device modeling is to trace electron trajectories from an emission site on the cathode surface to the position at which the electron is collected on the final anode. Account must be taken of the variation of the electric field along the trajectories because of the shapes and potentials of ancillary electrodes in the device and the space charge from the electron flux itself. Vacuum microelectronic devices present special problems for the models. The fact that the radius of curvature of a tip is usually very small compared with the electron path length and the electric fields close to a tip are so much higher than elsewhere along the trajectory, presents computational problems if the trajectories are to be calculated accurately. The crucial launching conditions are strongly modified by the electron path in the region close to the tip. The solution of axially symmetric problems is clearly simpler than for three-dimensional (3-D) systems, but the latter are required if devices using *arrays* of tips are to be properly addressed. On the assumption that the tip surface is smooth, of uniform work function, and radially symmetric, the electric field decreases from the center to the shank and the current density decreases very rapidly from the center of the tip because of its exponential dependency on field [1].

The first computations on microtips used uniform mesh sizes and were done by Brodie at SRI on a CDC 6400 computer using MAGLENS, a program developed at the Stanford Linear Accelerator Center (SLAC) to design electron guns for their linear accelerator. Field calculations near the tip were obtained with high resolution by repeatedly solving for the field using Laplace's equation on successively finer rectangular meshes. However, with uniform mesh size a curved cathode surface is un-realistically simulated by a jagged "staircase." These calculations were subsequently improved by Herrmannsfeldt et al. [60]. Computations using less versatile uniform mesh programs such as SIMION were carried out by a number of workers, for example Orvis et al. [61] and Zaidman [62].

When programs using the finite element method were developed, they soon became the preferred approach for analyzing field emission devices. This method allows the mesh to be both nonuniform and irregular. Thus it can accommodate a wide range of geometric scale as well as arbitrary electrode shapes (thus avoiding the "staircase" effect) and is now the preferred method used by the VME community, for example Zhu and Munro [63], Asano [64], True [65], Chang et al. [66], and Jensen et al. [17]. The Munro program takes into account space charge and relativistic effects and is available for solving 3-D problems. Kessling and Hunt [67] used the Maxwell 2-D field simulator program developed by the Ansoft Corporation (Pittsburgh, PA) for application to field emission, thin, flat panel displays.

High-resolution electron microscopy has shown that the molybdenum emitter cones in Spindt cathodes are an assemblage of microcrystals and the structure of the tip surface is not well defined or consistent from one tip to another [68]. Impurity atoms in the surface and strongly bound adsorbed atoms on the surface further complicate the situation. Field emission microscopy indicates that, in fact, the emission arises at only a few sites of atomic dimensions randomly situated around the tip center. Even if the tip surface was a crystal plane (the smoothest possible), the surface structure would have to be taken into account, since the diameter of the area of high field is typically about 10 nm compared with the atomic diameter of molybdenum of 0.28 nm. In cases where the tip peaks at a single atom, such as those used in scanning tunneling microscopy (STM) [69], electrons appear to be emitted uniformly into a narrow cone of about 1° half-angle so that in this case the computation of electron trajectories should be quantitative. Otherwise, until the emission distribution over the tip surface can be controlled, only approximate trajectories can be expected in the critical region near the tip surface. Unfortunately the launch velocity and current spatial distributions strongly affect the trajectories in the main parts of the device, making the results suspect even with the best programs. For arrays with very large numbers of tips the discrepancies may average out; however, this would require a 3-D program, and to date, this has not yet been done. We conclude that while 2-D computations can be used to show trends and as a guide for device performance, it should be recognized that they can only be used for detailed design if the specifications fall outside the range of the anticipated error.

4.4.2. Basic Spindt-Cathode Performance

When discussing FEA cathode performance, and especially when attempting to make comparisons between different emitter types, it is important to be certain that the conditions under which the data are taken are comparable. This includes the pretest processing, handling, and storage of the emitter arrays. The reason for this is the nature of the field emission process as illustrated by the well-known F-N equation [Eq. (1), repeated here for convenience]:

$$I = aV^2 e^{-b/V} \tag{1}$$

where $a = Cn\alpha\beta^2$ and $b = B(\phi^{3/2}/\beta)$, I is the emission current, V the applied voltage, n the number of tips, α the effective emitting area per tip, β is the geometric field-forming factor ($E = \beta V$), and ϕ is the effective work function of the emitting surface. For our purpose C and B can be treated as constants and can be approximated, with negligible error in regard to experimental results, by $C \cong 1.5 \times 10^{-6}\phi^{-1}(e^{10.4/\phi^{1/2}})$ and $B \cong 6.44 \times 10^7$, as shown by Brodie and Schwoebel [12].

These relationships illustrate the sensitivity of the field emission process to the field-forming factor and especially the work function of the emitting surface. With regard to work function it is important to realize that we are dealing with an atomic-level surface effect, and that surfaces in a vacuum system are dynamic and always moving toward some equilibrium state with their environment. The situation is further complicated by electron emission, as the environment changes because of heating and desorption resulting from the emission process at the emitter tip as well as the impact of electrons on anode surfaces — or tube surfaces in general when the cathode is operated at a high negative potential relative to ground, as is commonly done.

There is an extensive body of literature on FEEM and FIM that deals with adsorption, desorption, and surface migration on well-defined emitter tips under the influence of high electric fields [13, 19, 70–72]. In reality, under the conditions that one almost always operates microfabricated emitter arrays, the emitting surface is not clean in the strict sense that one associates with traditional FEEM studies. In classic FEEM work single-crystal emitter tips are thermally flashed for cleaning at temperatures of up to 1500–2000°C prior to taking observations. By contrast, microfabricated emitter arrays are rarely heated for cleaning at more than 450°C, and this is not sufficient to produce an atomically clean surface. As a result, we find ourselves working with an ill-defined emitting surface that can probably be best described as a combination of several microcrystalline surfaces, grain boundaries, and adsorbates. In addition, it is a dynamic situation as adsorbates diffuse about the surface and the surface evolves toward equilibrium with its environment.

It is well beyond the scope of this chapter to deal rigorously with the diffusion, adsorption, and desorption dynamics on an emitter tip, but a qualitative discussion of these phenomena is presented to illustrate the care that must be taken in gathering, analyzing, and comparing the results of emission tests with microfabricated emitter arrays.

The result of having an emitter tip that is not atomically clean is that at the onset of emission, there is liable to be a significant amount of emission-stimulated desorption from the tip, and related changes in the effective work function of the surface, as well as changes in the local field-forming β factor at the emission sites. In this case we refer to the local atomic level β factor as compared to the global β factor, which is a function of the emitter-tip radius, shank half-angle, gate-aperture diameter, and the height of the emitter tip relative to the bottom of the gate film. It has been clearly shown that the emission from emitter tips is an atomic-level effect, and is distributed over the surface of the emitter tip in relation to the local work function and β factor [13,19]. In the case of the Spindt-type microfabricated metal emitter, the emitter tip is normally polycrystalline and covered with adsorbates. This creates a dynamic surface condition because of diffusion and desorption during the emission process as well as adsorption from the operating environment. The result is unstable emission, at least during the initial stages of the process, as adsorbates migrate about the surface of the emitter and desorb under the stresses of the emission process. Furthermore, there is a supply of adsorbates on the entire cathode structure to replenish by diffusion any material that is desorbed from the tip by the emission process. Obviously, the details of this situation depend on the temperature, materials involved, the processing and handling of the emitters prior installation in the vacuum system, the vacuum

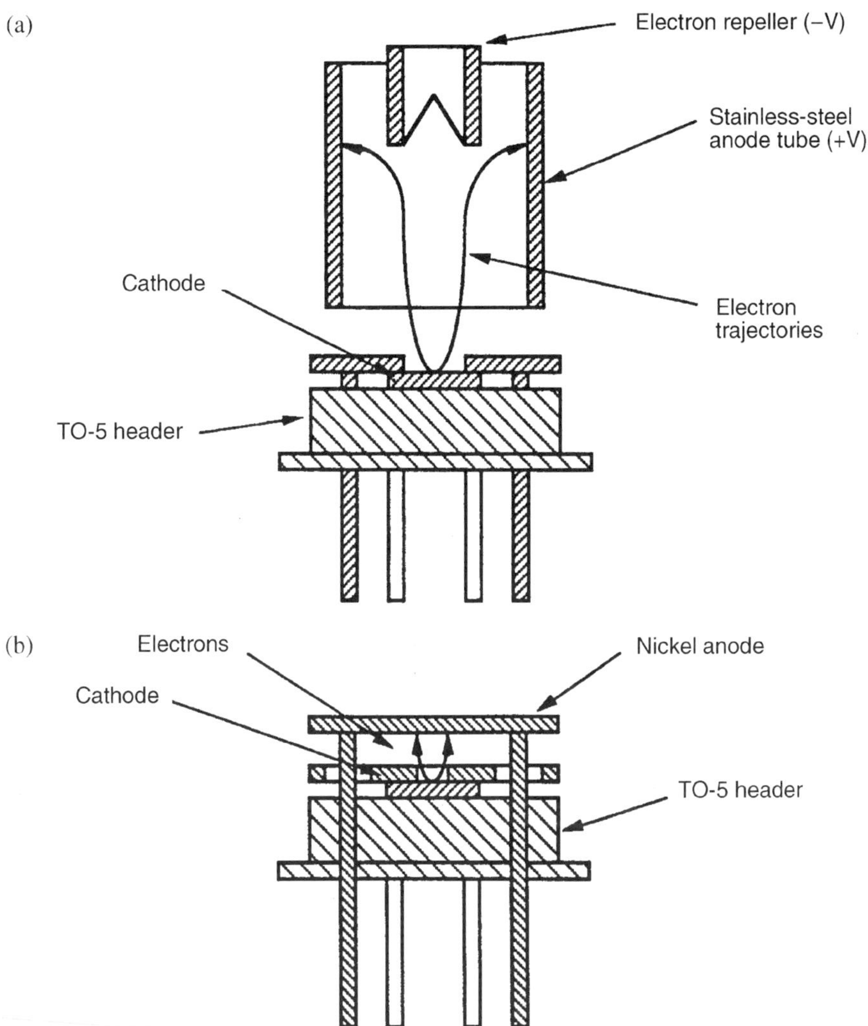

FIGURE 4.22. Anode configurations that have been used successfully for testing microfabricated FEA after rigorous processing, as described in the text.

processing, and the partial pressures and species of the various gases that make up the operating vacuum environment.

At this time we emphasize that the nature, geometry, and processing of the anode structure is also very important with regard to the performance that one obtains from the emitter. If the anode is not clean, the emitted electrons impacting the anode will cause electron-stimulated desorption from the surface, and some of them will find their way to emitter — usually with an impact on emission performance.

We have worked with a variety of anode configurations, and found that one of the best configurations is as shown in Fig. 4.22(a). The idea is that the electrons impact the surface of the anode in a field-free region on the inner walls of the tube and well within the tube as shown. Thus, reflected primary electrons and any ions formed are less likely to find their way back to the cathode. In addition, as the anode is a tube open at both ends, there is ample pump-out for any desorbed gases, and a large surface area for radiation cooling when operating at high emission levels. The anodes are often glowing at 800–900°C when operating cathodes in the tens of mA range with the anodes at 1 kV, and this sometimes produces interesting effects, which will be described in the next section.

The materials that we have used successfully for the anodes of this kind are carefully processed oxygen-free copper, 304 stainless steel, and electronic grade nickel. The anodes are thoroughly cleaned, hydrogen fired at 1100°C (900°C in the case of copper), mounted in the vacuum system, and heated by electron bombardment from a thermionic filament to about 900°C until there is no longer any indication of out-gassing from the anode structure. Then the system is vented to atmospheric pressure with dry nitrogen, and the emitter arrays to be tested are mounted as shown in Fig. 4.22(a). When this stringent procedure was implemented early in our cathode development program, we experienced an immediate and dramatic improvement in cathode reliability. Armed with this result, we later demonstrated that cathodes could also be operated with a close-spaced ($\sim$0.5 mm), flat-plate, electronic-grade nickel anode, as shown in Fig. 4.22(b), but only if the anode was preprocessed as described previously.

It is important to note that, in the light of the preceding discussions on emitter-tip surface conditions, vacuum partial pressures, and anode processing, the emitter tips will eventually come into some form of equilibrium with the operating environment. This equilibrium state will result in a net improvement in the emission with time, or a degradation with time depending on the environment and the condition of the emitter tips relative to the environment at the start of the experiment. It is also obvious that if the environment is not stable, the emitters will be continually changing in reaction to changes in environment. However, with well-processed anodes, and a stable vacuum in the low 10^{-9} torr range or better, long-term stability can be obtained. An example is shown in Fig. 4.23. The figure shows four I–V traces taken at intervals over approximately a year's continuous operation of a 1000-tip emitter array of molybdenum tips on a 0.01 Ω cm silicon substrate. The cathode was in a 1×10^{-9} torr ion-pumped vacuum and driven to a peak emission level of 15 mA (an average of 15 μA/tip) with a continuous 60-Hz, half-wave rectified, 75-V peak drive voltage. No long-term trend in the emission — up or down — was observed after the first few hours of turn-on, during which there was the usual emission improvement associated with the emission-stimulated desorption discussed earlier. After 8448 h of continuous operation, the test was terminated so that the apparatus could be made available for other work. An earlier life test under similar conditions with a 100-tip array driven using the same continuous 60-Hz pulse drive ran for over 8 years at an average emission of 20–50 μA/tip before an arc discharge in a small one-cell appendage ion pump on the tube ended the test [34]. This cathode was operated at an average emitter-tip current of 50 μA/tip during the last year of the experiment.

We did most of our testing with a continuous 60-Hz, half-wave pulse drive voltage. The initial reason was that it requires only a single transformer, diode, and a

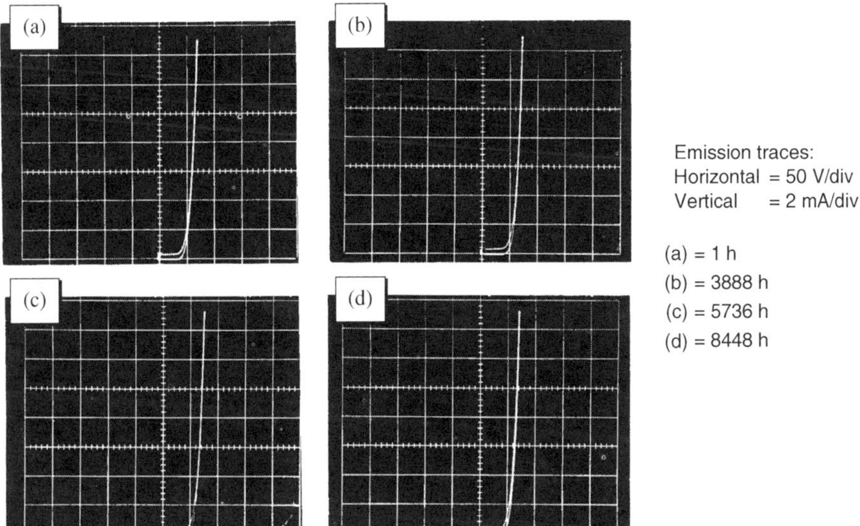

FIGURE 4.23. Voltage–current oscillographs taken during an emitter life test of a 1000-tip Spindt cathode with molybdenum emitter cones. The peak emission is 15 mA (and average emission of 15 μA/tip) with a continuous 60-Hz, half-wave rectified drive voltage. The anode was at 1 kV. (a) After 1 h; (b) after 3888 h; (c) after 5736 h; and (d) after 8448 h.

potentiometer for each cathode to drive many cathodes simultaneously. In addition, the ac emission allows us to monitor emission from several emitters simultaneously with only one dc anode supply by capacitively coupling out the emission pulse. The ac drive also allows us to simultaneously monitor the $I–V$ trace of the emission current and any gate current with an oscilloscope, which is helpful. Finally, it allows us to explore higher peak emission currents before heating of the collector electrode becomes an issue. We found no evidence that the 60-Hz drive has an influence on the long-term behavior of the emitters one way or another; however, we did not carry out exhaustive tests to firmly establish this.

These results can be taken as a demonstration that there is no fundamental limit to the lifetime of the microfabricated emitter array if it is properly processed and operated in a suitable environment. The challenge then becomes one of clearly defining these conditions and producing them reliably in a given tube application.

4.4.3. Emitter-Tip Shapes and Tip Build-Up During Operation

The emitter-tip radius is generally known to be one of the most important parameters with regard to the β factor of an emitter. Seemingly less well known in the VME community is the effect that the tip radius can have on the behavior of the emitters under the combination of high electric field and elevated temperature. This is a complex phenomenon involving parameters such as the characteristics of the tip material, crystal orientation, temperature, electric field, the effects of tip radius on effective surface tension. Tip build-up ("sharpening") may be due, at least in part, to the enlargement of principle crystal planes that cause the accentuation of the edges between them, resulting in field-enhancing microprotrusions on the tip [73]. Dulling due to excessive heating in the absence of an electric field is a consequence of the nature of materials to collapse into a sphere under the influence of surface tension and sufficient heat. Surface migration in the presence of high fields is discussed in depth by Dyke and Dolan [19] and Bettler and Charbonnier [73].

Thermal/field-forming of emitter tips has been well known and practiced by workers using classic etched-wire field emitters for many years, beginning with Benjamin and Jenkins in the late 1930s [19,74,75]. In relation to the etched-wire emitters, Herring [32] showed that the time rate of change in the length, $\partial z/\partial t$, of a field emitter tip because of diffusion of surface atoms is as follows:

$$\frac{\partial z}{\partial t} = \frac{2\sqrt{2}\Omega^2 D}{AkTR} \left(\frac{\gamma}{R^2} - \frac{F^2}{8\pi} \right) \tag{3}$$

where z is the length along the axis of the tip with the convention that decreasing length is in the positive direction, A is the surface area per atom in cm^2, k is Boltzmann's constant in ergs per atom per degree, T is the temperature in Kelvin, π is the atomic volume in cm^3, Ω is the atomic volume in cm^3, D is the surface diffusion coefficient in cm^2/s, R is the tip radius in cm, γ is the surface tension in ergs/cm^2, and F is the electric field in V/cm. Also, note from Gomer [13] that $F \approx V/5R$.

The electrostatic forces oppose the surface forces, and $\partial z/\partial t$ may have values that are positive, negative, or zero depending on the values of the parameters in the parentheses of the equation. Thus if $F^2/8\pi < \gamma/R^2$, $\partial z/\partial t$ is positive, and dulling occurs; conversely, if $F^2/8\pi > \gamma/R^2$, $\partial z/\partial t$ is negative, and tip build-up occurs [19]. Dulling can be used to enhance emission uniformity over an emitter array by reducing the emission of the sharper tips, whereas build-up leads to enhanced emission due to geometric accentuation of the localized high field-stress areas of the emitter tip. It is also important to note that the F^2 term indicates that the effect occurs with either positive or negative fields applied because the term is squared. Thus, in principle it is possible to perform build-up without drawing field emission from the emitter tip by applying a reverse bias to the structure, thereby avoiding the risk of excessive emission and a damaging vacuum breakdown during the forming process.

The emitter tip build-up phenomenon has been used extensively to good advantage with conventional etched-wire tungsten emitter tips, for example by Crewe et al. [74] at the University of Chicago in their work with ultrahigh-resolution, electron beam microscopy. The work of Benjamin and Jenkins [75] is of particular interest in regard to our work, as they showed that molybdenum exhibited the build-up effect at a temperature of only 450°C as compared to 1700°C for the more commonly used tungsten. The importance of the tip radius in this phenomenon is apparent from Herring's equation, and probably accounts for the results of the following experiment which dramatically illustrates the importance of surface effects such as diffusion on emitter array performance.

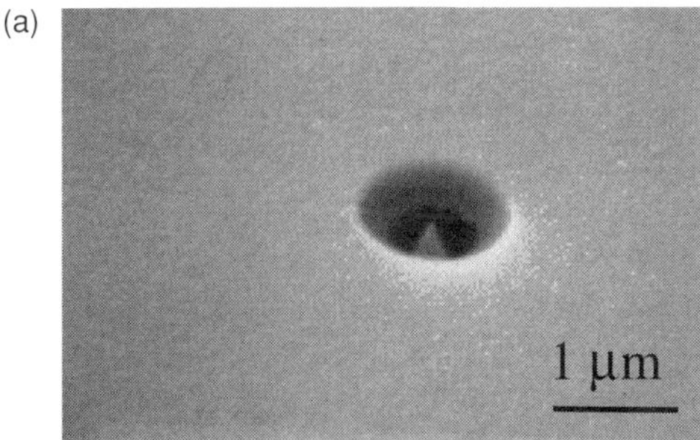
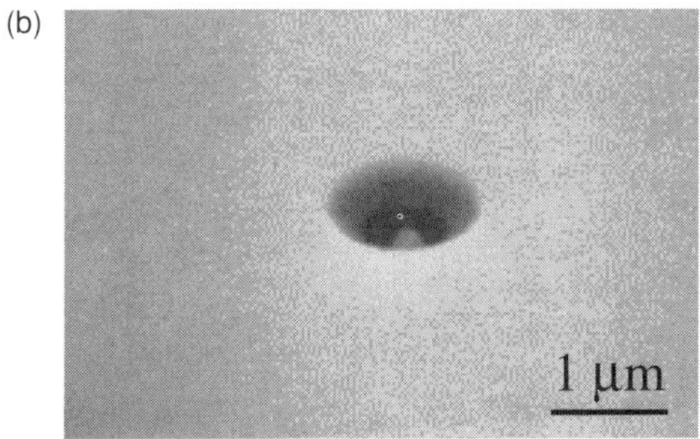

FIGURE 4.24. Scanning electron micrographs of typical tips from each of two groups of 10 000-tip arrays, one having tip radii of about 250 Å (a) and the other about 1200 Å (b), but all other features the same.

Six 10 000-tip emitter arrays from each of two fabrication batches having different emitter-tip radii, but all else the same within the accuracy of our measurements, were mounted in an ion-pumped vacuum chamber. Figure 4.24 shows scanning electron micrographs of the tip shapes, showing one with a tip radius of about 250 Å (a) and the other approximately 1200 Å (b). The difference was due to the amount of preclosure on the gate aperture prior to beginning the deposition of the cone material. The duller tips had less preclosure so that a larger, taller cone would have been formed because of the larger diameter gate aperture at the beginning of the cone deposition. However, the cone deposition was terminated when the cone was at the desired height, which was before the gate aperture was completely closed, with the result that the cone was not formed into a sharp point. Figures 4.4 and 4.5 help to visualize this effect.

The two groups of six emitter arrays were turned on and the emission increased to 10-mA peak emission from each array with a 60 Hz, half-wave rectified drive voltage applied between the base and gate. Each of the 12 emitter arrays had its own anode positioned about 3 mm away from the emitter and biased at +1200 V with respect to the gate electrodes, which were at ground potential. Under these conditions the temperature of the anodes was increased to 800–900°C by electron bombardment heating, and they, in turn, heated the emitters by radiation. Initially the six sharper emitter arrays required about 75 V to produce 10-mA of emission, while the duller emitters required about 150 V. This is consistent with what one would expect in view of the large difference in tip radii. The emission was maintained at the 10-mA level by adjusting the applied voltage as necessary. During a period of approximately 5 min, the emitter arrays with the sharper radii required progressively less voltage to maintain 10 mA, finally stabilizing at about 40 V. During the same period, the voltage required to maintain 10 mA with the arrays having blunter tips remained unchanged. After about 1 h of stable operation under these conditions, the drive voltage on all of the sharp-tip arrays was reduced to lower the emission current so that the anodes and cathodes would cool down. All of the sharp-tip arrays were observed to maintain the enhanced low-voltage emission at this reduced temperature. The blunter tips were held at the 10-mA level with the anodes hot for several hours, and during that time remained stable at their original emission level. Figure 4.25 shows the emission as a function of the applied voltage for these emitters, with the 250-Å radius tips having a reduction in required voltage for 10-mA emission from 75 to 40 V, and the 1200-Å radius tip arrays unchanged. The most likely explanation for this behavior is a form of the build-up phenomenon described earlier. Presumably, in the case of the 1200-Å radius tips the situation is such that $\gamma/R^2 \approx F^2/8\pi$, whereas with the 250-Å radius tips $\gamma/R^2 < F^2/8\pi$ as a higher field is needed to achieve the same emission as the 1200-Å tips because of the smaller emission area per tip. Then, as the temperature of the operating environment increased as a result of heating of the anodes at 10-mA peak emission, diffusion of surface atoms under the influence of the applied field resulted in localized build-up on the surface of the sharp tips. However, in the case of the duller tips the forces due to surface tension and applied field are in equilibrium. With the sharp tips, the voltage was reduced manually during the build-up process to maintain a constant emission, and thus presumably a nearly constant field (F) on the tips. Then, as build-up progressed, the effective value of R decreased and the value of γ/R^2 increased until it equaled $F^2/8\pi$ and equilibrium was established. Then as the system was cooled, the enhanced form was frozen in place.

This effect has been observed many times in our laboratory and found to be both repeatable and reversible. The intriguing aspect of the F^2 term, which suggests that the enhanced build-up condition can be produced by applying a reverse bias on an emitter array while baking it to the appropriate temperature for the emitter-tip material used, has yet to be demonstrated with microfabricated emitters. In principle this procedure could be incorporated into a tube processing routine as many tube manufacturers routinely bake at temperatures up to 600°C during tube pump-out. Thus it could be possible to apply a reverse bias to an emitter array during tube bakeout to promote tip build-up without the danger of producing a breakdown due to excessive emission as the build-up progressed. However, to our knowledge at the time of this writing, reverse-bias forming has not been demonstrated with microfabricated emitter tips, although it was used with good results by Crewe and coworkers at the University of Chicago [74].

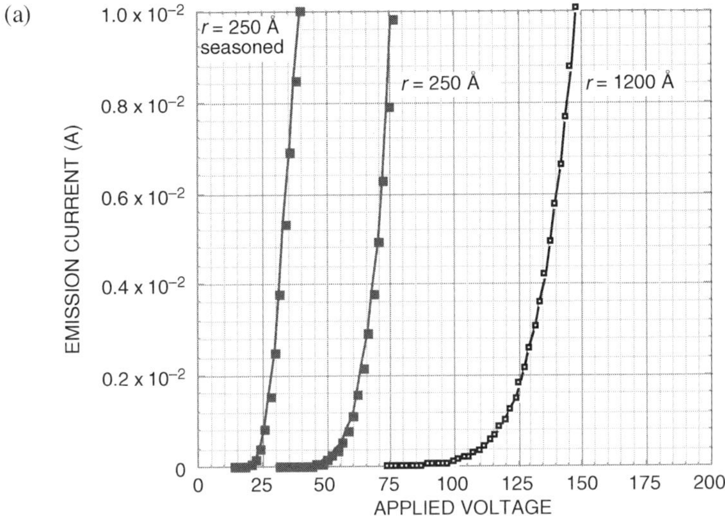

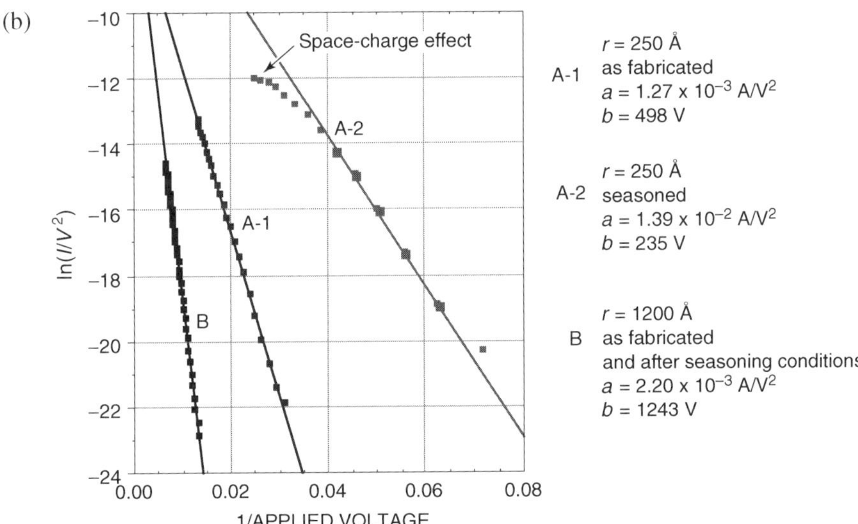

FIGURE 4.25. Voltage–current characteristics and F-N data for two 10 000-tip arrays, one with tip radii $\approx 250\,\text{Å}$ and the other with tip radii $\approx 1200\,\text{Å}$, showing radius-dependent tip-build-up-like "seasoning" that has been observed as the result of driving the cathodes to emission levels that heat the anodes to 800–$900\,°\text{C}$ [108].

As a very important final comment on the build-up-like phenomenon observed with microfabricated emitter arrays, we note that the effects described earlier may be due, at least in part, to field-assisted surface diffusion of adatoms deposited from the hot anodes rather than diffusion of surface atoms of the emitter-tip material.

Auger spectrometry of cathodes exhibiting the build-up-like behavior has shown small quantities of chromium, iron, and nickel on the surface of the cathode — materials that were almost surely deposited on the cathode from the hot (800–900°C) stainless-steel anodes during the experiment. See Gomer [76] for a discussion of adatom diffusion. In any case these build-up-like effects are a fertile field for additional study with microfabricated emitter arrays.

4.4.4. Emitter Materials

The material used to form a Spindt-type emitter array not only has an impact on the shape of the emitter as discussed previously, but it can also influence the emission characteristics by virtue of the inherent work function of the emitter surface. This is a complicated issue, as the tip is polycrystalline, and the work function of a microfabricated emitter tip relies on surface adsorbates as well as the basic emitter-tip material. Thus, the *effective* work function of an emitter tends to depend on the conditions of equilibrium between the emitter surface and the operating environment. However, there have been numerous reports of significantly improved emission from an emitter array by overcoating molybdenum tips with other materials such as DLC [57].

Because of the impact of materials selection on the Spindt-cathode fabrication process, the quickest way to explore alternate materials has been to overcoat standard microfabricated molybdenum emitters with a different material *in situ* and observe the voltage-current characteristics before and after coating. Mackie et al. [45] and Schwoebel et al. [77] reported experiments of this kind. In essentially every case, the overcoatings have resulted in improved performance in situ, but exposure to the atmosphere has caused the emitter performance to revert to the original values. Furthermore, experiment has shown that the emission characteristics after coating are also dependent on environmental and operating conditions such as temperature and emission level.

An experiment performed by Schwoebel in our laboratory illustrates these effects very well. A single-tip molybdenum Spindt emitter was installed in the emission microscope system, operated at 10-μA peak emission overnight to stabilize in the 10^{-10} torr vacuum environment, and shown to have a voltage-current characteristic with a peak emission of 10 μA at a little over 100 V, as shown in Fig. 4.26. The molybdenum emitter tip was then coated with approximately 100 Å of zirconium carbide (ZrC) by in situ electron-beam evaporation. The voltage-current characteristic was again measured and found to have improved to about 73 V for 10 μA of emission, as shown in Fig. 4.26. This was then followed by a series of processes listed below, and the resulting effects are shown in Fig. 4.27 that illustrate the nature of the field emission phenomena in general and the necessary caution that should be exercised when trying to compare emission data from various times and places. The complete test procedure was as follows:

1. The emitter was installed in an emission microscope and baked under vacuum at 250°C for 24 h. The system was then cooled to room temperature and

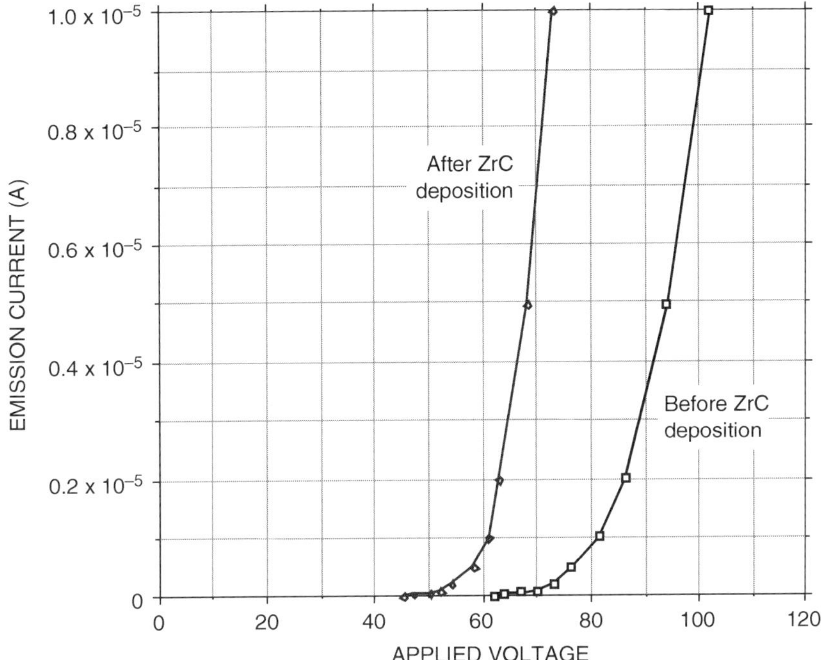

FIGURE 4.26. Voltage–current data for a single molybdenum Spindt tip before and after in situ coating with ~100Å of ZrC.

the cathode was turned on and brought up to 10-µA peak emission with our standard 60-Hz, half-wave rectified continuous-pulse drive voltage. The data shown in curve 1 of Fig. 4.27 were then obtained. System pressure was in the 10^{-10} torr range.

2. The emitter was left at 10-µA-peak emission overnight, and then the data for curve 2 were taken, showing a significant improvement in emission due to the overnight "burn-in." This is normal behavior for a new emitter in a good vacuum environment.

3. The emitter was turned off, and an estimated 100 Å of ZrC was deposited onto the cathode from a miniature electron-beam ZrC evaporation source mounted on a manipulater so that it could be positioned directly in front of the cathode for deposition and then moved back out of the way. The data for curve 3 were then taken, showing a marked improvement in emission.

4. A second 100 Å dose of ZrC was then added to the emitter surface, and the data for curve 4 were taken, showing a distinct decrease in emission, but performance still well above that of the uncoated molybdenum emitter tip.

5. The emitter was then heated to ~700°C in situ with the drive voltage off and cooled to room temperature. The data for curve 5 were then taken, showing an unexpected drop in emission to about the original value.

(a)

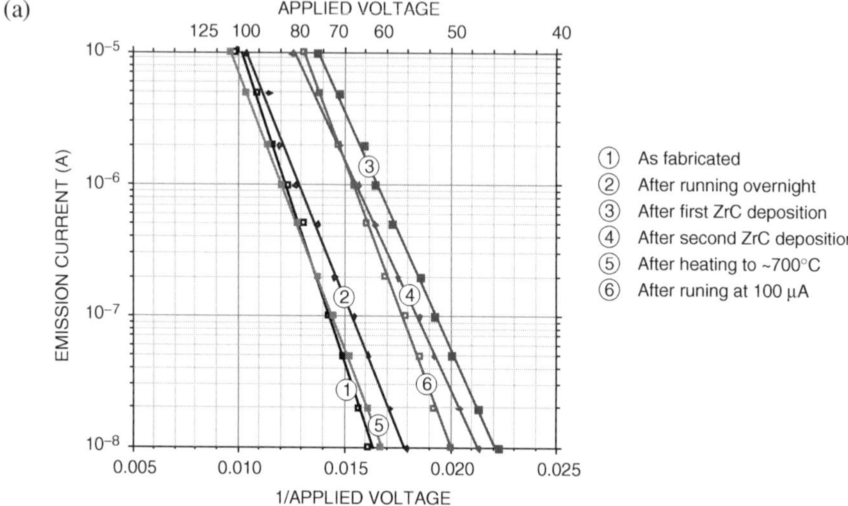

(b)

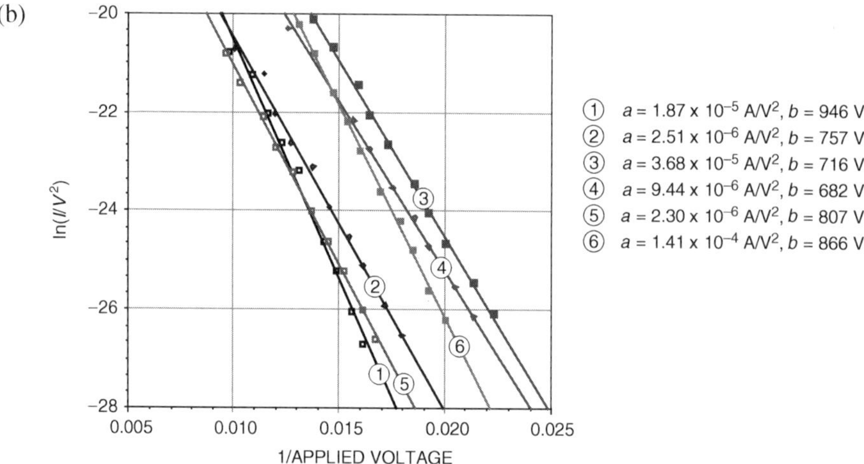

FIGURE 4.27. Emission data for a single molybdenum Spindt tip before and after coating with ZrC and various treatments as shown.

6. The emitter was then run up to 100 μA peak emission with the 60-Hz pulse drive and observed to improve in performance. Data for curve 6 were taken, showing that the emission had essentially returned to the original ZrC-coated value.

A host of effects are seen in this experiment that illustrate the importance of surface conditions on the emission process. The first is that even after a 250°C overnight bake in a 10^{-10} torr vacuum, the emitter "cleaned up" significantly because of emission-stimulated desorption in the ultrahigh vacuum environment of the emission microscope. This illustrates the importance of striving to obtain clean surfaces, and the caution that must be exercised with the first turn-on of an emitter that has not been "flashed" to very high temperature for cleaning as is commonly done with conventional etched tungsten wire emitters. Clearly emission-stimulated desorption cleaning was occurring during the overnight operation at 10 μA. It is important to note that, if left uncontrolled, this could lead to excessive emission and damage because of overheating and arcing.

The second is the effect of coating the surface with a material having an inherently lower work function such as ZrC. This is a straightforward way to obtain an emitting surface with a preferred material from a work function point of view, but often difficult to manage in situ with an actual application such as a microwave tube or field emission display panel. However, for first explorations of different emitter materials it is a very useful technique.

The third is the detrimental result of adding a second layer of ZrC to the first. It is well known that there is an effective work function minimum associated with sub-monolayer coverage by a low work function adsorbate on a cathode surface that is less than that of either the substrate material or the adsorbate [78]. Furthermore, with additional coverage beyond that minimum, the effective work function of the surface tends to approach the work function of the adsorbate material. It is likely that the coverage of about 100 Å is closer to the minimum for this substrate/adsorbate combination than the total of approximately 200 Å would be. In addition, the increase in the emitter-tip radius associated with the total ZrC coating of 200 Å could have a significant impact on the β factor for the emitter tip, thereby increasing the voltage required for a given emission level.

The fourth is the drop in the emission to the original bare molybdenum level as a result of the baking at 700°C. Certainly the vapor pressure of ZrC is negligible at 700°C, and in any case it is significantly lower than that of any of the other materials in the structure. The vapor pressure for ZrC is 10^{-4} torr at ~2500°C, and 2100°C for molybdenum (see Sloan's *Handbook of Thin-Film Materials*). Furthermore, there is no reaction between molybdenum and ZrC up to at least 2100°C [79]. This leaves the only explanation of this behavior as diffusion of the ZrC away from the emitter tip and down the emitter cone (the "blunting" effect predicted by Eq. (3) with heat in the absence of electric field) or contamination of the tip by material from the heater and other hot surfaces in the vicinity of the cathode. Finally, there is the return of the ZrC-coated emission characteristic because of the high-current pulse treatment and associated local high temperature at the tip. This behavior is also consistent with the notion of diffusion of ZrC in the high electric field, as well as the possibility of desorption of contaminants because of the emission process as is observed during the initial turn-on and overnight operation.

This series of experiments gives a good illustration of the need to be aware of the details of the surface of an emitter tip when attempting to characterize emission properties, and that the handling, processing, and operating environment all can have an effect on the results one obtains. Again, *these issues can be especially important when attempting to compare results from different times and places.*

4.4.5. The Effects of the Operating Environment and Changes in the Operating Vacuum Environment

Issues of concern with long-term emitter operation are back-ion sputtering, chemical reactions with partial pressure constituents in a normal vacuum system such as oxygen, water vapor, hydrogen, carbon monoxide, carbon dioxide, and any materials that may be peculiar to a specific application such as some spacecraft orbits that have relatively large amounts of atomic oxygen. Of particular concern are the materials that may be desorbed from an anode during bombardment with electrons. This has been a troubling issue with some of the low-voltage phosphors used in field emission display, for example.

Tests of the reaction of molybdenum emitters to high partial pressures of air, hydrogen, helium, argon, and neon have been reported, showing a variety of effects depending on the nature of the gas [34,80]. Figure 4.28 shows the effects of emitter operation in 10^{-5} torr of air as compared to 10^{-9} torr. In this test an emitter was operated for a few hundred hours at 10^{-9} torr in an ion-pumped system with an anode at 6 kV and shown to be stable under those conditions. The pump was then throttled down, and a leak valve was used to increase the pressure in the chamber to 10^{-5} torr of room air while maintaining a constant voltage applied between the cathode base and gate. The emission was monitored and observed to fall to about one tenth of the value maintained at 10^{-9} torr, and then stabilize. After about 100 h of stable operation under the high pressure conditions, the air leak was closed, the pump valve

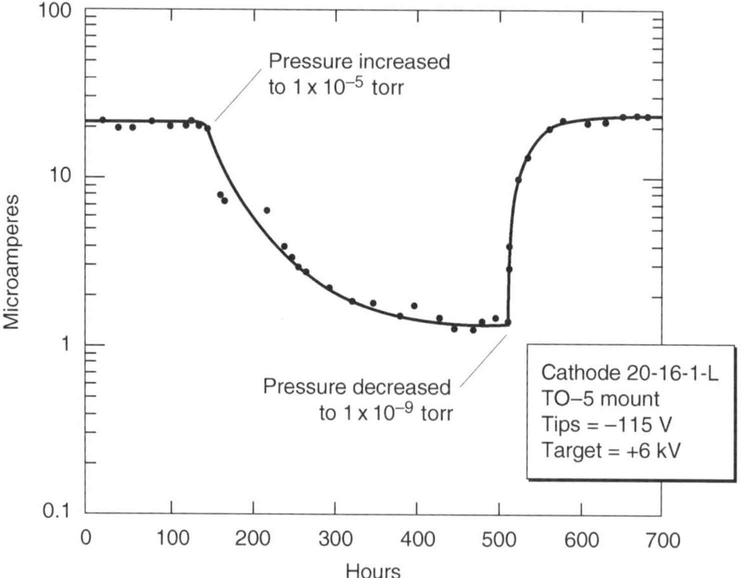

FIGURE 4.28. Emission current with a constant applied voltage at a pressure of 10^{-9} torr and at 10^{-5} torr of laboratory air introduced through a leak valve [34].

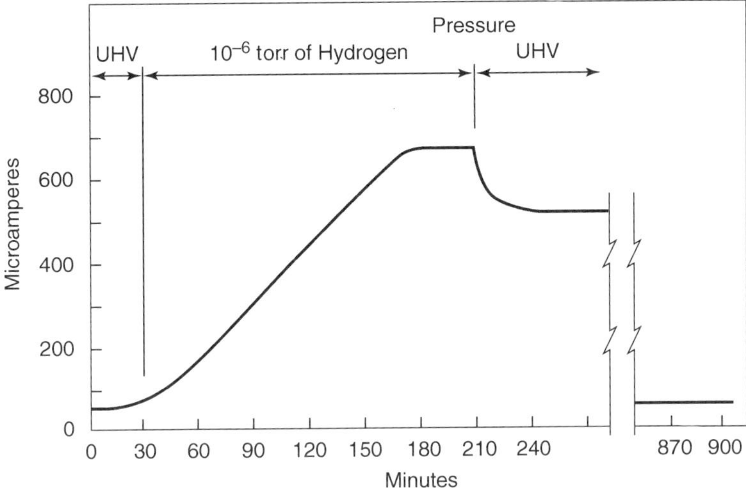

FIGURE 4.29. Emission current with a constant applied voltage at a pressure of 10^{-9} torr and at 10^{-6} torr of hydrogen introduced through a leak valve [34].

was fully reopened, and the emission was monitored with constant applied voltage as the pressure fell to the original value of 10^{-9} torr. As shown in Fig. 4.28, the emission returned to the original value in a relatively short time.

A similar test with hydrogen is shown in Fig. 4.29. In this case the emission increased with constant voltage applied between the base and gate and came to equilibrium with the hydrogen environment at an emission level of about a factor of 10 higher than it had been at 10^{-9} torr. Again the emission returned to the original value when the system was pumped back down to the original level of 10^{-9} torr. The same exercise was repeated with helium, argon, and neon. With these noble gases, no significant change in emission level was detected [34]. The fact that the emission returned to the original level when UHV conditions were restored suggests that the effect is due to changes in effective work function because of adsorbates from the operating environment, and not from changes in the β factor because of sputtering damage. Temple [81] reported qualitatively similar results for silicon emitter arrays.

The clear message is that the emitters come to equilibrium with the operating vacuum environment. A high partial pressure of air, for example, can decrease the emission for a given applied voltage by an order of magnitude, while high levels of hydrogen can improve performance by an order of magnitude for molybdenum emitter tips. This is yet another example of why it can be misleading to compare emission results without knowing the details of the operating environments.

Wei et al. [82] reported a more exhaustive test of aging as a function of environment. In this case, addressable molybdenum emitter arrays were operated in a sealed FED package with a ZnO:Zn phosphor for 1800 h. The residual gases in their display package were reported to be predominately O_2, CO_2, H_2O, and CO — all oxygen bearing species. The test arrays were divided into two groups, one of which was operated continuously for 1800 h while the other was dormant as a control group. The test was terminated when the emission fell to 50% of the original value, and the two cathode groups were examined with scanning Auger microscopy and X-ray photoelectron spectroscopy. The test group was found to have significantly higher quantities of oxygen than the control group, with the highest concentrations being at the emitter tips. Presumably this is the result of ionization of the residual gases in the vicinity of the emitters and subsequent reaction with the emitter surface. Similar effects have been reported by Itoh et al. [80]. These results show the importance of environment on lifetime, and that in circumstances where oxidizing species are likely to be present, alternate emitter-tip materials should be used. Iridium and transition-metal carbides are possible candidates.

4.4.5.1. Operation in Environments Containing Organics. While it is much less of a problem today than when the standard vacuum station was an oil-diffusion pumped chamber with o-ring gaskets, the issue of contamination from organics is very real in many practical applications and must be dealt with. The effect of operating an emitter array in a contaminated environment is dramatically illustrated by the following example. In the early 1980s, we explored the possibility of using an emitter array as a source of electrons for a mass spectrometer ionizer. The motivation was to eliminate pyrolyzation of the specimen by the hot filament normally used as a cathode in an ionizer—a notion recently confirmed by Felter [36]. At that time, the vacuum for a mass spectrometer was normally provided by an oil-diffusion pump and was typically about 10^{-7} torr, whereas we usually operated our emitters at 10^{-9} torr in an ion-pumped system with metal gaskets. Thus, the first test was to see how the emitters would behave in a 10^{-7} torr oil-diffusion pumped environment. A standard (for that time) 5000-tip Spindt cathode was mounted in a glass bell-jar, liquid-nitrogen trapped, oil-diffusion pumped station with a polished, prebaked, molybdenum anode plate positioned 1 cm from the cathode and pumped down to the 10^{-7} torr range. Emission was initiated, increased to about 100 μA, and observed to be stable. Over approximately an hour's time, the emission slowly improved, suggesting that emission-stimulated desorption was cleaning the tips. This was typical of the behavior we were accustomed to in the ion-pumped systems. However, shortly after that, the onset of intercepted gate current and emission instability was noted, along with a visually detectable darkening of the anode. With the realization that the darkening of the anode indicated electron-stimulated reactions with hydrocarbons on the anode surface, the cathode was turned off and removed from the system for examination in a scanning electron microscope.

Figure 4.30(a) is a low-magnification scanning electron micrograph of a large portion of the 5000-tip array, showing several emitters blown out and radial tracks emanating from the blow-out sites along with similar tracks from some tips that had not failed. This material appears to have been deposited by electron-stimulated reactions with the hydrocarbons on the gate surface in the same manner as the material deposited on the anode. Figure 4.30(b) is a close-up view of four of these emitters, showing how the radial tracks were formed by a curved "horn" of these reaction

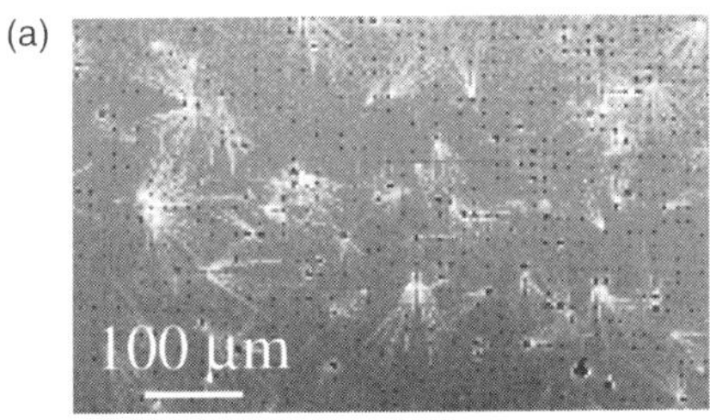

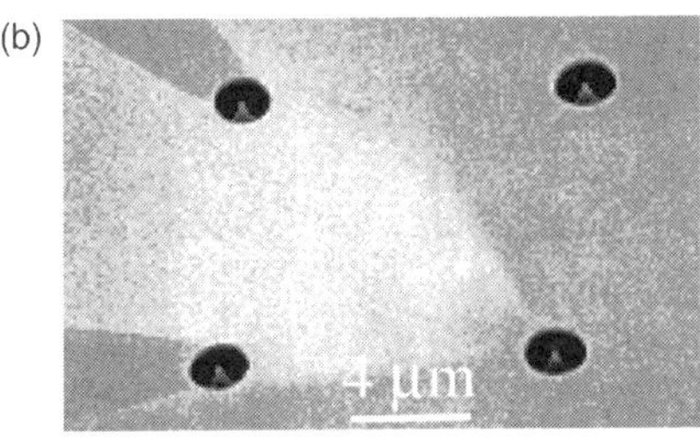

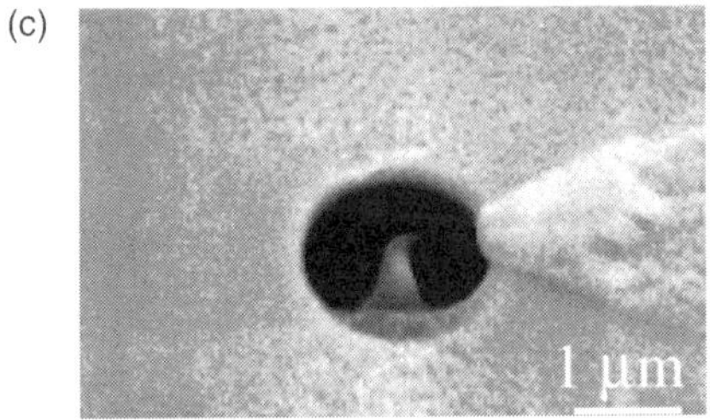

FIGURE 4.30. Scanning electron micrographs of an emitter array after operation in a hydrocarbon-contaminated vacuum system, showing electron-impact-stimulated build-up of contaminants on the emitter tips and the surface of the gate electrode.

products on the molybdenum emitter tip that grew in such a way as to form a new emitter tip directed along the surface of the gate film. This new tip spewed electrons across the gate surface, and they, in turn, built-up contamination on the surface of the gate. The fields due to the neighboring tips deflected the electrons skimming along the surface away from the gate, creating the radial-ray appearance in the deposited patterns. Figure 4.30(c) is a close-up of a single tip, showing heavy contamination near the edge of the gate aperture that surely would have created an arc if the test had not been terminated.

The obvious lesson from these results is that one should not operate emitter arrays in vacuum environments having a background of organics without taking steps to keep organics from building up on the cathode. This may be possible by heating the emitter array or performing periodic hydrogen-plasma cleaning treatments if the contamination is mild.

4.4.6. In Situ Processing with Plasmas

It is not possible to flash-heat the emitters Spindt cathodes to 1500°C for cleaning as is normally done with conventional tungsten emitters, but experience has shown that baking at 400–450°C for 48 h, followed by a careful first turn-on and "burn-in," can produce satisfactory results [34]. If integrated emission-buffering resistors are used, the first turn-on can be greatly accelerated. However, there are applications where temperatures are limited and long bakeout times are a difficulty (e.g., the production of display panels). We have found that a brief (i.e., ~2 min) hydrogen-plasma "scrubbing" of the vacuum surfaces followed by pump-out to operating pressures is a viable alternative to bakeout as a surface-cleaning process [83,84]. The idea for the hydrogen-plasma cleaning process grew out of our observation that operating an emitter array in relatively high partial pressures of hydrogen improves performance significantly [34]. Presumably, chemically active hydrogen ions formed near the emitter tips by the emitted electrons react with contaminants on the emitter tip to form gaseous species that are then desorbed from the tips.

The original experiments with hydrogen-plasma cleaning were performed in the FEEM shown schematically in Fig. 4.31. The apparatus is an ultra-high vacuum chamber with a base pressure low in the 10^{-11} torr range after a 12-h bakeout at 250°C. A molybdenum-tip Spindt cathode was installed and baseline data taken in the form of $I–V$ characteristics as well as by direct viewing of the emission pattern on a phosphor screen positioned about 5 cm away from the emitter. The emitter tips were then subjected to various treatments with glow discharges of hydrogen, hydrogen plus 10% neon, and hydrogen plus 10% xenon. The procedure was as follows:

After a standard pump-down and bakeout, $I–V$ data were taken to characterize the emitter. Plasma treatments were then performed by inserting a metal anode between the emitter tip and phosphor (about 1-cm from the cathode) and connecting (shorting) the base and gate of the cathode together. Matheson research-grade hydrogen gas was then bled into the chamber to a pressure of about 0.1–1.0 torr. At this spacing and pressure, a voltage applied between the cathode and anode of between 275 and 450 V was sufficient to sustain a glow discharge with the gases used. A dc current-regulated power supply was used to operate the glow discharge. Total ion-current densities were of the order of 10^{16} (ions/cm^2)/s, and typical doses were of the order of $10^{18}–10^{19}$ ions/cm^2. After each treatment the system was evacuated to UHV and emission data were taken.

F-N data and FEEM images are shown in Figs. 4.32 and 4.33 for a single-tip molybdenum Spindt cathode subjected to these procedures. Line 1 in Fig. 4.32 and FEEM image in Fig. 4.33(a) is the emission data taken following installation in the system and subsequent bakeout. The emitter was then treated with a hydrogen-plasma dose of 10^{18} ions/cm^2 (hydrogen pressure $\approx 5 \times 10^{-1}$ torr, and an applied voltage of ~300 V for about 2 min). The system was then pumped back down to UHV. The data for line 2 in Fig. 4.32 and the FEEM emission pattern shown in Fig. 4.33(b) were then

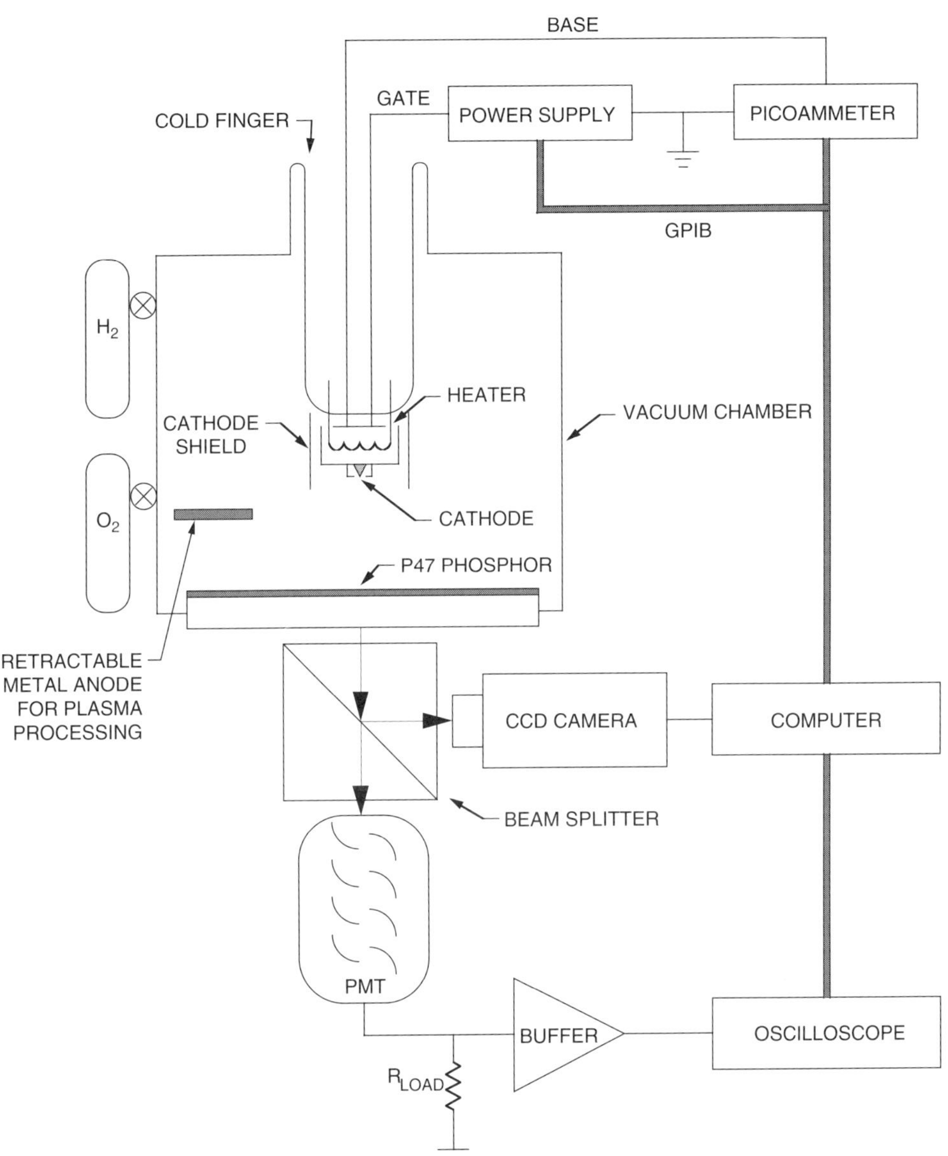

FIGURE 4.31. A schematic drawing of the FEEM used to study emitter-tip surfaces [88].

recorded showing a significant reduction in the voltage required for a given emission level and an apparent increase in the total effective emission area. We also note the elimination of an emission site on the lower right-hand corner of Fig. 4.33(a). Additional hydrogen-plasma treatments produced no significant added changes in the emission characteristics, suggesting that a 2-min (10^{18} ions/cm^2) treatment is sufficient to complete the hydrogen cleaning process.

1) $a = 7.60 \times 10^{-7}$ A/V^2, $b = 1759$ V
2) $a = 3.74 \times 10^{-6}$ A/V^2, $b = 1410$ V
3) $a = 4.40 \times 10^{-7}$ A/V^2, $b = 1103$ V
4) $a = 1.59 \times 10^{-6}$ A/V^2, $b = 1409$ V

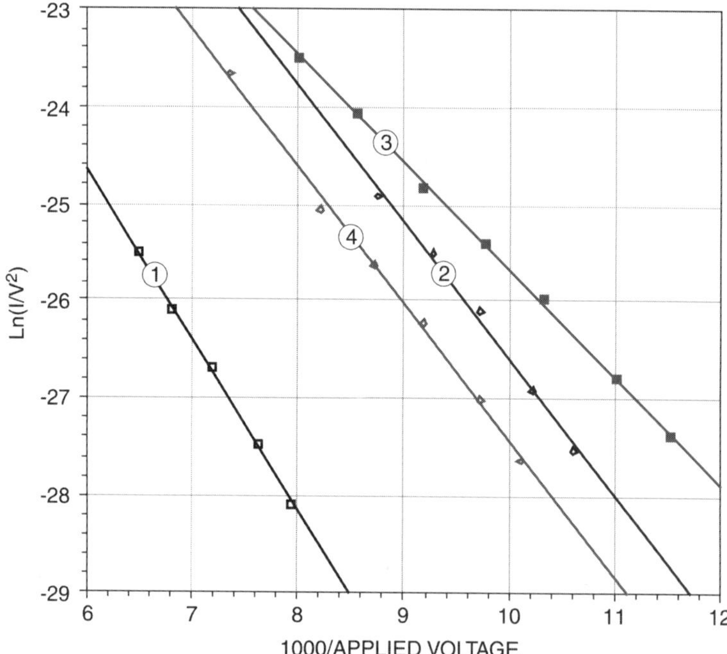

FIGURE 4.32. F-N data showing the effect of pure hydrogen and hydrogen + 10% neon plasma on a microfabricated single-tip field emitter. Line 1: F-N data prior to treatment. Line 2: F-N data following hydrogen-plasma treatment (dose $\approx 10^{18}$ ions/cm^2). Line 3: F-N data following hydrogen + 10% neon (dose $\approx 10^{18}$ ions/cm^2). Line 4: F-N data following an additional hydrogen + 10% neon treatment (total dose $\approx 7 \times 10^{18}$ ions/cm^2).

As a means of determining the role of sputtering in the cleaning process, the emitter was then subjected to a similar ion dose using hydrogen plus 10% neon. The data shown in line 3 of Fig. 4.32 were then acquired, showing an additional increase in emission for a given voltage. Additional doses of hydrogen plus 10% neon for a total of approximately 7×10^{18} ions/cm^2 reduced the F-N data to line 4 of Fig. 4.32 and the emission pattern similar to that shown in Fig. 4.33(c), from which a significant increase in effective emission area is evident.

In the case of the pure hydrogen treatment, it is reasonable to assume that the change in emission characteristics is due to the removal of contaminants and an accompanying decrease in effective work function as sputtering of molybdenum by hydrogen at the energies involved would be negligible [84]. However, with the hydrogen plus 10% neon there appears to be effects due to sputtering of the emitter surface. An

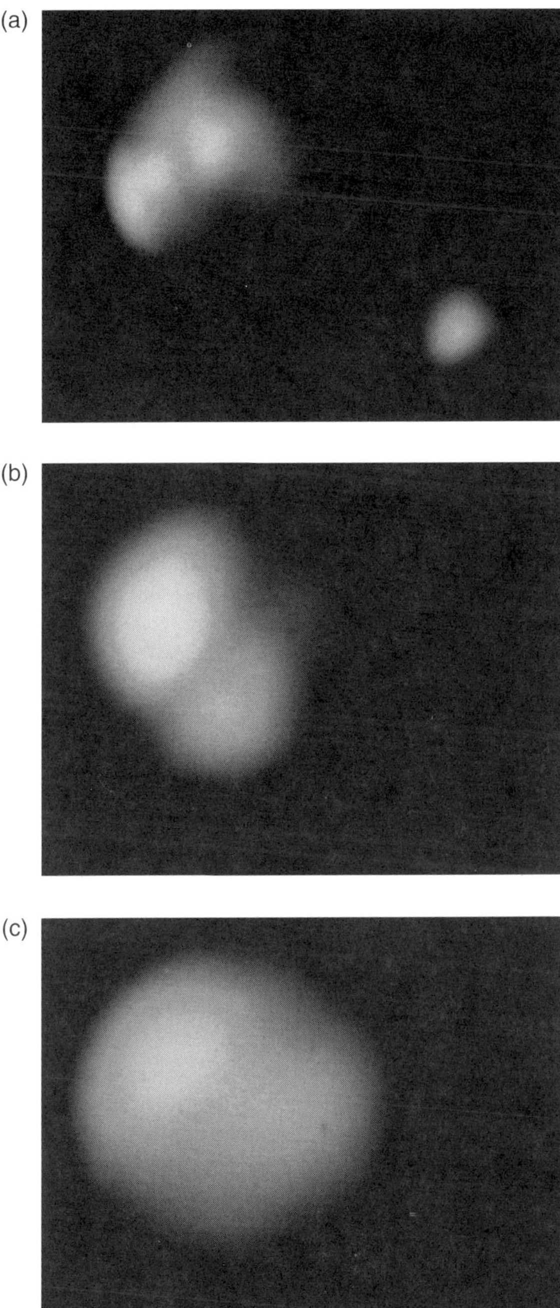

FIGURE 4.33. Field electron micrographs corresponding to the F-N data shown in Fig. 4.32. (a) Line 1 of Fig. 4.32—prior to treatment ($V = 175$ V, $I = 1\,\mu$A). (b) Line 2 of Fig. 4.32—after hydrogen treatment ($V = 133$ V, $I = 1\,\mu$A). (c) Line 4 of Fig. 4.32—after the second hydrogen + neon treatment ($V = 136$ V, $I = 1\,\mu$A) [84].

experiment to further characterize the effects of sputtering was done with hydrogen plus 10% xenon. A similar dose was again used; however, with the hydrogen 10% xenon plasma the cathode developed an electrical short between the base and gate. Lesser doses of hydrogen plus 10% xenon were then explored with a new cathode and found to produce reduced emission for a given voltage, followed again by base-to-gate shorting as the doses were increased. This result was taken as evidence that sputtering of the molybdenum is negligible with the hydrogen at $\sim$300 V, but can be significant at these voltages with heavier species. The result with xenon was that molybdenum sputtered from the sides of the cones and coated the silicon dioxide walls of the cavities, producing an electrical path that shorted the base and gate together. A paper by Schwoebel and Spindt [84] gives a detailed discussion of these results.

4.4.7. Integrated Resistance for Buffering and Stabilizing Emission

The initial onset of emission can be a time of sudden cathode failure, because it is not possible to flash-heat a microfabricated emitter array for surface cleaning prior to operation — as is traditionally done with conventional emitters. These failures are the result of localized emission-stimulated desorption from the emitters and/or the anode structure in the test system. The desorption of contaminants from the tip usually produces a rapid increase in emission as a result of lowering of the effective work function, and this, coupled with associated localized pressure bursts, can result in a local vacuum arc. In applications where it is permissible to integrate buffering resistors into the structure, burn-in times and the incidence of initial turn-on failures can be greatly reduced, and emission uniformity over large areas significantly improved. For example, buffering resistors are used to great advantage in display panels [85]. Unfortunately, the use of integrated resistors is not acceptable in some applications, as they introduce a series impedance to each tip and an energy spread in the emitted electron beam. However, in special cases it may be possible to have the resistors for initial burn-in and then use heat to effectively eliminate the resistors as outlined later.

For general purpose use, we have found that fabricating molybdenum emitter arrays on high-resistance silicon substrates (e.g., $\sim$200–500 $\Omega \cdot$ cm) greatly reduces the incidence of "infantile mortality" during initial turn-on and burn-in. In addition, the uniformity of emission over large areas is significantly improved; however, as mentioned previously, one must be able to accept the penalty of added series impedance and energy spread in the emitted electron beam. In some applications it is possible to have the benefit of a protective high-resistance silicon substrate during the initial turn-on and burn-in stages of operation, and then reduce the resistance of the silicon by heating the substrate. In principle, the heating can be either by operating at a level that produces joule heating of the silicon to reduce the resistance or by incorporating a heater into the cathode mounting structure. Experiments as described in the next section have shown this to be a workable technique.

4.4.7.1. Temperature Effects. Several experiments have been done using high-resistance silicon substrates for improved reliability. In a typical example, a 1-mm diameter, 50 000-tip Spindt-type molybdenum emitter array on a 2000 $\Omega \cdot$ cm n-type

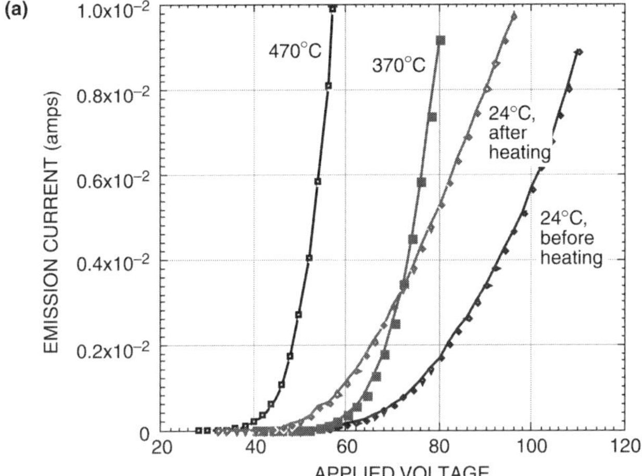

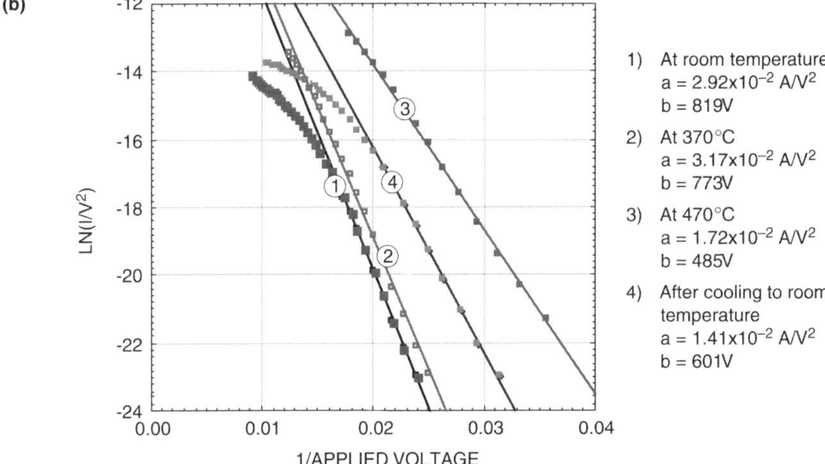

FIGURE 4.34. Emission data for a 50 000-tip Spindt cathode on a 2000 $\Omega \cdot$ cm, n-type silicon substrate before and after operating at temperatures up to 470°C, showing effects of the negative temperature coefficient of resistivity for the silicon and cleaning of the tip by high-temperature emission.

silicon substrate was operated at room temperature in 10^{-9} torr vacuum up to 10 mA with 112 V applied between the back of the high-resistance silicon substrate and the gate electrode. Analysis of the F-N curve 1 shown in Fig. 4.34 (with the problematic assumption that all tips are contributing) indicates that the effective resistance in series with each of the 50 000 tips was initially about 130 MΩ at 10 mA total emission. However, the curve 4 at room temperature after the temperature cycle shows an effective resistance per tip of about 90 MΩ. This change may be due to diffusion of gold into the silicon from the TO-5 transistor header on which the cathode was mounted. Gold diffusion into the silicon may well have been a factor with the initial measurements as well because of the vacuum bakeout of the system at ~400°C prior to testing. In any case, a calculation of the effective resistance to be expected from the substrate is complex because of the distributed 3-D nature of the resistance and non-uniform emission from the tips.

The system was then heated to 470°C while the cathode was emitting. At this temperature, the effective resistance of the silicon substrate was essentially eliminated because of the high negative temperature coefficient of the silicon. As a result, the emission (as a function of the voltage applied between the back of the substrate and the gate) exhibited F-N behavior up to a peak emission of 10 mA with an applied voltage of only 57 V. Figure 4.34 shows these data, along with the emission characteristics showing 10 mA emission with 97 V applied after the cathode was again cooled to room temperature. The voltage reduction from 112 to 97 V suggests that thermal desorption at 470°C reduced the effective work function of the tips. Over time, the emission characteristics reverted to the original values as the emitter tips came to equilibrium with the operating environment. Careful examination of this emitter array after it was removed from the test chamber showed that none of the 50 000 tips in the array suffered any damage during the experiment, suggesting that after the tips have been cleaned by desorption, they can be operated safely without protective resistors.

The notion of using a heater with field emitter arrays is offensive to many, as a major feature of field emission is that it requires no heat to emit electrons into the vacuum environment. Indeed, emitter arrays are often referred to as cold cathodes. Strictly speaking, emitter arrays are probably more precisely characterized as temperature-independent cathodes as it is generally accepted that the effect of temperature on the fundamental field emission process is small at temperatures below 1000 K [19]. As a practical matter, experiments have shown that Spindt cathodes can produce electrons over a range of at least 4 K [86,87] to 1100 K [88], and emitters are routinely operated at between 300 and 750 K in our laboratory. It should also be noted that, in general, better performance is obtained at elevated temperature. This is most likely due to maintaining a cleaner emitting surface by desorption at elevated temperature, as suggested by the data shown in Fig. 4.34.

4.4.8. The Nature and Management of Emission Fluctuations in Spindt Cathodes

The earliest workers in field emission recognized the cause of erratic emission behavior as the sensitivity of the emission process to adsorbates on the emitter surface. Armed with this insight, Millikan and Lauritsen [89] were among the first to achieve adequate emission stability for meaningful measurements through rigorous vacuum processing. To this day, those who work with conventional field emitters would not think of applying an electric field to an emitter tip without first achieving a vacuum of at least 10^{-10} torr and flashing the emitter tip to approximately 1500°C for forming and cleaning by desorption. Unfortunately, this is not possible when working with

Spindt cathodes, except when working with single tips where it is possible to heat the tip with high emission-current pulses. Nevertheless, in many applications, such as field emission display, fluctuations can be managed well enough by averaging the emission over many emitter tips and having current-limiting resistors integrated with the emitter tips to buffer emission from individual overactive emitters.

However, there are interesting potential applications for single-tip microfabricated emitters, such as electron beam probe tools, for which one is always striving to obtain the smallest, most intense electron source possible. For such applications to be practical with a microfabricated emitter, it is necessary to first reduce the emission fluctuations that are associated with microfabricated single-tip emitters. Furthermore, this must be done without the aid of buffering resistors, as series resistors introduce an energy spread to the emitted electrons that is unacceptable from an electron optical point of view.

There are three basic sources of short-term emission-current fluctuations from field emitters to be considered. These are shot noise, bistable "telegraph noise" resulting from one or more bistable emission sites on the emitter tip, and random fluctuations due to uncontrolled single events such as an ion impacting the emitter surface. Shot noise is the result of the discrete nature of the electron, and can be thought of as an irreducible minimum that dominates the noise spectrum for frequencies above 100 kHz. Bistable noise is a series of positive or negative pulses relative to a base emission level, and appears to be due to changes between two or more states by individual atoms (or molecules) on the surface of the emitter. Random-event noise is due to the temporal changes in the work function or β factor of the emitting surface caused by the interaction of the surface with the operating environment and surface diffusion of the substrate atoms. Flicker noise is due to the integration of bistable noise sites on the emitter over a large number of sites that occurs in a large array or the macroscopic surface of a thermionic cathode. Typically flicker noise has a spectral noise density that is proportional to $1/f^x$, where $0.5 < x < 2$. A detailed discussion of these phenomena is beyond the scope of this chapter, but these topics are well covered in the literature [76,90–95]. For our purposes here we will present an overview of investigations into the characteristics of random-event noise and bistable noise from single microfabricated emitter tips, and means to minimize their occurrence.

A field electron emission microscope, assembled by Paul Schwoebel in our laboratory for these studies, is shown schematically in Fig. 4.31. It is essentially a Müller microscope with provisions for heating and cooling the emitter and means for simultaneously recording the emission pattern on a phosphor screen as well as the amplitude of the emission current. A typical example of a burst of bistable emission current noise superimposed on a dc current level observed with this apparatus is shown in Fig. 4.35. This noisy mode of bistable bursts can consist of one to thousands of bistable transitions of varying length, interspaced with periods of quiescence that last from a few seconds to more than 10 minutes. Recordings of the FEEM images show that the changes in emission intensity with time are normally associated with localized regions rather than the whole emission pattern. These changes occasionally occur randomly over the emitter surface, but are far more often associated with fixed locations.

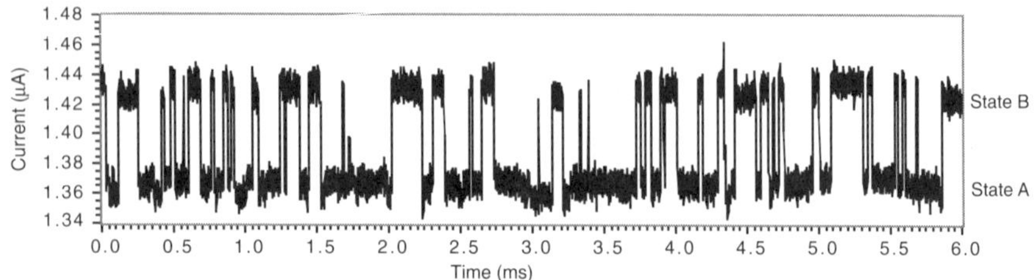

FIGURE 4.35. A typical example of bistable telegraph noise with an amplitude of about 5% of the peak emission [88].

Time-correlated FEEM images and bistable noise measurements were made by Paul Schwoebel of SRI and Russ Olson, Gary Condon, and John Panitz of the University of New Mexico. Figure 4.36 shows the results that illustrate the localized nature of these bistable fluctuations [88]. The total emission current on an expanded scale is shown as a function of time. The dc emission level is nominally 1.5 µA and the fluctuations are 20–40 nA in amplitude, or roughly ±1% of the total. Recorded FEEM images A, B, and C correlated to the emission levels are not sensitive enough to distinguish changes in intensity associated with the emission fluctuations; however, subtracting and amplifying the stored images (B–A and C–B) clearly shows that the emission changes occurred in highly localized areas.

Collecting many time-correlated images and emission noise data in this way has shown that, at any given time, there are only a few (three to five) active bistable sites on a tip, each with its characteristic burst amplitude generating the noise. These results suggest that this kind of bistable noise is due to adsorbates residing at localized sites on the emitter surface rather than adsorption/desorption from the surrounding environment. There are only a small number of sites, and the bistable pulses are repeatedly observed from the same sites, whereas fluctuations resulting from ion bombardment of the emitter surface or adsorption/desorption from the environment would be randomly distributed over the emitter surface. The implication of this result is very important as it means that, in an adequate vacuum environment, it may be possible to achieve acceptable emission stability for electron beam probe applications if these bistable sites can be deactivated or removed from the emitting surface.

4.4.8.1. *Processing Microfabricated Emitter Tips to Reduce Bistable Noise.*

Since bistable noise appears to be due to adsorbates on the emitter surface jumping between two modes, the obvious solution would be to remove these adsorbates from the surface. Thermal desorption of the offending species, as is done with conventional field emitters, has been investigated with mixed results. We recall an early experiment done in a 10^{-9} torr ion-pumped UHV system, during which a microfabricated single-tip molybdenum Spindt cathode was heated to about 1000 K on a special-made molybdenum cathode mount having an integrated heater. After cooling, the cathode exhibited very stable emission for several hours, and then experienced a gradual reoccurrence of the bistable noise.

Unfortunately, those data and apparatus were lost in a recent laboratory fire. Therefore, a new experiment was done by Schwoebel in our laboratory, in which a Spindt

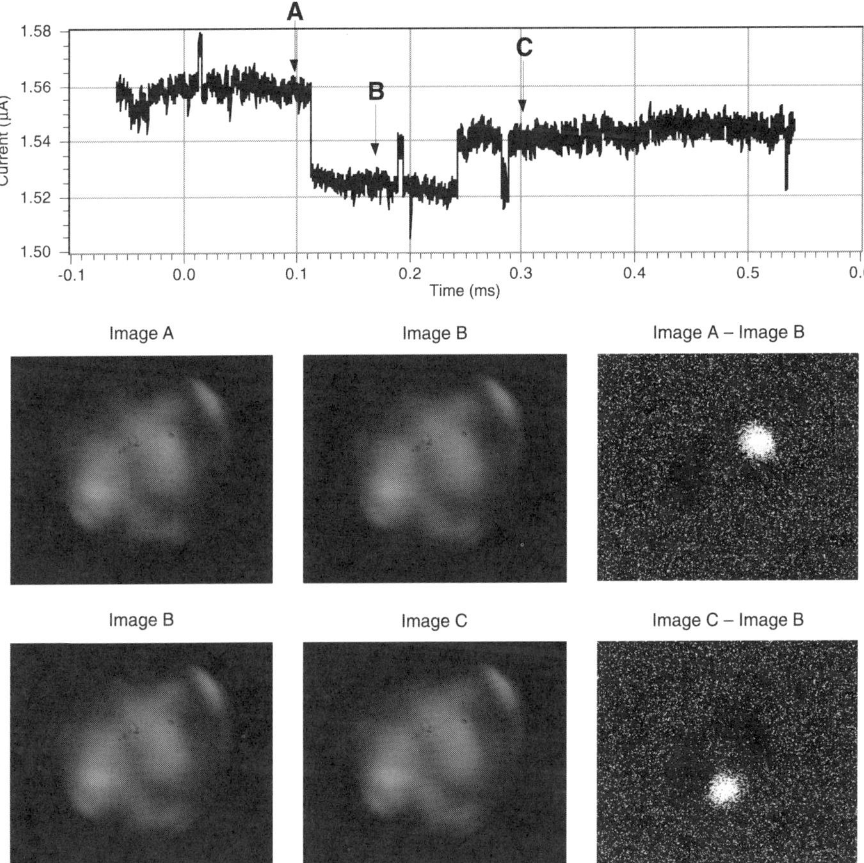

FIGURE 4.36. Time-correlated emission and FEEM images for a single molybdenum Spindt tip, showing the localized nature of the emission transitions. Images A and B show the emission pattern at times A and B on the current vs. time plot, and amplified image A–B reveals the site responsible for the change in emission amplitude. Similarly with images B and C and C–B showing a second bistable site [88].

cathode with a molybdenum emitter tip on a silicon substrate was mounted on a TO-5 header (which has glass insulators), heated to 1100 K, and then cooled to 300 K in the FEEM. In this case an increased incidence of flicker noise was noted after cooling. We theorize that this unexpected result was most likely caused by materials desorbed from the glass in the TO-5 header used in the experiment, which found their way onto the emitter tip, resulting in additional bistable emission sites. This increased noise subsided to the original preheating level with operation over several days, again demonstrating that emission-stimulated desorption tends to bring the emitter tip to equilibrium with the operating environment.

Cooling the emitter tip to approximately 77 K with liquid nitrogen reduced both the number of active bistable sites and the frequency of the bistable transitions. However, the magnitude of the current fluctuations remained the same. This leads us to the intuitively satisfying conclusion that the mechanism responsible for the bistable transitions is temperature dependent.

Clearly, the best solution to the noise issue is to find a practical means for removing the offending adsorbates from the emitter surface. Two possibilities, other than baking or the previously discussed plasma treatments, are emission-stimulated desorption by tip heating with high-current pulses and field desorption. Neither is a simple matter, as emission-stimulated desorption requires unusually high emission currents and is, therefore, risky, and field desorption of even weakly bound adsorbates requires the application of an electric field at least five times greater than that required for field electron emission. However, field desorption is done with a reverse bias, so the risk of emission-initiated arcing is essentially eliminated as long as there are no field-stress-rising sharp edges or microprotrusions on the edge of the gate aperture. An additional concern is that, with most microfabricated emitter structures, a voltage of at least five times the normal electron-emission potential would exceed the breakdown strength of the dielectric between the base and gate. However, the high aspect ratio cone fabrication methods described earlier may well make it possible to build emitters with a thick enough dielectric to hold off sufficient voltage for some field desorption to be achieved. If this can be accomplished, the very exciting prospect of atom-probe studies of Spindt cathodes to determine the identity of the adsorbates responsible for flicker noise presents itself [96].

In the meantime, emission-stimulated desorption has been shown by Schwoebel and Olson to be an effective means for reducing flicker noise. Recent ongoing experiments have shown that high-current pulse conditioning of an emitter tip can significantly reduce emission noise. Figure 4.37 illustrates the effect. The first plot in the figure shows the noise as a function of time for single emitter tip that has been relatively heavily contaminated by heating prior to any other treatment. The second shows performance after an aggressive 1-mA, 100-µs single pulse; and the third shows the noise after applying less frightening repetitive 100-µA, 50-µs pulses with a 50% duty cycle to the cathode for 2 min. Very recent ongoing work — at the time of this writing — has shown that, with this thicker oxide (1.5–2.0 µm) and high aspect ratio cones, emission in the mA range can be achieved more or less routinely with 100-µs, 10-Hz pulses in the 275-V range.

These high-current pulses decreased the flicker noise from 50–100% to 10%, clearly demonstrating that emission-stimulated desorption can be an effective means for reducing emission-current fluctuations. It may be that in many single-tip applications, periodic pulse cleaning of the tips will produce emission that is stable enough for any particular application.

4.4.9. Using F-N Analysis to Design Emitter Arrays to Meet Performance Specifications

Attempts to extract details of emitter properties such as work function, emitting area, and β factor from the well-known F-N equation [1] invariably require assumptions and estimates that qualify the results. However, it is a common practice to make use of an F-N plot to verify that experimental data are consistent with the accepted

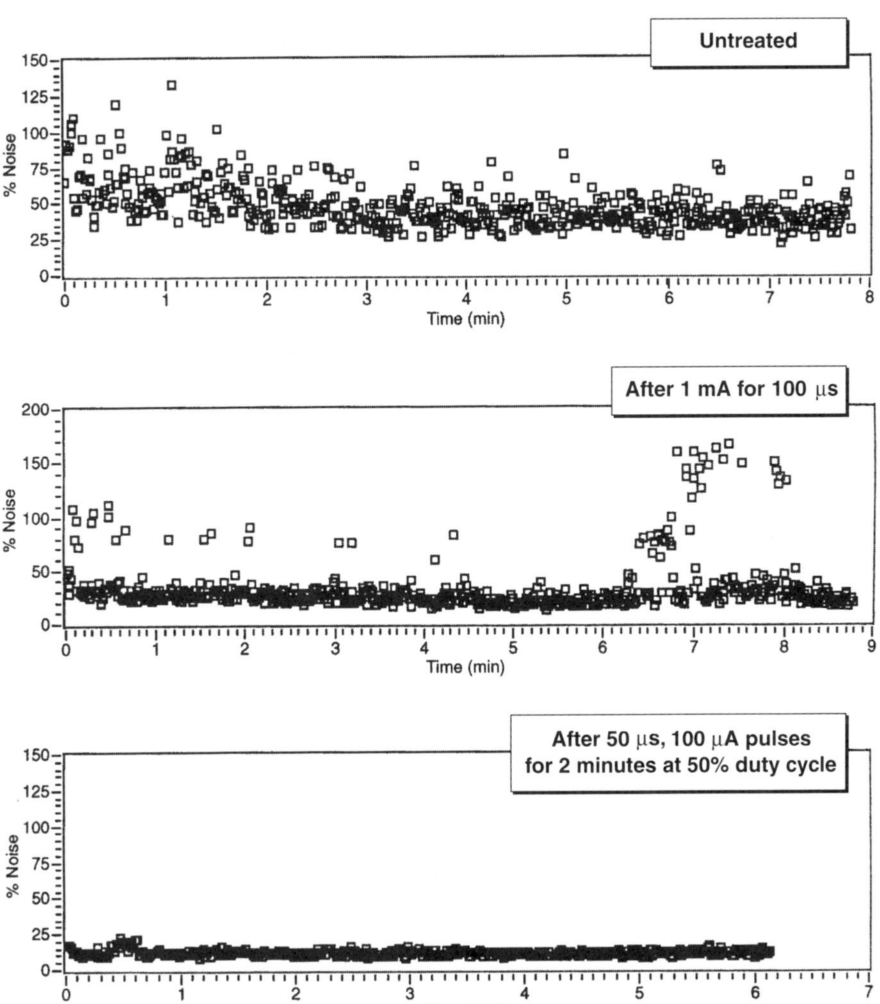

FIGURE 4.37. Noise reduction on a single Spindt tip by emission-stimulated desorption due to high-current pulses by Schwoebel and Olson. First by a single 1 mA, 100-μs pulse, and then by a series of 100 μA, 50 μs pulses (Cathode 39C-102F-15H).

field emission theory, and indeed, carefully obtained field emission data give a very good fit to the F-N plot over many orders of magnitude of emission current. It is also very convenient to extract the a and b coefficients for the F-N equation from a plot of experimental data (calculated from the y intercept and slope of the plot, respectively). These values are useful as figures of merit for characterizing cathodes, and for calculating cathode transconductance (g_m) obtained by differentiating Eq. (1):

$$g_m = \frac{\partial I}{\partial V} = ae^{-b/V}(2V + b) \tag{4}$$

Transconductance in conjunction with the capacitance of the cathode is generally used as a figure of merit (F_t) for high-frequency devices $(F_t = g_m/2\pi C)$, where C is the capacitance of the device. Thus, for example, if one is designing a cathode to operate in the 10-GHz range, the value of F_t must be of the order of 10^{10} or higher, and in general, it is desirable to have the highest possible transconductance and lowest possible capacitance. The way in which F-N analysis can be used to design emitter arrays for specific applications is best illustrated by the example of a recent program to design an emitter array to meet given specifications for a microwave amplifier tube.

The task was to design an emitter array in the form of an annulus with an outer diameter of 600 μm that could produce 160 mA of peak emission and be modulated at 10 GHz to prebunch electrons in an experimental klystrode amplifier [42,97]. Our approach was to first perform tests using available state-of-the-art, standard, low-frequency emitter arrays to obtain data on high-current operation, F-N parameters, and cathode capacitance. Using F-N analysis, these experimental data were then used to design emitter arrays to meet the program's specifications with a reasonable degree of assurance, as the cathode design parameters were supported by experimental results.

First, a standard 10 000-tip array was driven up to 180-mA peak, demonstrating that emission levels that satisfied the tube specifications were possible. Then the capacitance of an array having an oxide thickness of 1 μm was measured as 6 nF/cm^2 for emitter arrays having a 1-μm pitch (10^8 tips/cm^2). Finally, as a means of obtaining F-N data for designing an array to satisfy the high-frequency specifications of the Klystrode tube, emission tests were performed with a standard 100-tip Spindt cathode having 0.4-μm diameter gate apertures formed using electron-beam lithography. These are the smallest gate apertures that could be planned for, given the tools available to the program at that time. The importance of using small dimensions is that the β factor and the emitter-tip packing density are both increased by reducing the size of the emitter structures. These are both important parameters with respect to maximizing the transconductance and minimizing capacitance. The 100-tip cathode was driven up to a peak emission of 5 mA (50 μA/tip) with a 60-Hz, half-wave rectified drive voltage and current/voltage was recorded for F-N analysis. Figure 4.38 shows the data.

The a and b coefficients ($a = 1.35 \times 10^{-5}$ A/V^2, $b = 530$ V) on a per tip basis were extracted from the data and used to calculate the transconductance per tip as a function of emission current. Figure 4.39 shows the result. These data were then available to help determine a cathode array configuration that would satisfy the tube designer's specifications for peak emission current, current density, total transconductance, and capacitance. Additional concerns involved acceptable emitter-tip loading, and lithography issues associated with cathode fabrication.

Experience has shown that small emitter arrays can be driven up to an average emitter-tip current loading of 100 μA/tip [34]; however, an average emission of 1–10 μA/tip is regarded as a more comfortable design goal. The constraints associated with the fabrication issues center around the smallest diameter gate aperture that can be formed with acceptable uniformity, and the minimum thickness of the insulating oxide layer that can be used without defect density becoming an issue. The gate diameter is an important parameter impacting the β factor of the structure and

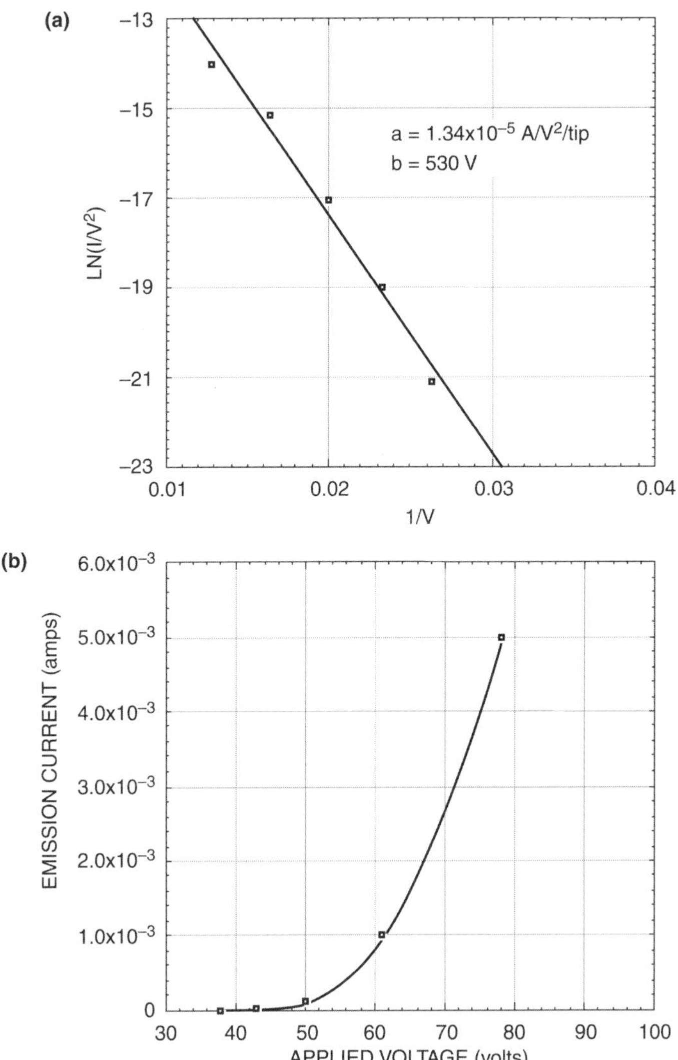

FIGURE 4.38. Voltage–current and F-N data for a 100-tip Spindt cathode (Cathode 56C-S2-10L).

the emitter-tip packing density that can be achieved. The thickness of the oxide layer is important with regard to the cathode capacitance, and also has an impact on the cone-fabrication process with regard to the aspect ration and tip sharpness of the cones as discussed in Section 4.3 on fabrication. The tip radius is one of the most important parameters determining the emitter's β factor. Thus for high-transconductance emitters, we always strive to produce the sharpest possible tip radius by forming the smallest possible gate apertures. Using the multiple-deposition process for high

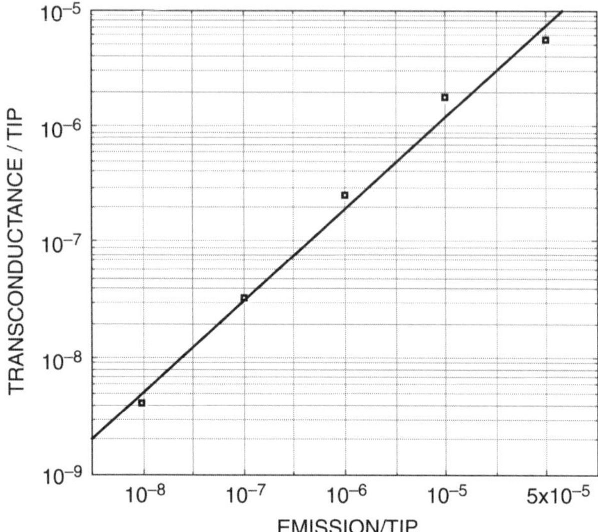

FIGURE 4.39. Transconductance per tip as a function of emission per tip for a 100-tip Spindt cathode (Cathode 56C-S2-10L).

aspect ratio cones also reduces the cone's half-angle, which, in turn, reduces the k factor in the relationship $\beta = R/kr(R - r)$ as discussed in Section 4.3.1.

The lithography tools available to the program at that time limited the gate diameter to a minimum of about 0.4 μm, and the tip pitch to 1 μm (10^8 tips/cm^2). Given a 0.4-μm gate diameter, and the double-deposition cone fabrication process for high aspect ratio emitter cones (Section 4.3.3.1), the maximum oxide thickness that could be used comfortably was 1 μm. With these issues in mind, and the given set of specifications for an annular array with a 600-μm outer diameter, 160-mA peak emission, 10-GHz operation and a maximum emitter-tip packing density of 10^8 tips/cm^2, a spreadsheet was constructed.

The spreadsheet (Table 4.1) shows the available emitting area as a function of a set of inner annulus diameters, the current density at a peak emission of 160 mA, the number of emitter tips, the emitter-tip loading, the transconductance per tip at that loading, the total transconductance, the cathode capacitance for the given area, and finally, the ratio g_m/C.

Recalling that $F_t = g_m/2\pi C$ should be of the order of 10^{10} or higher for operation at 10 GHz, we see from the table that the inner diameter of the annulus must be at least 560 μm. A 580-μm inner diameter gives a 10-μm wide annulus with the outer diameter set at 600 μm, and a total of 18 500 emitter tips. This should be adequate, as we recall that our standard test emitter array produced 180-mA peak emission with only 10 000 emitter tips (18 μA/tip). With the 10-μm wide, 600-μm diameter annulus, we see that we have an average peak emitter-tip loading of 8.65 μA/tip, and an emission-current density of 865 A/cm^2. Both are also well within previously experimentally

TABLE 4.1. Calculated Characteristics for a 600-μm Outer Diameter Annular Cathode with Various Inner Diameters, 0.4-μm Diameter Gate Apertures on a 1-μm Pitch, and Operating at 160-mA Peak Emission

ID (microns)	AREA (cm^2)	J (A/cm^2)	TIPS	I/TIP (A/tip)	Gm/TIP (Siemens/tip)	Gm (Siemens)	C (farads)	Gm/C (Siemens/farad)
0.00E + 00	2.83E − 03	5.66E + 01	2.83E + 05	5.66E − 07	1.41E − 07	4.00E − 02	1.70E − 11	2.36E + 09
1.00E − 02	2.75E − 03	5.82E + 01	2.75E + 05	5.82E − 07	1.45E − 07	3.98E − 02	1.65E − 11	2.41E + 09
2.00E − 02	2.51E − 03	6.37E + 01	2.51E + 05	6.37E − 07	1.56E − 07	3.93E − 02	1.51E − 11	2.60E + 09
3.00E − 02	2.12E − 03	7.55E + 01	2.12E + 05	7.55E − 07	1.81E − 07	3.83E − 02	1.27E − 11	3.01E + 09
4.00E − 02	1.57E − 03	1.02E + 02	1.57E + 05	1.02E − 06	2.33E − 07	3.66E − 02	9.42E − 12	3.88E + 09
5.00E − 02	8.64E − 04	1.85E + 02	8.64E + 04	1.85E − 06	3.88E − 07	3.35E − 02	5.18E − 12	6.46E + 09
5.60E − 02	3.64E − 04	4.40E + 02	3.64E + 04	4.40E − 06	8.08E − 07	2.94E − 02	2.18E − 12	1.35E + 10
5.80E − 02	1.85E − 04	8.65E + 02	1.85E + 04	8.65E − 06	1.44E − 06	2.66E − 02	1.11E − 12	2.40E + 10
5.92E − 02	7.45E − 05	2.15E + 03	7.45E + 03	2.15E − 05	3.12E − 06	2.32E − 02	4.47E − 13	5.19E + 10

FIGURE 4.40. A scanning electron micrograph of a portion of a 10-µm wide annular array having an outer diameter of 600 µm. Gate diameters $\approx$0.5 µm. (Cathode 104L-E18F-16M) [29].

demonstrated values of up to 100 µA/tip [34] and 2000 A/cm^2 averaged over the array [25,33], as well as consistent with our design goal of 1–10 µA/tip.

Figure 4.40 is a scanning electron micrograph showing a portion of an annular emitter array fabricated to these specifications with gate-aperture diameters of 0.4 µm, tip pitch of 1 µm, an oxide thickness of 1 µm, and molybdenum emitter cones formed by the double-deposition process. Low-frequency emission tests were performed to characterize the emission properties of the annular cathode prior to tests in the klystrode tube. The average current per tip was calculated and the experimental data were plotted on a F-N graph in Fig. 4.41, along with the data from the test of the 10 000-tip standard array used to obtain emission data up to 180 mA. Remarkably, the data points fall on essentially the same line for the average current-per-tip data. This is most certainly better agreement than can reasonably be expected; however, given that the *exact* match is a fluke, the results validate the method of designing emitter arrays as described earlier, and lend credibility to the understanding of the technology.

4.4.10. Performance Summary

Although at the time of this writing it has been over 30 years since the first report of emission from microfabricated FEA, the overall level of effort on the development of the basic emitter technology has been very low. By far the major portion of the work has been in the last 10 years, and that has been almost entirely directed toward

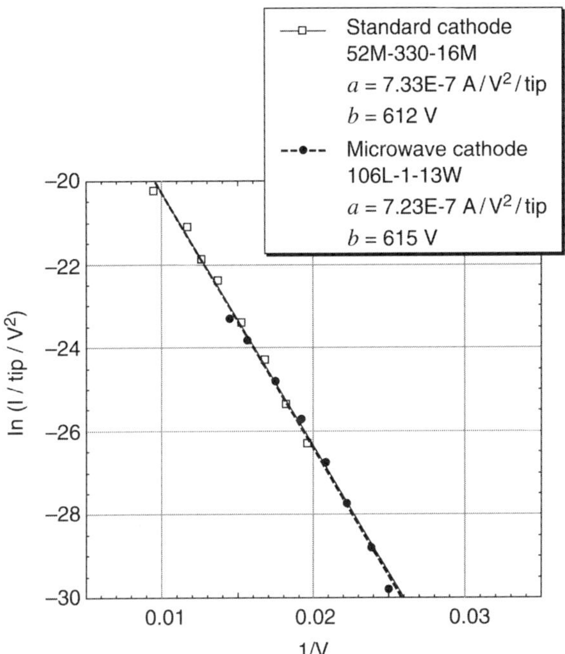

FIGURE 4.41. An F-N plot of a standard emitter array (52M-330-16M) and a low-capacitance microwave cathode array (106L-1-13W) [29].

field emission display development, which is covered in Chapter 7. Suffice is to say that, while the FED technology is very difficult overall, it places very limited demands on the emitter tips with regard to the emission current. Indeed, emission currents of nA/tip are adequate for most display designs, and the primary concern regarding the emitter tips is longevity in the display environment — which is really a vacuum-tube packaging and surface processing issue rather than a basic cathode issue. As a result, relatively little effort has gone into researching high-level emission performance with Spindt-type cathodes.

In any case, while there has been significant progress in the development of emitter arrays, there is much that can still be done by way of improving performance. Further reductions in dimensions are surely possible, although Candescent may be approaching the limits of miniaturization of emitter arrays with 0.15-μm diameter gate apertures formed using particle-tracking methods, which have produced a 10–20 V operating range for their display [23].

To date, molybdenum and silicon have been by far the most commonly used emitter materials in microfabricated field emitter arrays. However, there has been a growing move to investigate alternate materials for lower effective work functions for lower voltage operation, improved resistance to less than optimum environments, higher aspect-ratio cone fabrication, cathode architectures and processing methods for improved reliability, etc. These are all fertile fields that, up to this time, have had relatively little effort expended in their development, but hold a great deal of promise for further improvements in performance.

Takemure et al. [98] demonstrated a novel vertical current limiter (VECTL) configuration that appears to greatly reduce the incidence of arc damage. Several workers have reported an emitter with an integrated MOSFET having excellent controllability and emission stability [99]. Mackie et al. [44] demonstrated that ZrC emitters are capable of very large emission currents, and are also relatively resistant to poor environments. A host of workers have published data on emission from diamond-like carbon (DLC) in various forms, but mostly from planar surfaces (see Chapter 6). However, some recent results have shown both reduction (of about 30%) in required drive voltage and significantly improved emission stability with conformal, diamond-like coatings on standard molybdenum Spindt cathodes [57]. Pflug et al. [26] and Choi et al. [27] reported emitter-cone packing densities of 2.5×10^9 tips/cm^2 using laser interferometric lithography. These studies of materials, along with advances in lithography technology, will surely lead to significant improvements in emitter array performance. Carbon and other exotic materials are covered extensively in Chapter 6.

As of this writing some of the experimentally demonstrated performance characteristics of Spindt-type cathodes with molybdenum cones are as follows:

1. Over 2000 A/cm^2 averaged over the area occupied by an emitter array [25,33];
2. Over 100 μA per microfabricated emitter tip [34], and 1-mA pulsed emission (Fig. 4.37);
3. Eight years of continuous operation with a 60-Hz, half-wave rectified drive voltage at an average tip loading of over 20 μA/tip [34];
4. Emitter array packing densities of 10^8 tips/cm^2 are routine [29], and 10^9 tips/cm^2 has been demonstrated [100];
5. Transconductances of up to 5 μS/tip have been demonstrated [34];
6. Integrated emission buffering resistors have been developed to significantly improve reliability and uniformity of emission over large areas for applications such as spacecraft charge management and field emission display [59];
7. Processes such as in situ hydrogen plasma cleaning and overcoating with selected materials have been shown to reduce the operation voltage by up to 50% [77,84,100];
8. Field emission monochrome display lifetimes of over 25 000 h have been reported [59].

Finally, it has been clearly demonstrated that emitter performance relies on the nature of the operating environment and the interaction of the emitter surface with that environment. For example, the evidence to date indicates that a high partial pressure of oxygen is detrimental, while a hydrogen-rich environment can be beneficial. It is also likely that other materials such as iridium or some of the carbides that are more resistant to oxidation than molybdenum would be better suited for applications where it is difficult to tailor the environment to suit the emitter's needs. Lastly, it is clear that environments containing organics are best avoided in any case.

4.5. APPLICATIONS AND ONGOING DEVELOPMENTS

Flat panel display and microwave tube applications have been the focus of most emitter array development, and these are covered in Chapters 7 and 8. However, there are several other applications that are of interest, and for which the fundamental features of microfabricated FEAs, such as high efficiency, cold emission, small and light-weight structures, high current density, and the potential for high total currents, are attractive. In addition, the technology lends itself to integrating emitters with control circuitry for switching and closed-loop control of emission for stability, and results have been reported by many workers (e.g., Yokoo et al. [101], and Itoh [102]).

The cold emission characteristics have been a critical feature in at least three applications aside from display. Buchman et al. [86] reported using an emitter array for charge control in a Gravity Probe-B Gyroscope Experiment where emission must be obtained at temperatures as low as 2 K. In their laboratory experiments they have demonstrated adequate emission for their purposes at 4 K with a 10 000-tip Spindt cathode on a silicon substrate [87].

In another example, Felter [36] reported obtaining significant differences in the spectrum obtained with a quadrupole mass spectrometer when electron-impact ionization with a hot-filament cathode in the ionizer section was alternated in situ with electron-impact ionization using electrons from a cold Spindt cathode. Effects associated with the hot-filament cathode, such as pyrolysis of sample gas, outgassing of the filament and nearby surfaces, light, magnetic fields, etc., are all eliminated with the cold cathode. Felter found that the residual gas spectrum obtained with the cold electron source had significantly less carbon monoxide and carbon dioxide than it did with the hot filament. In addition, with the filament off, the O_2 and C_2H_2 (acetylene) peaks are larger by factors of 2 and 3, respectively. He theorizes that the hot filament catalyzes the reaction of acetylene and oxygen to form carbon monoxide, and hydrogen released by the process forms methane. In addition, it was suggested that the efficiency, small size, and cold operation of the cathode are all features that lend themselves to miniaturization of the instrument for portable operation. Finally, Baptist et al. [37] and Baker and Outlaw [38] reported replacing the hot filament in a Bayard–Alpert-type pressure gauge with a Spindt-type cold emitter. As with the ionizer for mass spectrometry, the cold emitter has the advantage of avoiding outgassing issues associated with the heat of a conventional hot-filament cathode. These applications for cold emission have clear advantages; however, it must be noted that field emission is a surface work-function related effect, and the performance of an emitter array is sensitive to its operating environment. For example, in high partial pressures of oxygen, oxygen ions are formed and have been shown to chemically react with the molybdenum emitter tips and increase the effective work function of the emitter surface [81]. This effect may be manageable through the use of less reactive emitter surfaces such as iridium, DLC, or a carbide.

In addition, hydrocarbons of the kind found in many ordinary vacuum environments can build up on the emitter tips and lead to failure. Under these conditions, it would be necessary to periodically clean the surface by some means such as heating or hydrogen-plasma cleaning. The practicality of such an approach would of course depend on the application and on how frequently cleaning would be required.

4.5.1. Integrated Focus Electrodes

For many applications, a parallel or focused electron beam is required. By the nature of the microfabrication technology used to fabricate emitter arrays, it is a relatively simply matter to integrate focus electrodes with the microfabricated field emitter. There are three ways in which this can be done: One is to build a third electrode over the gate separated by an insulating film described by Spindt et al. [34]. A second method is to pattern a focus electrode that is co-planar with the gate electrode as outlined by Itoh et al. [103] and Tang et al. [104]. The third is to pattern the gate film so that it is in the form of a narrow ring around the gate aperture with only a few (e.g., four) fine connecting lines extending radially from the ring for electrical contact to adjacent emitter-tip gate rings in an array of emitters or to a contact pad [105]. In this configuration the fringing electric field due to the base layer protrudes through open areas in the gate electrode into the space above the gate, and turns electrons that are emitted off-axis to the emitter cone back toward the axis of the cone, or perpendicular to the surface of the cathode.

All three focusing configurations work. The first has the disadvantage of being the most difficult to fabricate and increasing the capacitance of the gate electrode, but has the advantage of providing the most effective focusing. The other two configurations invariably have some asymmetry in the focusing field that they form because of the need for contact leads for the gate film. Figure 4.42 shows an example of a single

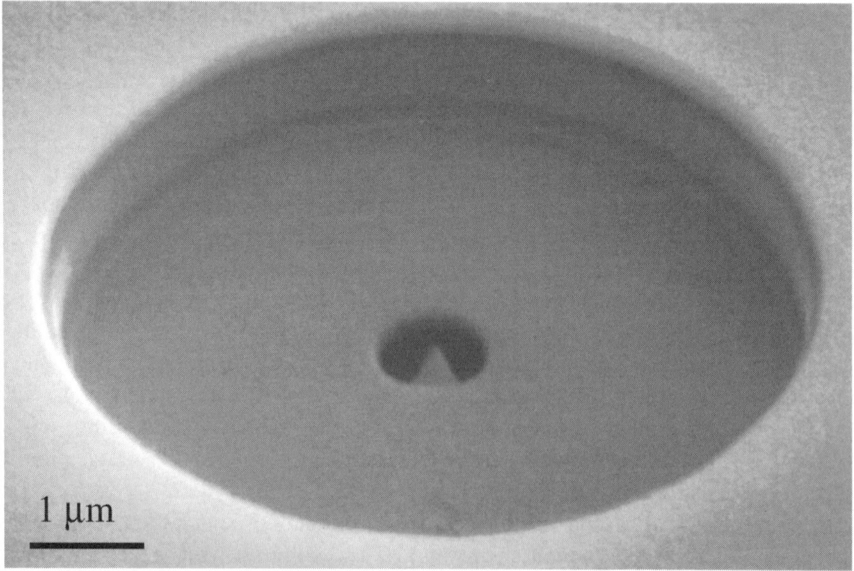

FIGURE 4.42. A scanning electron micrograph of a Spindt cathode with an integrated focus electrode.

emitter tip with an integrated focus electrode fabricated at SRI, and presently under study as a source for focused electron beams.

4.6. SUMMARY

There has been a significant amount of progress in the development and characterization of microfabricated FEA to date; however, this is still very much an emerging technology. We have attempted not only to illustrate the development and characterization of Spindt-type emitter arrays, but also to give the reader a sense of the very significant opportunities that exist for ongoing development with the basic technology. The FEA has been shown to be a very efficient, high current density, temperature-independent, electron emitter with a fabrication technology that lends itself to cost-reducing, large-scale, parallel-processing methods of the kind developed for the integrated circuit industry. Research for new materials such as DLC and carbides has indicated that improved, durable emitters may well be possible by using these materials, and emission-controlling techniques such as integrated resistors, MOSFETs, and VECTLs have been demonstrated to effectively improve reliability and performance. These ongoing developments, along with emerging applications in microwave, spacecraft, electronics for harsh environments with respect to temperature and radiation, and electron sources for established vacuum-based systems such as mass spectrometers, cathode-ray tubes, and scanning electron microscopes, create prospects for an exciting future for VME technology.

ACKNOWLEDGMENTS

We are pleased to have this opportunity to acknowledge our friend and colleague, Ken Shoulders, the universally acclaimed "Father of VME," along with our many colleagues here at SRI International who, over the past 30 years, have had an impact on the development and characterization of the Spindt cathode (in alphabetical order): Leonor Almada, Bill Chu, Betsy Harris, Cliff Hartelius, Earl Heydon, Lou Heynick, Hazel Pakka, Purobi Phillips, J. D. Reed, George Reich, Ken Rogers, Charlie Rosen, Arne Rosengreen, Shari Shepherd, Marv Simkins, Bob Stowell, Joe Suarez, Jan Terry, David Thibert, and Gene Westerberg.

We are also pleased to make special mention of the very fruitful FEEM studies performed in close collaboration with Prof. John Panitz, Dr. Gary Condon and Dr. Russ Olson of the University of New Mexico. This work has greatly advanced our knowledge and understanding of the intricacies of the emission process, as well as means to improve emission behavior.

In addition, we gratefully acknowledge support for much of the work reported here by the NASA Lewis Research Center (now the NASA John H. Glenn Research Center at Lewis Field), the DARPA/NRL VME Initiative, the DARPA High-Definition Systems Program, and Dr. Bill Parker through Ion Diagnostics, Inc. Dr. Jim Dayton of NASA (now with Hughes Aircraft) has been particularly supportive over the years.

Lastly, we are most grateful for the long-term support of this work in the early years by the late Dr. Ralph Forman of the NASA Lewis Research Center in Cleveland, Ohio. He will be always remembered as a good friend and respected colleague.

REFERENCES

1. R. H. Fowler and L. W. Nordheim, *Proc. R. Soc. Lond. A* **119**, 173 (1928).

2. D. A. Buck and K. R. Shoulders, An approach to microminiature systems, in *Proceedings of the Eastern Joint Computer Conference*, Amer. Inst. of Elect. Engrs.: New York, 1958, p. 55.

3. K. R. Shoulders, Microelectronics using electron beam activated machining techniques, in *Advances in Computers, Vol. 2* (F. L. Alt, Ed.), p. 135, 1961.

4. M. E. Crost, K. Shoulders, and M. E. Zinn, Thin electron tube with electron emitters at the intersection of crossed conductors, US Patent 3,500,102 (1970).

5. C. A. Spindt and K. R. Shoulders, Research in micron-size Field-emission tubes, in *IEEE Conference Record, 1966 Eighth Conference on Tube Techniques*, 1966, p. 143.

6. C. A. Spindt, *J. Appl. Phys.* **39**, 3504 (1968).

7. C. A. Spindt, I. Brodie, L. Humphrey, and E. R. Westerberg, *J. Appl. Phys.* **47**, 5248 (1976).

8. R. N. Thomas, R. A. Wickstrom, D. K. Schroder, and H. C. Nathanson, *Solid State Electron.* **17**, 155 (1974).

9. D. V. Geppert, Integrated vacuum circuits, US Patent 3,701,919 (1972).

10. R. Meyer, A. Ghis, P. Rambaud, and F. Muller, Microchip florescent display, in *Proc. Japan Display*, 1985, p. 513.

11. C. E. Holland, C. A. Spindt, I. Brodie, J. Mooney, and E. R. Westerberg, Matrix addressed cathodoluminescent display, in *Int. Display Conf.*, London, UK, 1987.

12. I. Brodie and P. R. Schwoebel, *Proc. IEEE* **82**, 1006 (1994).

13. R. Gomer, *Field Emission and Field Ionization*, Harvard University Press: Cambridge, MA, p. 54, 1961.

14. R. B. Markus, K. K. Chin, Y. Lyuan, H. Wang, and W. N. Carr, *IEEE Trans. Electron Devices* **37**, 1545 (1990).

15. E. G. Zaidman, K. L. Jensen, and M. A. Kodis, *J. Vac. Sci. Technol. B* **14**, 1994 (1996).

16. D. Temple, C. A. Ball, W. D. Palmer, L. N. Yadon, D. Vellenga, J. Mancusi, G. E. McGuire, and H. F. Gray, *J. Vac. Sci. Technol. B* **13**, 150 (1995).

17. K. L. Jensen, E. G. Zaidman, M. A. Kodis, B. Goplen, and D. N. Smythe, *J. Vac. Sci. Technol. B* **10**, 1942 (1996).

18. J. H. Kang, J. W. Cho, J. W. Kim, and J. M. Jim, *J. Vac. Sci. Technol. B* **14**, 124 (1996).

19. W. P. Dyke and W. W. Dolan, Field emission, in *Advances in Electronics and Electron Physics, Vol. 8*, (L. Martin, Ed.), 1956, p. 89.

20. C. A. Neugebauer, in *Handbook of Thin Film Technology* (L. I. Massel and R. Glang, Eds.), McGraw-Hill: NY, 1970.

21. D. S. Campbell, in *Handbook of Thin Film Technology* (L. I. Maissel and R. Glang, Eds.), McGraw-Hill: NY, 1970.

22. I. Brodie, E. R. Westerberg, D. R. Cone, J. J. Muray, N. Williams, and L. Gasiorek, *IEEE Trans. Electron Devices* **ED28**, 1422 (1981).

23. T. S. Fahlen, Performance advantages and fabrication of small gate openings in Candescent's thin CRT, in *Proc. 12th Int'l Vacuum Microelectronics Conf.*, Darmstadt, Germany, 1999, p. 56.

24. X. Chen, S. H. Zaidi, D. J. Devine, and S. R. J. Brueck, *J. Vac. Sci. Technol. B* **14**, 3339 (1996).

25. L. Parameswaran, R. A. Murphy, C. T. Harris, R. H. Mathews, C. A. Graves, M. A. Hollis, J. Shaw, M. A. Kodis, M. Garven, M. T. Ngo, K. L. Jensen, M. Green, J. Legarra, L. Garbini, and S. Bandy, Field emitter arrays for inductive output amplifiers, in *Vacuum Electronics Annual Review Abstracts*, San Diego, CA, May 2, 1997, p. IV-7.

26. D. G. Pflug, M. Schattenburg, A. I. Akinwande, and H. I. Smith, 100nm Gate aperture field emitter arrays, in *Proc. 11th Int'l Vacuum Microelectronics Conf.*, Asheville, NC, 1998, p. 130.

27. J. O. Choi, A. I. Akinwande, and H. I. Smith, 100nm gate hole openings for low voltage driving field emission display applications, in *Proc. 13th Int'l Vacuum Microelectronics Conf.*, Guangzhou, P. R. China, 2000, p. 61.

28. C. H. Oh, *J. Vac. Sci. Technol. B* **16**, 807 (1998).

29. C. A. Spindt, C. E. Holland, P. R. Schwoebel, and I. Brodie, *J. Vac. Sci. Technol. B* **16**, 758 (1998).

30. C. A. Spindt, Method of fabricating a funnel-shaped miniature electrode for use as a field-ionization source, US Patent 4,141,405 (1979).

31. G. N. A. van Veen, K. Theunissen, K. van de Heuvel, R. Horne, and A. L. J. Burgmans, *J. Vac. Sci. Technol. B* **13**, 478 (1995).

32. C. Herring, *Structures and Properties of Solid Surfaces*, University of Chicago Press: Chicago, p. 5, 1953.

33. C. A. Spindt, C. E. Holland, P. R. Schwoebel, and I. Brodie, *J. Vac. Sci. Technol. B* **14**, 1986 (1996).

34. C. A. Spindt, C. E. Holland, A. Rosengreen, and I. Brodie, *IEEE Trans. Electron Devices*, **38**, 2355 (1991).

35. C. C. Curtis and K. C. Hsieh, *Rev. Sci. Instr.* **57**, 989 (1986).

36. T. E. Felter, *J. Vac. Sci. Technol. B* **17**, 1993 (1999).

37. R. Baptist, C. Beith, and J. Py, *J. Vac. Sci. Technol. B* **14**, 2119 (1996).

38. D. H. Baker and R. A. Outlaw, in *Proceedings of the 44th National Symposium of the American Vacuum Society*, Philadelphia, PA, 1996.

39. K. L. Jensen, R. H. Abrams, and R. K. Parker, *J. Vac. Sci. Technol. B* **16**, 749 (1998).

40. H. Makishima, H. Imura, M. Takahashi, H. Fukui, and A. Okamoto, Remarkable improvements of microwave electron tubes through the development of the cathode materials, in *Proc. 10th Int. Vacuum Microelectronics Conf.*, Kyongju, Korea, 1997, p. 194.

41. S. G. Bandy, M. C. Green, C. A. Spindt, M. A. Hollis, W. D. Palmer, B. Goplen, and E. G. Wintucky, Application of gated field emitter arrays in microwave amplifier tubes, in *Proc. 11th Int. Vacuum Microelectronics Conf.*, Asheville, NC, 1998.

42. D. R. Whaley, B. Gannon, C. R. Smith, C. M. Armstrong, and C. A. Spindt, *IEEE Trans. Plasma Sci.*, **28**, 727 (2000).

43. R. Adamo and V. Aguero, Space applications of Spindt cathode field-emission devices, in *The USAF/NASA Spacecraft Charging Technology Conference*, Hanscom AFB, Lexington, MA, Nov. 1998.

44. W. A. Mackie, X. Tianbao, and P. R. Davis, Transition metal carbide emitters for FEA devices and high current applications, in *Proc. 11th Int. Vacuum Microelectronics Conf.*, Asheville, NC, 1998, p. 84.

45. W. A. Mackie, J. E. Blackwood, S. C. Williams, and P. R. Davis, *J. Vac. Sci. Technol. B* **16**, 1215 (1998).

46. J. H. Orloff and L.W. Swanson, *J. Vac. Sci. Technol.* **12**, 1209 (1976).

47. O. A. Weinreich and H. B. Bleecher, *Rev. Sci. Inst.* **28**, 56 (1952).

48. C. A. Spindt, Method of manufacturing an electric field producing structure including a field emission cathode, US Patent 5,064,396 (1991).

49. H. H. Busta, Lateral cold cathode triode structures fabricated on insulating substrates, in *Vacuum Microelectronics 89 (Inst. Phys. Conf. Ser. 99)* (R. Turner, Ed.), 1989.

50. C. Nureki and M. Araragi, Planar field emission devices with three-dimensional gate structures, in *Proc. 4th Int. Vacuum Microelectronics Conf.*, Nagahama, Japan, 1991, p. 78.

51. J. G. Fleming, D. A. A. Ohlberg, T. Felter, and M. Malinowski, *J. Vac. Sci. Technol.*, **14**, 1958 (1996).

52. D. S. Y. Hsu and H. F. Gray, A low-voltage, low-capacitance, vertical multi-layer thin-film-edge dispenser field emitter arrays electron source, in *Proc. 11th Int'l Vacuum Microelectronics Conf.*, Asheville, NC, 1998, p. 82.

53. C. A. Spindt, *Surf. Sci.* **266**, 145 (1992).

54. H. D. Beckey, *Principles of Field Ionization and Field Desorption Mass Spectrometry*, Pergamon Press: Oxford, 1977.

55. W. Aberth, R. Marcuson, R. Barth, U. Emond, and W. B. Donham, *Biomed. Mass. Spec.* **10**, 89 (1983).

56. J. Mitterauer, *Surf. Sci.* **246**, 107 (1991).

57. J. H. Jung, N. Y. Lee, J. Jang, M. H. Oh, and S. Ahn, Electron emission performance of Mo-tip field emitter arrays's with nitrogen-doped hydrogen-free DLC coating, in *Proc. 12th Int. Vacuum Microelectronics Conf.*, Darmstadt, Germany, 1999, p. 110.

58. I. Brodie and C. A. Spindt, *Adv. Electron. Electron Phys.*, vol. **83**, p. 1 (1992).

59. R. Meyer, in *Proceedings of the International Display Research Conference*, Strasbourg, France, 1993, p. 189.

60. W. B. Herrmannsfeldt, R. Becker, I. Brodie, A. Rosengreen, and C. A. Spindt, *Nucl. Instrum. Methods Phys. Res. Section A* **298**, 39 (1990).

61. W. J. Orvis, C. F. McConaghy, D. R. Ciarlo, J. H. Yee, and E. W. Hee, *IEEE Trans. Electron Devices* **38**, 2383 (1989).

62. E. G. Zaidman, *IEEE Trans. Electron Devices* **40**, 1009 (1993).

63. X. Zhu and E. Munro, *J. Vac. Sci. Technol. B* **7**, p. 1862 (1989).

64. T. Asano, Simulation of geometrical change effects on electrical characteristic of micron-size vacuum triode with field emitter, in *Proc. 3rd Int'l Vacuum Microelectronics Conf.*, Monterey, CA, 1990, p. 27.

65. R. True, Simulation of thin film field emitter arrays cathodes, in *Technical Digest of IEEE International Electron Devices Meeting*, p. 379, 1992.

66. T. P. H. Chang, D. P. Kern, and M. A. McCord, *J. Vac. Sci. Technol. B* **7**, 754 (1994).

67. W. D. Kessling and C. E. Hunt, *IEEE Trans. Electron Devices* **42**, 340 (1995).

68. P. R. Schwoebel and I. Brodie, *J. Vac. Sci. Technol. B* **13**, 1391 (1995).

69. C. J. Chen, *Introduction to Scanning Tunneling Microscopy*, Oxford University Press: Oxford, 1993.

70. E. W. Müller and T. T. Tsong, *Field Ion Microscopy, Principles and Applications*, American Elsevier Publishing Company: New York, 1969.

71. M. K. Miller and G. W. D. Smith, *Atom Probe Microanalysis, Principles and Applications to Materials Problems*, Materials Research Society: Pittsburgh, 1989.

72. G. Fursey, The forming of submicrogeometry on the solid and liquid surface I the strong electric fields, in *Proc. 3rd Int'l Conf. On Vacuum Microelectronics,* Monterey, CA, 1990.

73. P. C. Bettler and F. M. Charbonnier, *Phys. Rev.* **119**, 85 (1960).

74. A. V. Crewe, D. N. Eggenberger, J. Wall, and L. M. Welter, *Rev. Sci. Inst.* **39**, 576 (1968).

75. M. Benjamin and R. O. Jenkins, *Proc. R. Soc. Lond. A* **176**, 262 (1940).

76. R. Gomer, *Surf. Sci.*, **299/300**, 129 (1994).

77. P. R. Schwoebel, C. A. Spindt, and I. Brodie, *J. Vac. Sci. Technol. B* **13**, 338 (1995).

78. J. M. Macaulay, I. Brodie, C. A. Spindt, and C. E. Holland, *Appl. Phys. Lett.*, **61**, 997 (1992).

79. W. H. Kohl, *Handbook of Materials and Techniques for Vacuum Devices*, AIP: New York, p. 103, 1960.

80. J. Itoh, T. Niiyama, and M. Yokoyama, *J. Vac. Sci. Technol. B* **11**, 647 (1993).

81. D. Temple, *Mater. Sci. Eng.* **R24**, 185 (1999).

82. Y. Wei, B. R. Chalamala, B. G. Smith, and C. W. Penn, *J. Vac. Sci. Technol. B* **17**, 233 (1999).

83. P. R. Schwoebel and C. A. Spindt, *Appl. Phy. Lett.*, **63**, 33 (1993).

84. P. R. Schwoebel and C. A. Spindt, *J. Vac. Sci. Technol. B* **12**, 2414 (1994).

85. A. Ghis, R. Meyer, P. Rampaud, F. Levy, and T. Leroux, Sealed vacuum devices: microtips fluorescent display, in *Proc. 3rd Int'l Vacuum Microelectronics Conf.*, Monterey, CA, 1990.

86. S. Buchman, T. Quinn, G. M. Keiser, and D. Gill, *J. Vac. Sci. Technol. B* **11**, 407 (1993).

87. D. Monstavicius, Stanford University Gravity Probe-B Program, Private communication, 1999.

88. R. T. Olson, G. R. Condon, J. A. Panitz, and P. R. Schwoebel, *J. Appl. Phys.*, **84**, 2031 (2000).

89. R. A. Milikan and C. C. Lauritsen, *Proc. Natl. Acad. Sci. U.S.A.* **14**, 45 (1928).

90. D. J. Rose, *J. Appl. Phys.*, **27**, 215 (1956).

91. C. B. Duke and M. E. Alferieff, *J. Chem. Phys.*, **46**, 923 (1967).

92. I. Giaever, *Surf. Sci.*, **29**, 1 (1972).

93. L. W. Swanson, *Surf. Sci.* **70**, 165 (1978).

94. I. Brodie, *Surf. Sci.*, **70**, 186 (1978).

95. J. R. Chen and R. Gomer, *Surf. Sci.*, **79**, 413 (1979).

96. J. A. Panitz, *Prog. Surf. Sci.*, **8**, 219 (1978).

97. NASA Contract NAS3-27362, 1994.

98. H. Y. Takemura, T. N. Furutake, F. Matsuno, M. Yoshiki, N. Takada, A. Okamoto, and S. Miyano, A novel vertical current limiter fabricated with a deep trench forming technology for highly reliable field emitter arrays, in *Technical Digest of the IEEE International Electron Devices Meeting*, 1997, p. 709.

99. S. Kanemaru, K. Ozawa, K. Ehara, T. Hirano, and J. Itoh, MOSFET-structured Si field emitter tip, in *Proc. 10th Vacuum Microelectronics Conf.*, Kyongju, Korea, 1997.

100. C. O. Bozler, C. T. Harris, S. Rabe, D. D. Rathman, M. A. Hollis, and H. I. Smith, *J. Vac. Sci. Technol. B* **12**, 629 (1994).

101. K. Yokoo, M. Arai, M. Mori, J. Bae, and J. Ono, *J. Vac. Sci. Technol. B* **13**, 491 (1995).

102. J. Itoh, *Appl. Surf. Sci.*, **111**, 194 (1997).

103. J. Itoh, Y. Tohma, K. Morikawa, S. Kanemaru, and K. Shimizu, *J. Vac. Sci. Technol. B*, **13**, 1968 (1995).

104. C. M. Tang, T. A. Swyden, K. A. Thomason, L. N. Yadon, D. Temple, C. A. Ball, W. D. Palmer, J. Mancusie, D. Vellenga, and G. E. McGuire, *J. Vac. Sci. Technol. B* **14**, 3455 (1996).

105. C. A. Spindt, Automatically focusing field emission electrode, US Patent 4,874,981 (1989).

CHAPTER 5

Silicon Field Emitter Arrays

JONATHAN SHAW

Naval Research Laboratory, 4555 Overlook Avenue SW, Washington, District of Columbia 20375-5347

JUNJI ITOH

National Institute of Advanced Industrial Science and Technology, Tsukuba, Japan

5.1. INTRODUCTION

Field emitter array (FEA) technology [1,2] promises to provide a new free electron source with many enticing features, including small size, easy manufacture, high speed, and long lifetime. Such a source would allow a variety of new or improved devices to be fabricated using vacuum electrons. Some devices (e.g., x-ray sources) require the high kinetic energy that only free electrons can achieve. Other devices (e.g., high voltage or microwave power tubes) utilize the high speed of free electrons and/or the high electric field of vacuum. Many devices (e.g., displays) benefit from having a large number of independently addressable sources operating in parallel. A number of prototype devices have been demonstrated.

FEA structures provide a means of increasing the electric field at emitting sites geometrically (by forming sharp points or edges inside apertures). Issues other than geometry, e.g., the density and energy of electrons at the surface relative to the vacuum level, charge trapped near the emission site, inhomogeneous or discontinuous work function, etc., can also affect the emission current obtained at a given gate potential. One objective in designing an FEA is to achieve a sufficiently large field enhancement (geometric or otherwise) to produce field emission with a low gate potential. Since there are drawbacks to excessively high field enhancement, the application requirements should be considered. The potential applied between the emitting sites and corresponding gate apertures should be low for several reasons:

(1) Excessive gate voltages can cause leakage current (and perhaps breakdown) in the surface or bulk of the dielectric isolating the gate, and also produce large mechanical stresses;

(2) The energy stored in the arrays ($CV^2/2$) and the gate modulation power increase with the average gate voltage;

Vacuum Microelectronics, Edited by Wei Zhu.
ISBN 0-471-32244-X. 2001 John Wiley & Sons, Inc.

187

(3) Both the electronics controlling the gate voltage and any current-limiting structures in series with the gate-emitter circuit must withstand higher voltages and thus may be more difficult to fabricate or require more area when large gate voltages are used;

(4) The possibility of damage caused by ions created by the electron beam is reduced in devices where the electron energy is low, and ionization might be eliminated if the voltages were kept below 20–30 V.

Assuming that little or no emitted current arrives at the gate electrode, the gate drive power is negligible in low frequency or dc applications and thus *not* affected by the gate voltage. Higher gate voltages are favored in some applications where higher initial electron velocities are needed. In such cases, it may be possible to apply the higher voltage to a second integrated gate or suspended grid [3].

For high frequency applications, it is important to achieve very high emission current and low capacitance. For example, the FEA current gain, anode current/gate current, is reduced at high frequency when the displacement current in the gate circuit becomes significant. The frequency of unity-current-gain is the ratio $g_{\mathrm{m}}/2\pi C$, where $g_{\mathrm{m}} = \mathrm{d}I/\mathrm{d}V$ is the transconductance of the array [3]. Resonant gate circuits can be built, improving the unity-current-gain frequency by factors of at least 10, which is limited by losses in the gate material. Because field emission is an exceptionally fast process [4] and electrons travel ballistically in vacuum, transconductance is the only practical limit on the modulation frequency. Since the emission current rises exponentially with gate voltage, the transconductance also increases rapidly with emission current. In addition, the transconductance can be increased by reducing the gate voltage required for a given emission current, and by reducing parasitic capacitance (i.e., the capacitance with portions of the array where high electric fields are not required). Since the emission currents demonstrated from individual emission sites far exceed the average emission current from large arrays, and since nearly all of the capacitance in a typical FEA is parasitic, large increases in modulation frequency over the current state-of-the-art are obtainable.

FEAs are readily fabricated from crystalline silicon using oxidation to create very sharp emitters. The processing steps can be carried out using standard fabrication tools. Emission has been demonstrated at gate voltages as low as 25 V. However, at this writing, silicon FEA components are not commercially available. One reason is inadequate reliability and scalability. Although silicon FEAs having only a few tips have produced currents in excess of 10 μA/tip [5], the average current per site typically declines with the total number of emission sites in a single array. That is, the total current does not scale with the number of tips in large arrays, although progress is being made as fabrication techniques improve [6]. Non-uniform emission, combined with local failures at emission sites with the highest emission, is thought to be responsible for the failure to scale. Some form of local emission current control or feedback is required to achieve large total emission currents and more reliable operation.

The emitting surfaces of an FEA are generally contaminated with a layer of molecules that could be adsorbed from the ambient, desorbed from an anode, transported across the tip shank, or segregated from the bulk. The properties of this surface layer, together with the emission history, can affect the emission properties. For example, the emission current from many FEAs typically increases with time when the emitter is first turned on. After stable emission current is established by operating for some time in good vacuum, the total emission from an array degrades if the pressure of a reactive gas such as H_2O, CO, or O_2 is increased [7]. However, gas exposure (even to atmosphere) may have little or no effect on the emission properties if the exposure occurs while the array is not emitting and good vacuum is re-established before emission resumes. Exposure to some gases may even increase the emission [8]. On the other hand, clean silicon is quite reactive and can be contaminated within hours, even at pressures near 10^{-10} torr [9,10]. In some cases oxidation and other surface reactions can change the shape of the emitter as well.

FEAs commonly fail when local arcing destroys individual structures or creates emitter–gate shorts. Arcs are initiated by power dissipated in a specific emitter, and this power increases with the emitter current (among other parameters). The current necessary to initiate an arc is dependant on the electronic properties of the emitter surface, which can vary due to processing, environment, and emission history as well as the emitter material.

Apart from high frequency issues, it is important to reduce the stored energy ($CV^2/2$) available to an arc (thus limiting damage). The energy stored in FEAs not containing current-limiting structures at operating potential is typically sufficient to create a short by melting parts of the array. Arc damage can be limited by reducing either the total available energy or the rate the energy can be transferred to the arc [11,12]. Increasing the gate insulator thickness [13], or reducing its average dielectric constant can reduce the total capacitance of an array. Reducing the distance between emission sites also reduces the capacitance per emission site [14]. However, reducing some dimensions such as the gate thickness also reduces the power needed to create a short or ruin an emitter, and thus may produce excessively fragile structures.

Arc damage is much less severe in FEAs that contain current-limiting circuit elements (e.g., resistors) in series with each tip. The elements can reduce arc initiation by limiting the maximum emission current at each site. They can limit the power dissipated in an arc, and if a local short does occur, current-limiting elements can prevent overall failure. The voltage across each circuit element makes the connected emission site(s) more positive, thereby reducing the electric field and emission current. To the extent that such emission-limiting occurs, nonuniformity in *current* is traded for nonuniformity in *potential* [1]. Thus applications demanding uniform emission potential can rely on current-limiting elements only as a fail-safe mechanism. When high current as well as uniform potential is required, the current-limiting circuit element should be nonlinear (change potential only at the highest current). For applications where uniformity is paramount and beam quality is less important (such as displays), the arrays can be run with all the emitters current-limited; however, the gate voltage and modulation power will be increased, and power will be dissipated in the circuit elements. If large variations in emitter performance are corrected using current-limiting elements, large potentials will be placed across some of the elements. The elements must be designed to avoid breakdown problems [15,16]. Thus the use of circuit elements does not completely remove the need for good uniformity in un-regulated emitters.

In the case of emitters made from silicon or other semiconductors, the electronic properties of the surface can also affect transport through the semiconductor. A number of semiconductor structures can provide nonlinear current-limiting. One current-limiting structure is the emitter tip itself. Because the electron velocity in semiconductors saturates at high electric field, a given electron density within a given structure (post, cone, pyramid, wall, or wedge) will allow a finite maximum current [17]. For example, current saturation may occur when the surface of an n-type emitter shank is depleted. However, surface charge can heavily influence the electron density near semiconductor surfaces. Thus using the emitter shaft or apex as the current-limiting element will require a means of controlling the surface charge density in the applicable regions.

A p–n junction or p-type Schottky barrier under each emitter can also provide a nonlinear current-limiting element. The emission current is limited to the reverse bias diode current. Electrons can also be regulated using bipolar or field effect transistors [18]. Standard planar transistors and vertical field effect transistors [19] have been realized.

To produce nearly uniform emission from an array of emitters, an ideal FEA fabrication method would produce emitters with uniform shape (including small and uniform radius of curvature) centered within gate apertures with uniform diameter. Perhaps most importantly, the electronic properties of each emitting site should also be uniform. Fabrication of emitting sites that maintain uniform electronic properties as well as high and uniform field enhancement is one of the main challenges of FEA technology. Designing a successful FEA requires careful attention to the various mechanisms that can degrade performance or lead to failure. The lowest work functions and smallest geometries only will probably not provide optimal performance.

5.2. FABRICATION OF SILICON FEAs

Crystalline silicon is a natural choice for FEA development for many reasons, including its excellent electronic, crystalline, and mechanical properties, ready availability, and well-developed fabrication techniques and equipment. Fabricating emitter tips having apex radii below 10 nm *and* with small standard deviation calls for fabrication techniques that are largely independent of errors introduced by lithography and etching.

5.2.1. Orientation-Dependant Etching

One of the initial silicon FEA fabrication methods utilized orientation-dependant etching (ODE) to form convex pyramidal structures [20,21]. The ODE method is attractive since pyramids with (111) crystal faces extending nearly to the last atom would provide a highly uniform tip geometry with well-defined surfaces.

The fabrication process begins by lithographic definition of a set of dot masks on the (100) face of a silicon wafer over the points where tips are to be placed. The mask

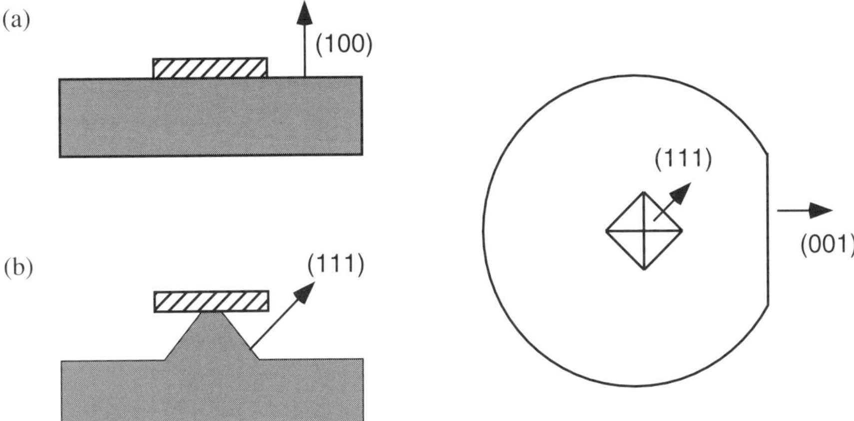

FIGURE 5.1. Use of ODE to produce pyramids with sides terminated in (111) crystal planes. (a) A mask is patterned lithographically to inhibit etching in locations where pyramids are desired. (b) Silicon is etched more quickly at the (100) planes that at the (111) planes, forming (111) bounded pyramids.

dots are typically 1 μm in diameter or less. Then the wafer is etched in a solvent such as KOH that dissolves the (100) crystal face much more quickly than the (111) face. This etching undercuts the etch mask, leaving a pyramid bounded by (111) crystal planes. The process is illustrated in Fig. 5.1. Gate apertures can be formed by coating the tips with conformal oxide and metal layers, then opening holes in the metal and oxide at the tip apex. To produce a dielectric layer with high breakdown potential, the silicon/dielectric interface is best formed by thermal oxidation of the silicon. Additional dielectric can be deposited on top of the thermal oxide. To reduce the gate-substrate capacitance, the oxide should be reasonably thick, typically 0.5–2 μm. One way to define the gate apertures is to use a planarization technique. Planarization can be used to create gates on tips defined by many methods. The traditional method of planarization using photoresist or similar spin-on material is illustrated in Fig. 5.2. Recently chemical–mechanical polishing (CMP) has become an attractive alternative to the traditional method. An example of a gate aperture defined by CMP is shown in Fig. 5.3.

In practice, the orientation-dependant etch becomes difficult to control once the emitter apex is formed. To achieve reasonable uniformity, considerable care must be exercised to keep the liquid etching solutions at the proper concentration and temperature. The pyramids do not always remain terminated in the (111) face, sometimes inverting to become a concave pit [22]. For this reason the ODE method has seldom been used successfully. However, the process can be inverted to form concave pyramidal pits. The concave apex regions are stable and very sharp. The pits can be used as molds for emitter tips [23,24]. Oxidizing the pits before depositing the emitter material can make the apex region of the molds more acute and provides a convenient release layer [25]. This process is especially attractive for forming tips from experimental materials such as polycrystalline diamond, diamond-like carbon, or metal carbides,

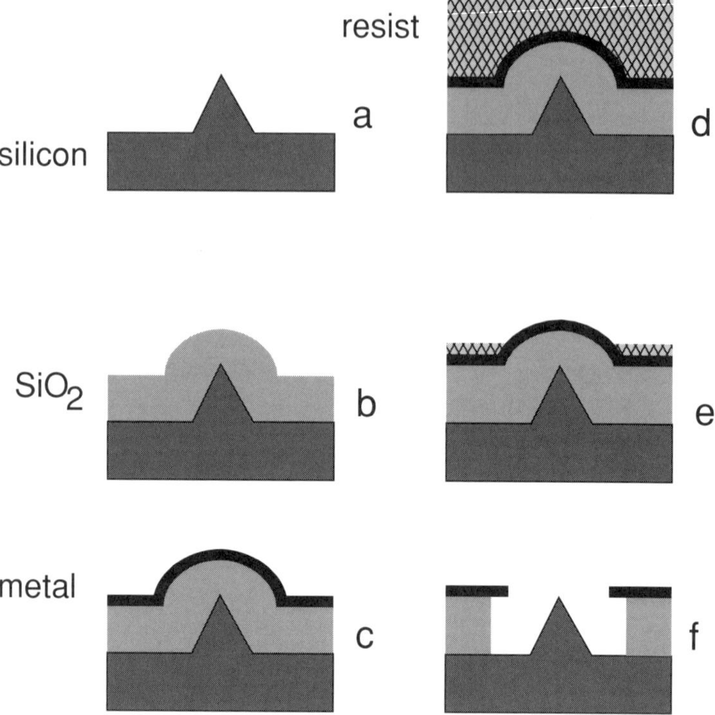

FIGURE 5.2. Silicon FEA fabrication steps using planarization. (a) A tip is formed under a mask, e.g., by ODE. (b) Dielectric is deposited, e.g., using CVD. (c) Gate metal deposition. (d) Application of photoresist or other planarizing coating. (e) Etch back the coating to reveal the nonplanar metal. (f) Etch the exposed metal, then the exposed oxide. (Steps d–f can be replaced by simply polishing the wafer, i.e., using CMP.)

especially when the materials are difficult to etch with sufficient control to make sharp tips. Gates can be formed over such tips using planarization, or by a novel technique [26] making use of doping-dependant etching (commonly used in micro machining). Before etching the (111) pit, the surface of a lightly doped silicon wafer is doped n^+, and the pit is thermally oxidized to form an effective gate dielectric. After filling the pit with emitter material and bonding to a substrate, the lightly doped silicon bulk is etched electrochemically, which stops at the n^+ layer. This n^+ layer thus forms the gate.

Orientation-dependant *growth* processes [e.g., molecular beam epitaxy (MBE) or chemical vapor deposition (CVD)] may be a useful alternative to etching for creating pyramids with near atomic sharpness and possibly very high uniformity. MBE growth on (100) terminated posts can produce pyramids on top of the posts, terminated in a combination of (113) and (111) facets [27,28]. The TEM lattice images shown in Fig. 5.4 demonstrate that the (113) facets terminate at the apex with precision within a few atomic layers. Gates can be added by sequential deposition of oxide and metal, followed by planarization. The flat-top posts used as starting structures may

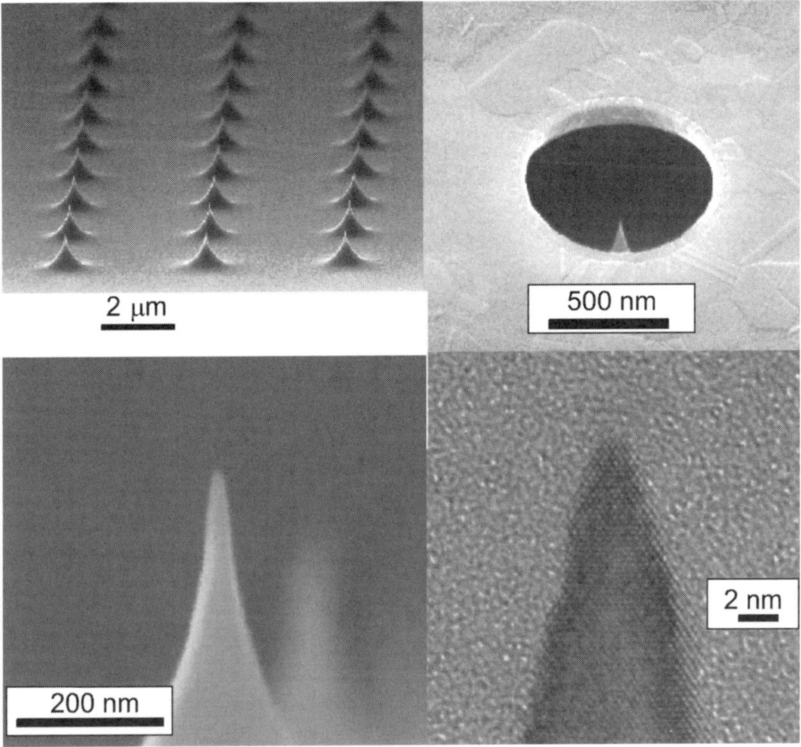

FIGURE 5.3. Oxidation sharpened silicon field emitter tips gated by CMP planarization. The photo on the lower right is a transmission electron micrograph, the others are scanning electron micrographs. (Courtesy of M. Ding and A. I. Akinwande, MIT.)

be fabricated with high aspect ratio using, e.g., reactive ion etching, such that FEAs with tall tips and high density can be produced. It appears feasible to use MBE in a similar manner to grow sharp apex structures of other crystalline materials, including many of the group IV, III–V, and II–VI semiconductors and combinations or alloys thereof. In addition to their structure, the chemical and electronic properties of these alternative tip materials offer many possibilities for improved emitters such as graded electron affinity [29].

Selected area vapor phase eptitaxy (SA-VPE) has also been used to grow faceted pyramids. This process starts with a flat crystalline surface, masked with an inert material such as SiO_2 in areas where growth should not occur. Under the appropriate conditions, growth often proceeds more quickly at a specific crystal face, e.g., the (100) face. Pyramids of GaAs [30] and GaN [31] have been formed in this way. SA-VPE growth of pyramids made of many other materials including silicon is also possible. Although the (111) pyramids show very smooth and uniform walls, SEM micrographs and emission tests of initial structures indicate that the apex regions are not especially sharp. Perhaps future work will identify changes in the growth conditions or additional processing to improve the tip apex shape.

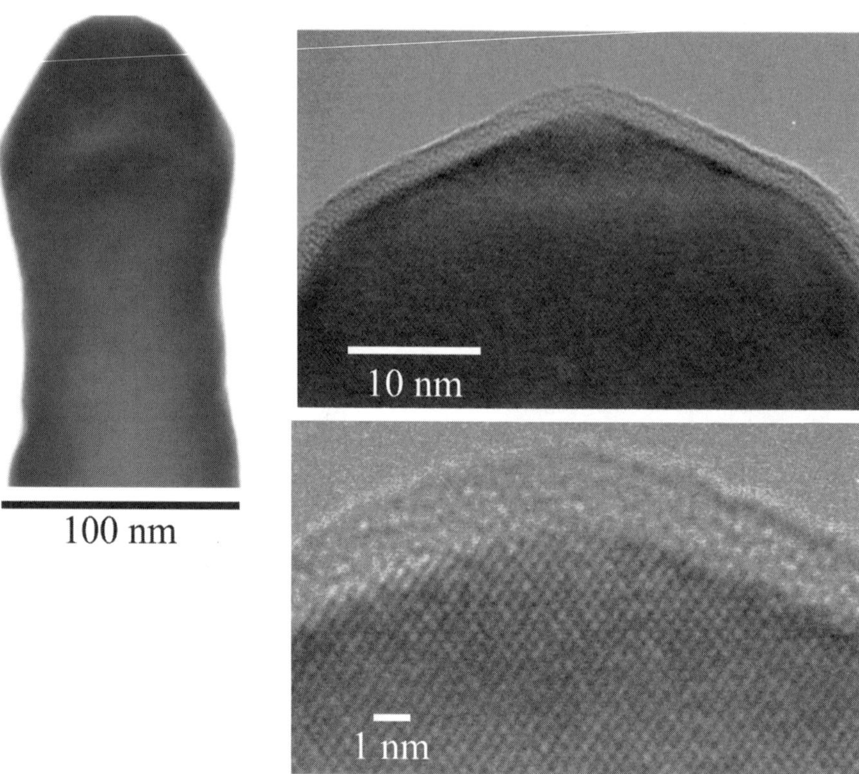

FIGURE 5.4. TEM micrographs of a silicon post after creating a faceted tip apex via MBE. (Courtesy of M. Twigg, NRL.)

5.2.2. Oxidation Sharpening

The most often-used method of creating sharp tips utilizes the oxidation properties of silicon [32]. When oxidized at low to moderate temperature, silicon does not oxidize as quickly at asperities (such as the edge of a mesa), as it does at flat surfaces. Thus oxidizing a tip or edge can serve to sharpen it. A typical process based on this property is illustrated in Fig. 5.5. The process typically starts by forming dot masks and then undercutting the masks (as in the ODE method), using any of a variety of etching techniques. Plasma etching is typically preferred. In this case etching is stopped short of the point where the masks are completely undercut. If a tall tip structure is desired (to produce an array with low tip-gate capacitance), an anisotropic (vertical) etch (such as reactive ion etching) step can be performed to form a "tip-on-post" structure [33]. To form the tip apex, the structures are thermally oxidized at about 950°C or less. The silicon remaining under the oxide becomes extremely sharp. Apex radii below the resolution of many scanning electron micrographs (<10 nm) are typically realized [13]. One TEM analysis of oxidation sharpened tips showed that the tips had a standard deviation in tip radius of 3 nm and average tip radius of 9 nm [7].

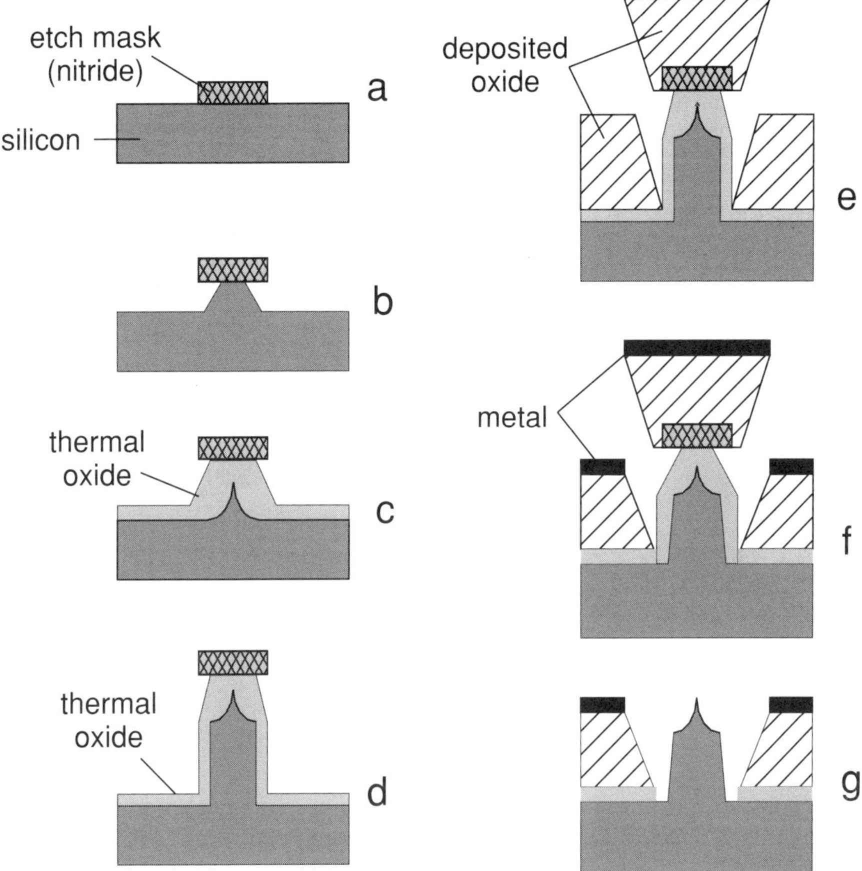

FIGURE 5.5. Silicon FEA fabrication steps using oxidation sharpening. (a) Define mask dot. (b) Form tip under mask by either ODE (wet etching) or isotropic (e.g., plasma) etching. (c) Sharpen tip by thermal oxidation. (d) (Optional) Form post under tip using directional etch (e.g., reactive ion etching) followed by thermal oxidation. (e) Directional deposition (e.g., e-beam) of thicker dielectric (may be done in multiple steps with planarization). (f) Directional deposition (e.g., e-beam) of gate metal. (g) Etch (e.g., HF etch) exposed oxide to remove cap and expose silicon tip.

More recent efforts (starting with smaller dot masks) have produced arrays with even smaller tip radii [34].

Because the oxidation rate slows as the oxide becomes thicker, the oxidation step is more tolerant of process variations in the initial structures than etching processes. Diagrams of a process simulation and micrographs of typical structures before and after oxidation are shown in Fig. 5.6 [35]. The resulting tips are typically very tall and narrow near the apex, a geometry that will provide very high field enhancement. The thermal oxidation step also serves to form the high quality silicon/oxide interface necessary for good breakdown characteristics. In addition, the sharpened tips can

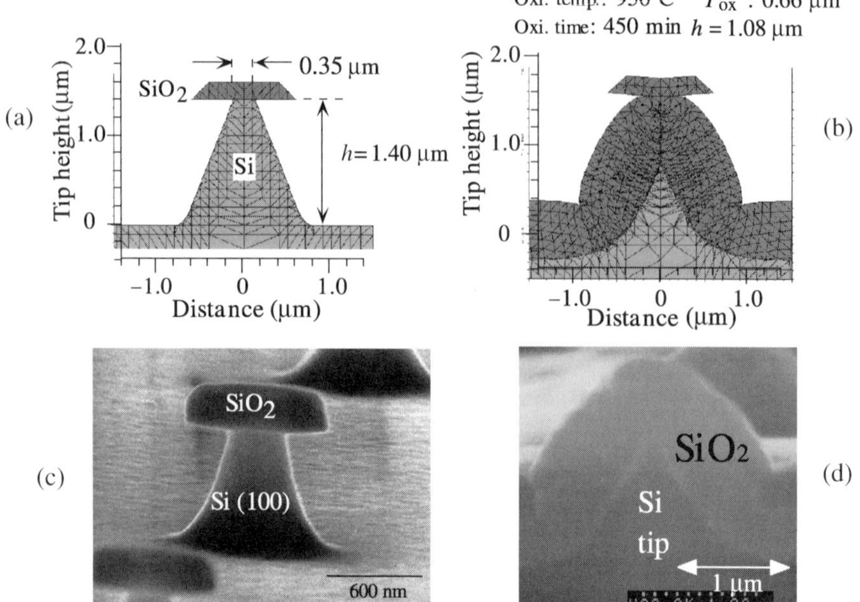

FIGURE 5.6. Oxidation sharpening process to form sharp silicon tips. (a) Schematic diagram of the initial structure. (b) The shape of the oxidized structure as modeled. (c) An SEM micrograph of the initial structure. (d) An SEM micrograph of the oxidized structure [35].

remain protected by the oxide during subsequent processing steps. For example, the oxidized tip, protected by the mask, can be used to define ion implants and gate apertures.

To increase the thickness of the gate oxide, additional dielectric (typically SiO$_2$) can be deposited over the oxidized tip using a directional method such as e-beam. E-beam deposition decomposes the silicon dioxide to some extent, forming a silicon-rich oxide. Annealing in an oxidizing ambient can restore the density to that of thermal oxide. During e-beam deposition, vapor phase atoms landing on the surface typically move about laterally while accommodating to the surface, causing the diameter of the oxide deposited on top of the mask to increase relative to the original mask. This property is undesirable in this case, because the deposited oxide forms a lift-off mask for the gate metal. Thus it is difficult to deposit thick oxides (needed for tall "tip-on-post" structures) without also increasing the gate aperture diameter. To reduce this effect, some of the dielectric deposited on the mask can be removed (via planarization techniques [33]) before the gate metal is deposited. In another process [36], silicon posts and the gate oxide can be formed *before* the tips are oxidation sharpened. Etch masks for the tip-definition step can be formed on the exposed post tops by depositing a Pt layer, heating to form platinum silicide, and selectively etching the remaining elemental platinum. Most recently, processes have been developed, which use oxidation-sharpened tips but remove the mask after the initial etch and

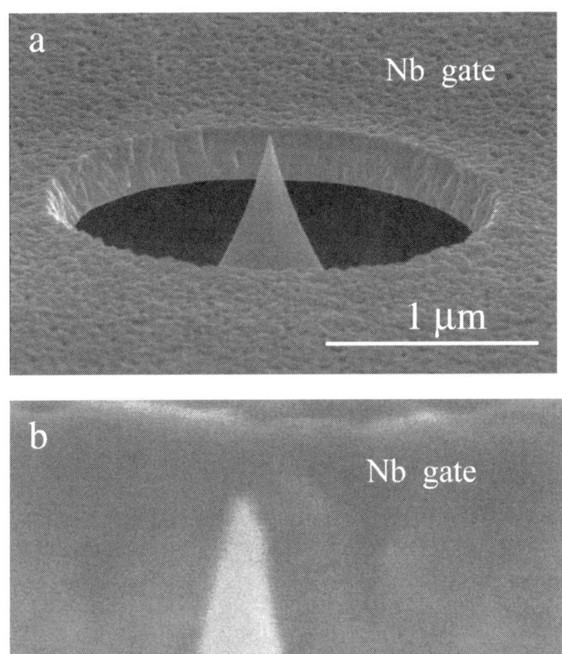

FIGURE 5.7. (a) An SEM micrograph of a finished FEA tip inside its gate aperture. (b) Close-up of the apex region [40].

form the gates using either traditional planarization techniques [37] (see Fig. 5.2) or CMP [6,38,39]. Using these methods, the gate aperture diameter can be made smaller than the original mask. However, uniformity can suffer. Figure 5.7 shows SEM images of a finished oxidation-sharpened tip and the adjacent aperture [40–42]. Note that the cone angle near the tip apex is very acute in this case.

The gate thickness, material(s), and deposition method(s) must be selected carefully to provide good adhesion to the dielectric, provide sufficient mechanical strength to withstand the electrostatic force created when the gate is biased, avoid stress, and provide the desired conductivity. Gate materials that have been used successfully include Nb, Cr/Pt, WSi, and polycrystalline silicon.

5.2.3. Variations

A second dielectric layer and second gate layer can be deposited to form an upper aperture. This is accomplished in a straightforward manner using lift-off gate definition [43] or using planarization methods. If the emitter tips can be grown or

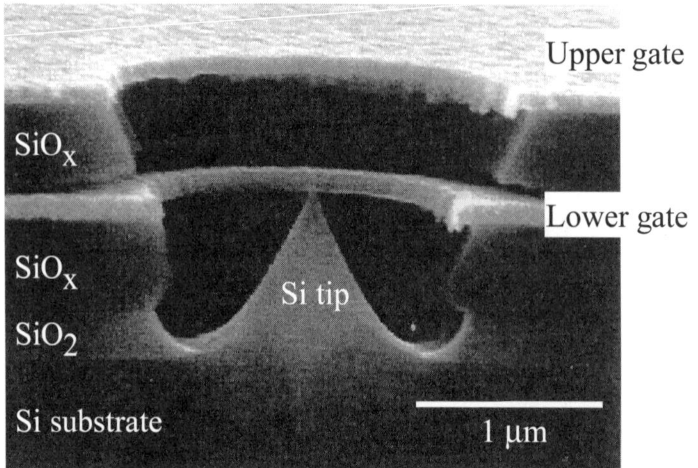

FIGURE 5.8. FEA cell having both an upper and lower gate fabricated by using the tip mask to define both gate apertures by lift-off [43].

deposited, multiple gate apertures can be fabricated on the flat substrate. The second aperture can focus the emitted beam from each emitter. Such focusing of each emission site is the best way to improve the brightness of the overall beam [44–46]. Using a double gate requires much less lateral space and is more effective than co-planar lenses. A finished double gated FEA cell is shown in Fig. 5.8. In this case, reducing the upper gate potential significantly reduced the emission, suggesting that the upper aperture diameter might optimally be increased or the aperture plane raised relative to the lower aperture [47]. Additional processing of finished arrays is often possible. For example, the gate can be used as a mask to coat or implant the emitter tips. Gate metal adhesion issues and/or the formation of leakage paths across the gate oxide can limit such processing.

The use of silicon wafers is not feasible for some applications, such as those requiring very large arrays and/or very low fabrication cost. In such cases, polycrystalline silicon can be deposited onto other substrates. The resulting silicon can be processed in much the same way as single crystal material, but with some loss in uniformity. The electrical resistivity of the polycrystalline silicon is also much higher, which may be an advantage or a disadvantage, depending on the application.

The surface of a silicon tip can be processed at high temperatures by removing the thermal oxide before the gates are added. For example, the silicon surface can be coated with a variety of materials including metals, metal silicides, nitrides, silicon carbide or metal carbides, various forms of carbon, etc. Forming these surface layers before adding the gates precludes any physical or electrical degradation of the gate that might otherwise occur during the coating process (e.g., delamination, melting, gate-tip leakage, etc.). On the other hand, removing the thermal oxide from the silicon typically results in somewhat inferior dielectric performance relative to thermally grown SiO$_2$. Coatings may well prove to be an excellent way to take advantage of the sharp tips that can be formed by silicon oxidation as well as the electronic and chemical properties of the coating material. The electronic properties of the interface between the coating and the silicon tips can also influence transport to the emission site, and thus have important effects on the current–voltage characteristics.

A large and growing body of literature exists on field emission from silicon and other FEAs fabricated in a variety of ways, including a variety of coatings and treatments. The performance of the new process is often reported with the uncoated silicon tips as a standard, in the form of I–V or Fowler-Nordheim (F-N) plots. However, in most cases FEA I–V characteristics vary with many factors which are difficult to control, including any cleaning or "burn-in" procedures, the previous emission history of the device, the temperature, the partial pressures of various gases, and perhaps other effects still not recognized. Knowing that such variations exist, it is impossible to evaluate an alternative material or coating based on a single measurement.

5.3. FREE ELECTRON THEORY OF FIELD EMISSION

Basic field emission theory assumes that the emitting surface can be described as a one-dimensional (1-D) free-electron gas. Early experiments on atomically clean single-crystal tungsten tips showed that this model was a good first approximation for the (111) surface, but less accurate for the (110) surface. Exposing these surfaces to oxygen and other gases causes additional changes in the field emission. More recent experiments have shown that three-dimensional (3-D) effects on an atomic scale can cause dramatic changes. None of these effects is accounted for within the F-N theory, but the basic form of the field emission I–V characteristic is typically observed nonetheless. To better understand the strengths and limitations of the basic theory, we briefly review it below. More detailed accounts can be found in the review article by Gadzuk [48], the book by Moldinos [49], and references therein.

5.3.1. Current Density

Electron emission from a solid is generally stated as Eq. (1), the integral of the supply function N, and the barrier transmission probability D over energy W, multiplied by the electron charge e:

$$J = e \int_0^\infty N(W)D(W)\,\mathrm{d}W \tag{1}$$

Simple theories of electron emission make several assumptions. First, the emitting material is approximated as a free electron gas. The number of electrons available for field emission is determined by the bulk properties of the free electron gas. Electrons in the gas move toward the emitting surface, such that the supply function N is the product of the Fermi function, the density of states, and the group velocity in the direction of the emitting surface $N = f(E)\rho(E)v_z(E)$. The density of states and velocity are inversely proportional, and $f(E)$ is assumed to be independent of emission current, i.e., *the emitted electrons are assumed to be replaced immediately and without perturbing the number or distribution of electrons in any way*. Although this approximation is quite accurate for clean metal surfaces, it is clearly not strictly correct, and ignores a significant part of the emission physics, i.e., how the emitted electron is replaced. This is a significant issue for many contaminated surfaces. Second, most theories make a 1-D approximation, as if the emitter tip was flat and all the relevant parameters are constant across the emitting surfaces. From a geometric point of view, the 1-D approximation is reasonably accurate only for smooth tips with radii much greater than the tunneling distance. In a 1-D approximation, only the velocity component traveling toward the surface contributes to the emission current. For a metal surface that obeys the free-electron theory, integrating the Fermi function results in Eq. (2), where m is the electron mass, k is Boltzman's constant, h is Plank's constant, and T is the temperature (in Kelvin):

$$N(W, T) = \frac{mkT}{2\pi^2 h^3} \ln\left[1 + \exp\left(-\frac{W - E_F}{kT}\right)\right] dW \tag{2}$$

This is the expression for N used to derive the F-N field emission current. Of course, most surfaces will not behave exactly like free electron gases, and are not 1-D. The solid/vacuum interface itself often creates a major change in N, and differences in the exposed crystal faces, the surface texture or protruding atoms, and foreign atoms also change N. The density of electrons near the Fermi level inside a semiconductor will typically be much lower than a metal. However, the density of states at an air-exposed metal surface may be much lower than in the bulk, or may not be metallic at all (due to a reacted layer). Conversely, the density of states at a semiconductor surface may be much higher than in the bulk. Thus expression (2) can easily be in error by several orders of magnitude and should not be taken too seriously when attempting to predict emission performance or calculate emitting area based on I–V data.

In the case of emission from a semiconductor, the density of electrons near the Fermi level just below the surface will depend on the Fermi level position relative to the band edges, which is determined primarily by the applied electric field and the charge density in surface states. Stratton's theory describes these issues. Transport to the near-surface region may also be affected by the Fermi level position. The density of states at the surface will in general be a combination of the bulk states near the surface and the surface states.

The tunneling transmission function D also requires a number of approximations. In general, the tunneling probability decreases exponentially with the area under the curve in a plot of potential vs. distance from the surface (between the Fermi energy and the potential barrier). The exponential dependence is common to most forms of the potential barrier and to most methods of calculating the transmission. The most common assumption is a triangular barrier modified by an image potential, given by Eq. (3):

$$V(z) = E_F + \phi \frac{e^2}{4z} - eF \tag{3}$$

where z is the distance from the surface, E_F is the Fermi potential, ϕ is the work function, e is the electron charge, and F is the electric field. The so-called Jeffreys–Wentzel–Kramers–Brillouin (JWKB) solution of transmission function for tunneling through such a barrier is

$$D(W) = \exp\left\{-c + \left[\frac{(W - E_F)}{d}\right]\right\} \tag{4}$$

where c and d are functions of field F and work function ϕ. The exponential form of the solution is also derived for other potential functions, but different potentials do change the terms c and d. Substituting Eqs. (2) and (4) in Eq. (1) and integrating results in the F-N equation:

$$J(F) = AF^2\phi^{-1} \exp\left(-\frac{B\phi^{3/2}}{F}\right) \tag{5}$$

The values A and B vary slightly with ϕ and $F (A \sim 1.5 \times 10^{10}, B \sim 0.7$ where J is in A/cm^2 and F in V/Å). For the reasons noted above, the derived values of A and B are probably far from correct except in special cases. If Eq. (5) is valid, a plot of $\ln(J/F^2)$ vs. $1/F$ yields a straight line. Such a plot is referred to as a F-N plot. In practice, the field at the emitter tip *cannot* be measured, but the field is presumed to vary linearly with the potential V (applied between the emitting surface and an anode or aperture). Thus $F = \beta V$, where the geometrical field enhancement factor β is based on the cathode, gate, and anode geometry. Since it is generally not possible to determine the exact shape of the field emitter tips, it is not possible to check the precise values of A and B against their derived values. Thus the linearity of the F-N plot is usually the only way to check on the validity of Eq. (5) based on I–V data.

The field emission I–V characteristics of most surfaces produce roughly straight lines when presented as a F-N plot. However, this qualitative agreement should not be taken as proof that all of the assumptions required to derive Eq. (5) are valid. Similarly, deviations from a straight line do not prove that external factors are altering the emission characteristic. Approximate F-N plot linearity means only that the exponential variation of the tunneling probability with gate voltage dominates the emission function. Any slowly varying preexponential function could be substituted for Eq. (2) and still get reasonably straight lines on the F-N plot. Figure 5.9 shows a comparison of two exponential functions of $(1/V)$, one includes a V^2 factor and one does not. The F-N plots of both functions look reasonably linear, especially at lower voltages where the exponent is varying most rapidly.

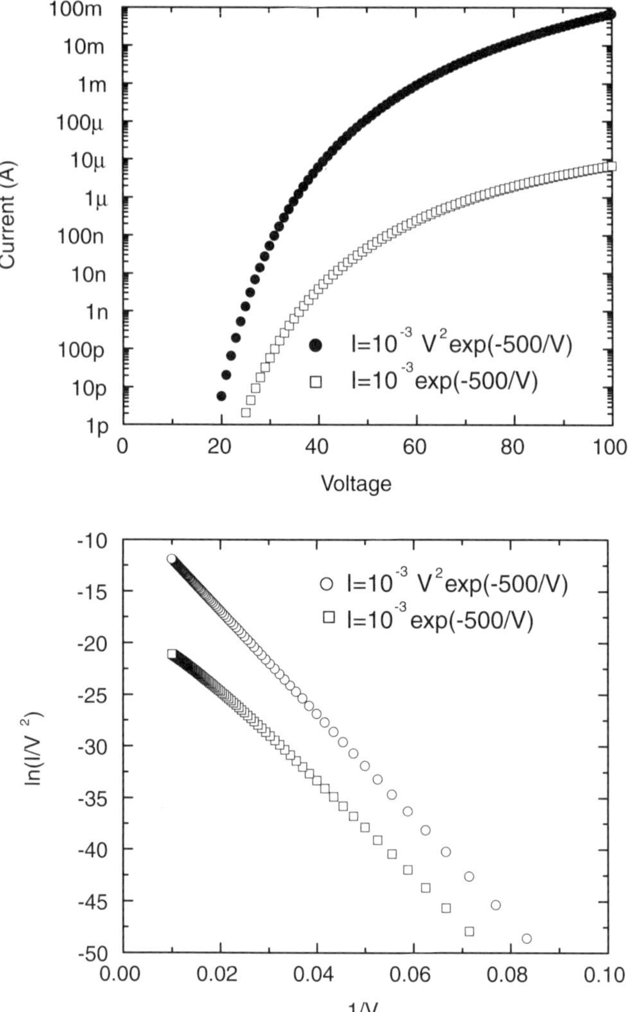

FIGURE 5.9. *I–V* characteristics generated from the standard F-N expression and a similar expression without the V^2 term. The calculated currents are plotted on a semi-log scale and as a F-N plot.

5.3.2. Energy Distribution

The energy distribution of the emitted current is a more complete diagnostic of the emission physics than the linearity of the F-N plot obtained from *I–V* data. The energy distribution from a conducting surface is most generally given by

$$\frac{\mathrm{d}J}{\mathrm{d}E} \propto N(E)D(E) \tag{6}$$

Again assuming free electron theory and a rounded triangular potential barrier, the transmission function D is given by Eq. (4). If the emitter were actually 1-D, the emitted current entering an analyzer located at normal angle to the surface would come from bulk electrons headed toward the surface at normal incidence. In that case, the source function N would be given by Eq. (2). Using Eqs. (2) and (4) in Eq. (6) gives the "normal energy distribution". However, because the emitters are curved, current emitted at many angles with respect to the local surface normal may enter the analyzer. Therefore, the source function includes more bulk electrons. The usual theory is to integrate over all angles, giving the total energy distribution (TED). Here, the source function returns to a Fermi function:

$$N(E) = \frac{mk_\mathrm{b}T}{2\pi^2 h^3}\left\{1 + \exp\left(\frac{E - E_\mathrm{F}}{kT}\right)\right\}^{-1} \tag{7}$$

For large vacuum barriers, the transmission function is an exponential as in Eq. (4). If the vacuum barrier is not large with respect to the electron energy, the transmission function saturates at high field, as may be modeled with a Fermi function [50]. Thus these very simple approximations give the form

$$\frac{\mathrm{d}J}{\mathrm{d}E} = A\left\{1 + \exp\left[\frac{(E - E_\mathrm{F})}{kT}\right]\right\}^{-1}\left\{1 + \exp\left(\frac{-c + (E - E_\mathrm{F})}{d}\right)\right\}^{-1} \tag{8}$$

Classic field emission spectra from a clean tungsten (111) surface [51] are reproduced in Fig. 5.10. To first order, the data and Eq. (8) are in good agreement. However,

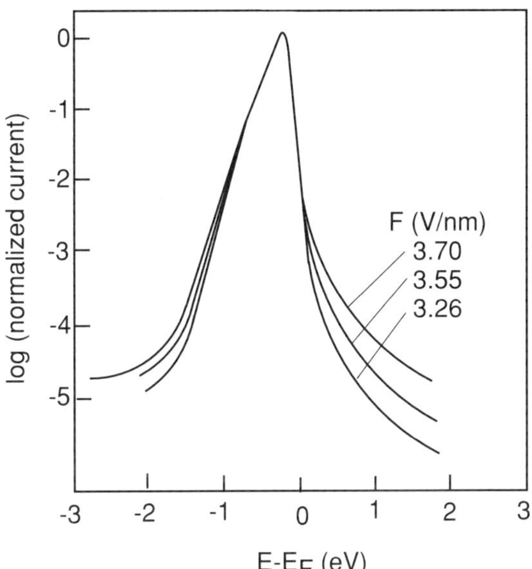

FIGURE 5.10. A field emission energy spectrum typical of a clean metal surface [51].

this theory fails to predict some features. On the high-energy side of E_F, there is a low current "tail" that extends several volts above E_F. This tail has been attributed to many body effects [51], i.e., energy produced when electrons are emitted from below E_F is transferred to nearby electrons at E_F, creating "hot" electrons. If these hot electrons are created within a few angstroms of the surface of a metal, they are transmitted to the surface with little loss and emitted readily. The hot electrons can interact with adsorbed molecules on the surface, stimulating desorption or surface chemistry [52]. This mechanism may be responsible for some of the emission changes that occur over time.

The low energy side of the distribution is typically not perfectly exponential. Such non-exponential behavior is prominent in emission spectra from the (100) plane of tungsten (where it was first observed [53]), and from a variety of other metal surfaces [48]. However, the deviation from exponential behavior is typically less than a factor of 10. Non-exponential behavior may be caused either by peaks in the density of states $N(E)$, or deviations in the transmission function $D(E)$. Structure in the density of states function can be created by surface states and/or bulk band structure. Structure in the transmission function can be caused by "resonant" tunneling [48] through an atomic-like state on the surface (such as might be produced by an adsorbed atom or molecule). Resonance can occur when a surface state exists in close proximity to the substrate free electron gas, but isolated from the substrate by a potential barrier. An electron wave function at the same energy as the surface state can tunnel out with higher probability than waves with higher or lower energy. *Inelastic* resonant tunneling occurs when the tunneling electron is scattered into lower energy states in the adjacent potential well (losing energy) before being emitted. This often occurs when the tunneling electron excites vibrations in absorbed atoms or molecules. Field emission spectra from tungsten tips coated with organic molecules showed large peaks in the energy distribution attributed to inelastic tunneling [54]. Field electron microscope images created by molecules adsorbed on tips are often highly symmetrical, a consequence of scattering [48].

The surface potential barrier can be 3-D. Fink [55,56] reported field emission patterns showing a single small spot, implying emission from a single atom with a very narrow angle ($\sim4°$). Explanations of this narrow emission angle include the angular dependence of the potential profile [57] and the angular dependence of a resonant surface state [58]. Similar narrow-angle, single spot emission patterns have been reported by Binh et al. after a thermal-field build-up treatment [59]. In that work, the energy distribution of the electrons in the narrow spot showed multiple peaks widely separated in energy. However, other groups [60] could not reproduce the multi-peak distributions, finding identical energy spectra for the blunt tip, the 3-atom tips, and the single-atom tips. Other multi-peak emission spectra and narrow-angle emission were obtained from tips made from metal carbides and from clean tungsten tips *only* after carburization [61]. Binh et al. [59] postulated that the peaked emission spectra they observed were caused by resonant tunneling through states produced by the single atom tip apex. However, it is clear from Ref. [60] that single atom field emission alone does not necessarily produce multi-peak spectra. It seems more likely that the thermal-field build-up process added some additional atoms such as C or O near the apex, and that the electronic properties of the contaminated surface caused the multi-peak emission [62]. Thus it appears that such contamination can produce narrow angle emission.

Fink et al. [63] have also produced narrow emission beams by adsorption of Cs, in which case the width of the energy distribution became narrower, but maintained a similar shape in agreement with free electron theory for a reduced work function. Similar narrow-energy/narrow-angle emission patterns have been obtained from thermal-field build-up tips made on W(111) and Pt(001) [64].

5.4. EMISSION CHARACTERIZATION OF SILICON FEAs

5.4.1. Experimental Apparatus

Energy distributions can be measured with the same energy analysis equipment commonly used for surface analysis such as photoemission spectroscopy. Unlike surface analysis, no external photon or electron source is required, but some means of making contact to the emitters and gates must be provided. An example of the system in one of the author's laboratories (JS) using a hemispherical analyzer is drawn in Fig. 5.11. The chamber is equipped with probe wires mounted on manipulators, used

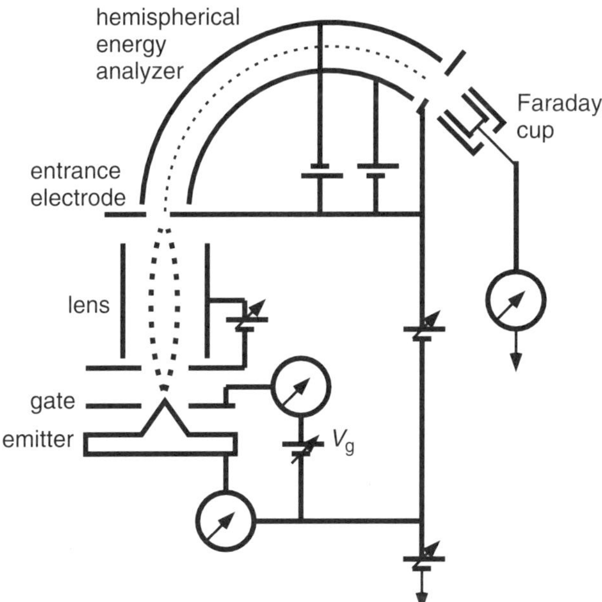

FIGURE 5.11. Apparatus for simultaneously measuring the emission current, gate current, and emission energy spectrum of an FEA.

for contacting the arrays and serving as anodes. The currents from the cathode, gate, and anode are measured independently. The cathode potential can be biased negative or positive with respect to ground to encourage or discourage the beam from hitting the chamber walls. The hemispherical analyzer allows only electrons with a specific energy to pass through the circular path to the detector. The potential applied to the slit in front of the hemispheres adds or subtracts energy from the incoming electrons, so that scanning the slit potential while monitoring the detector current produces an energy distribution. (It is worth noting that the potential seen by the electrons in space is equal to the voltage applied to the slit plus the work function of the slit surface. Thus the measured energy distribution must be shifted by the work function of the slit when it is plotted with respect to the Fermi level. The work function of the *emitter* does *not* shift the energy distribution.) In this case a Faraday cup is used to detect the analyzer current in order to sample a significant fraction of the large currents produced by FEAs (typical capacitively coupled electron multipliers cannot detect large currents).

One very useful method of evaluating the emission uniformity of an array is to image the FEA surface using an electron lens, i.e., make an emission microscope [65,66]. Either magnetostatic or electrostatic lenses can be used. An example of an electrostatic lens design is shown in Fig. 5.12 [65]. Using this type of lens, the emission initially launched parallel to the cylinder axis is focussed to an enlarged image on the phosphor screen at the right. The cathodoluminescence intensity can be detected

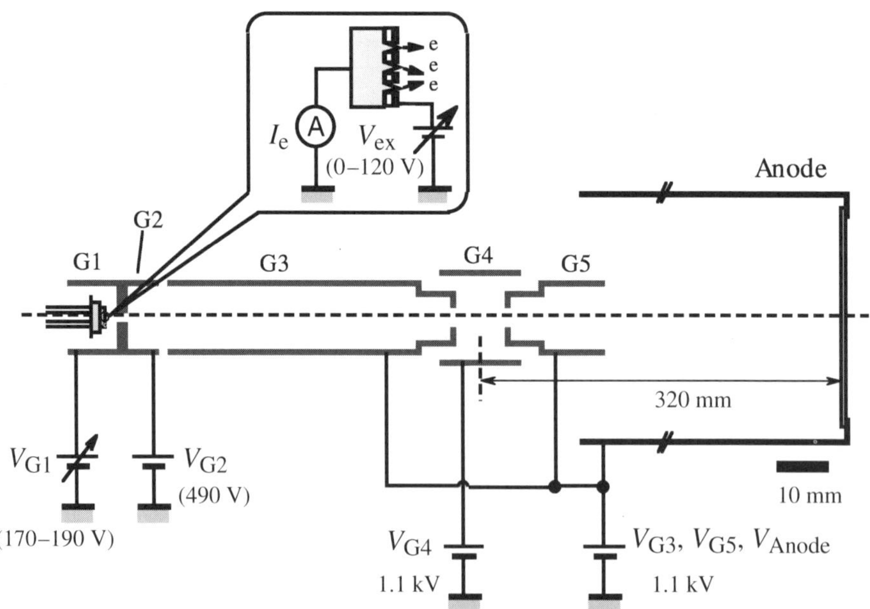

FIGURE 5.12. An electron lens system suitable for imaging the emission from individual emitters in an FEA [65].

$I_e = 48$ nA, $V_{ex} = 35$ V $I_e = 190$ nA, $V_g = 38$ V

$I_e = 1.0$ nA, $V_g = 42$ V $I_e = 1.9$ μA, $V_g = 45$ V

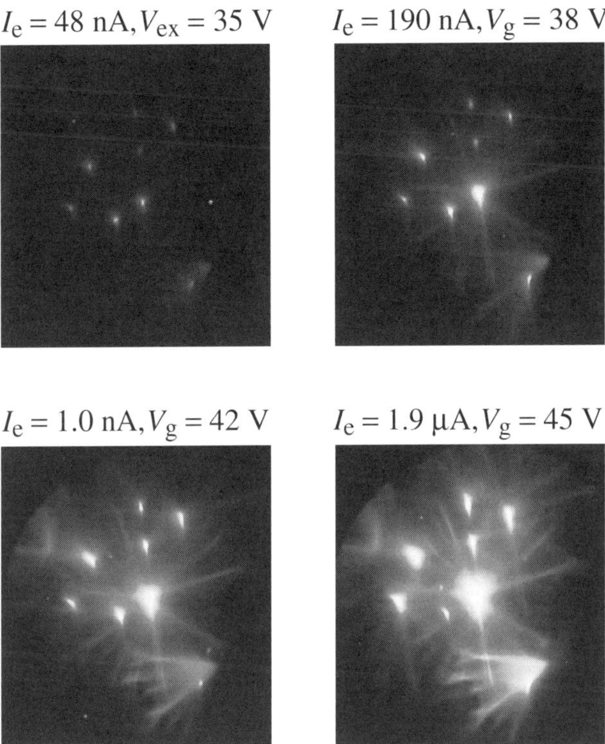

FIGURE 5.13. Images generated from the focused emission of an n-type silicon FEA [65].

and related to the current density impacting the screen at each point. Thus the system images part of the original emission pattern, information that is difficult to obtain in any other way.

An example of an emission microscope image is shown in Fig. 5.13. The image indicates that only a small fraction of the many tips in the FEA are emitting. There are limitations to the system that must be understood. Only the on-axis emission is imaged. Current launched from the array with significant off-axis energy will not pass through the lens apertures, and thus will not contribute to the image. Traditional field emission images of single points often produce emission patterns that show considerable variation over the solid angle of emission and often produce nearly all the emission in a small solid angle that does not intersect the tip axis. Thus it is likely that more tips are emitting than the image implies. Since the apertures have a finite size, emission with some small off-axis energy will pass through the apertures, but will not be well focussed. A good example of this effect is evident in the lower right corner of the images, where a pointed or comet-like spot with multiple tails is evident. Observing these images in real time, the comet-like spots often rotate (sometimes called the "searchlight effect" [67]), and any of the spots can fade/brighten, grow/shrink, or flicker. The searchlight effect must be caused by changes in the initial emission angle, which can also cause flickering. The properties of the detection system should also be considered when interpreting emission images. Most light detection systems have a limited sensitivity and dynamic range, i.e., currents that are too low are not detected and currents that are too high are not resolved. The response can be limited by either the light detection system, (e.g., film or CCD camera) or the phosphor.

5.4.2. Effects not Accounted for by Free Electron Theory

The field emission $I–V$ curve of most contaminated surfaces (i.e., FEAs) is not static; it can vary with time, temperature, the partial pressure of various gases, etc. These variations must reflect changes in the relevant transport and emission properties. Because the only variables in the theory of field emission from free electron gases are the electric field and surface work function, changes in these parameters (which can occur via atomic motion, adsorption, and desorption) are thus suggested. However, since no practical surface is a free electron gas, changes in the shape of the tunneling barrier, in the density and energy of initial states (and intermediate states), and in transport of electrons into the initial states are also potential issues. Understanding these issues should prove useful in directing future efforts aimed at obtaining more reproducible and robust emission as well as better performance for specific applications.

Figure 5.14 illustrates the change in emission current that can occur with time soon after an FEA is first activated [68]. To gather the data, the gate voltage of an FEA was held steady for several minutes at a time, then a quick $I–V$ measurement was made by reducing the gate voltage. The highest emission current and gate voltage are plotted in the top part of the figure. The slope and intercept of the F-N plot are shown in the bottom part of the figure. The data show that the emission current increased by several orders of magnitude during the first four hours of operation, but finally saturated. This general behavior, a rapid rise in current followed by a slower rise and finally saturation, is typically observed from silicon FEAs (similar behavior has been reported for molybdenum FEAs as well [1]); the time required to reach a steady emission current varies between specimens and with other factors. Most of the change in emission is due to a change in the intercept (A) of the F-N plot. The slope (B) also increased slightly in this case, but such an increase was not consistently observed. The large increase in A suggests that the emission area and/or the density of initial state electrons increased during the measurement period.

Figure 5.15 shows how temperature can affect the emission characteristics after the initial increase to saturation. The emission current, and the slope and intercept of the F-N plot are plotted vs. time. A heater under the specimen was turned on during the time indicated, causing a temperature rise of roughly 200°C. After the heater was turned on, both the slope and intercept decreased; the current first decreased then increased. After the heater was turned off, the current returned to its initial value monotonically, and much more quickly than did the slope and intercept.

Figure 5.16 shows that the emission current is influenced by the value of the gate voltage at a previous time (hysteresis effect). The emission current was measured

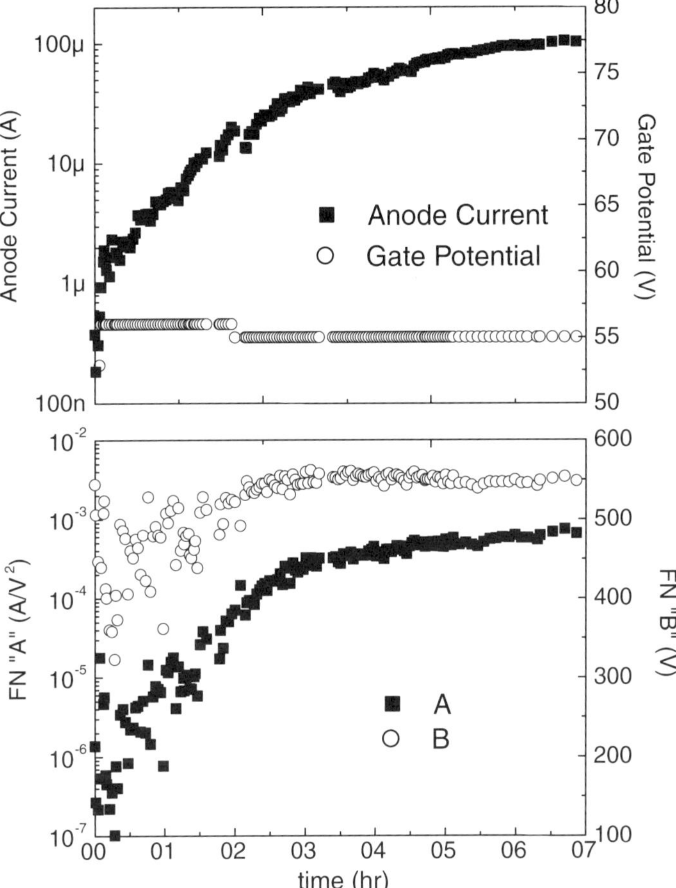

FIGURE 5.14. Plot of the anode current (top), and the intercept (*A*) and slope (*B*) of the F-N plots (bottom) measured from a silicon FEA as a function of time. The gate voltage was held at the voltage plotted on the upper right axis (56 or 55 V) except during the *I–V* measurements made at lower gate voltages [68].

periodically at $V_g = +55$ V, but the gate was held at a different voltage, called the "soak voltage" at other times. In this case the soak voltage was either 0, +55 V, or −55 V. The emission current changed by more than 10% depending on the soak voltage. The FEA produced the highest emission currents for negative soak voltage. Somewhat lower currents occurred at zero soak voltage, and the lowest currents were produced with positive soak voltage. The changes in current took several minutes to fully develop and were reversible.

The changes in emission characteristics shown in Figs. 5.15 and 5.16 cannot be explained simply by changes in the parameters of the F-N equation such as the density of states or the work function. The *correlated* changes in *A* and *B* shown in Fig. 5.15 suggest that a different physical parameter, that affects both *A* and *B*, is changing.

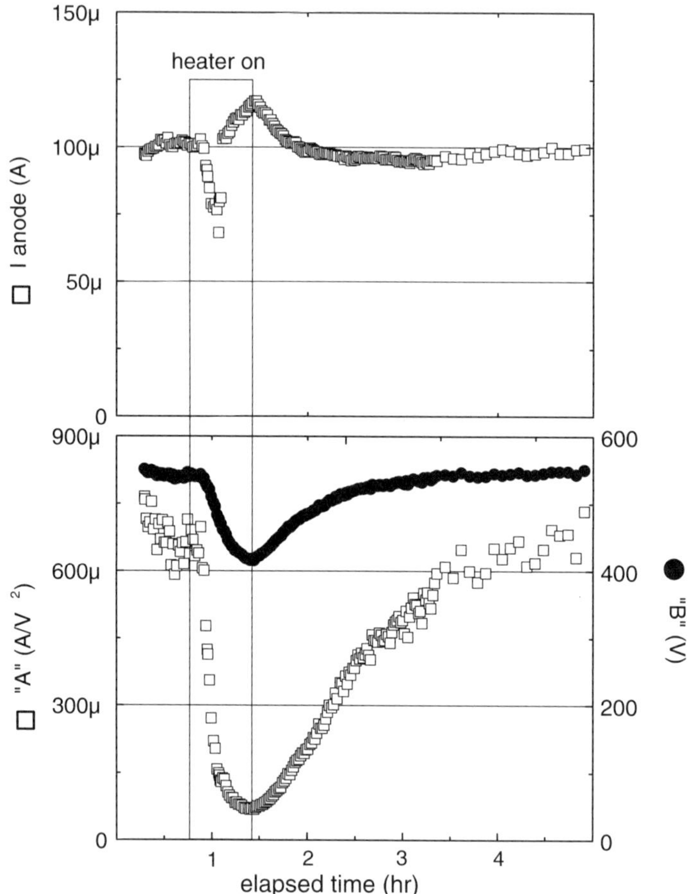

FIGURE 5.15. Plot of the emission current (top), and the intercept (*A*) and slope (*B*) of the F-N plots (bottom) measured from a silicon FEA as a function of time. The gate voltage was held steady at 55 V except while making *I–V* measurements at 55 V or less. A heater under the FEA was turned on during the time indicated, increasing the temperature by perhaps 200°C [68].

The hysteresis with gate voltage shown in Fig. 5.16, particularly the sensitivity that occurs independent of emission current (i.e., zero vs. negative gate voltage), suggests that *charge* (embedded in dielectric material) is attracted or repelled by the gate voltage. The temperature may affect the embedded charge because of increased thermal promotion of charge from the valence band into gap states, from gap states to a band or into vacuum, and perhaps by aiding transport or diffusion of the charged species. The only dielectric materials in the FEA are the oxide layer under the gate and the thin oxide on the surface of the tip. The gate will largely screen any charge in the oxide under the gate, thus charge in the oxide on the tip surface is implied. This oxide

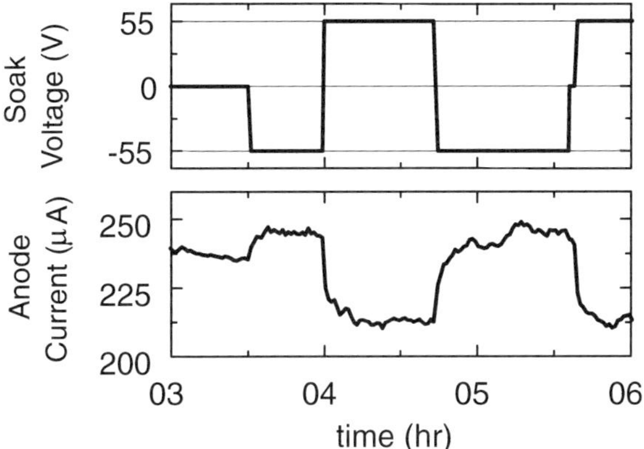

FIGURE 5.16. The top plot of the "soak voltage" is the voltage applied to the FEA gate during at least 59.9 s of each minute. At the end of each minute, the gate voltage was placed at 55 V for a few milliseconds, and the resulting emission current is plotted in the bottom plot. Holding the gate voltage negative prior to the emission measurement substantially increased the emission [68].

charge can create an electric field that adds to the field created by the gate. Figure 5.17 illustrates a simple model of an additional field created by fixed charge, calculated by adding a fixed voltage V' to V_g in the F-N equation. The F-N plot becomes curved as a result of the added charge. Depending on the voltage range used to calculate the slope and intercept, both quantities are changed by the fixed charge. In many cases

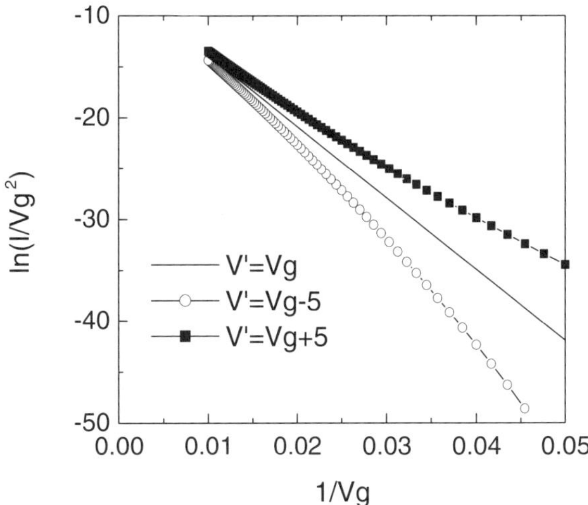

FIGURE 5.17. F-N plots $\ln(I/V_g^2)$ vs. $1/V_g$, where $I = AV'^2 \exp(-B/V')$ and $V' = V_g$, $V' = V_g + 5$, and $V' = V_g - 5$. These plots illustrate the effect of adding to or subtracting a small voltage from the applied gate voltage V_g. The additional voltage is a simple approximation of the effect of charge in the oxide on the effective electric field at the emission site.

measured F-N plots are also slightly nonlinear and can be fit by adjusting V' in this way.

5.4.3. Energy Distributions from Silicon FEAs

Energy distributions obtained from silicon FEAs [10,69] as well as from other semiconductors [62] and molybdenum FEAs [70] show multiple peaks and emission at energies up to several volts below E_F. The peak of the emission distribution typically shifts to lower electron energy linearly with increasing gate voltage. These peaked emission spectra immediately demonstrate that the emission from FEAs does not originate from the bulk band structure of the emitters. The peaks have been interpreted as either direct [10] or resonant [70,71] tunneling through states in a surface dielectric layer. Both models can qualitatively account for the multiple peaks and shifts to lower energy. Field penetration through a dielectric surface layer containing gap states will also shift the states to lower energy in this manner. (Resistive potential drops in the bulk are ruled out, since the peak shifts do not correlate linearly with emitted current.) The surface potential of semiconductors can also shift due to band bending below the oxide interface.

Two band diagrams for an n-type semiconductor emitter with a dielectric coating, illustrating different applied fields, are drawn in Fig. 5.18. The lateral dimension in

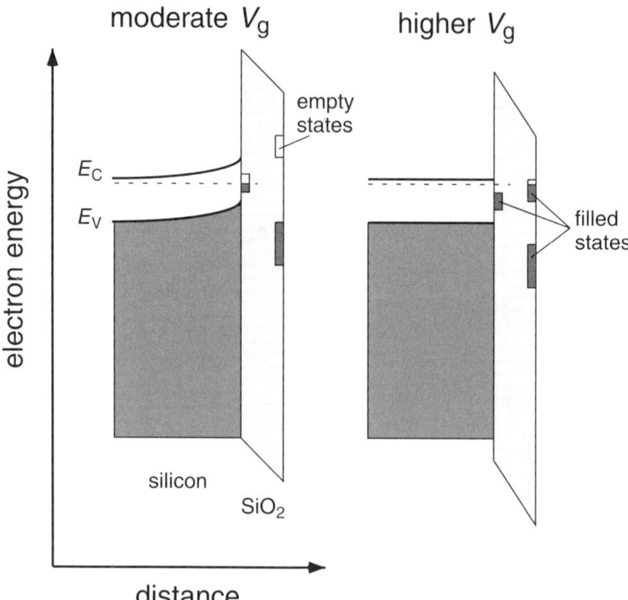

FIGURE 5.18. Band diagram of an n-type semiconductor surface covered with a thin oxide. Examples of filled and empty states at the semiconductor/oxide interface and within the oxide are indicated with filled and empty boxes.

the figure is distorted for clarity, but the vertical dimension is drawn to scale. The surface layer is assumed to be SiO_2, having 9 eV band gap, where the conduction band minimum of the oxide occurs about 3.5 eV above E_c (the silicon conduction band minimum) [72]. The native oxide may contain significant amounts of carbon and other species, which can change these properties. Native oxide on flat surfaces is typically 2–3 nm thick. Significant concentrations of allowed states within the SiO_2 band gap are typically present at the Si–SiO_2 interface and within the SiO_2 layer. Two types of surface-related states are drawn in the figure: silicon–oxide interface states, and bulk oxide states. The net charge in interface states causes band bending in the silicon below the interface, and moves the potential of the oxide relative to the Fermi level (dashed line). Charge within the oxide causes band bending in the oxide as well as the silicon, so that the potential of the oxide surface can change with respect to the silicon band edges at the interface. To simplify the figure, oxide states are drawn only at the oxide surface, such that the oxide band edges are straight.

Interface and oxide states, as well as the bulk bands, can serve as initial states in a tunneling process. Emission from the states at the oxide surface will be much more probable than emission from bulk states at the same energy, since electrons in the bulk must tunnel through the oxide as well as the potential barrier in vacuum. However, emission of an electron from an oxide state will leave the state positively charged, and the state must be refilled before emission from the same state can take place again. Thus most oxide states at energies above the Fermi level will become positively charged. Oxide states near or below the Fermi level may be refilled with charge from the bulk silicon. Transport from the bulk bands into oxide states could take place by tunneling through the oxide, either in a single step or in multiple steps via intermediate states. Alternatively, resonant tunneling can occur from an initial state (e.g., in the bulk silicon) through an intermediate oxide state in a single-step process.

In Fig. 5.18, the silicon bands are drawn bending upward on the left (moderate V_g case), indicating that the interface has net negative charge. The upward band bending moves electrons away from the surface and exposes positive donors, such that the number of donors is equal to the negative surface charge. Surface depletion such as this often occurs at silicon and other semiconductor surfaces. Moving additional negative charge into interface states can screen an electric field outside the surface (such as produced by making the FEA gate positive). The additional negative charge will move the Fermi level closer to the silicon conduction band and reduce the band bending in the silicon. If the density of surface states is large enough, the change in Fermi level position will be small. However, if the gate potential is increased to a point where the density of oxide and interface states is insufficient to terminate the gate field, the bands will flatten (as drawn on the right side of the figure). Electrons that accumulate in the conduction band near the surface will terminate additional gate field.

The electric field needed to create a significant tunneling probability may also produce accumulation. Electric fields normally associated with field emission near threshold, 2×10^7 V/cm, correspond to surface charge densities near 10^{13} cm^{-2}. Thus if the density of interface and oxide charge at the tip apex is below 10^{13} cm^{-2}, the bulk silicon just below the surface in the apex region will accumulate when the electric field adjacent to the surface is strong enough to produce field emission. The charge density at a flat native oxide interface is typically 10^{12} cm^{-2}, well below the density required to screen the electric field applied by the gate. However, higher densities of interface and oxide states are quite possible. High interface state concentrations can be created by impurities (such as metals), ion or electron impact, and the high curvature at the apex. In addition, states can be created when significant energy is transferred from a tunneling electron to the oxide [73]. Thus it is quite possible for the interface state density near the tip apex to be large enough to prevent accumulation. It is also possible that the interface state density can change due to either environmental factors or the emission history, and that such changes could move the surface from accumulation to depletion (or vice versa).

If the interface state concentration is not too high, most of the negative charge needed to screen the gate field is accumulated in the conduction band. Because of the modest density of states near the bottom of the conduction band, the Fermi level position may move several tenths of an electron volt above the conduction band minimum. Thus accumulation can significantly reduce the emission barrier potential for electrons near the Fermi level at the surface as well as drastically increase the density of initial states.

Figure 5.19 plots the electric field along the surface of a model field emitter tip starting from the apex (assuming a tip having a 10° cone angle, 5 nm apex radius, and 1 μm gate diameter). The surface charge density needed to screen the same electric

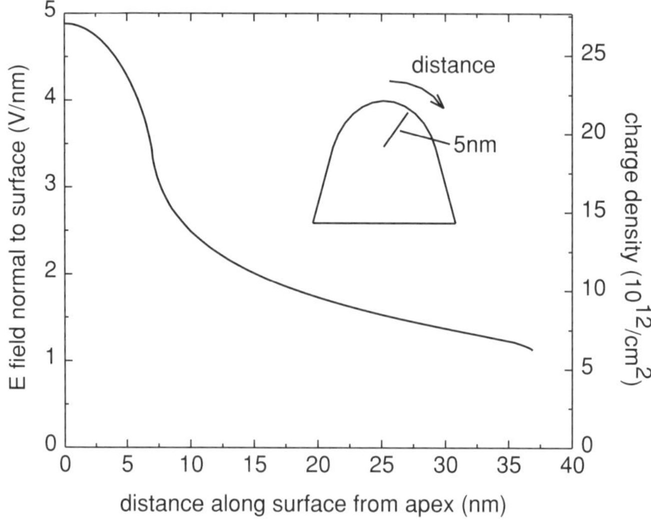

FIGURE 5.19. Plot of the electric field and induced charge density as a function of distance from the apex of a field emitter tip.

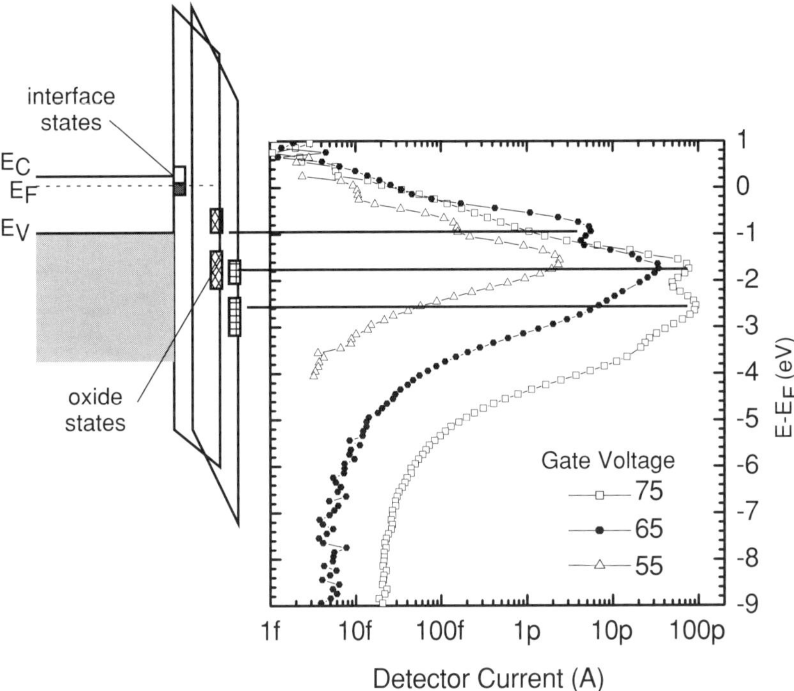

FIGURE 5.20. Energy spectra of a single silicon field emitter measured at three different gate voltages. A band diagram drawn to the same energy scale is indicated to the left. In the current axis at the bottom, f stands for femto (10^{-15}), and p stands for pico (10^{-12}) [10].

field ($\sigma = \varepsilon E$) is plotted on the right. Since the electric field falls off quickly along the tip shank below the apex, the silicon surface along the shank may be depleted even when the apex is accumulated. This result was confirmed via a detailed calculation [74] where the surface state density was assumed to be $3 \times 10^{13}\,\mathrm{cm^{-2}\,eV^{-1}}$ (uniform in energy within the band gap, charge neutrality level at midgap).

Energy spectra of the emission from a single n-type silicon tip indicate that nearly identical emitters can have either accumulated or depleted surfaces. Figure 5.20 shows energy spectra of the emission from a single n-type ($0.02\,\Omega$ cm) silicon tip, measured at three gate voltages [10]. Each of the spectra show emission at energies near the Fermi level (zero), indicating that some initial states near the Fermi level are filled. The spectra show peaks, and some of these spectral features shift with gate voltage, indicating that these features come from states in the surface oxide. However, the emission threshold energy does not shift, showing that the band bending in the silicon does not change. This behavior is consistent with flat bands (or accumulation) in the silicon. Two bands of oxide states consistent with the peaks in the spectra are drawn in the figure with and without a field across the oxide layer.

Figure 5.21 shows energy spectra from a second single-tip array, located adjacent to the first. All the spectra measured at different gate potentials have emission

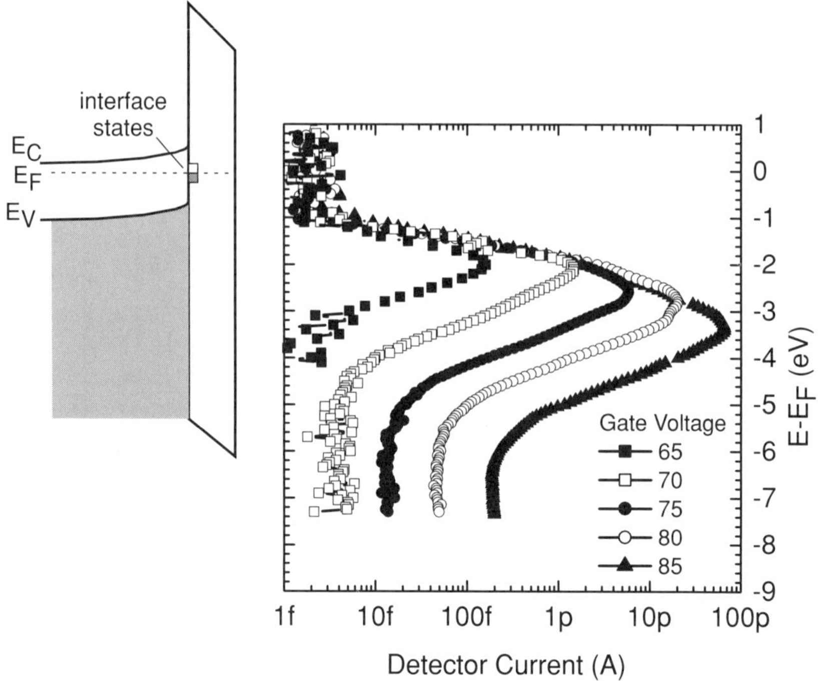

FIGURE 5.21. Energy spectra of a single silicon field emitter, produced with the gate voltage set at the values indicated. A band diagram with the same energy scale showing surface depletion is indicated to the left [10].

thresholds about 1 eV below E_F. The lack of any emission closer to E_F is consistent with upward band bending near the surface (as shown in the band diagram), since the band bending would prevent electrons in the conduction band from reaching the surface. The consistent threshold energies indicate that the interface state density is high enough to keep the Fermi level position nearly constant.

In Fig. 5.21, the emission current at energies just below threshold does not increase with gate voltage. Instead, the emission at a given energy saturates with increasing gate potential. Field emission from a free electron gas, with or without a resonant process, would increase indefinitely as the external field is increased. In contrast, the observed current saturation implies that the emission current is limited by transport into the initial states, rather than the tunneling probability. The transport limit can be explained by emission from states on the oxide surface, which must be refilled by transport through the oxide. The oxide states appear to be distributed over a wide energy range. Increasing the gate potential increases emission from progressively lower energy states, and states at lower energies emit more current because they are refilled more quickly. (Transport limits in the silicon below the emission area would shift the entire distribution, and so do not explain the data.)

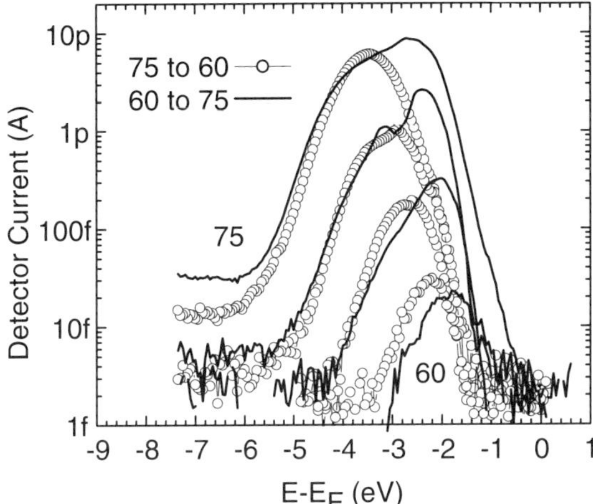

FIGURE 5.22. Energy spectra of a single silicon field emitter, measured sequentially as the gate potential was increased from 60 to 75 V in 5 V increments after holding it at 55 V, and as the gate potential was reduced after holding it at 75 V [10].

Figure 5.22 shows a hysteresis effect in emission spectra of a third single silicon tip. Distinct sets of spectra were obtained, depending on whether the gate voltage had been held high or low prior to the measurements. This hysteresis is similar to the effect recorded in Fig. 5.16, again indicating the presence of oxide charge and a change in that charge induced by the gate voltage and/or emission current. The energy spectra have consistent threshold energies below E_F, again indicating that the bulk bands are depleted near the surface. The spectra measured after the gate voltage has been held low for several minutes are different than the spectra measured after the gate voltage has been held high. The spectra measured after holding the gate low seem to have two emission bands, one being very similar to the spectra made after holding the gate voltage high, plus a second band at higher energy. Thus holding the gate potential low appears to have "turned on" an additional emission site.

The physics behind this "turn-on" may become more apparent with reference to Fig. 5.23. This figure shows the potentials calculated near a model emitter covered with a dielectric layer containing a discrete number of positive charges. The modeled emitter structure had gate radius 1 μm, a conical 45° tip terminated in a spherical section with 5-nm radius, and gate voltages 50–100 V. A 2-nm thick dielectric coating with relative dielectric constant 3.9 increases the radius to 7 nm. Charge is assumed to be uniformly distributed in a small volume (0.8-nm radius cylinder, 0.8-nm high) on the symmetry axis at the apex. The apex region of the tip is drawn at the top of the figure. A vertical line on the axis of rotation and a line 30° off-axis are shown. The potentials calculated [75] along these lines, starting from the conductor surface and extending through the dielectric into vacuum, are plotted. The left side and center plots show the potentials calculated for no charge, and for 1, 2, and 3 positive charges. Removing a single electron from the oxide (center) reduces the oxide surface potential by 0.8 eV, a

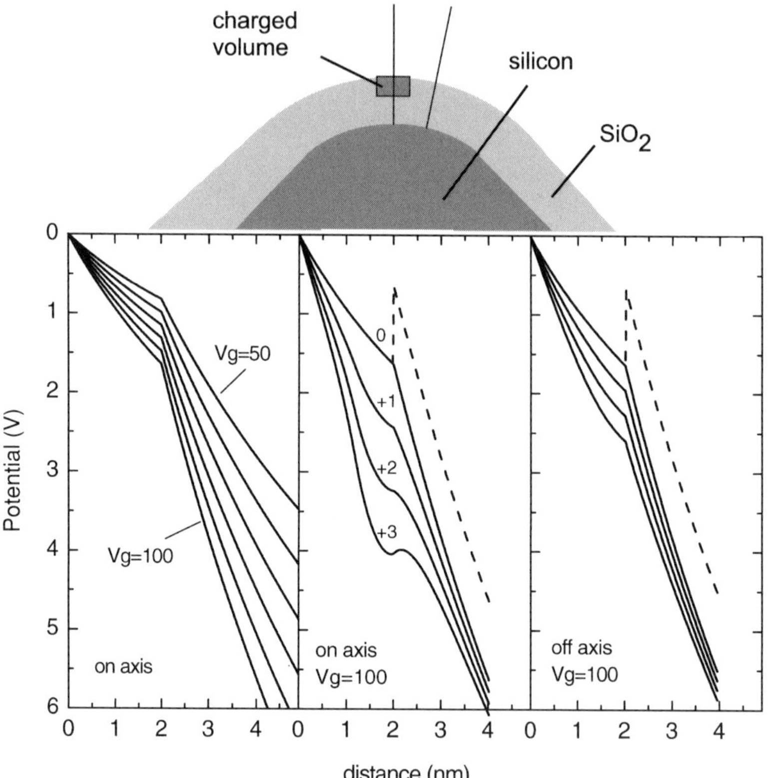

FIGURE 5.23. Potentials at the surface of a model field emitter tip. (Top) Drawing of the apex region of the same model emitter tip, showing the hemispherical portion of the conductor tip with 5 nm radius, covered with a dielectric coating 2 nm thick. (Bottom left) The potentials calculated on axis without charge in the dielectric. (Center) Potentials on axis with 0, +1, +2, and +3 charges in the dielectric. (Right) Potentials along the line drawn 22° off-axis assuming the same charge [10].

similar effect as increasing the gate potential from 50 to 100 V (left plot); however, the effect is reduced off-axis (right). Thus the additional emission band observed (shown in Fig. 5.22) after holding the gate voltage low is consistent with the net loss of an electron from the surface oxide. The similarity of the lower energy portion of both sets of emission spectra suggests that the emission came from a separate portion of the tip.

Figure 5.24 shows a set of emission spectra that shift to lower energy with higher gate potential, but shift back up after holding the gate potential high for several minutes. The spectra measured at the lower gate potentials show current saturation at energies just below threshold (similar to Fig. 5.21), and the threshold energies shift to more positive potential (lower electron energy) with more positive gate voltage. The spectra measured at 90 V gate potential shifts to higher energy after the gate

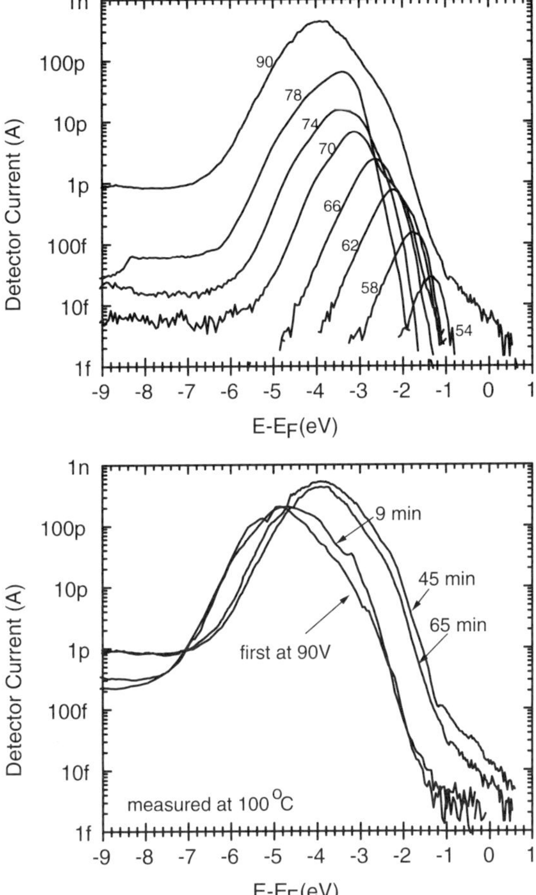

FIGURE 5.24. Energy spectra of a silicon field emitter array with a single emitter tip. The spectra in the above were measured while the gate voltage was held at potentials increasing from 54 to 90 V. The spectra in the below were made at time intervals with constant gate potential of 90 V [10].

voltage was held high for 45 min. The shift occurred discretely, suggesting that a single charge or state was responsible.

5.4.4. Fluctuations

Fluctuations in field emission current from field emitters are well known. Their cause has generally been described as changes in the tunneling probability (e.g., due to increased work function) accompanying adsorption and desorption [1,7], consistent with a model of emission from a clean metal surface similar to a free electron gas. Silicon FEAs also produce large emission changes after gas doses of the order of 100

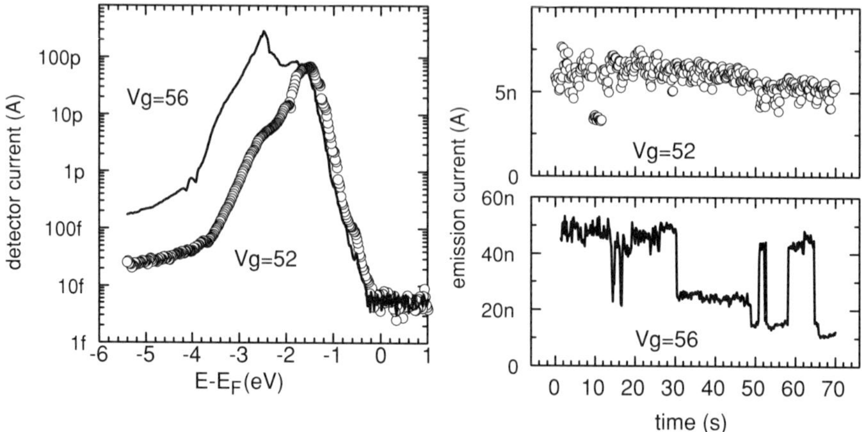

FIGURE 5.25. Plots of the energy spectra and simultaneous current measurements made at two gate voltages from a single silicon tip.

Langmuirs (1 Langmuir $= 10^{-6}$ torr $\cdot$ second) [7]. However, the reduced emission is observed only when gas is present while the FEAs are emitting; the emission recovers when the gas is removed. In most cases the arrays can be exposed to large gas doses or room air without significant changes in the emission characteristics if ultra high vacuum is re-established before the arrays are turned on. Relative insensitivity to gas is expected; one assumes that the silicon surface is oxidized, and the work function of SiO_2 changes little at pressures below 10^{-3} torr, and changes less than 100 meV even at 10 torr [76]. Thus simple molecular adsorption does not appear to provide a complete explanation for the change in emission. Very similar fluctuations also occur in solid-state devices (e.g., MOS structures) where adsorption cannot occur, and have been explained as changes in interface (oxide) charge [77].

Changes in oxide charge can also account for the fluctuations in emission. Emission fluctuations and correlated energy distributions are shown in Fig. 5.25. There, two energy distributions recorded at different gate voltages and emission currents measured simultaneously are plotted. The energy distributions are identical at energies above -1.5 eV, showing saturated emission. At the higher gate voltage, the additional current emitted occurs in a band centered 2.5 eV below E_F. The total current fluctuated between three discrete levels, i.e., the same current obtained at 52 V and two higher values. The reason for the fluctuation is illustrated in Fig. 5.26. Application of a higher gate potential may initially increase the emission current, but the additional potential across the oxide may also move an oxide state below the Fermi level or below the valence band maximum such that an electron is captured in the oxide. After electron capture, the electric field across the oxide is reduced, reducing transport and/or tunneling across the oxide. However, the reduced field across the dielectric increases the electron potential, so that it may be released again. Thus the oxide surface energy and consequently the emission current are unstable and fluctuate.

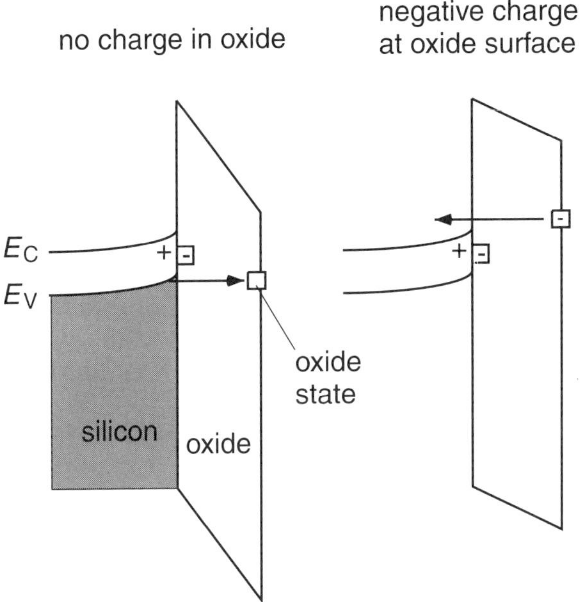

FIGURE 5.26. Energy band diagram illustrating how charge in an oxide state can create an unstable potential profile.

5.4.5. Surface Chemistry

The density of interface and oxide states can be changed by hot electron induced surface chemistry. Bonds can be broken by the energy released when energetic electrons impact a surface. SiO_2 can be reduced to SiO by bombardment with electrons or photons, and the SiO can be thermally desorbed at much lower temperatures than SiO_2. The impacting particles are thought to create ("hot") secondary electrons which interact with the oxide bonds. Energy to create hot electrons can also be generated when electrons are emitted from states well below E_F. Tunneling through the oxide in a silicon–SiO_2–metal (or polysilicon) structure can also create hot electrons. This mechanism is thought to cause "stress induced leakage current" (SILC), the increase in leakage current observed at a low voltage after current is forced through the devices at higher voltages. The additional states increase transport through the oxide, presumably by allowing multi-step tunneling. Field emission from silicon can produce a similar energy loss if the emission occurs at energies well below the Fermi level, and presumably will have similar effects in increasing the density of oxide states and thus transport and emission.

Figure 5.27 shows energy spectra of a silicon tip measured after the tip had been emitting for many hours at V_g between 32 V and 50 V (2 V increments) and $T{\sim}300°C$ [10]. The emission current increases exponentially at all energies with very little shift in the peak potential, showing that the emission was not limited by transport to the surface. The peak position was nearly 0.5 eV below E_F, showing that the emission

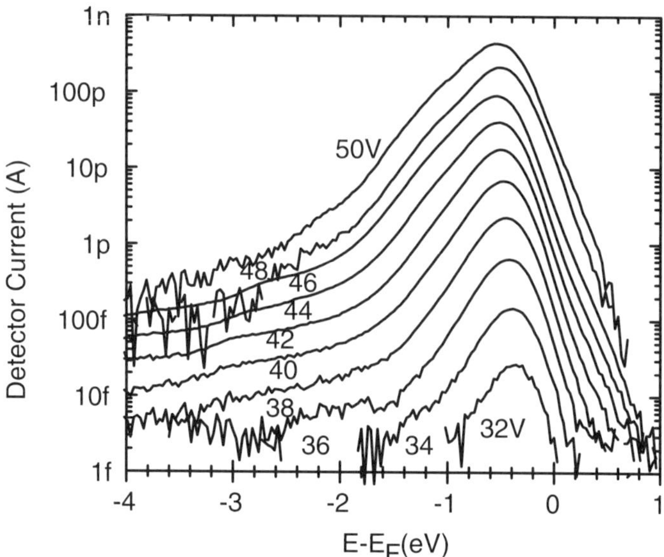

FIGURE 5.27. Energy spectra from a silicon field emitter after operating the emitter for several days at 300°C [10].

came from surface states rather than the bulk conduction band. The energy spectra extended to energies at least 1 eV above the Fermi level, indicating that electrons were sufficiently concentrated at the emission site to cause significant electronic energy transfer between electrons.

After allowing the emitter of Fig. 5.27 to rest in UHV (with the gate disconnected) for several days, the minimum gate voltage required to produce emission increased to more than 100 V. Presumably, the change in emission occurred due to oxidation of the clean surface. In contrast, arrays still displaying emission saturation effects can be exposed to air without a significant change in the *I–V* curve (provided that they are not emitting while exposed). Apparently adsorbed molecules such as O_2 or H_2O or CO_2 do not normally react with the silicon native oxide surface, and are thus free to desorb when the pressure is reduced. However, gas exposure while the emitters are operating does reduce the emission. In this case the energy released during field emission apparently stimulates the oxidation in a manner similar to photo-stimulated oxidation [78]. These reactions can heal un-terminated oxide bonds, thus reducing the state density and consequently the emission current. Thus it appears that surface bonds are concurrently healed and broken while the emitters are operated.

The steady state surface condition will depend on a rate balance between bond breaking and reoxidation. The oxidation rate depends on the density of reactive molecules on the surface, the density of dangling bonds, and the flux and energy of emitted electrons. Relatively clean surfaces will oxidize spontaneously (without emission), whereas oxidized surfaces require the energy released by field emission to stimulate reaction. The rate oxide states are created (by breaking oxide bonds) depends on the emission current and emission energy. The range of available energy varies with the condition of the surface and the gate voltage. In general, less energy is released at lower gate voltages. Surfaces with high densities of oxide states behave more like clean metal surfaces and emit at average energies closer to E_F (releasing less energy). In the spectra of Fig. 5.27, most of the electrons are emitted no more that 2 eV below E_F. Stress-induced leakage current measurements [79,80] indicate 1.5–2 eV is the minimum energy required to create oxide states. Thus the surface has come to an equilibrium condition where the energy released by emission is just enough to create oxide states. SILC measurements show that oxide states are created more efficiently when greater energy is released by the electrons; thus a higher rate of bond breaking (for a given emission current) is implied by emission at energies further below E_F, such as the spectra in Fig. 5.22. This process explains the initial increase in emission current (as oxide bonds are broken by energy released as electrons are emitted far below E_F) followed by saturation (when reoxidation of a relatively high density of broken oxide states balances bond breaking), such as shown in Fig. 5.14.

5.4.6. Arc-Initiation Mechanism

Clean tungsten etched wire field emitters have been shown to fail (explode) because of Joule heating at high current densities [81,82]. This mechanism is often assumed to be responsible for arc-initiation in FEAs as well. However, arcs at contaminated surfaces occur over a wide range of current, including rather low currents, indicating that the initiation mechanism is not the Joule and Nottingham heating that is typical of clean surfaces. The preceding discussion shows that oxidized surfaces emit significant current from states several volts below E_F, and the energy distribution may move to lower energies at higher currents. Significant emission can occur as much as 10 V below E_F. This large energy probably causes dielectric breakdown via direct electronic excitation of antibonding states, which could initiate an arc. Measurements of FEA failure events have confirmed that plasma is formed between the tip and gate during the failure event [83] and correlated the presence of SiO_2 with the incidence of arcs [84]. The incidence of arcs also increases when the gas pressure is higher [85]. However, simulation of the ion flux generated by the field emitted electrons striking gas molecules showed that damage due to this mechanism would require significantly higher pressures than observed experimentally [86]. Instead, the adsorbed molecules probably react with the oxide to heal gap states and thus increase the potential across the oxide. The increased potential will produce more energetic ("hotter") electrons when emission occurs at energies further below E_F. If these hot electrons stimulate decomposition and desorption of surface atoms, the chance of breakdown would be increased in agreement with experiment.

This mechanism is consistent with the experimental work by Latham et al. [87,88] on breakdown at high voltage electrodes, showing that arcs are often initiated at microscopic dielectric particles on large flat metal electrodes. Emission from the particles could be obtained at macroscopic fields two orders of magnitude lower than expected for field emission. That group reported energy distributions similar to those plotted in Fig. 5.24, i.e., shifting to lower energy at higher gate voltage, and more symmetric than typical of clean metals. Drawing on the early work on oxide breakdown available at the time, they assumed that excessive temperature causes the oxide to break down and initiate an arc, and that electron scattering heated the oxide. The group worked out a three-part model of field-induced injection into the oxide followed by bulk transport, and finally quasi-ballistic transport and classical emission at the surface. Fixed positive charge in the oxide (with sufficient density), perhaps in combination with geometrical field enhancement at fine conducting channels in the dielectric particles, were proposed as the mechanism producing the high fields required for injection into the oxide. This model accounts for the observed exponential increase in current with anode voltage, the low macroscopic fields required, and the symmetric energy distributions [87].

The emission model proposed by the Latham group does not account for some of the results reviewed above. For example, emission at energies significantly above E_F would not be possible, and there is no mechanism to account for the spontaneous and discrete shifts in the emission energy to more positive potentials (as shown in Fig. 5.24). Nevertheless, the similarity of their data to the more recent work suggests that similar mechanisms were at work in both cases.

5.4.7. Emission Uniformity

Assuming that emission-induced oxide states are in fact responsible for most of the emission from silicon surfaces, it follows that some significant initial emission current from any given tip is required to begin the process leading to the steady-state emission level. That is, an emission site having few or no oxide states, and/or relatively low geometric field enhancement, might not emit at all when the gate potential is first increased. Thus the process for creating additional gap states in the native oxide would not be operative at such sites. To improve the fraction of sites that emit at a significant level, a minimal initial surface state density must be created at all sites (assuming the required geometry exits). Gap states in the native oxide can potentially be created with a number of external processes. Oxygen vacancies can be created by thermal or chemical reduction of the oxide, bombardment with charged particles (electrons or light ions), or UV radiation. Alternatively, the gate potential could be increased sufficiently to get all of the tips to emit. Having created a significant density of initial states, the gate potential could be returned to a normal level. The latter option requires some sort of current-limiting element in series with each tip to prevent excessive emission.

5.5. LOCAL CIRCUIT ELEMENTS

Placing a circuit element in series with each emission site or each gate aperture can prevent local shorts from causing overall array failure. Elements in series with each emission site or group of sites can also reduce the emission current variations with time and from site-to-site, prevent or reduce the damage done by arcs, and allow faster and more successful turn-on. However, such elements can add complexity to the fabrication process and are not necessarily compatible with all applications.

5.5.1. Energy Spread

A circuit element in series with an emitter limits the emission current by reducing the potential between the gate electrode and the emitter tip, so that the electric field at the emitter surface is reduced. Some of the gate potential falls across the circuit instead of the vacuum, making the potential at the emitter surface positive with respect to the emitter contact. Thus some of the nonuniformity in current is traded for nonuniformity in potential, i.e., a wider energy spread in the emitted beam. Some applications (e.g., displays) require excellent uniformity in current and do not require uniformity in potential, while the opposite is true for other applications (where electron optics are important). The power dissipated in the current-limiting elements may be an issue in high current applications or where high efficiency is required. The emission current should be limited only as a fail-safe mechanism if uniform emission potential is desired, i.e., the emitter potential should not be a function of current until some critical (excessive) current is reached. To avoid changing the emission potential at currents just below the maximum current set-point, a nonlinear circuit may be required (rather than a resistor).

5.5.2. Design Considerations

- Breakdown voltage: To be useful for protection against short-circuits between the emitters and the gate, a current-limiting element must be able to hold off the gate voltage, which could potentially be applied across the transistor in the event of a short or an arc. Crystalline silicon can begin to avalanche at fields of several hundred thousand volts per centimeter, thus diode or transistor designs capable of tolerating gate voltages of 100 V must distribute the potential over several microns.
- Stored energy: Even when the dc emission current is limited to a low value, the energy available to an arc can be significant. This is due to energy stored in the capacitance between the gate and the portion of the substrate in good contact with the current-limiting element. Thus to minimize the stored energy, each current-limiting element should ideally contact only a small emitting area and as little as possible of the surrounding area.
- Low resistance path for displacement current: If the gate voltage is to be modulated at frequencies near 1 GHz or more, the displacement current may be much larger than the emission current. In such cases, a low resistance path for the displacement current (separate from the emission current) should be provided.
- Ease of fabrication: The current-limiting elements must be fabricated in a manner compatible with the FEAs, adding as little complexity as possible.

5.5.3. Resistive Elements

Resistive elements can be fabricated from thin deposited films of polycrystalline silicon. "Spindt-type" molybdenum FEAs deposited on resistive films exhibit little or no arcing and improved uniformity. The vertical silicon post of a tip-on-post style FEA could be used as a vertical resistor. As discussed earlier, the surface conductivity of a silicon tip near its apex can be a strong function of the surface state density, since an unusually large surface state density can prevent the surface from accumulating negative charges. Provided the surface near the apex does accumulate, the apex area may have a low resistance even though its cross section is small. Metal coatings or other surface treatments could be applied to ensure that the apex region is conductive. Normal surface state densities will typically prevent surface charge accumulation further down the cone and post. Thus the post resistance will probably be somewhat more than expected from the bulk resistivity ρ. Cylindrical posts of the order of 5 μm high and 1 μm in diameter made from 10 Ω cm silicon would have a resistance near 1 MΩ. This post resistance would serve to isolate shorts, limit the local emission current to a few microamperes, and reduce the damage done in the event of an arc [11]. If the application requires gate-voltage modulation at RF frequencies, the displacement current associated with the gate-substrate capacitance may exceed the peak emission current. In this case, a high conductivity path such as a metal layer must be provided under the gate dielectric. No such path is required through the emitter because the capacitance of the emitter tip is a small fraction of the total capacitance [12].

Another way of dealing with the displacement current is to create a capacitance in parallel with the resistance. This would allow larger resistors to be used. However, the capacitor must hold off the gate voltage and fit within the area near each emission site.

Additional arc protection might be achieved with a resistive gate layer, or a resistive coating over the exposed portions of the gate [11]. Resistance in the gate circuit would limit the current available in the event of an arc or a short. For low-frequency devices, the gate can be prepared conveniently using polycrystalline (deposited) silicon, perhaps patterned to provide local resistors for groups of emitters. High-frequency devices require a metal gate. However, patterned local polysilicon resistors with RF bypass capacitors could be prepared in a manner similar to the low-frequency devices.

5.5.4. Diode Elements

One simple nonlinear current-limiting circuit is a reverse bias p–n junction [89]. The emission current is limited to the reverse current though the junction, typically caused by a combination of thermal and optical generation.

5.5.4.1. P-Type Silicon Field Emitter Tips. The interface between vacuum and a p-type emitter forms one type of p–n junction, where vacuum is considered to be an n-type medium. Figure 5.28 shows a band diagram for a p-type field emitter. A dipole exists at the surface with negatively charged acceptors within a depleted region,

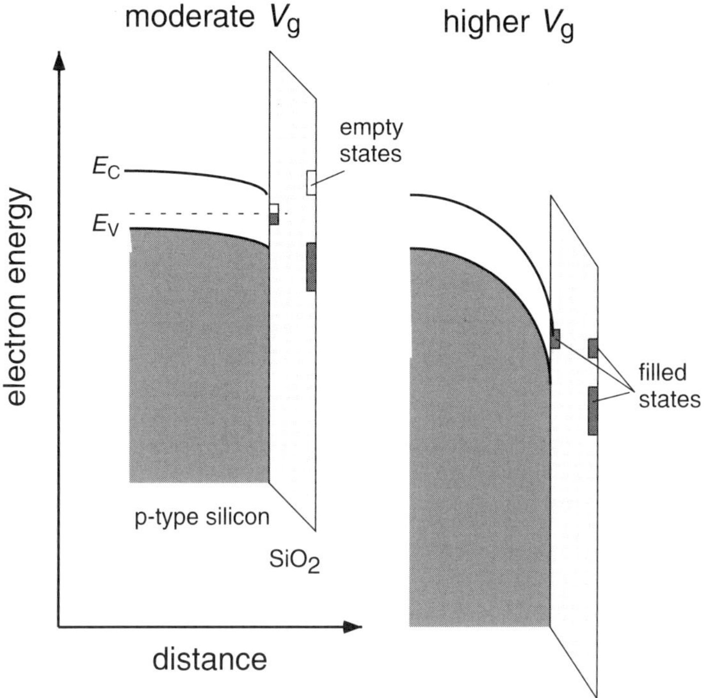

FIGURE 5.28. Band diagram of a p-type semiconductor surface with low and high electric fields.

balancing interface states having net positive charge. The number of acceptors in the depletion region is constant for a given interface charge, hence the volume of the depleted region will increase in more lightly doped material as in any semiconductor diode. If the net positive charge in interface states increases, the surface potential will move down and the thickness of the depleted region will increase. When the tip apex begins to emit, electrons have to move to the surface to replace the emitted charge. At low emission current, these extra electrons can be supplied by thermal and optical generation. Generation can take place at the interface, populating interface states directly from the valence band, and in the bulk, where valence electrons are moved to the conduction band. The conduction band electrons that are generated in the depleted volume and that diffuse into the depleted volume will be swept toward the surface by the internal electric field. This thermal generation current increases with the defect density in the silicon bulk or at the surface and can be significant.

If the density of interface states is large enough, the Fermi level position at the interface will be in the center of the energy gap, regardless of the bulk doping. Thus the bands will bend downward at the p-type surface (Fig. 5.28), and upward at the n-type surface (Fig. 5.18). As a result of this band bending, p-type tips may have more electrons in the conduction band near the emitting surface than n-type tips. In

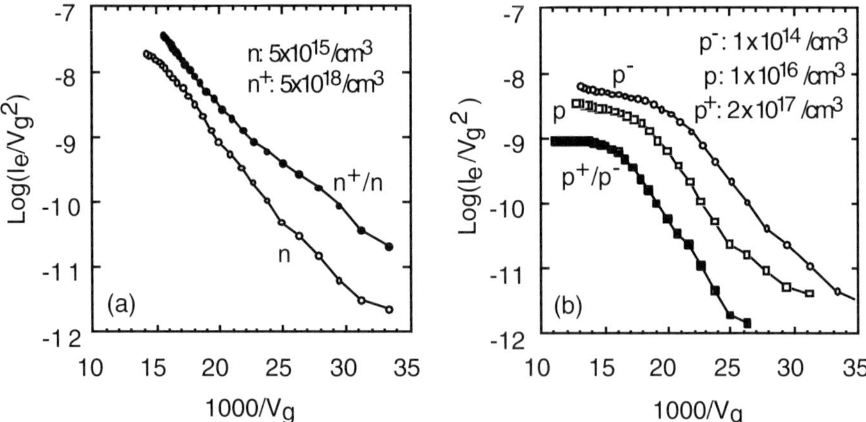

FIGURE 5.29. F-N plots of field emission from 1000 tip n-type and p-type silicon tips, doped by ion implantation technique [60 keV, P^+ ions for (a) and 50 keV, B^+ ions for (b)]. The p-type emission is limited by the supply of electrons [74].

such a case, p-type tips can produce more current at low gate voltages than similar n-type tips. This is illustrated in Fig. 5.29, which shows *I–V* curves and F-N plots of the emission from p-type (0.6 Ω cm) and n-type (4–6 Ω cm) field emitters [74].

Thermal generation can supply only a limited current. If the emission current becomes too large, some of the electrons emitted from interface states will not be replaced, so the net positive charge in the interface states will increase. Thus some of the potential applied to the gate will fall between the surface and the bulk silicon, rather than across the vacuum. This additional potential will increase the volume and surface area from which electrons can be drawn to the emission site. The current from thermal generation will be larger for more lightly doped material since the depleted volume will be larger. Thus as the gate voltage is increased, the emission current will initially be limited by the tunneling probability and increase exponentially, then will become limited by the thermal generation current and increase much more slowly.

Figure 5.29 illustrates this behavior. The figure shows F-N plots of the emission from n-type and p-type Si emitters [74,90,91]. The F-N plots of current emitted from n and n^+/n emitters are linear, as expected, when the supply of conduction band electrons is larger than the emission current. On the other hand, the p-Si emitters show nonlinear dependence on $1/V_g$ in the high electric filed. This confirms that thermal generation current limits the emission at higher gate voltage. At lower gate voltage, where the p-type emitters are not limited by the thermal generation current, the p-type emitters produce more current than the similar n-type emitters at the same voltage. Field emission *I–V* curves and F-N plots from n-type and p-type single-tip emitters, measured in detail in the low field region, are shown in Fig. 5.30 [74]. The p-type emitters produced more current than the n-type emitters. This result is due

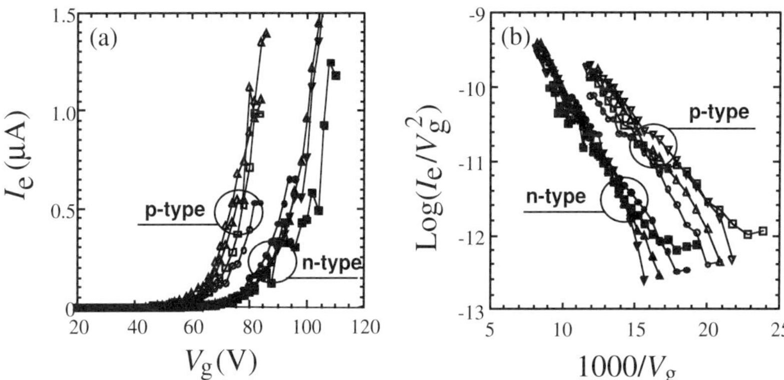

FIGURE 5.30. Field emission *I–V* curves and F-N plots from single tip n-type and p-type field emitters [74].

to the relative abundance of electrons at the apex of the p-type tips compared to the n-type tips. Downward bending at the p-type surface produces a significant current of thermally generated conduction band electrons at the surface, whereas upward band bending at the n-type surface pushes electrons away from the surface. The more lightly doped surfaces saturate at higher currents, consistent with the expected dependence on the depletion volume.

Figure 5.31(a) shows the emission current stability of an n-type array and a similar p-type array each with 1000 tips [92]. The p-type array has much less high frequency noise than the n-type array, as confirmed by the noise spectra plotted in part (b) of the figure. Illuminating the p-type arrays increases the noise level nearly to that of

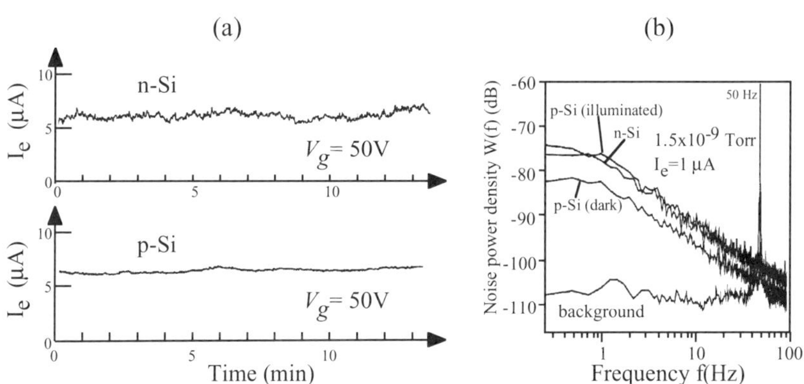

FIGURE 5.31. Current vs. time plots and noise power spectra showing lower noise from p-type vs. n-type field emitters [92].

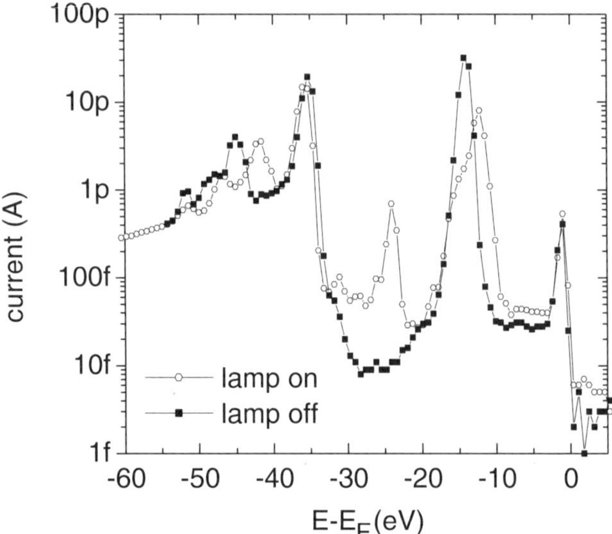

FIGURE 5.32. An energy spectrum of the emission produced by a p-type silicon FEA. The potential drop inside the emitter tips causes the emission peaks many volts below the contact Fermi level.

the n-type arrays, confirming that the current stabilization effect is due to current saturation in the p-Si emitters.

The current that can be extracted from p-type tips can vary considerably between individual arrays. Figure 5.32 shows an energy spectrum of the emission produced by a p-type silicon FEA fabricated at MCNC. N-type arrays prepared in the same fabrication run emitted well at gate voltages in the 60–80 V range. However, the p-type arrays emitted lower currents and required much higher gate voltages (over 100 V). The energy spectrum shows that most of the increased gate potential was due to large potentials across the p-type silicon. Most of the electrons are emitted 20-40 V below the Fermi level, and potentials over 60 V were developed within the silicon at some points. Only a small fraction of the emission came near the Fermi level. Although illumination (from an incandescent microscope lamp) slightly changed the emission intensity and/or shifted some of the peaks, the total current was not significantly changed. This shows that electrons generated in the bulk silicon are not likely to contribute to the emission in this case. The difference in behavior between the p-type arrays fabricated by different means in different facilities shows that the surface properties and/or the shape of the emitter tip play an important role in determining the emission properties.

The transport process from the p-type bulk silicon to the emission site is illustrated in Fig. 5.33 [40]. A depletion layer will form in the tip shank and under the gate. Conduction band electrons generated near or within the depletion region can potentially contribute to the emission current. The depletion layer width, and thus the

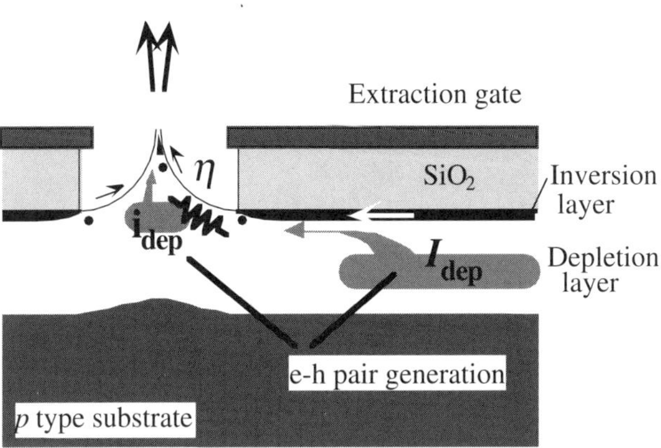

FIGURE 5.33. Schematic diagram of the expected regions of depletion and inversion in a p-type field emitter. Thermal generation under the gate can contribute to the saturation current [40].

thermal generation current, is reduced for higher doping levels. The interface with thermal oxide under the gate typically contains a low density of interface states, so positive potential applied to the gate will probably form an inverted layer under the gate. This inverted layer will largely screen the electric field created by additional gate potential, so the bulk generation rate I_{dep} is basically independent of the gate voltage at gate potentials where emission typically occurs. The electrons generated in the depletion layer must move to the tip apex before they can be emitted. Before arriving at the tip apex, most of the conduction band electrons will encounter the silicon surface exposed along the tip shank, where they may experience increased scattering and recombination. The probability (η) that a generated electron will arrive at the apex will thus depend on a variety of surface-related issues such as surface charge, roughness, and surface mobility. Any potential across the tip shank will change the electron trajectory, including the number of scattering events at the surface. Thus η can depend strongly on V_g as well as the tip shape and its surface condition. Some additional conduction band electrons may be generated in the silicon just below the tip apex. This current i_{dep} will be much smaller than I_{dep}, but transport to the tip will be more probable. The current i_{dep} may also be a function of V_g. The total saturated emission current, i_s, may be expressed as the summation of the near-surface and bulk generation currents i_{dep} and I_{dep}.

$$i_s = i_{dep}(V_g) + \eta(V_g)I_{dep} \tag{9}$$

5.5.4.2. Fabricated p–n and n–p Junction Diodes. Effects related to surface scattering can be removed by moving the p–n junction below the surface. A solid-state interface can be fabricated with little or no interface states, and hence the junction characteristics are more stable and predictable than the p-vacuum interface. In addition, a large fraction of the bulk generation current is transported to the n-type layer and

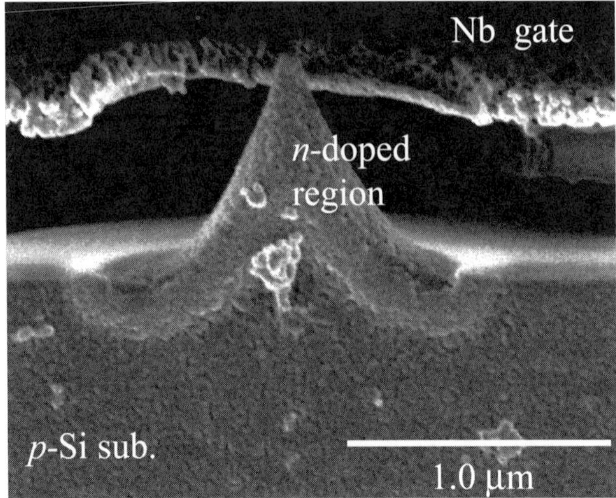

FIGURE 5.34. An SEM photograph of the n-type region created by ion implantation of a p-type silicon FEA [40].

emitted. As mentioned previously, ion implantation can be used to change the doping of finished silicon FEAs, using the gate as an implant mask. This technique can be used to conveniently fabricate isolated p–n or n–p junctions at each tip. An SEM image of a silicon tip fabricated from a p-type substrate but doped n-type on the surface is shown in Fig. 5.34 [40]. To better show where the tip was doped, the n-type material was removed with a selective etch. In this case phosphorous ions were implanted at 60 keV to a density of 5×10^{15} cm^{-2}, then activated in vacuum at 800°C. As is evident from the figure, the n-type impurity is uniformly doped through the extraction gate opening and forms a p–n junction under the entire tip surface. Figure 5.35 shows

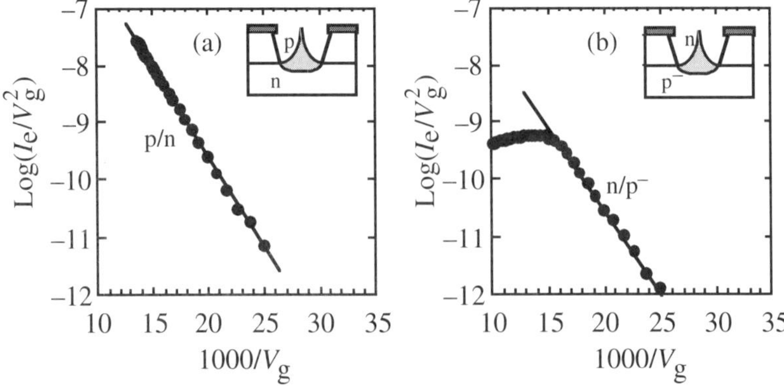

FIGURE 5.35. F-N plots of the emission from p–n and n–p emitter structures fabricated with ion implantation [93].

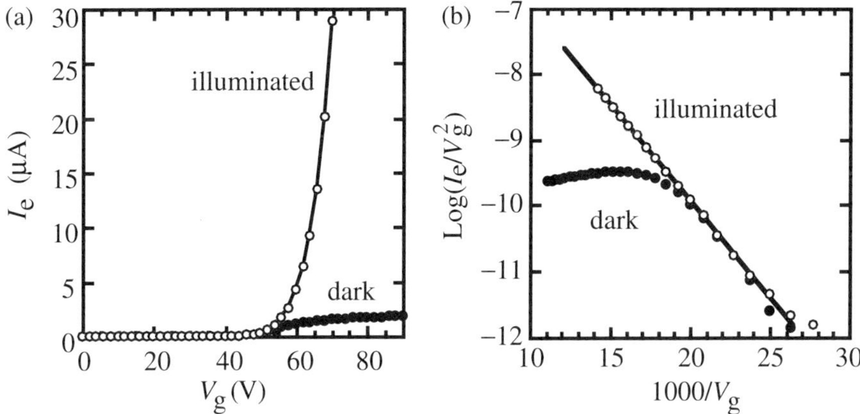

FIGURE 5.36. *I–V* and F-N plots of the current emitted from a n/p emitter structure with and without illumination from a 100 W incandescent lamp mounted external to the test chamber [91].

F-N plots of emitters with built-in p–n and n–p⁻ diodes, respectively [93]. The p–n diode is forward biased, and the emitted current produces a linear F-N plot. The n–p⁻ diode is reverse biased, so the emission current is limited to the reverse bias saturation current. This buried junction provides more consistent saturation current than the p-type emitters. The current supplied to the surface is still limited by the generation rate of carriers in the n–p junction depletion layer. Unlike the p-type emitters, the emission current in the saturation region is nearly independent of gate voltage.

The n–p diode was photo-sensitive [91]. Illumination with a 100 W incandescent lamp through the vacuum chamber window was sufficient to eliminate the current saturation observed in the dark. This effect is illustrated in Fig. 5.36. Figure 5.37

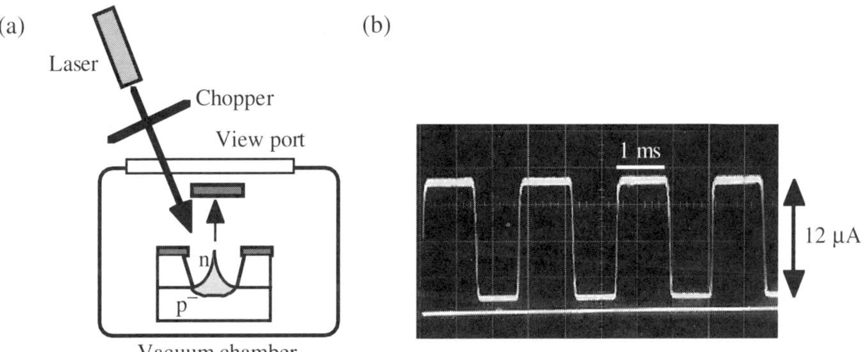

FIGURE 5.37. Schematic of the photo-modulated emission and chopped laser used in the experiment [91].

illustrates modulation of the emitted current achieved by mechanically chopping a HeNe laser trained on the FEA. By using a faster device (e.g., electro-optical) in place of the chopper, the maximum photomodulation frequency would be limited only by the transit time of the photoelectrons through the volume of silicon where optical generation occurs. This frequency could exceed 1 GHz, independent of the FEA transconductance and capacitance.

5.5.5. Transistor Structures

The reverse saturation current of a diode is linked to the density of gap states in the junction area and the diode geometry as well as the temperature and illumination. Although the reverse saturation current can be adjusted with these properties, it may be more practical to adjust the emission electronically, i.e., by replacing the diode with a transistor. Such electronic control also creates new device opportunities.

5.5.5.1. Simple MOSFET. A straightforward transistor configuration would place a planar transistor such as a MOSFET in series with and adjacent to each emission site or group of sites, with separate voltages applied to the transistor gate and the field-emitter gate apertures [18]. The electronic control of the maximum (saturation) current afforded by a transistor also allows the saturation level to be adjusted independently of the FEA gate voltage. Improved emission stability has been demonstrated from FEAs connected to external transistors [94]. Several fabrication processes have been demonstrated for FEAs with integrated transistors at each emission site (or group of sites).

5.5.5.2. MOSFET-Structured Emitters

5.5.5.2.1. Single Gate MOSFET Emitters. In most cases it will be desirable to reduce the complexity and expense of fabricating a completely separate transistor next to each emitter. A very simple transistor structure can be fabricated by using the field emission gate as the transistor gate as well. Such a structure is illustrated in Fig. 5.38

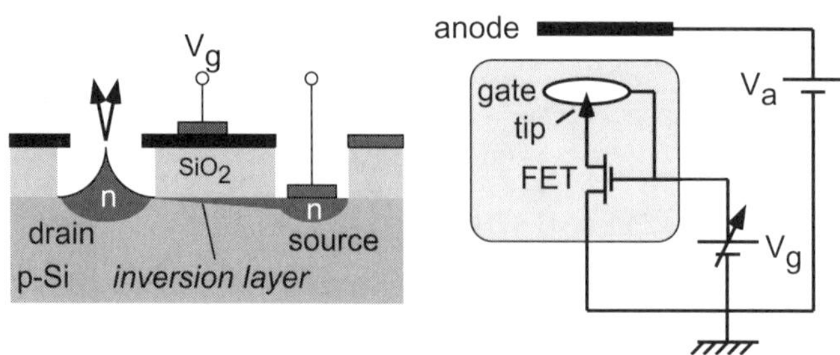

FIGURE 5.38. A MOSFET-structured emitter and its equivalent circuit [41].

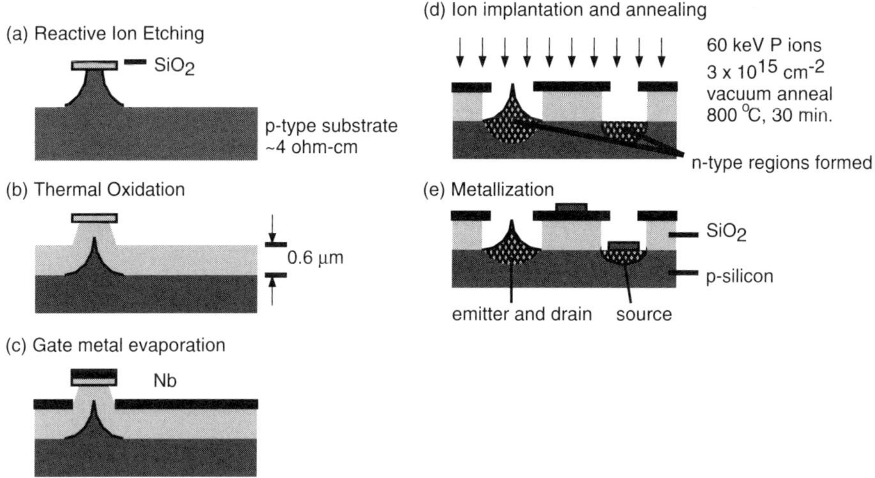

FIGURE 5.39. Fabrication process for the MOSFET-structured emitter [41].

[41]. The device is very similar to the n–p diode FEA, except that a ground contact is made adjacent to the emitter. As shown in the equivalent circuit, the resulting structure is an *n*-channel MOSFET, where the emitter tip is the drain. Single tip emitters using this structure have shown much improved controllability and stability.

The fabrication process of the MOSFET-structured emitter is shown in Fig. 5.39 [40,41]. In order to form n-type regions for the drain and source of the MOSFET, only two steps (steps (d) and (e) in the figure) are added to the conventional fabrication process based on RIE and thermal oxidation sharpening. Typically, a p-type (100)-oriented Si substrate with a resistivity of about 4 Ω cm is used. A 300-nm-thick SiO_2 disk with a diameter of 1.5 μm is made by wet etching. The tip outline is made by RIE with a mixture gas of SF_6 and oxygen (6:1) (step a), and then a 0.63-μm-thick thermally oxidized SiO_2 layer is formed at 1000°C (step b). A 0.3 μm thick niobium layer is deposited with vacuum evaporation (step c). After etching the source pattern, the conventional emitter structure and the source contact hole are formed by the lift-off process simultaneously. p^+ ions are implanted into both the tip and the source with a dose of 3×10^{15} cm^{-2} at 60 keV (step d). Then the substrate is annealed at 800°C for 30 min in vacuum for activation. Aluminum pads of 0.3 μm in thickness are used as the gate and source contacts (step e).

Figure 5.40 shows typical emission characteristics of a single MOSFET-structured emitter tip. In the figure, dots represent the emission current (I_e) and a solid line represents the drain current (I_d). At lower voltages (<74 V), the emission current increases exponentially as is typical. Above $V_g = 74$ V, the emission current is limited by the drain current I_d rather than the tunneling probability. Variations in the emission current are thus reduced relative to a conventional n-Si emitter tip, as shown in Fig. 5.41.

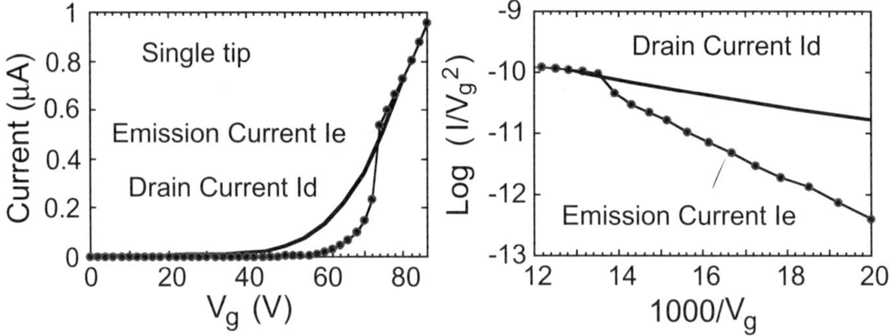

FIGURE 5.40. Typical *I–V* characteristics of a MOSFET-structured emitter [41].

Some of the initial proof-of-concept studies that showed external transistors can achieve active control of the emission current [95,96] indicated some problems with impact ionization. To avoid high-field breakdown effects, a simulation code (TMA MEDICI V.4.0) was used to design the MOSFET-structured emitter [15]. This program solves the Poisson and current-continuity equations to estimate carrier transport in the device structure.

The same concept and fabrication process may be used to form thin film transistors (TFT) in amorphous or polycrystalline silicon [97]. These materials can be deposited on glass, enabling large area devices such as displays. The emission current is precisely controlled and stabilized with this structure, similar to the MOSFET-structured emitters.

5.5.5.2.2. Multi-gate MOSFET Tips. The basic structure of the MOSFET-emitter can be easily extended to a dual-gate MOSFET structure as shown in Fig. 5.42 [98,99].

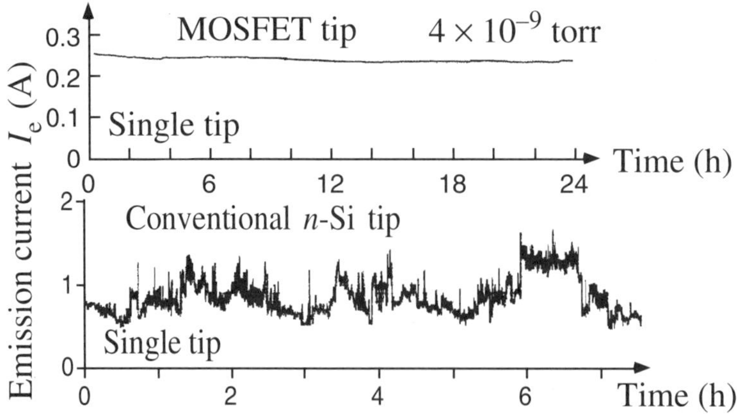

FIGURE 5.41. Emission current stability of a single MOSFET-structured tip vs. a single conventional n-type emitter tip [41].

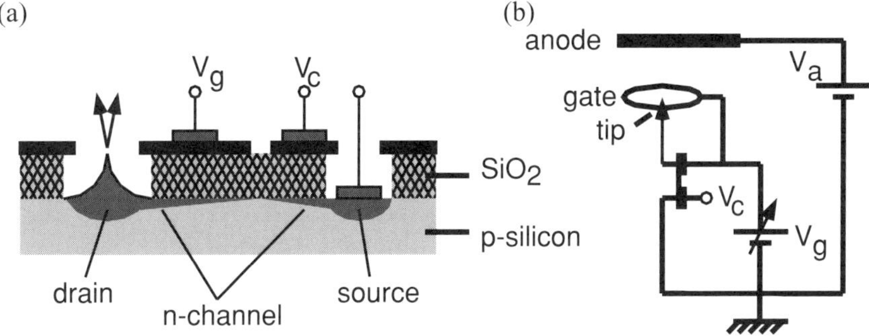

FIGURE 5.42. Illustration of a dual-gate MOSFET structure and its equivalent circuit [98].

The gate is divided into two co-planar pieces. Each of the gates makes an inversion layer (*n*-channel) in the adjacent p-type Si substrate. The left gate plays the same role as that of the single-gate MOSFET structure, while the right gate independently controls the channel conductance. The fabrication sequence is very similar to that of the single-gate MOSFET tip. Emission characteristics of the dual-gate MOSFET tip are shown in Fig. 5.43. In the figure, the emission current I_e was measured as a function of the left-gate voltage V_g at various right-gate voltages V_c that varied from 19 to 25 V as marked. The current I_e is limited by the tunneling probability and increases with V_g at low V_g, but saturates after I_e reaches the channel current determined by V_c, for example, 1.2 μA for $V_c = 25$ V. The emission current is stabilized in the saturated condition. Compared with conventional n-Si tips, the dual-gate MOSFET tips exhibit much better emission current stability, even at pressures up to 10^{-5} torr of ambient gases such as H_2, N_2, He and CO.

Figure 5.44 shows a structure and equivalent circuit of a tri-gate MOSFET tip. The gates labeled V_x and V_y can be used as *AND* logic device. The tip emits electrons

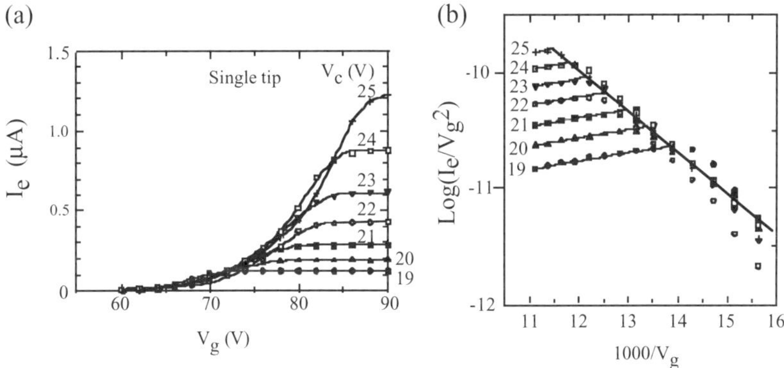

FIGURE 5.43. Emission characteristics of a dual-gate MOSFET-structured emitter [98].

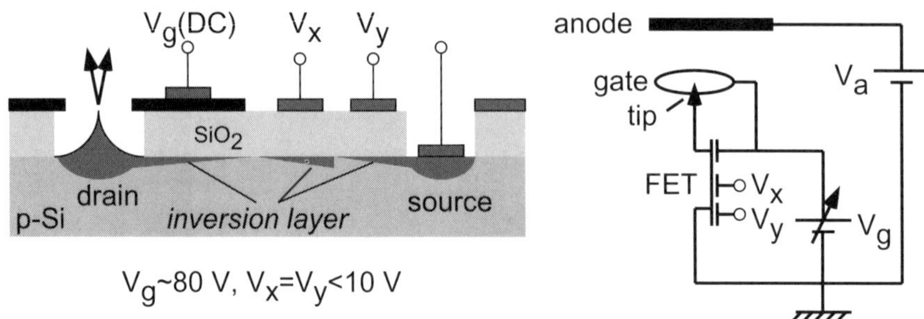

FIGURE 5.44. A tri-gate MOSFET emitter and its equivalent circuit [101].

only when both signals become positive. This feature enables active-matrix operation in addressed devices such as field emission displays (FED). This tri-gate structure lowers the driving voltages, V_x and V_y, down to TTL (transistor-transistor-logic) level and therefore, significantly lowers the consumption power compared to the conventional passive-matrix FED. Furthermore, the two control gates, V_x and V_y, can be directly connected with other TTL devices such as CPU, memories and even with solar cells monolithically integrated in the same Si substrate. This technology enables the "system-on-panel" concept in FEDs and therefore, a new-generation FED that are much more intelligent than the previous ones [100,101].

5.5.5.2.3. MOSFET-Tip with Focusing Electrode. Focused beams are important for many applications; for example, a focused beam would allow the anode in a FED to be separated from the cathode without loss of resolution. The emitted electron beam can be focused using a double-gated silicon emitter tip [43,102]. In the double-gated structure, the lower gate acts as an extraction electrode of electrons from the tip and the upper gate acts as an electrostatic lens to focus the electron trajectory. The emitted beam is focused when the vacuum level of the upper gate surface is near the initial potential of the emitted electrons. Normally this means that the upper gate potential will be a few volts more positive than the tip potential for best resolution. If the potential of the emitted electrons is more positive than ground potential, the focus potential will become equally more positive.

To prevent excessive emission, the MOSFET structure can be integrated with the double-gated tip structure, as shown in Fig. 5.45 [103]. In this structure, the lower gate is divided into two parts. The extraction gate is biased to a voltage (V_e) sufficient for electron emission and makes an inversion layer in a p-type Si substrate surface. The control gate is biased to a voltage (V_c) sufficient to create the n-channel in the MOSFET.

Figure 5.46 shows F-N plots of the beam current measured with a double-gated MOSFET tip. By reducing V_c, saturation in the current appears clearly. The beam current stability is also shown. The stability is excellent compared to that of the conventional double-gated tips. The beam focusing effect is also demonstrated at

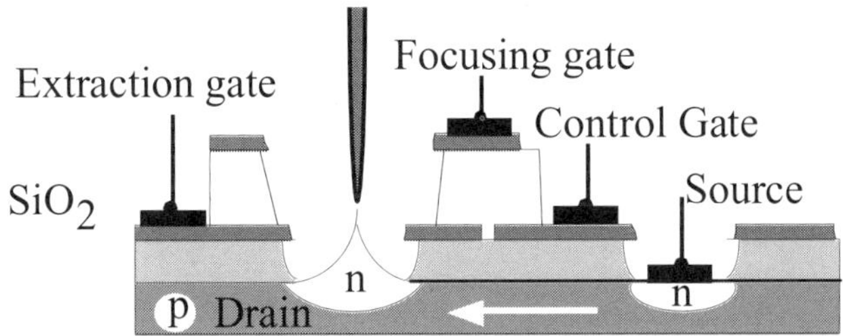

FIGURE 5.45. A MOSFET emitter including focusing lens [103].

$V_\mathrm{f} = 20$ V vs. 80 V. The current is reduced at lower focus voltages, in part because the upper gate reduces the field at the tip apex, and in part because the MOSFET causes some dispersion in tip potential.

Besides the above double-gate beam focusing, various co-planar lens focusing structures have been demonstrated [47]. The co-planar lenses easily generate cross-shaped beams and linear-shaped beams. The emitter generating cross-shaped electron beam has been applied for a 3-D vacuum magnetic sensor [104,105]. The concept of MOSFET-tip is suitable for use in these devices.

5.5.5.3. Vertical Transistor Structures.

For applications where large current densities are desired, the surface area needed to fabricate lateral resistors and transistors adjacent to each emitter or small cluster of emitters can be a liability. Vertical resistors and transistors provide an alternative.

One suggested structure [11,12] consists of a tip-on-post FEA where a metal layer under the gate extends up the sides of the posts, perhaps insulated from the silicon with thermal oxide. The metal layer serves as a gate in a vertical Schottky junction

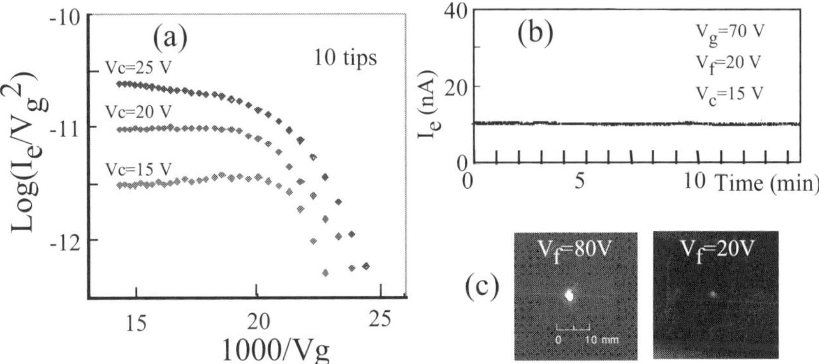

FIGURE 5.46. Emission characteristics, emission stability, and focussed emission from a MOSFET emitter with focussing lens [103].

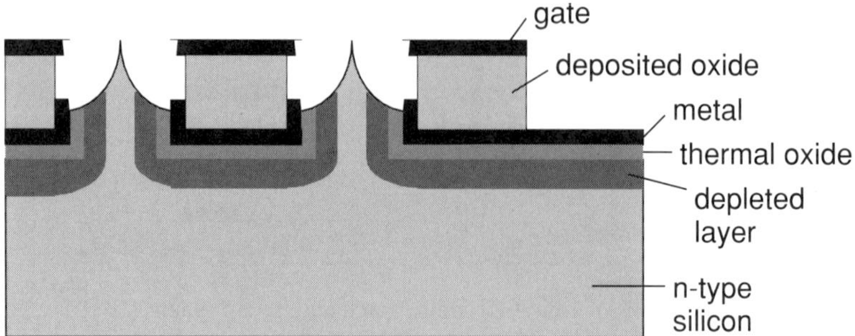

FIGURE 5.47. A tip-on-post design FEA, where the post is also a vertical MOSFET. The metal layer also serves as a low resistance ground path for RF modulation [12].

FET (JFET) or MOSFET and provides a low resistance path for the RF displacement current. The MOSFET structure is illustrated in Fig. 5.47.

A similar vertical JFET [106] can be prepared by replacing the metal layer with a p-type layer. This can be accomplished by forming tips in a manner similar to that illustrated in Fig. 5.5, but modified to include boron implantation and annealing before the oxide and gate metal are deposited (steps c and d). Thus a p-type layer is formed under the gate, but not on the tip surface or under the tip as shown in Fig. 5.48. By using the tip cap as an implant mask in this way, additional lithography is required only to contact the p-type layer.

Relatively large vertical transistors may be fabricated by etching deep ($\sim$10 μm) trenches in n-Si and refilling the trenches with oxide [19], forming roughly cubic columns in the silicon separated by oxide walls. The structure, called a "VErtical CurrenT Limiter (VECTL)" is illustrated in Fig. 5.49. Most of the usual FEA structures, including deposited metal and etched silicon, may be formed on top of the vertical transistors. In the event of a short between the gate and the top of a vertical transistor, the silicon columns neighboring the shorted column will act as the gate conductor in an MOS transistor, forcing the shorted column to become depleted and thus limit the short circuit current. One drawback of such a structure is that it limits the

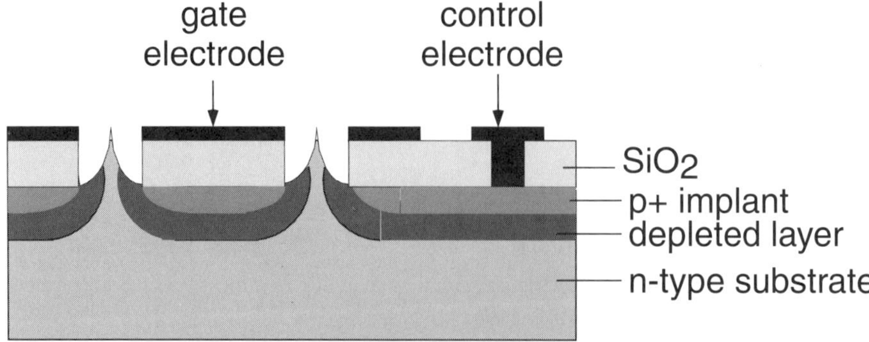

FIGURE 5.48. A JFET-based FEA [106].

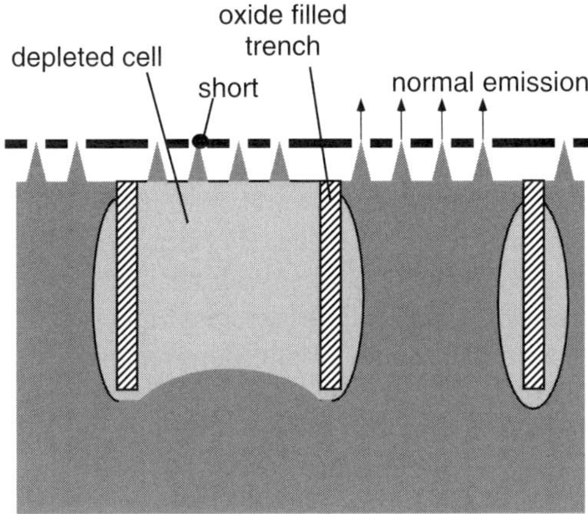

FIGURE 5.49. A VECTL structure with an FEA fabricated on top [19].

current from a subgroup of emission sites, typically only after a short has occurred, and the device will not function if any of the emitters on the adjacent columns have also shorted.

5.6. SUMMARY

Single crystal silicon provides an excellent material for producing structures that are both mechanically and electrically consistent and robust. Oxidation currently provides a simple means of sharpening field emission sites. Epitaxial growth offers a means to further improve the uniformity and perhaps tailor the surface electronic properties at the emission sites. Current silicon FEAs commonly produce dc currents of several hundred microamperes for indefinite periods. To increase the practical size and total currents of the arrays, the maximum current at each emission site or group of sites must be limited. The semiconductor properties of the silicon substrate and the emitter tips themselves can be used to achieve this control. Integrated resistor, diode, and transistor structures are easily fabricated and prevent excessive emission at individual sites, thus improving stability, robustness, and scalability. Although the presence of native silicon oxide at the emitting surface is not ideal for demanding high current applications, it provides a chemically robust surface and works reasonably well as an emitter without difficult cleaning procedures.

The physics of field emission from "practical" surfaces such as silicon covered with native oxide is not accounted for by theories that neglect transport and assume a simple free electron gas (such as the F-N theory). The variable charge and thickness of the silicon native oxide make the emission characteristics unstable and probably cause arcs in most cases. The native oxide creates an initial state density much different than the bulk silicon and inhibits charge transport to and from those initial states. The resulting surface charge may affect the band bending in the silicon along the tip shank and even at the apex. The band bending can affect transport to the tip apex as well as the low current electron distribution. The properties of the native oxide can be altered by emission and by the environment. Despite these complications, the silicon native oxide works reasonably well as a chemically inert yet electron permeable vacuum interface. Further, most of these issues cause no problems if only small amounts of current with relatively wide energy distribution are required, since resistor or transistor structures can be placed in series with each tip to prevent excessive local currents. Applications requiring modest current density ($\sim$10 A/cm^2), robust operation, and scaleable total current can be addressed with current silicon technology when a local-current-limiting structure is used. Robust surfaces with better transport and stability than silicon native oxide undoubtedly exist and provide a worthy research goal.

DEDICATION

We dedicate this chapter to Henry Gray, who died of cancer on July 21, 1999. Henry wished for a more robust field emitter source, and so began to experiment with alternative materials and fabrication techniques. These experiments first centered on building the emitter tips from silicon, which he hoped could prevent destructive arcs by limiting transport to the tip apex. He also took part in building and characterizing a variety of other emitter structures and materials. Henry was always ready to discuss emission and fabrication issues with anyone. He always took time to listen to anyone who sought his help, tried to help improve their understanding, and (when he agreed with them) help promote their ideas. He was thrilled to debate new ideas and new applications when interesting and unexplained results were reported. He was a constant source of encouragement and enthusisam. Henry co-organized the International Vacuum Microelectronics Conferences and served as the steering committee chairman for the first 11 conferences.

REFERENCES

1. I. Brodie and C. Spindt, *Adv. Electron. Electron Phys.* **83**, 1 (1992).
2. D. Temple, *Mater. Sci. Eng. R* **24**, 185 (1999).
3. J. P. Calame, H. F. Gray, and J. L. Shaw, *J. Appl. Phys.* **73**, 1485 (1993).
4. G. Fursey, in *J. Vac. Sci. Technol. B* **13**, 558 (1995).
5. K. Betsui, in *Proc. 4th Int. Vacuum Microelectronics Conf.*, Nagahama, Japan, 1991, p. 26.
6. M. Ding, H. Kim, and A. I. Akinwande, *IEEE Electron Devices Lett.* **21**, 741 (2000).
7. D. Temple, W. D. Palmer, L. N. Yadon, J. Mancusi, D. Vellenga, and G. E. McGuire, *J. Vac. Sci. Technol. A* **16**, 1980 (1998).

8. B. R. Chalamala, R. M. Wallace, and B. E. Gnade, *J. Vac. Sci. Technol. B* **16**, 3073 (1998).

9. J. Miller and R. Johnston, *J. Phys. Cond. Mater.* **3**, S231 (1992).

10. J. Shaw, *J. Vac. Sci. Technol. B* **18**, 1817 (2000).

11. J. L. Shaw and H. F. Gray, in *Proc. 9th Int. Vacuum Microelectronics Conf.*, St. Petersburg, Russia, 1996, p. 547.

12. H. F. Gray and J. L. Shaw, in *Proc. 10th Int. Vacuum Microelectronics Conf.*, Kyongju, South Korea, 1997, p. 220.

13. D. Temple, H. F. Gray, C. A. Ball, J. E. Mancusi, W. D. Palmer, G. E. McGuire, and J. L. Shaw, in *Proc. 8th Int. Vacuum Microelectronics Conf.*, Portland, OR, 1995, p. 113.

14. C. O. Bozler, C. T. Harris, S. Rabe, D. D. Rathman, M. A. Hollis, and H. I. Smith, *J. Vac. Sci. Technol. B* **12**, 629 (1994).

15. T. Matsukawa, K. Koga, S. Kanemaru, and J. Itoh, *IEEE Trans. Electron Devices* **46**, 2261 (1999).

16. L. Parameswaran, C. T. Harris, C. A. Graves, R. A. Murphy, and M. A. Hollis, in *Proc. 11th Int. Vacuum Microelectronics Conf.*, Ashville, NC, 1998, p. 136.

17. M. Ding, H. Kim, and A. Akinwande, *IEEE Electron Devices Lett.* **21**, 66–69 (2000).

18. H. F. Gray, Regulatable field emitter device and method of production thereof, US Patent 5359256 (1994).

19. H. Takemura, Y. Tomihari, N. Furutake, F. Matsuno, M. Yoshiki, N. Takada, A. Okamoto, and S. Miyano, in *Tech. Dig. IEDM*, 1997, p. 709.

20. H. F. Gray, in *Proc. 29th Int. Field Emission Symp.*, Stockholm, Sweden, 1982, p. 111.

21. G. J. Campisi and H. F. Gray, R. Greene, H. Gray, and G. Campisi, in *Tech. Dig. IEDM*, 1985, p. 175.

22. J. W. Faust and E. D. Palick, *J. Electrochem. Soc.* **130**, p. 1413–20 (1983); E. D. Palick, H. F. Gray, and P. B. Klein, *J. Electrochem. Soc.* **130**, 956 (1983).

23. H. F. Gray and R. F. Greene, Method of manufacturing a field emission cathode structure, US Patent 4307507 (1981).

24. H. F. Gray and R. F. Greene, Field emitter array, US Patent 5150192 (1992).

25. W. P. Kang, A. Wisitsora-at, J. L. Davidson, Q. Li, J. F. Xu, and D. V. Kerns, in *Proc. 11th Int. Vacuum Microelectronics Conf.*, Ashville, NC, 1998, p. 168.

26. T. Sakai, T. Ono, M. Nakamoto, and N. Sakuma, *J. Vac. Sci. Technol. B* **16**, 770 (1998).

27. K. D. Hobart, M. E. Twigg, H. F. Gray, J. L. Shaw, P. E. Thompson, F. J. Kub, and D. Park, in *Proc. 9th Int. Vacuum Microelectronics Conf.*, St. Petersburg, Russia, 1996, p. 463.

28. K. D. Hobart, F. J. Kub, H. F. Gray, M. E. Twigg, D. Park, and P. E. Thompson, *J. Crystl. Growth* **157**, 338 (1995).

29. J. L. Shaw, H. F. Gray, K. L. Jensen, and T. M. Jung, *J. Vac. Sci. Technol. B* **14**, 2072 (1996).

30. J. L. Shaw, R. S. Silmon, D. W. Park, H. F. Gray, and A. C. Constant, in *Proc. 9th Int. Vacuum Microelectronics Conf.*, Portland, OR, 1995, p. 408.

31. S. Kitamura, K. Hiramatsu, and N. Sawaki, *Jpn. J. Appl. Phys.* **34**, 1184 (1995).

32. R. B. Marcus and T. T. Sheng, *J. Electrochem. Soc.* **129**, 1278 (1982); R. B. Marcus, R. Soave, and H. F. Gray, in *Proc. 1st Int. Vacuum Microelectronics Conf.*, Williamsburg,

VA, 1988, p. 136; R. B. Marcus et al. *Appl. Phys. Lett* **56**, 236 (1990); T. S. Ravi, R. B. Marcus, and D. Liu, *J. Vac. Sci. Technol. B* **9**, 2733 (1991).

33. D. Temple, H. F. Gray, C. A. Ball, J. Mancusi, W. D. Palmer, G. E. McGuire, and J. L. Shaw, in *Proc. 9th Int. Vacuum Microelectronics Conf.*, Portland, OR, 1995, p. 113.

34. A. I. Akinwande, Private communication.

35. J. Itoh, S. Kanemaru, and T. Matsukawa, *Oyo Butsuri* **67**, 1390 (1998) (in Japanese).

36. H. F. Gray, J. L. Shaw, and D. Temple, in *Proc. 9th Int. Vacuum Microelectronics Conf.*, Portland, OR, 1995, p. 31.

37. H. Takemura, N. Furutake, M. Nisimura, S. Tsuida, M. Yoshiki, A. Okamoto, and S. Miyano, *J. Vac. Sci. Technol. B* **15**, 488 (1997).

38. H. Takemura, M. Yoshiki, N. Furutake, Y. Tomihari, A. Okamoto, and S. Miyano, in *Tech. Dig. IEDM*, 1998, p. 859.

39. D. G. Pflug, M. Schattenburg, H. I. Smith, and A. I. Akinwande, in *Tech. Dig. IEDM*, 1998, p. 855.

40. T. Hirano, S. Kanemaru, H. Tanoue, and J. Itoh, in *Tech. Dig. IEDM*, 1996, p. 309.

41. T. Hirano, S. Kanemaru, H. Tanoue, and J. Itoh, *Jpn. J. Appl. Phys.* **35**, 6637 (1996).

42. Y. Toma, S. Kanemaru, and J. Itoh, *J. Vac. Sci. Technol. B* **14**, 1902 (1996).

43. J. Itoh, Y. Tohma, K. Morikawa, S. Kanemaru, and K. Shimizu, *J. Vac. Sci. Technol. B* **13**, 1968 (1995).

44. Y. Toma, S. Kanemaru, and J. Itoh, *J. Vac. Sci. Technol. B* **14**, 1902 (1996).

45. J. Itoh, Y. Tohma, K. Morikawa, S. Kanemaru, and K. Shimizu, *J. Vac. Sci. Technol. B* **13**, 1968 (1995).

46. D. Nicolaescu, V. Filip and J. Itoh, *Jpn. J. Appl. Phys.*, **40**, 83 (2001).

47. C. Py, J. Itoh, T. Hirano, and S. Kanemaru, *IEEE Trans. Elec. Dev.* **44**, 498 (1997)

48. J. W. Gadzuk and E. W. Plummer, *Rev. Mod. Phys.* **45**, 487 (1973).

49. A. Moldinos, *Field, Thermionic, and Secondary Emission Spectroscopy*, Plenum: New York, 1984.

50. S. C. Miller and R. H. Good, *Phys. Rev.* **91**, 174 (1953).

51. J. W. Gadzuk and E. W. Plummer, *Phys. Rev Lett.* **26**, 92 (1971).

52. J. W. Gadzuk, *Chem. Phys.* **251**, 87 (2000).

53. L. W. Swanson and L. C. Crouser, *Phys. Rev. Lett.* **16**, 389 (1966).

54. L. W. Swanson and L. C. Crouser, *Surf. Sci.* **23**, 1 (1970).

55. H. W. Fink, *Phys. Scr.* **38**, 260 (1988).

56. H. W. Fink and H. Schmidt, *Phys. Rev. Lett.* **65**, 1204 (1990).

57. N. D. Lang, A. Yacoby, and Y. Imry, *Phys. Rev. Lett.* **63**, 1499 (1989).

58. J. W. Gadzuk, *Phys. Rev. B* **47**, 12832 (1993).

59. V. T. Binh, S. T. Purcell, N. Garcia, and J. Doglioni, *Phys. Rev. Lett.* **69**, 2527 (1992).

60. N. Ernst, J. Unger, H. W. Fink, M. Grunze, H. W. Muller, B. Volkl, M. Hofmann, and C. Woll, *Phys. Rev. Lett.* **70**, 2503 (1993).

61. Ming L. Yu, N. D. Lang, B. W. Hussey, and T. H. P. Chang, *Phys. Rev. Lett.* **77**, 1636 (1996).

62. R. G. Forbes, in *Proc. 9th Int. Vacuum Microelectronics Conf.*, St. Petersburg, Russia, 1996, p. 58.

63. R. Morin and H. W. Fink, *Appl. Phys. Lett.* **65**, 2363 (1994).

64. S. T. Purcell, V. T. Binh, and N. Garcia, *Appl. Phys. Lett.* **67**, 436 (1995).

65. T. Matsukawa, S. Kanemaru, K. Tokunaga, and J. Itoh, *J. Vac. Sci. Technol. B* **18**, 952 (2000).

66. H. Nakane, Y. Muto, K. Yamane, and H. Adachi, *J. Vac. Sci. Technol. B* **18**, 1003 (2000).

67. H. F. Gray, Private communication.

68. J. L. Shaw and H. F. Gray, in *Proc. 10th Int. Vacuum Microelectronics Conf.*, Kyongju, South Korea, 1997, p. 743.

69. A. J. Miller and R. Johnston, *J. Phys. Condens. Matt.* **3**, 231 (1991).

70. S. T. Purcell, V. T. Binh, and R. Baptist, *J. Vac. Sci. Technol.* **15**, 1666 (1997).

71. R. Johnston, *J. Phys. Condens. Matter* **3**, 187 (1991).

72. J. W. Keister, J. E. Rowe, J. J. Kolodziej, H. Niimi, T. E. Madey, and G. Lucovsky, *J. Vac. Sci. Technol. B* **17**, 1831 (1999).

73. K. R. Farmer, C. P. Debauche, A. R. Giordano, P. Lundgren, M. O. Anderson, and D. A. Buchanan, *Appl. Surf. Sci.* **104/105**, 369 (1996). D. J. DiMaria and E. Cartier, *J. Appl. Phys.* **78**, 3883 (1995).

74. T. Matsukawa, S. Kanemaru, K. Tokunaga, and J. Itoh, *J. Vac. Sci. Technol. B* **18**, 1111 (2000).

75. Electro Boundary Element Potential Solver, Integrated Engineering Software, 300 Cree Cresent, Winnipeg, Manitoba Canada R3J 3W9.

76. N. Shamir, J. G. Mihaychuk, H. M. van Driel, and H. J. Kreuzer, *Phys. Rev. Lett.* **82**, 359 (1999).

77. G. B. Alers, K. S. Krisch, D. Monroe, B. E. Weir, and A. M. Chang, *Appl. Phys. Lett.* **69**, 2885 (1996).

78. H. Richter, T. E. Orlowski, M. Kelly, and G. Margaritondo, *J. Appl. Phys.* **56**, 2351 (1984); T. E. Orlowski and D. A. Mantell, *J. Appl. Phys.* **64**, 4410 (1988).

79. D. J. DiMaria and E. Cartier, *J. Appl. Phys.* **78**, 3883 (1995).

80. Shin-ichi Takagi, N. Yasuda, A. Toriumi, *IEEE Trans. Electron Devices* **46**, 348 (1999).

81. W. P. Dyke and J. K. Trolan, *Phys. Rev.* **89**, 799 (1953).

82. G. N. Fursey, D. V. Glazanov, and S. A. Polezhaev, *IEEE Trans. Diel. Electr. Insulation* **2**, 281 (1995).

83. J. Browning, N. McGruer, W. J. Bintz, and M. Gilmore, *IEEE Electron Device Lett.* **13**, 167 (1992).

84. W. J. Bintz and N. E. McGruer, *J. Vac. Sci. Technol. B* **12**(2), 697 (1994).

85. S. Meassick and H. Champaign, *J. Vac. Sci Technol. B* **14**, 1914 (1996).

86. N. E. McGruer, J. Browning, S. Meassick, M. Gilmore, W. J. Bintz, and C. Chan, *J. Vac. Sci. Technol. B.* **11**, 441 (1993).

87. K. H. Bayliss and R. V. Latham, *Proc. R. Soc. Lond. A* **403**, 285 (1986).

88. N. S. Xu, in *High Voltage Vacuum Insulation* (R. V. Latham, Ed.), Academic Press: New York, Sec. 4.3, 1995.

89. R. F. Greene and H. F. Gray, p-n junction controlled field emitter array cathode, US Patent 4513308 (1985).

90. T. Hirano, S. Kanemaru, H. Tanoue, and J. Itoh, *Jpn. J. Appl. Phys.* **34**, 6907 (1995).

91. S. Kanemaru, T. Hirano, H. Tanoue, and J. Itoh, *J. Vac. Sci. Technol. B* **14**, 1885 (1996).

92. J. Itoh and S. Kanemaru, Hyomen Kagaku **17**, 718 (1996) (in Japanese).

93. J. Itoh, *Appl. Surf. Sci.* **111**, 194 (1997).

94. K. Yokoo, M. Arai, M. Mori, J. Bae, and S. Ono, *J. Vac. Sci. Technol. B* **13**, 491 (1995).

95. G. Hashiguchi, H. Mimura, and H. Fujita, *Jpn. J. Appl. Phys.* **35**, L84(1996).

96. D. Kim, S. J. Kwon, and J. D. Lee, in *Proc. 9th Int. Vacuum Microelectronics Conf.*, St. Petersburg, Russia, 1996, p. 534.

97. H. Gamo, S. Kanemaru, and J. Itoh, *Appl. Phys. Lett.* **31**, 1301 (1998).

98. J. Itoh, T. Hirano, and S. Kanemaru, *Appl. Phys. Lett.* **69**, 1577 (1996).

99. S. Kanemaru, K. Ozawa, K. Ehara, T. Hirano, H. Tanoue, and J. Itoh, *Jpn. J. Appl. Phys.* **36**, 7736 (1997).

100. K. Werner, *Inf. Display* **14**, 128 (1998).

101. J. Itoh and S. Kanemaru, *Inf. Display* **14**, 12 (1998).

102. A. Hosono, S. Kawabuchi, S. Horibata, and S. Okuda, *J. Vac. Sci. Technol. B* **17**, 575 (1999).

103. J. Itoh, Y. Kagawa, and S. Kanemaru, in *Proc. 11th Int. Vacuum Microelectronics Conf.*, Asheville, NC, 1998, p. 128.

104. K. Uemura, S. Kanemaru, and J. Itoh, *Jpn. J. Appl. Phys.* **35**, 6629 (1996).

105. J. Itoh, K. Uemura, and S. Kanemaru, *J. Vac. Sci. Technol. B* **16**, 1233 (1998).

106. M. Arai, N. Kitano, H. Shimawaki, H. Mimura, and K. Yokoo, in *Proc. 10th International Vacuum Microelectronics Conf.*, Kyongju, South Korea, 1997, p. 38.

CHAPTER 6

Novel Cold Cathode Materials

WEI ZHU

Bell Laboratories, Lucent Technologies, 600 Mountain Avenue, Murray Hill, New Jersey 07974-0636

PETER K. BAUMANN

AIXTRON AG, Kackertstr. 15-17, 52072 Aachen, Germany

CHRISTOPHER A. BOWER

Bell Laboratories, Lucent Technologies, 600 Mountain Avenue, Murray Hill, New Jersey 07974-0636

6.1. INTRODUCTION

Ever since microtip field emitter arrays (FEAs) were invented, there have been efforts in exploring novel cold cathode materials for better and more reliable performance. The search for new materials is also motivated by the desire to have a more manufacturing-friendly process for making the cathodes. The fabrication of uniform metal or semiconductor-tip FEAs over a large area has proven to be a difficult and expensive process. The control voltage required for emission from these FEAs is also relatively high ($\sim$50–100 V) because of their high work functions. The high voltage operation increases the possibility of damaging the tips due to ion bombardment and surface diffusion. It also necessitates high external power to be supplied to produce the required emission current. Additionally, FEAs are highly susceptible to surface contamination and have poor environmental stability. For these reasons, there is a need to develop more robust, low voltage field emitters. Field emitters operating at low voltages are important to the development of many vacuum microelectronic devices, in particular displays, as they will allow simpler driver circuitry to be used, help reduce the overall power consumption, and increase the device lifetime.

This chapter primarily addresses a new class of carbon-based cathode materials that has emerged recently as promising field emitters. They include diamond, diamond-like or amorphous carbon, and carbon nanotubes. Early in the 1970s, investigators already discovered that graphite fibers show better stability than refractory metals in a number of environments [1,2], suggesting that carbon-based materials are uniquely suited as stable field emitters. The discovery of negative electron affinity (NEA) on

Vacuum Microelectronics, Edited by Wei Zhu.
ISBN 0-471-32244-X. 2001 John Wiley & Sons, Inc.

hydrogen-terminated diamond surfaces, coupled with the new ability of depositing diamond in thin film forms on a variety of substrates by low pressure chemical vapor deposition (CVD), has excited great interest in the use of diamond and diamond-like carbon materials as field emitters since the early 1990s. Carbon nanotubes have diameters in the range of 1–30 nm with aspect ratios greater than 1000. This one-dimensional nanometer-scale structure consisting of cylindrical graphitic sheets makes nanotubes naturally attractive as field emitters, as high field concentrations can be induced locally at the tips of nanotubes and allows electrons to be emitted at low voltages. Both diamond and carbon nanotube materials have been experimentally shown to emit electrons at relatively low electric fields and generate useful current densities. In particular, nanotube emitters can deliver macroscopic current densities exceeding 1 A/cm^2. Prototype field emission displays based on these carbon materials have been demonstrated.

In this chapter, we will introduce the concept of NEA of diamond surfaces and experimental techniques useful for field emission measurements. We will review field emission data obtained from many different types of carbon materials, and discuss physical mechanisms involved and potential device applications that are being explored.

In addition, we will also briefly describe three other non-carbon based cathodes that have shown promise for vacuum microelectronic applications. They are surface conduction emitters (SCEs), ferroelectric emitters, and metal–insulator–metal (MIM) emitters. While these cathodes clearly belong to different classes of electron emitters, they are similar to both the microtip FEAs and carbon-based emitters in that the electron emission process is controlled principally by the application and manipulation of voltage or electric field. Although specific schemes of applying the fields to activate the emission vary, ranging from fields perpendicular to the emitting surface in both FEAs and carbon emitters, to those parallel to the emitting surface in SCEs, and to those created within the cathodes in both ferroelectric and MIM emitters, it is the electrostatic force exerted by the external voltages that is extracting electrons from these cathode surfaces. We will provide sufficient references for readers who are interested in more detailed and in-depth understanding of the electron emission phenomena from these non-carbon based cathodes.

6.2. DIAMOND EMITTERS

Diamond is one of the main crystalline allotropes of carbon, as shown in Fig. 6.1 [3]. It is the high-pressure form of carbon with an sp^3 tetrahedral bonded cubic structure. This is in contrast to graphite, the high-temperature form of carbon that is composed of hexagonal layers of carbon atoms with sp^2 trigonal bonds. Among the many unusual and fascinating characteristics that diamond possesses is the presence of a very small barrier to its electrons leaving the surface and emitting into vacuum. When diamond surfaces are terminated with hydrogen atoms, the electron affinity, which is a measure of the energy barrier that electrons must overcome to escape into vacuum, can become negative [4]. In contrast to other materials such cesium (Cs) and barium (Ba) that also have low energy barriers to the emission of electrons but

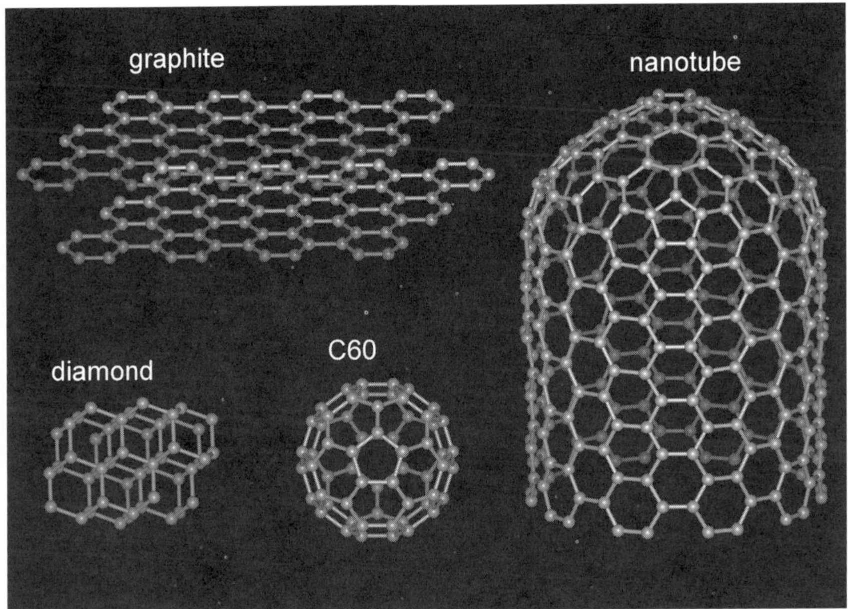

FIGURE 6.1. The allotropes of carbon, graphite, diamond, C_{60}, and carbon nanotubes [3]. (Courtesy of Daniel Colbert, Rice University, Houston, TX)

are chemically reactive, the diamond surfaces are chemically stable and mechanically strong. In fact, diamond is the only known material with NEA that is stable in air. This combination of low surface barrier to electron emission in an otherwise robust material has attracted attention to diamond's promise as a high performance cold cathode material. The low barrier to electron emission allows diamond, in principle, to emit electrons efficiently and at low applied fields, even from planar surfaces, without the need of fabricating sharp microtips to provide geometric field enhancement. This offers potentially important advantages in ease of fabrication and low cost manufacturing. In addition, the excellent mechanical and chemical stability of diamond leads to highly durable and reliable emitters.

6.2.1. Negative Electron Affinity

The electron affinity χ of a semiconductor is defined as the energy required to remove an electron from the conduction band minimum to a distance macroscopically far from the semiconductor (i.e. away from image charge effects). At the surface, this energy is simply the difference between the vacuum level and the conduction band minimum, as shown schematically in the band diagrams of a semiconductor surface in Fig. 6.2. The electron affinity is not dependent on the Fermi level of the semiconductor. Thus, while doping can change both the Fermi level and the work function (difference between the vacuum level and the Fermi level), it will generally not affect the electron affinity.

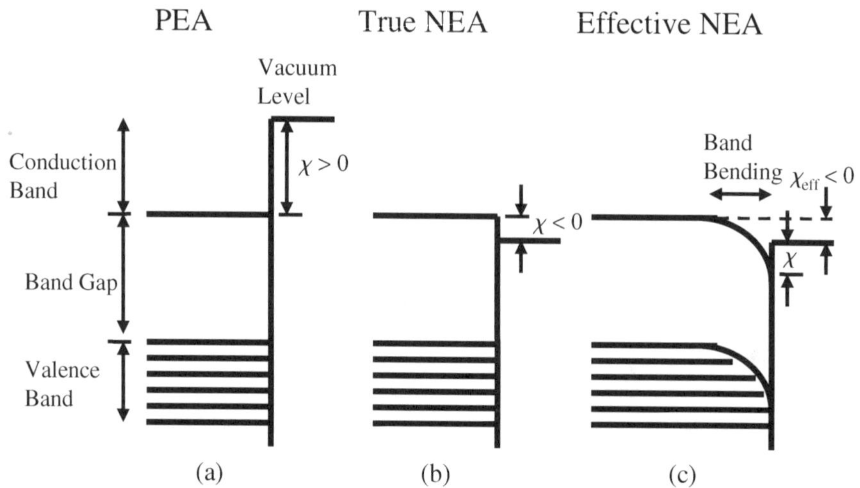

FIGURE 6.2. Band schemes of a semiconductor surface. (a) PEA, (b) true NEA, and (c) effective NEA.

The electron affinity of a material is closely related to both the fundamental atomic structure of the material and surface dipoles created by surface terminations, including steps, reconstructions, and adsorbates [5]. While the atomic arrangements are an intrinsic property of the material, the surface dipoles can readily be altered and manipulated to affect the electron affinity. As an example to illustrate the magnitude of this effect, let's consider a surface terminated with hydrogen. If the average nuclear and electronic charges are assumed to be point charges separated by 0.5 Å, then, for a surface density of 1×10^{15} cm^{-2} for hydrogen atoms, we would find a $\sim$9 eV effect on the energy levels due to the surface dipoles [6]. Certainly a complete charge transfer is never a reasonable possibility, but this simple calculation demonstrates the significance of the surface dipoles.

For most semiconductors, the conduction band minimum is below the vacuum level as shown in Fig. 6.2(a), and the electron affinity is positive. Electrons in the conduction band are bound to the semiconductor by an energy equal to the electron affinity. In some cases, surface conditions can be obtained in which the conduction band minimum is above the vacuum level as shown in Fig. 6.2(b), and the electron affinity is thus negative. Under such conditions, the conduction electrons are not bound to the sample but can escape with a kinetic energy equal to the difference in energy between the conduction band minimum and the vacuum level, although the electrons are still bound to the vicinity of the surface by coulomb forces. These true NEA surfaces are rare, however. NEA semiconductor surfaces (e.g. Si, GaAs, GaP) are more often prepared by a combination of heavy p-type doping together with Cs or Cs/O surface coatings [7–11]. Both Cs and Cs/O deposits are ionic species that form affinity-lowering surface dipoles. When coupled with a characteristically steep downward band bending within the heavily p-type doped semiconductors, these dipoles lead to the formation of an effective NEA surface as shown in Fig. 6.2(c). It

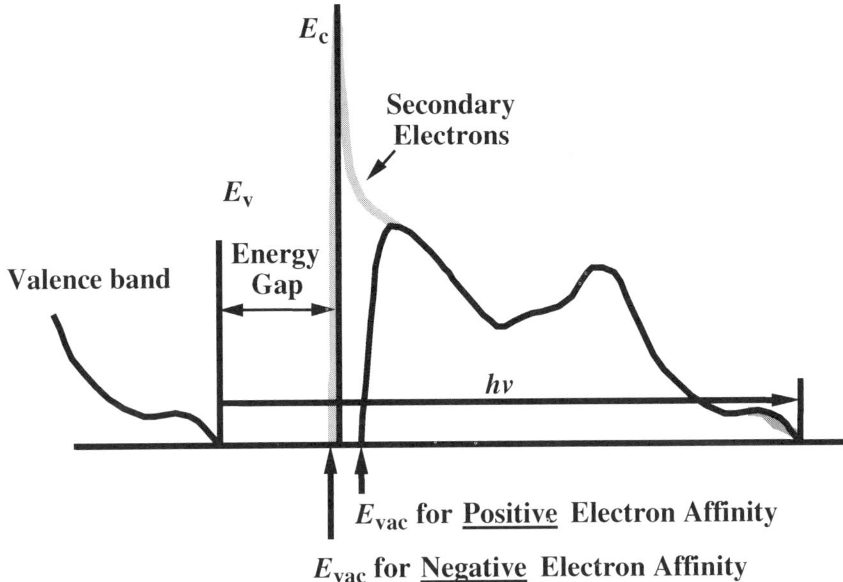

FIGURE 6.3. Photoemission spectra of PEA and NEA semiconductor surfaces. The NEA spectrum is broadened to lower kinetic energy with a peak at the lowest kinetic energy due to quasi-thermalized electrons [12].

is termed effective NEA, because, in fact, there is still a positive barrier at the very surface. The short range band bending is a critical component of these conventional (effective) NEA semiconductor surfaces.

The electron affinity can be conveniently determined by photoemission spectroscopy. An incident light (e.g. an ultraviolet light) excites electrons from the valence band into states in the conduction band, and some of these electrons quasi-thermalize to the conduction band minimum. For an NEA surface, these electrons can emit into vacuum and are detected as a sharp feature at the low energy end of a photoemission spectrum. However, for a surface with positive electron affinity (PEA), emission from the conduction band minimum will not occur, and the value of the electron affinity can thus be deduced. Figure 6.3 shows schematically a photoemission spectrum from a semiconductor [12]. The electron affinity χ or the presence of NEA can be deduced from the width of the spectrum W as follows:

$$\chi = h\nu - E_g - W \quad \text{for a positive electron affinity}$$

$$0 = h\nu - E_g - W \quad \text{for a negative electron affinity}$$

where $h\nu$ is the photon energy, and E_g is the bandgap.

Another method to study surface emission is secondary electron emission (SEE). When a semiconducting material is sexposed to high-energy (1–5 keV) electrons, electron–hole pairs are created in the conduction and valence bands. A gain is obtained when one incident (high-energy) electron generates several electron–hole pairs. The electrons in the conduction band can move to the surface and emit into vacuum, the process of which can be enhanced by the presence of an NEA surface. However, the electrons generated in SEE reside deeper inside the sample than those from ultraviolet photoemission spectroscopy (UPS), which makes SEE not as surface sensitive. Although an energy spectrum of emitted SEE electrons can be similarly obtained, it is more difficult to draw conclusions from spectral width measurements, because the energy of the incident electrons depends on the work function difference between the electron gun and the sample. Additionally, band bending at the surface can influence the gain in SEE, because electron–hole pairs are created deep inside the sample and will be either facilitated by downward band bending or impeded by upward band bending when they try to reach and escape the surface. Therefore, UPS is generally a more suitable technique than SEE to perform electron affinity measurements.

Himpsel et al. [4] was the first to report from photoemission measurements that the unreconstructed diamond (111) surface of a boron doped, type IIb single crystal exhibited NEA. They found that the photoemission yield started to increase at a threshold equal to the fundamental band gap of 5.5 eV, and the SEE was detected at energies down to the conduction band minimum, both of which pointed directly to the existence of an NEA surface. Pate [13,14] later studied the effects of hydrogen on the electronic and atomic structures of diamond surfaces. He found that the unreconstructed NEA diamond (111) surface was usually hydrogen-terminated, and the removal of hydrogen resulted in a $2 \times 1/2 \times 2$ reconstructed diamond surface that showed PEA. This finding is remarkable because, unlike Cs activated conventional NEA semiconductor surfaces that are extremely chemically sensitive and unstable [15], only covalently bonded species (C, H) are involved in the NEA diamond surface. Such hydrogenated diamond surfaces are stable over air exposure, and no heavy p-type doping and resultant sharp downward band bending are involved. In this sense, hydrogen-terminated diamond surfaces are true NEA surfaces. Nemanich and his coworkers [12,16–18] further confirmed that NEA could be induced or removed repeatedly by the addition to or removal of hydrogen atoms from the diamond surfaces. They found that the role of hydrogen in inducing NEA could be extended to all the major index surfaces, including diamond (111), (100), and (110). Hydrogenation of diamond surfaces was also reported to significantly increase the surface conductance by 10 orders of magnitude and enhance the secondary electron emission yield by a factor of 30 compared to clean or oxygenated surfaces [19,20].

While hydrogen induces NEA and can lead to a change of 1.65 eV in electron affinity between a hydrogenated and "clean" diamond surface [21], oxygen is found to form a PEA diamond surface [18]. The difference in electronegativity between C-H and C-O bonds in the surface dipoles certainly plays an important role. Additionally, different and asymmetric charge distributions associated with specific hydrogen and oxygen adsorption sites on diamond surfaces could result in different dipole effects and thus affect the electron affinity. For example, Rutter and Robertson [22] calculated a smaller PEA for "bridge bonded" oxygen on the (100) diamond surface than for "double bonded" oxygen. Similarly, different electron affinities were measured for oxygenated diamond (100) surfaces that were prepared by different surface cleaning procedures [18]. Because of the large effect of the surface dipoles, it is important to identify the type of surface termination in describing the electron affinity of diamond.

Thin metallic films can also lead to the formation of NEA diamond surfaces. Cs deposition on diamond has been shown to induce NEA [23,24]. However, since diamond has a large band gap, other relatively higher work function metals may also be suitable to establish an NEA surface. Working examples include Ti, Ni, Co, Cu, and Zr on various surfaces of diamond [17,18, 25–30], and the NEA property induced by these metals was again found to be stable after air exposure. A submonolayer of TiO deposited on a hydrogen-free, reconstructed diamond (111) surface was also effective in modifying the electron affinity from positive to negative [31]. Baumann and Nemanich [26,30] correlated the electron affinity of metal-coated diamond with the Schottky barrier height at the metal–diamond interface and found that a lower Schottky barrier height would result in a lower electron affinity.

6.2.2. Emission from Diamond

Although the NEA property of diamond was reported by Himpsel et al. [4] in 1979, investigations of field emission properties of diamond did not start in earnest until the early 1990s when diamond films and coatings with controlled quality on many different substrates became widely available due to the emergence of low pressure CVD techniques. Early reports on diamond field emission include Wang et al. [32] who reported low-field (<3 V/μm) electron emission from undoped CVD polycrystalline diamond films, Geis et al. [33] who measured emission currents from a diamond diode structure consisting of a p-type diamond substrate and a carbon ion implanted diamond surface, and Djubua et al. [34] who showed that arrays of diamond-like carbon (or DLC, a form of amorphous carbon) cones formed by plasma polymerization required the lowest operating voltage for emission when compared to arrays of Mo and Hf tips. Since then, a large body of literature has developed on field emission from diamond. Although discrepancies in results exist among the many published reports, there is irrefutable evidence that indicates that many diamond and DLC materials are good field emitters, with low turn-on fields and useful emission current densities.

6.2.2.1. Emission Measurements. The field emission properties of diamond are typically measured in a parallel plate apparatus by bringing a probe (i.e., anode) close to a sample surface (i.e., cathode) and applying a voltage between the two. The emission current is measured as a function of the applied voltage. The anode can be a small-size metallic probe that can be moved either vertically towards or laterally across the cathode surface. It can also be a large-size fluorescent screen maintained at a certain distance from the cathode by electrically insulating spacers. In the case of a metallic probe, the anode often assumes a spherical shape with a sufficiently large diameter (on the order of a millimeter) to avoid nonuniform field distributions at edges, especially at small anode–cathode spacings. With an anode that can be stepped vertically towards or away from the surface, the current–voltage (*I–V*) characteristics can be determined at different anode–cathode spacings that typically range from a few micrometers to hundreds of micrometers, as shown in Fig. 6.4(a). This allows both the

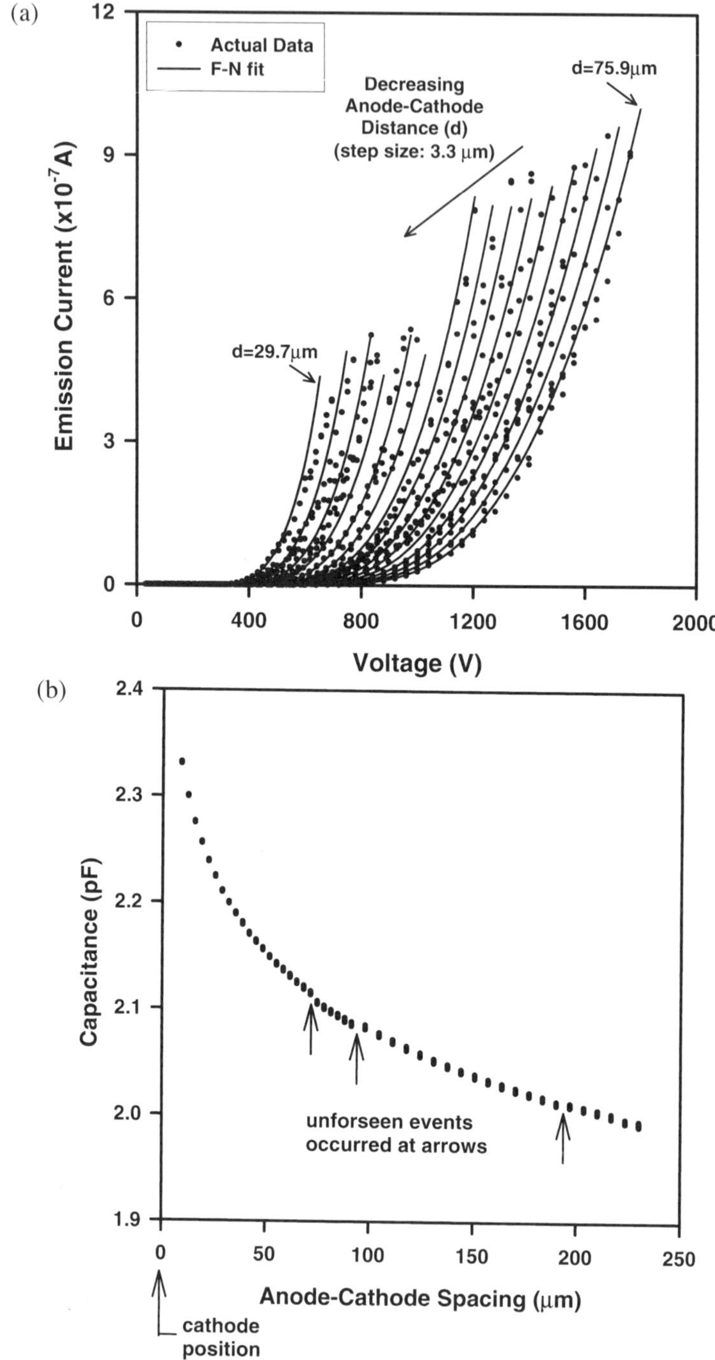

FIGURE 6.4. (a) Field emission current versus applied voltage at different cathode–anode distances for diamond emitters, and (b) capacitance as a function of cathode–anode distance. The arrows indicate that unpredicted events occurred at these positions during the measurements [35].

stability and uniformity of the emitters to be estimated. As the probe is moved away from the cathode surface, the area of relatively uniform electric field under the probe expands, and more emitters can contribute to the measured current. If all the emitters were identical, the current would scale proportionally with the area. However, if the physical conditions of the emitters vary, the current is likely to increase faster than the area, as a larger area has a larger chance of containing an exceptionally good emitter. Often, in arrays of field emitters, a single exceptionally good emitter dominates the current from the entire array.

Another useful aspect of measuring I–V characteristics at different anode–cathode spacings is to determine if the emission properties are distance-sensitive, as highly resistive samples or surfaces with significant morphologies might produce data that are seemingly independent of the nominal anode position. The actual distance between the anode and cathode can be determined by simple mechanical means such as using a micrometer, but it potentially involves substantial uncertainties, particularly when significant surface morphologies are involved. It is equally unreliable to simply use the position at which the current becomes infinite and presume that the anode touches the cathode surface, as large currents can occur as the result of an arc while the anode is still more than tens of micrometers above the surface. A method reported by Zhu et al. [35] that measures the capacitance between the anode and cathode as their spacing changes and fits the capacitance data to a sphere-to-plane model allows the position of the cathode surface and thus the anode–cathode distance to be better determined. As shown in Fig. 6.4(b), the capacitance measurements also indicate problems that occur during the measurements, such as arcing, missteps of the probe, moving dust particles, or loose samples, because they introduce discontinuities in an otherwise smooth capacitance-distance curve. The error involved in estimating the distance from the capacitance data was reported to be 3 μm out of a 100 μm spacing or 3% [36].

It should be noted that it is generally difficult to extract physically meaningful parameters from a fit of the Fowler-Nordheim (F-N) equation [37] to the I–V data. This is especially true for data obtained from randomly distributed multiple emitting tips such as those on the surfaces of CVD diamond films, where the best that can be hoped for is to extract an average of the distribution of tip parameters. The F-N equation itself is degenerate, because it is described almost exactly with two parameters (offset a and slope b when plotted as $\log(J/E^2)$ vs. $1/E$), yet there are three physically interesting parameters to be extracted (work function ϕ, tip area α, and field-enhancement factor β). As a result, it is difficult to make many claims about the details of the emission physics from the I–V data alone, since the electronic properties of the tips cannot be easily separated from the geometric field-enhancement factor. Nevertheless, it is typical to calculate the field required to produce a certain current density from the I–V data and use them as a figure of merit to rank the various emitting samples. These are not the local fields on the tips of the emitters; instead, all the field values thus derived are for the macroscopic electric field, far above the surface, where the field is uniform and independent of surface roughness. This macroscopic field, which represents both electronic and geometric contributions, is actually the relevant parameter for building any devices using a field emission cathode.

A laterally movable probe placed at a certain distance from the surface (e.g., a few micrometers) will allow a scanned image of an emitting surface to be obtained. This technique is useful in getting a measure of the emission site density of a surface as well as in correlating the spatial distribution of emission sites with the surface morphological features. A large fluorescent screen (i.e., a phosphor screen) can similarly be used for examining the emission uniformity and determining the location of emission sites. This experiment actually serves as a field emission microscope (FEM) that can be used to monitor the dynamics of the field emission process and study atomic interactions on cathode surfaces during the emission. Such measurements complement the I–V characteristics in many important ways, because they provide critical information on the emission uniformity and stability, two of the most critical parameters for device applications. I–V data alone without corresponding FEM images could often be misleading in that emission currents might be dominated by loose particles, sharp edges, or surface contaminants. A problem associated with the use of phosphor screens is that it has poor spatial resolutions at high currents because of the phosphor saturation. The phosphor particles can also be easily pulled away by the intense electric field or bombarded loose by the electron beams and subsequently contaminate the cathode.

It is evident that performing field emission measurement requires great care, and in fact, some early reports might have been dominated by artifacts attributable to poor vacuum, surface contaminations, or other effects related to the high electric fields. The residual gas in the vacuum chamber could result in a background ion current. Additionally, the strong fields can sometimes cause a discharge to ignite and severly damage the surface. Following arc discharges, crater formations and molten areas with debris were observed on the surface of CVD diamond and amorphous carbon films deposited on silicon substrates [38]. The emitted electrons impacting the anode can lead to desorption or removal of material. This will result in the deposition of materials on the cathode or the formation of a destructive discharge. Positive ions can be accelerated towards the cathode and may damage the surface, even if no discharge froms. One way to avoid these damaging effects is the use of lower voltages. This implies that smaller distances between anode and cathode need to be employed. At smaller distances, however, effects due to the surface roughness of the cathode become more significant.

6.2.2.2. Emission Characteristics. For a semiconducting field emitter such as diamond, the emitting electrons can originate from either the conduction band, the valence band, and/or surface states. Diamond has a wide bandgap ($\sim$5.5 eV), and undoped diamond is thus generally thought to be unable to produce sustained electron emission because of its insulating nature. Although electrons emitting from surface states in diamond can occur [13], there are no obvious mechanisms by which electrons can be transported through the undoped bulk to the surface states. Either the bulk or the surface of diamond must be made conductive in order to sustain the emission process. Furthermore, to fully take advantage of diamond's NEA to realize electron emission at low applied fields, the Fermi level must be close to the conduction band.

This would require the diamond to be n-type doped. Although p-type diamond can be readily made available by boron doping, very high electric fields are needed for emission to occur, because the emitting electrons reside deep (>5 eV) below the vacuum level in p-type diamond.

An ideal emitting structure may, therefore, consist of an n-type doped semiconducting diamond with a true NEA surface. However, making n-type diamond effectively and reliably and thus supplying electrons to the conduction band remains a significant challenge. Prins reported that n-type characteristics can be obtained by ion implanting diamond [39,40]. This method was employed by Geis et al. [33] to create an all-diamond p–n junction for electron emission. Carbon ions were used to create an n-type layer in a p-type diamond substrate. Dopants such as nitrogen, phosphorus, and lithium have also been explored to produce n-type diamond, but with limited success because of either their donor levels being too deep or their incorporation in the diamond structure being unstable [41]. Reported improvements in emission from these "doped" diamond [24,42–47] are likely due to the creation of in-gap defects by the ion implantation or the heavy doping, giving rise to defect enhanced electron emission.

It is not surprising, therefore, that published reports have mostly dealt with undoped or p-type doped diamond emitters. Electron emission has been observed from many different types of diamond materials, including synthetic single crystals and powders, vapor-deposited islands and films with varying surface morphologies and crystallite sizes, and nanocrystalline coatings, as shown in Fig. 6.5. They are generally considered to be in a "planar" geometry because the diamonds are applied or deposited on nominally flat surfaces. Nevertheless, there are many distinct crystallographic facets and edges that protrude out of the cathode surface at the micrometer or nanometer scale that could provide local field enhancements. As the data in Fig. 6.6 show, most of these diamond materials are found to emit electrons efficiently under applied fields. Diamond materials with small grain sizes and high defect densities generally emit better than those with large crystallite sizes and low defect contents. Heavy doping, whether it is n-type or p-type in nature, enhances emission. Outstanding emission properties are discovered in both ultrafine diamond powders containing 1–20 nm crystallites produced by explosive synthesis and nanocrystalline or ultrananocrystalline diamond films (composed of ~1–100 nm crystallites) [36,48–52]. Emission has been found to originate from sites that are associated with defect structures in diamond rather than sharp features on the surface [32,52–57]. Compared with conventional Si or metal microtip emitters, diamond emitters show lower threshold fields, improved emission stability, and robustness in low/medium vacuum environments. Attempts have also been made to apply diamond coatings on tips of silicon or metal-emitter arrays, as shown in Fig. 6.7, to further enhance the emission characteristics [49,58–63]. High emission currents of 60–100 μA per tip have been measured for Si tips conformally coated with nanocrystalline diamond films [64].

6.2.2.3. Emission Mechanisms. There are a number of mechanistic models proposed to account for the emission characteristics of diamond. While the NEA property

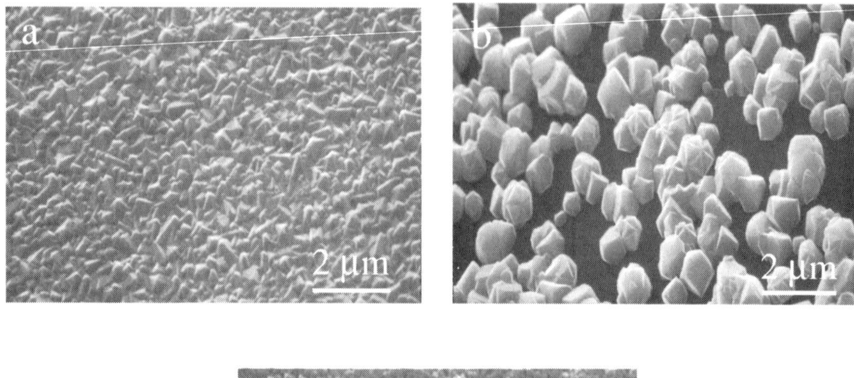

FIGURE 6.5. Scanning electron micrographs of diamond cold cathodes of (a) continuous film-type, (b) island-type, and (c) nanostructured diamond clusters.

of diamond surfaces may be important and can make diamond an efficient photoemitter, it is not adequate by simply invoking NEA to explain why diamond is a good field emitter. As shown in Fig. 6.8, for field emission to occur, and more importantly to sustain, there must be a continuous supply of electrons and a sustainable transport mechanism for the electrons to reach the surface. Moreover, the energy levels of these electrons relative to the vacuum level are critical in determining the threshold field required for emission. An NEA surface is practically useful in reducing the barrier for electron emission only when the energy levels of some occupied states or bands, including surface states, are positioned sufficiently close to the conduction band minimum in diamond.

Among the various models, the simplest and most obvious argument cites the classic field emission theory by noting that there are local field enhancements on sharp morphological features protruding on the diamond surface. Because the surfaces of CVD diamond films and coatings are typically full of facets and ridges, these rough crystallographic features can function as conventional field emitters in which electrons tunnel into vacuum at the protruding small features, facilitated by the enhanced local

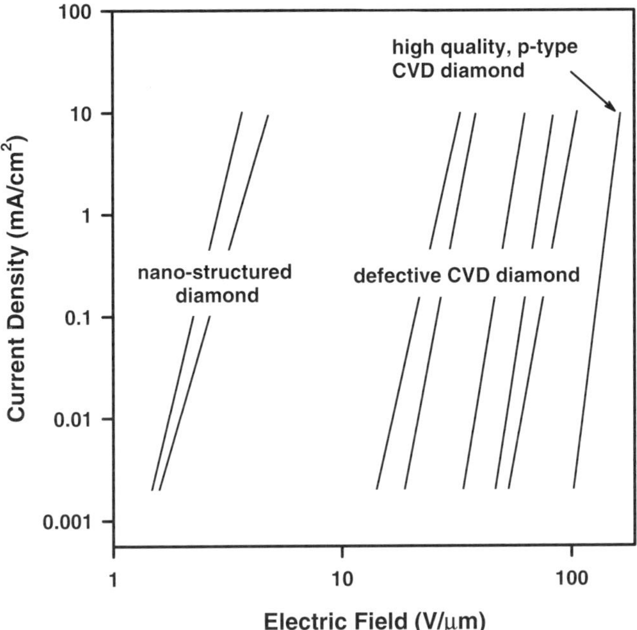

FIGURE 6.6. Emission current density vs. applied electric field for various types of diamond emitters [36].

electric fields. This could certainly be a working and contributing mechanism, but it addresses neither the source of electrons in the wide-bandgap diamond nor the issue of effective electron transport to the surface of diamond.

The defect/impurity theory suggests that structural defects and impurities can form energy states within the band gap of diamond [24,47,65–67]. When the defect density is sufficiently high, the electronic states of various defects can interact and form energy bands. If these bands are wide enough or closely spaced, the electron hopping mechanism within the bands, similar to the Poole–Frenkel conduction mechanism or the Hill type conduction [68,69], could easily provide a steady flow of electrons to the surface and sustain a stable electron emission. The electrons can either be excited into the conduction band or unoccupied surface states from these defect/impurity bands and emit, or tunnel directly from the defect/impurity bands and emit. For example, recent photoelectron yield spectroscopy detected sub-bandgap emission associated with presence of graphite in diamond [21]. The defects/impurities essentially raise the Fermi level by acting as donors of electrons and thus reduce the tunneling barrier. This theory is supported by overwhelming experimental data indicating that defective or lower quality diamonds have better emission properties. It also appears to explain why emission characteristics are enhanced in many "doped" diamonds, not necessarily

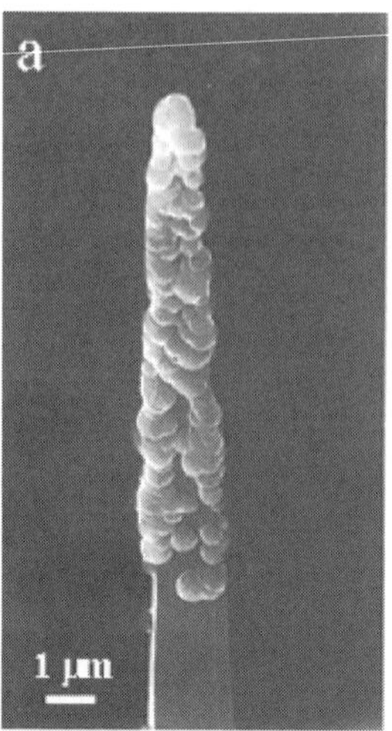

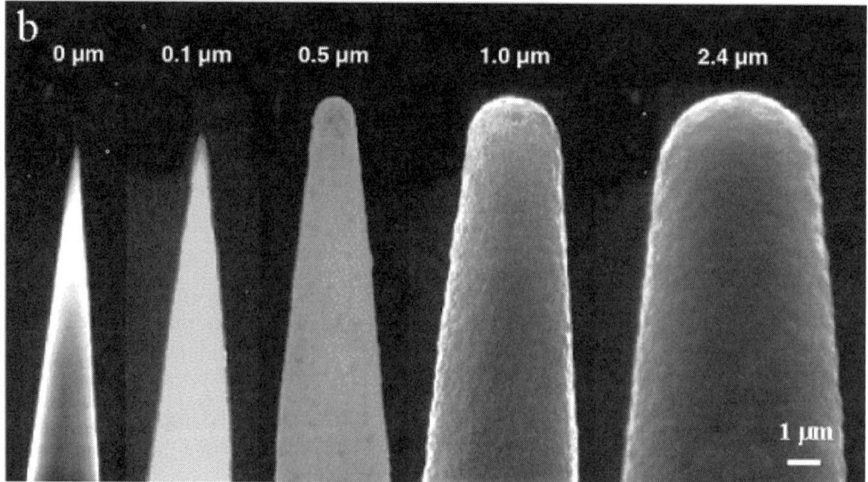

FIGURE 6.7. Scanning electron micrographs of silicon tips coated with (a) conglomerated diamond particles [62] and (b) ultra-nanocrystalline diamond films of varying thickness [64].

because of the electrical doping effect, but rather by the creation of defects in the doping process. However, the exact nature of the responsible defects has not been identified, and the very existence of many defect/impurity energy bands and their locations within diamond's band gap have not been positively confirmed or linked to field emission. The increasing use of FEED (field emission energy distribution), first employed by Henderson and Badgley [70], is helpful in determining the origin of the emitting electrons [53,54,71–75].

In regard to the role of graphite in the electron emission from diamond, Athwal et al. [76] and Xu and Latham [77] suggested a field-induced hot electron emission

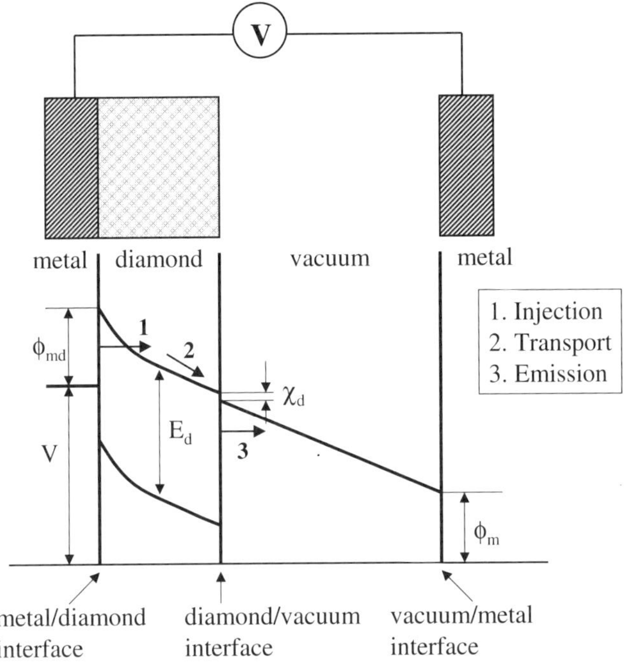

FIGURE 6.8. A schematic band diagram of metal–diamond–vacuum interfaces, showing the three steps involved in stable electron field emission. V is externally applied voltage, ϕ_{md} is metal-diamond Schottky barrier height, E_d is bandgap of diamond, χ_d is electron affinity of diamond, and ϕ_m is surface work function of metal.

process from isolated graphitic inclusions in diamond, citing an antenna effect that leads to field concentrations on a "floating" conductive particle (i.e., graphite) embedded in an insulator matrix (i.e., diamond). This model is based on the observation that active emission sites correspond to discrete locations of defects or graphite inclusions on the diamond surface [32,53–55,78]. To sustain a continuous emission current, electrons are assumed to be supplied to the emitting surface through conduction channels formed in diamond via an electroforming process at high electric fields [79,80], similar to the emission process described for composite resin–carbon materials by Bajic and Latham [81]. Grain boundaries in diamond films [82,83] and hydrogenated diamond surfaces [78,84] have also been suggested to function as conduction channels. Similarly along this mechanism, Cui et al. [21] concluded from their photoelectron sub-band gap emission study that the diamond phase provides a thermally and mechanically stable matrix with a comparatively low work function, and graphitic phases provide the transport path for electrons to reach the surface and emit.

Another model considers the injection of electrons over a Schottky barrier at the back-contact interface between a metallic substrate and diamond as the controlling mechanism [42,62,71,85]. The emission current here is limited by tunneling of electrons through the metal–diamond Schottky diode into the conduction band of diamond, not by electron emission from the diamond surface into vacuum. For very thin diamond films such as those coated on sharp Si or Mo tips [63], the injected electrons could directly reach the diamond surface and emit into vacuum. This electron injection role is supported by the observation that a roughened interface between the nickel contact and nitrogen-doped diamond improved electron emission due to local field enhancement effects at the interface that facilitate tunneling through the Schottky barrier [42,85]. Electron injection facilitated by band bending at the interface related to space charge build-up has also been proposed [79,86–88]. It is worth noting that the tunneling current density at the interface as a function of applied field follows a numerical formula similar to the F-N equation, and because of this, it is difficult to differentiate between this interface barrier mechanism and the surface barrier mechanism from I-V data alone. Additionally, it is not clear what kind of Schottky barrier structure forms at the interface of metal and diamond that is highly defective. The internal emission through the Schottky barrier at the metal-diamond interface could lead to enhanced Nottingham cooling effect [89] as suggested by Miskovsky and Cutler [90].

Other proposed models include field concentrations induced by chemical inhomogeneity (such as hydrogen termination) on the surface [84,91], dielectric breakdown that provides conductive channels in diamond [80], surface arcing that results in the formation of tips and protrusions and thus provide additional geometric field enhancement [72,73, 92], space charge limited conduction current through the bulk of diamond [93], and surface conduction enhanced electron emission [78,94]. All these models are not necessarily mutually exclusive, because each addresses a particular part of an overall complex field emission process that includes the critical steps of supplying electrons to diamond, transporting them through the bulk to the surface, and emitting them into vacuum. Overall, the physics remains ambiguous, and because of the wide variety of diamond materials available, particularly of those grown by CVD processes, it has been difficult to establish a quantitative relationship between the emission and material properties. The optimal set of material parameters for the best diamond emitters remains to be better defined.

6.2.3. Emission from DLC

There has also been substantial work done on many types of nondiamond carbon emitters, including amorphous carbon [95,96], DLC [97–99], carbon fibers [1,2,100–103], graphite [104,105], "nanocoralline" graphitic carbon [106], resin–graphite composite materials [81], and even polymers [107]. Of particular interest is DLC, a three-dimensional network of sp^3 and sp^2 coordinated carbon atoms, the properties of which resemble those of diamond. Interest in exploring DLC and other carbon emitters is motivated by the combination of their good field emission properties and the convenience of low temperature deposition. The latter allows DLC or other forms of carbon emitters to be applied directly onto temperature-sensitive materials such as glass substrates for device applications. In contrast, CVD diamond is usually deposited at temperatures of 700°C or higher. Depending on the deposition conditions, DLC materials can be made to contain a substantial amount of hydrogen (a-C:H) or be hydrogen-free (a-C).

They can also be tailored to have different portions of sp^2 and sp^3 bonded carbon in their structures. These structural parameters dictate the field emission properties of these materials [108,109]. For example, the lowest emission threshold field was found for DLC materials with the highest sp^3 content deposited by a cathodic arc process [95]. Field emission from hydrogen-free DLC films was found to be more stable than from hydrogenated DLC materials [95,110,111]. Emission was also found to improve with increasing nitrogen content in a-C:H films [86,87]. Films with nitrogen up to 0.4% showed a reduction of threshold field down to 3–5 V/um. However, for higher amounts of nitrogen, an increase in the threshold field was observed, because the sp^3 content was reduced by nitrogen [94]. It is interesting to note that the emission properties of these amorphous carbon or DLC emitters improve with increasing sp^3 content, while those of diamond materials deteriorate. This suggests that there is an optimal amount of sp^3/sp^2 content in an amorphous carbon structure in order to yield a low-field electron emitter. If the sp^3 content is too high, problems could arise in the effective supply and transport of electrons to the diamond surface for sustaining a stable emission. On the other hand, if the sp^2 content is too high, the NEA property that is characteristic of a diamond surface would be suppressed, and the tunneling barrier would then become too high for low-field electron emission.

6.2.4. Device Applications and Issues

There have been great efforts in exploring emitters of diamond, DLC, and some other forms of carbon for building useful devices. Researchers at SI Diamond Technology (Austin, TX) have been at the forefront of pursuing applications for diamond field emission technology. Over the years, they have made tremendous progress in improving the diamond cathode performance, currently achieving an emission site density of 2×10^5 cm^{-2} and a current density of 100 mA/cm^2 while operating at 10 V/μm. Cathode lifetime has reached 6900 hours under continuous operation at 10^{-9} torr and 1–2 mA/cm^2, a projected equivalent of 20 000 hours in display operation [112]. The company made 1-in. (50×50 pixel array) monochrome diode-structured displays in 1993, using DLC emitters fabricated by laser ablation [113], and demonstrated high brightness field emission picture elements useful for large outdoor displays and specialty lighting in 1996 using diamond emitters deposited by hot filament CVD [114]. More recently, the company announced the development of a hybrid FED (HyFED) that combines matrix-addressed CRT electron optics with a diamond thin film cathode to produce a thin, flat display [115]. A panel structure consisting of 8 electron emitters illuminating 720 pixels on a screen was built and tested. The operation of a monochrome 1/16 VGA (640×480) HyFED device having 405 integrated electron emitters was also reported most recently [116].

For high-end products such as high-resolution displays and high-power microwave tube amplifiers, there are formidable problems to overcome. The emission site density from diamond cathodes is not sufficiently high to achieve the high-emission uniformity ($>10^6$–10^7cm^{-2}) necessary for displaying high-resolution

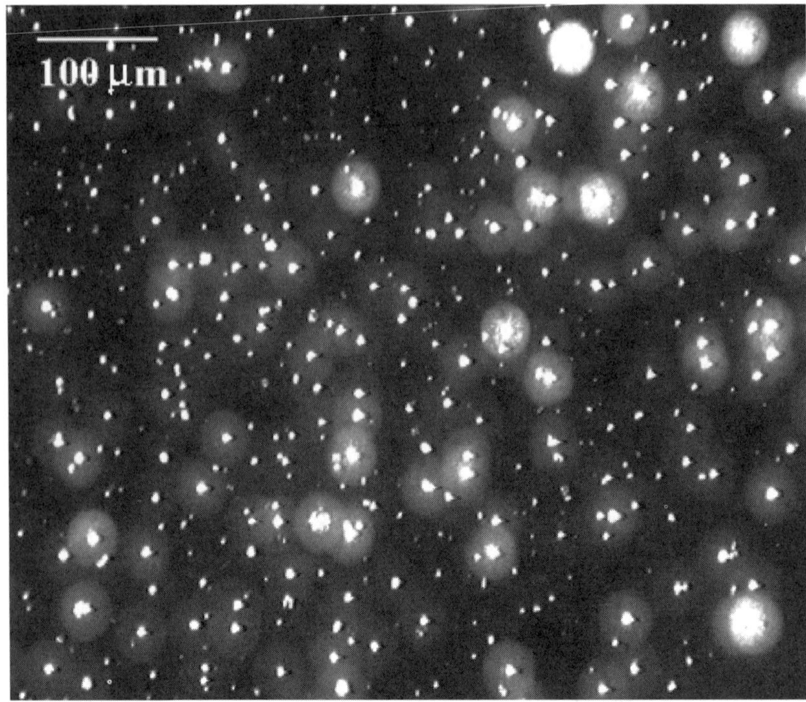

FIGURE 6.9. A phosphor screen image of the emission pattern from a diamond cathode showing the presence of "hot" spots and the nonuniform nature of emission.

images (>80 pixels/in.). As is often observed, the emission from diamond is typically from a set of spots as shown in Fig. 6.9, each of different efficiency, possibly related to the variations in surface morphologies and adsorbates. The non-uniform distribution of emission current was even found within individual grains on a nanometer scale, closely associated with the local surface conductivity established by the hydrogen termination [84]. Those few highly efficient, "hot" spots dominate the total emission current and limit the emission site density in a range of 10^2–10^5 cm^{-2}, far below that required for high-resolution displays. Worse yet, these "hot" spots are susceptible for destruction and failure, especially when operated at relatively high current densities (e.g., >50 mA/cm^2), because current overloading at these spots causes local overheating, arcing, or explosion, as shown in Fig. 6.10. Internal structural changes, such as the phase transformation of diamond into graphite in these local "hot" spots, could be a contributing factor in the failure mechanism. As in FEAs, good emission uniformities hold the key in enabling the generation of useful emission current densities while preventing premature failures of diamond emitters. Without further improvements in the emitter durability associated with the low emission site density, diamond emitters would be likely restricted in applications where requirements for current

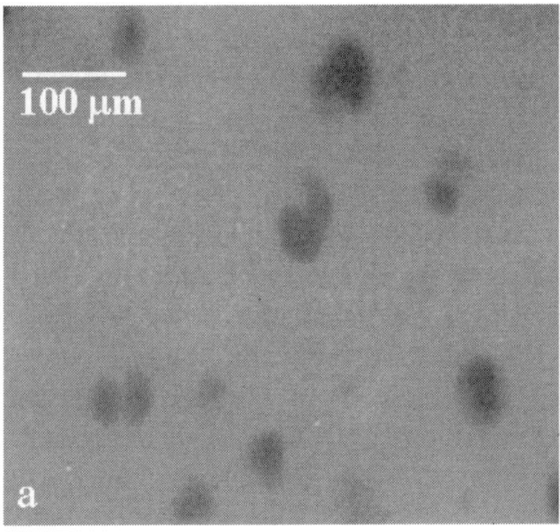

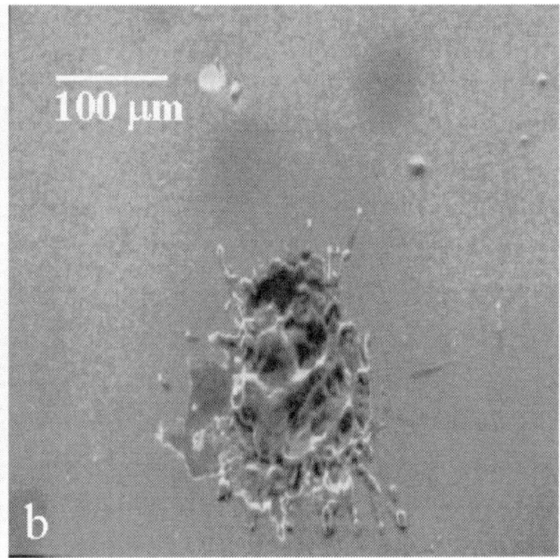

FIGURE 6.10. Scanning electron micrographs showing (a) overheated spots and (b) explosive failure on a diamond cold cathode.

density and emission uniformity are moderate, such as in low- or medum-resolution displays, LCD (liquid crystal displays) backlights, image sensors, or distributed X-ray sources.

6.3. CARBON NANOTUBE EMITTERS

Carbon nanotubes are a new, stable form of carbon consisting of long (>1 μm) graphitic cylinders with nanometer-scale diameters (<50 nm) [117], as illustrated in Fig. 6.1. They are generally classified into single-walled nanotubes (SWNT) and multiwalled nanotubes (MWNT) and can be produced using several different techniques, including electric are discharges between two carbon electrodes, pulsed laser ablation from a carbon target containing Co/Ni catalysts, and chemical vapor deposition from carbon-containing gaseous species [118,119]. Both theoretical predictions and experimental observations have shown that these one-dimensional carbon nanotubes possess unique physical properties [120–125]. In particular, the extraordinary geometric property of high aspect ratio, coupled with their high mechanical strength and chemical stability, makes carbon nanotubes natually attractive as electron field emitters. Despite the large work function (~ 5 eV) associated with graphite and nanotubes [126,127], high field concentrations can be effectively induced at the sharp nanotube tips that allow electrons to overcome the surface barrier and emit into vacuum at low macroscopic fields. Although the tips of nanotubes can be terminated with a variety of structures with different radii of curvature, large field enhancement factors in the range of 10^2-10^3 can generally be achieved, exceeding those associated with metal tip FEAs.

6.3.1. Emitter Preparation and Properties

There are several methods of preparing nanotube emitters and measuring their emission properties. Rinzler et al. [128] attached a single MWNT to a graphite fiber and found that the emission current increased 100-fold when the nanotube cap was opened by either laser heating or annealing in oxygen. De Heer and co-workers [129] made films of partially aligned nanotubes by drawing nanotube suspensions through 200 nm pore ceramic filters and measured emission current densities in excess of $100 \, mA/cm^2$. Wang et al. [130] and Collins and Zettl [131] prepared nanotube cathodes by mixing nanotubes in epoxy resin and then exposing the nanotubes by either polishing or oxygen plasma etching the surface. Bower et al. [132] used an air-spraying technique to apply purified and dispersed nanotube suspensions onto heated substrates to make thin film cathodes of randomly oriented nanotubes, as shown in Fig. 6.11(a). Xu et al. [133] and Kuttel et al. [134] made similar randomly oriented MWNT cathodes by CVD. Finally, Ren et al. [135], Fan et al. [136], Bower et al. [137,138] and Murakami et al. [139] fabricated cathodes consisting of highly oriented nanotubes grown by CVD. As shown in Fig. 6.11(b), this type of aligned nanotubes, if properly spaced [140], is considered by many to be an ideal field emitting structure.

Although the emission properties reported for nanotubes vary, depending on the nanotube content and size distribution in the specific samples measured, carbon nanotubes, in general, exhibit excellent emission properties, regardless of their structures (single-wall vs. multi-wall), orientations (randomly distributed vs. highly aligned), or production techniques (are discharge, laser ablation or CVD). Emission is identified to originate from the tips of nanotubes. Loose ends of randomly oriented ranotubes can align themselves in applied electric fields, emitting electrons in a current level similar to that of oriented nanotubes. As shown in Fig. 6.12, the nanotubes rank among the best carbon-based field emitters, with field emission occurring at low applied fields and generating high current densities. Individual nanotubes can stably emit up to $1 \, \mu A$ [141], and the current density from randomly oriented films routinely reaches $1 \, A/cm^2$ and can be as high as $4 \, A/cm^2$ [142]. These are the highest emission current densities ever reported for any carbon-based emitters. Upon operating at such high current densities, the emission remains robust with no apparent structural degradation or surface damages occurring to the nanotube emitters.

There are some interesting and unique I–V characteristics of nanotube emitters that have not been observed in other carbon-based emitters. As shown in Fig. 6.13, the I–V behavior typically follows the F-N tunneling mechanism at low voltages and currents (region I). However, at high voltages and increased current ($>100 \, nA$ for individual nanotube emitter or $>100 \, mA/cm^2$ for groups of emitters), the I–V data exhibit a current saturation region that sharply deviates from the tunneling mechanism (region II). At further higher voltages and currents, the I–V characteristics return to the electron tunnelling behavior governed by the F-N theory (region III). These distinct I–V regions have been observed in emissions from both SWNTs and MWNTs in the form of either an individual emitter or large groups of nanotubes (such as in a film) [131,133,143–145]. Saturation in emission current is of great interest for device

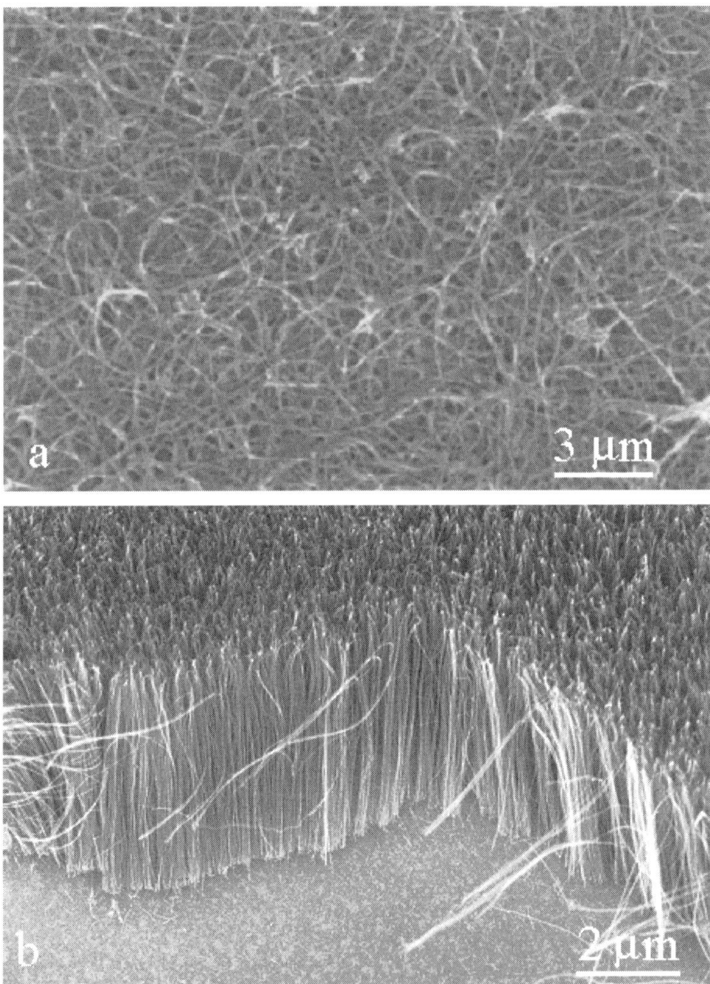

FIGURE 6.11. Scanning electron micrographs of (a) spray-coated, randomly oriented SWNT ropes, and (b) highly oriented MWNTs grown using CVD techniques [137].

applications, because devices operating in the current saturation region are more stable and less influenced by external factors. The current saturation characteristics of nanotubes, if proven and understood, may, therefore, make them better suited for building stable devices.

Possible explanations for the saturation phenomenon in nanotube emitters range from space charge effects [146,147] to interactions between neighboring nanotubes [144] and to the existence of nonmetallic, localized states at the nanotube tips [143]. However, an investigation by Dean and Chalamala [141] indicated that the current saturation is actually a surface adsorbate effect that is not observed with

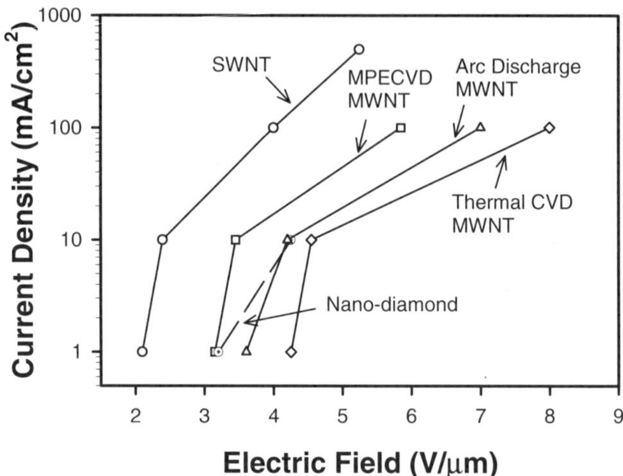

FIGURE 6.12. Emission current density vs. applied electric field for a variety of carbon nanotube emitters as well as the nanostructured diamond emitter.

clean nanotubes. As was reported [141,148,149], nanotubes emit electrons through adsorbate states at room temperature. These adsorbate states, mostly water-related, enhance the field emission current of nanotubes by 2–4 orders of magnitude. As the field and emission current increase, the adsorbate states are perturbed, resulting in the reduction of tunneling enhancement at the adsorbates and an accompanying current plateau. The current saturation was found to be concurrent with rapid fluctuations in emission current (a 100-fold increase in fluctuation, compared to a typical 5–10% [143]) and distinctive changes in field emission patterns from lobed configurations to circular ones, consistent with physical changes occurring at the adsorbate sites. At even higher fields, the adsorbate states are completely removed, and the $I-V$ behavior now represents that of clean nanotubes, which shows no evidence of current saturation for emission current reaching 1 μA per tip. In non-ideal vacuum conditions where the adsorbates return to the emitter surface when the applied field is reduced, the $I-V$ characteristics, including the saturation region, are completely reversible. However, this adsorbate effect, while it is likely present, does not fully explain why the current saturation data consistently follow a certain slope. In the authors' laboratory at Bell Labs, we found that the $I-V$ data in the saturation region fit nicely with the Child–Langmuir law [150,151] that governs the vacuum space charge phenomenon. It is, therefore, tempting to attribute the current saturation to the local space charges established by the adsorbates. Considering the molecular nature of these adsorbates that can induce very high-field concentrations and generate very high current densities locally, such a space charge effect appears plausible.

Dean and Chalamala [149] further used FEM images to corroborate their findings about the roles of surface adsorbates. As shown in Fig. 6.14, images obtained at room temperatures on SWNT emitters are usually symmetrical and contain 1–4

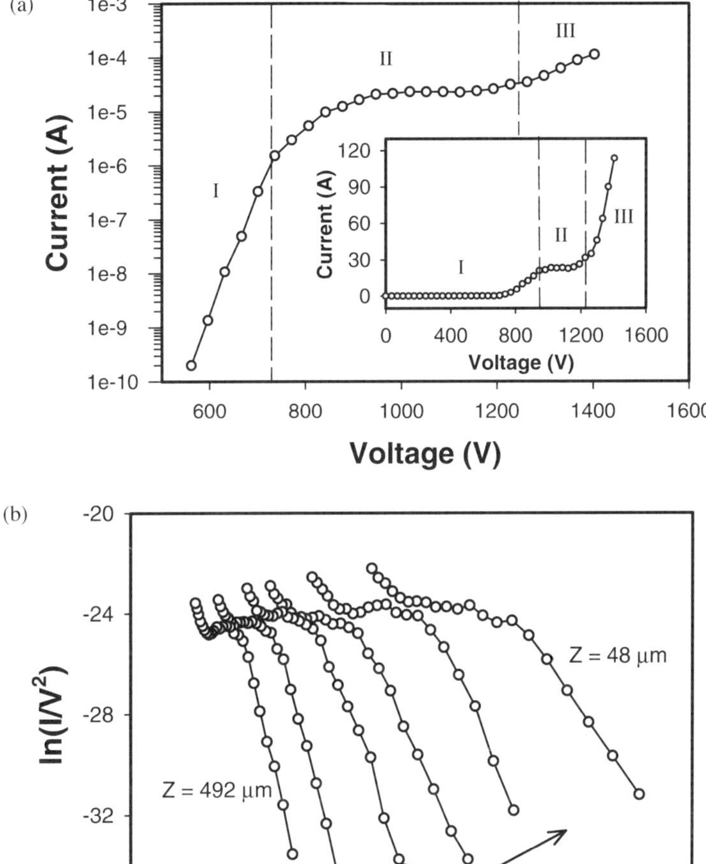

FIGURE 6.13. (a) A single emmission *I–V* curve from a SWNT sample shows three distinct regions, adsorbate enhanced emission (region I), current saturation associated with the removal of the adsorbates (region II), and emission from the clean SWNTs (region III). (b) A series of *I–V* curves taken at different anode–cathode spacings (*Z*) from a SWNT cathode. The data are plotted as $\ln(I/V^2)$ vs. $1/V$ and should fall on a straight line if the emission obeys the F-N equation.

lobes, identical to those commonly produced on metals by adsorbates [152,153]. These lobed images are explained as scattering patterns of elections passing through the electronic states of the adsorbed molecules [154]. The lobed patterns are dynamic, changing from one lobed pattern to another with time because of the mobility of adsorbates. The rate at which the images switch between patterns increases with

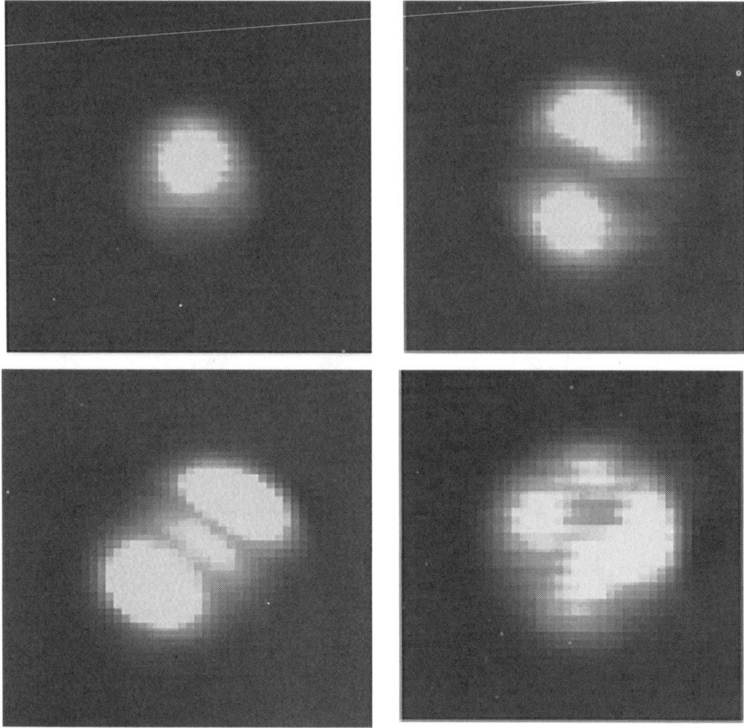

FIGURE 6.14. Field emission patterns observed from carbon nanotubes, which exhibited the one, two, three, and four lobe patterns, associated with adsorbates [149].

temperature. When the nanotubes are heated to temperatures between 700–900 K, lobed patterns abruptly disappeared, revealing significantly dimmer circular patterns, and the emission current drops by 2 orders of magnitude. Finer structures, as shown in Fig. 6.15, can be discerned within each circular pattern, showing high levels of symmetry, including 5-fold and 6-fold symmetries, that are consistent with the atomic configurations at the nanotube tips as well as their associated electronic structures. When the sample is cooled to room temperature, the brighter lobed patterns return over a period of time with an accompanying increase in emission current. At further higher emission currents (>2 μA from individual emitter) or higher temperatures (>900 K), rapid circular motion occurs in the circular patterns of clean nanotubes, and rings form around the primary pattern, as shown in Fig. 6.16, suggesting atomic as well as electronic rearrangements at this emission level, possibly similar to the atomic unraveling mechanism proposed by Rinzler et al. [128] and Lee et al. [155]. Saito et al. [156] and Zhu et al. [142] have also reported similar circular patterns in both MWNT and SWNT films. Saito et al. [156] suggested that the ring structure corresponded to emission from open cap MWNTs, while the filled circular

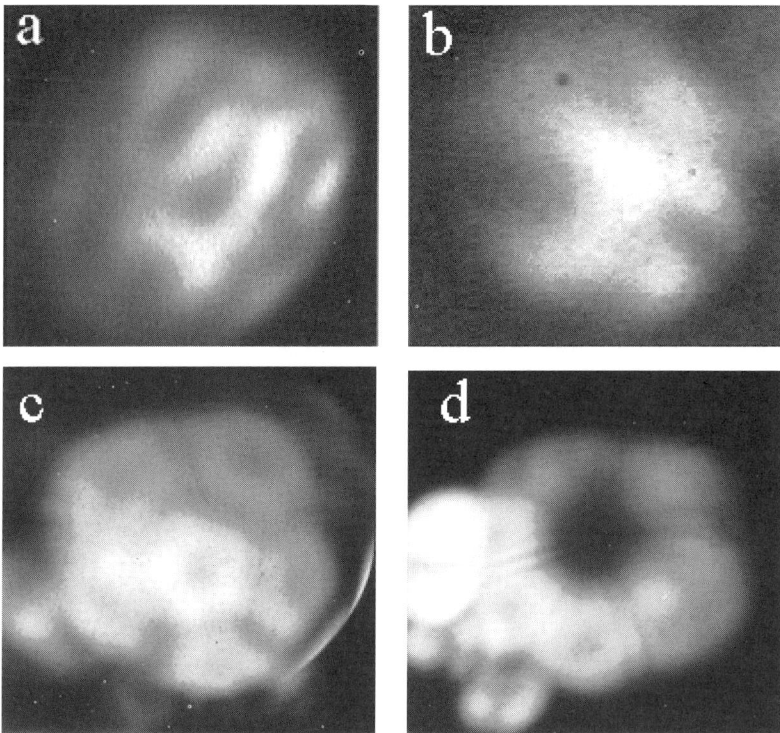

FIGURE 6.15. Field emission patterns observed after removing the adsorbates from (a) and (b) SWNTs [149], and (c) and (d) MWNTs [157].

pattern was associated with closed cap MWNTs. More recently, Saito et al. [157] have identified field-emission patterns originating from pentagons located at the tips of closed MWNTs after removal of the surface adsorbates, as shown in Figs. 6.15c and 6.15d.

Another interesting observation was the field-emission-induced luminescence from carbon nanotube emitters. As reported by Bonard et al. [158], light emission was detected from MWNT nanotube emitters at a current density of 2 mA/cm^2 or higher. By analyzing the spectra of the emitted light, they concluded that the luminescence does not come from blackbody radiation or current-induced heating, rather it results from electronic transitions between different electronic states participating in the field emission at the tips of nanotubes. The electronic structure at the nanotube tips is known to be different from that of the bulk [159,160], and electron energy distribution measurements have confirmed the existence of nonmetallic, localized electronic states above the Fermi level at the tips of SWNTs [161].

The stability of field emitters in various vacuum environments is of critical importance to device applications. The lack of environmental stability of metal-FEAs

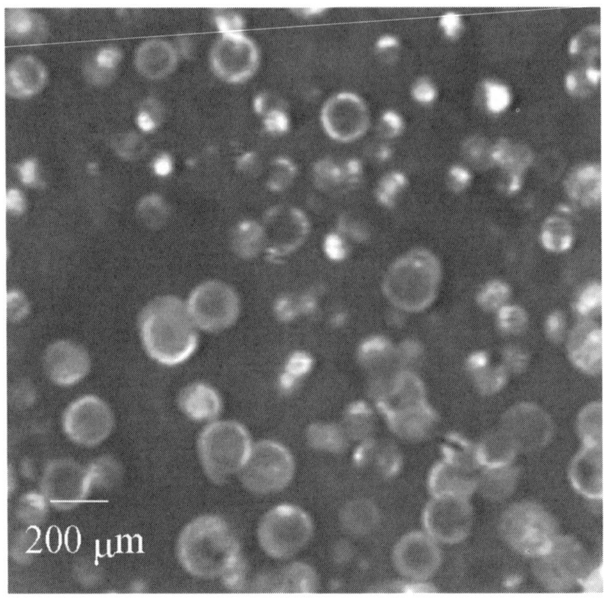

FIGURE 6.16. Ring shaped field emission patterns observed from SWNTs at high currents [142].

has been one of the primary reasons that field emitters have not reached widespread uses. The stability is mostly associated with the resistance of emitter to sputtering and to oxidation. In metal emitters, the easy formation of microprotrusions on the emitter surface under the bombardment of residual gas species often leads to runaway emission currents, resulting in the destruction of the emitter through vacuum arcs [162,163]. Nanotube emitters do not show this type of emitter destruction, even though the emission is from adsorbate states at room temperature. They are also generally more durable and stable than other carbon emitters such as diamond and carbon fibers, possibly because nanotubes have a relatively defect-free graphite structure with small sputtering yield and low carbon atom mobility. They do degrade, however, depending on the emission current level and the environments. For example, Bonard et al. [143] reported significant degradations of nanotube emitters in uncharacterized vacuum environments and found that SWNTs degrade faster than MWNTs. Zhu et al. [142] found SWNT emitters to be stable and robust at 20 mA/cm^2, but observed degradation over time at current densities in excess of 500 mA/cm^2. They attributed the emitter failure primarily to the "uprooting" of nanotubes from substrate surfaces due to the poor adhesion strength, rather than to the intrinsic structural degradation and failure as found in diamond emitters. Dean and Chalamala [164] studied the environmental stability of SWNT emitters in a number of different gases. As shown in Fig. 6.17, an exposure to 10^{-6} torr of hydrogen showed no significant effects on the emission. When water (10^{-7} torr) was introduced, the emission current experienced a rapid

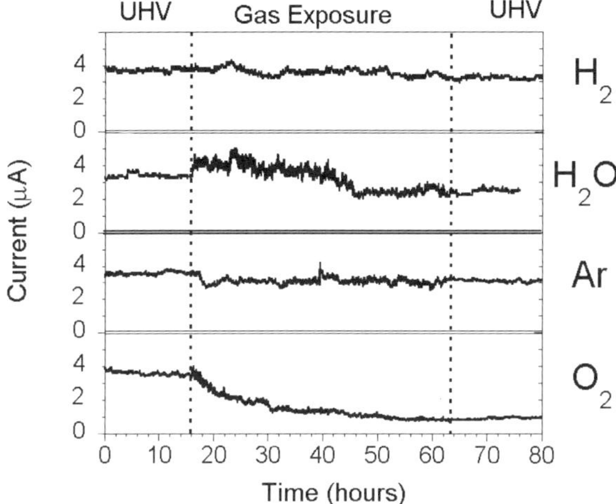

FIGURE 6.17. Stability tests of SWNT emitters showing emission current as a function of time in different environments [164].

increase due to the establishment of the adsorbate tunneling states. During extended exposure (around 45 h), the current started to decay due to ion bombardment that removed this state. Operation of nanotubes in 10^{-7} torr of argon showed little effect aside from a small initial decrease in current, consistent with the sputter cleaning of some adsorbates by argon. In 10^{-7} torr of oxygen, the current steadily decreased over the duration of the exposure with a drop of 75% over 48 h. This decrease in current was related to chemical interactions at the surfaces such as the formation of C-O dipoles. Oxygen dipoles are known to reduce emission currents from metal and diamond emitters [18,165].

6.3.2. Device Prospects

Carbon nanotube emitters are still relatively new, but researchers have already begun incorporating them into prototype field emission displays [166]. Saito and coworkers fabricated triode-type CRTs based on MWNT emitters produced by arc discharges [167]. As shown in Fig. 6.18, these cold cathode CRT lighting elements, 20 mm in diameter and 74 mm in length, were reported to be twice as bright as conventional CRTs and had a lifetime greater than 10000 hours. They are currently being explored for use in giant outdoor displays. Wang et al. [168] built a matrix addressable, 32×32 pixel, diode-type display that utilized a 10×10 mm^2 MWNT/epoxy composite cathode. Researchers from Samsung demonstrated a fully sealed 4.5 in. color field emission display, shown in Fig. 6.19, that made use of aligned SWNT emitters prepared by squeezing a nanotube "paste" through a metal mesh of 20 μm in size and rubbing the surface to expose the nanotubes [169]. They observed low turn-on

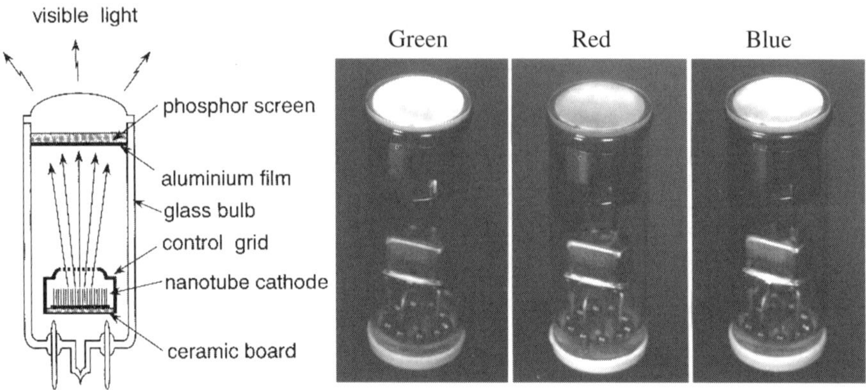

FIGURE 6.18. A schematic and photographs of field emission lamps based on carbon nanotube emitters [167].

fields ($\sim$1–3 V/μm), high brightness (1800 cd/m^2 at 4 V/μm), and good stability (7% current fluctuation), all of which demonstrated the potential applicability of carbon nanotubes in display devices.

There are other devices, such as microwave power amplifier tubes, that can be contemplated by taking advantage of the high emission current capability of carbon nanotubes. However, the quality factors of electron beams from carbon nanotubes, such as the energy spread and beam divergence, are largely unknown. The long-term stability of these nanotube emitters is also uncertain, especially at high current densities, even though the emission process has been found to be relatively robust in a short-term scale. The emission uniformity remains poor, and like any carbon-based field emitters, nanotubes are susceptible to oxidation damage. The sensitivity of field emission to surface adsorbates further raises concerns in emission noise and environmental stability, which will, at the minimum, make the cathode processing, including the disposition, cleaning, and conditioning of nanotube emitters, more critical and challenging. As a result of these outstanding issues, commercialization of carbon nanotube field emission devices remains to be seen.

6.4. OTHER COLD CATHODES

6.4.1. Surface Conduction Emitters

Surface conduction emission (SCE) is a phenomenon in which electrons are emitted from a cathode when electric current flows through the cathode in parallel with the cathode surface. It has attracted attention only recently when researchers at Cannon of Japan succeeded in building a 10-in. full color display incorporating a thin film PdO cathode based on the SCE mechanism [170]. The basic properties of surface conduction emitters were reported much earlier on thin films of tin oxide (SnO_2)

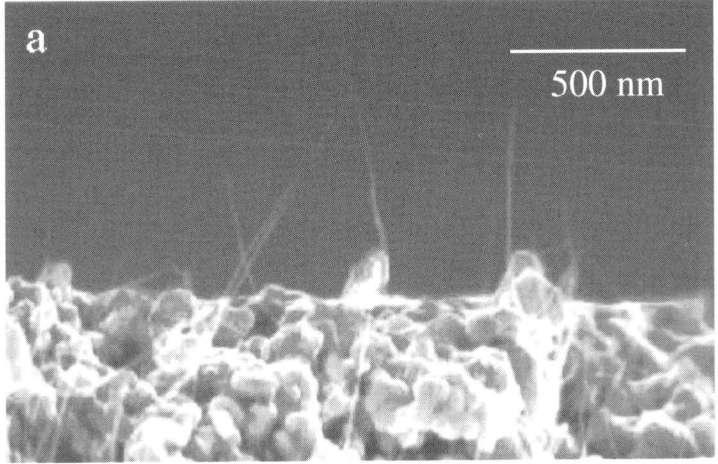

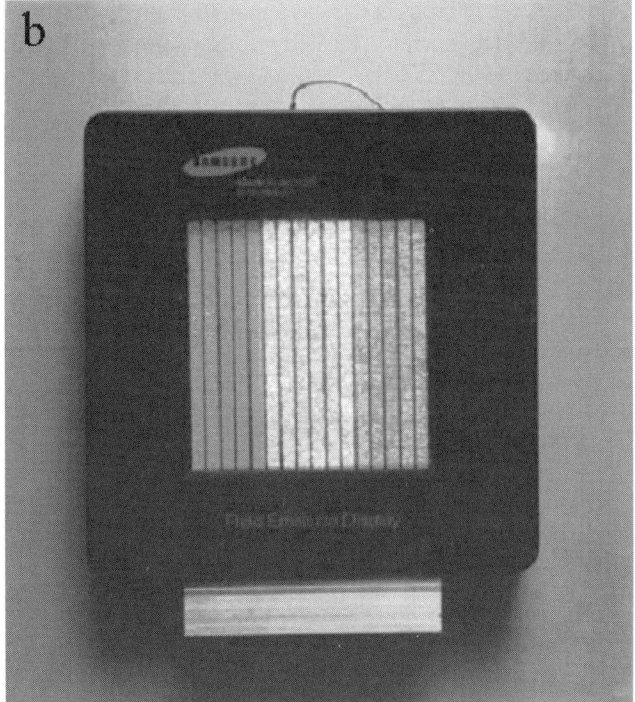

FIGURE 6.19. (a) An SEM micrograph of the SWNT cathode used in the FED shown in (b) [169].

[171], gold (Au) [172], indium tin oxide (ITO) [173], and carbon (C) [174], but the unstable and irreproducible emission characteristics have impeded progress in this area.

Researchers at Cannon found solutions to these problems by developing innovative techniques of energetically forming and activating the ultrafine (5–10 nm) PdO particulate films. As shown in Fig. 6.20, an electroforming process is employed to activate a surface conduction film, in which a voltage is applied between two electrodes on the cathode so that an electric current flows through the film in parallel with the surface. The thin film generates Joule heating that locally modifies the microstructure of the film by generating microscopic cracks or fissures (0.1–5 μm wide). These cracks or fissures lead to the formation of island structures that cause structural discontinuity and increase the electrical resistance. The dimensions of these fine island structures range from several nanometers to several micrometers, and they exist in a spatially discontinuous but electrically continuous state. Field emission occurs at the microcracks because of the high fields established across the cracks along the surface, and the emitted electrons, after multiple scattering on the cathode, can be collected at an anode spaced apart from the surface of the cathode. Because it is difficult to accurately and predictably control the electroforming process and create the island structure and fissure configurations in a reproducible and deliberate manner, surface conduction emitters had not been seriously considered for device applications. The

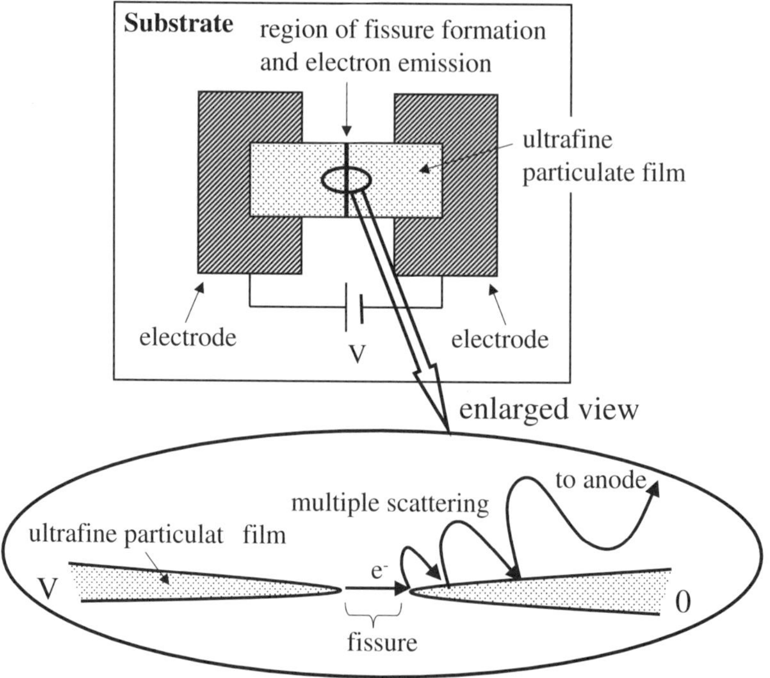

FIGURE 6.20. Schematic illustrating the surface conduction emission mechanism [170].

success at Cannon clearly established this technology as a viable solution to building practical cold cathode devices.

6.4.2. Ferroelectric Emitters

Electron emission from the surfaces of ferroelectric materials during polarization reversal was first observed almost 40 years ago [175, 176]. It was recognized that the polarization induces macroscopic charge separation on the two opposite surfaces of a ferroelectric sample. As illustrated in Fig. 6.21(a), screening charges are then developed to compensate the net charges developed on the surfaces. A fast reversal (on the time scale of sub-microseconds) of the polarization leads to the build-up of a large electric field that ejects the electrons from the negatively charged surface, as shown in Fig. 6.21(b). In contrast to "conventional" field electron emission, no external extraction field is required to overcome the surface work function and obtain electron emission from ferroelectric emitters [177]. The emission depends on the polarization fields within the ferroelectric material, and only an excitation energy sufficient to overcome the coercive fields and obtain the polarization reversal is needed for the emission. On a broader term, ferroelectric emission can occur under any external perturbation (electrical, optical, thermal, or mechanical) that leads to the disturbance of the initial charge equilibrium and to the appearance of an unscreened charge as well as an electrostatic field at the free polar surface. Ferroelectric emission is thus a transient unipolar effect generated from a nonequilibrium charged ferroelectric surface.

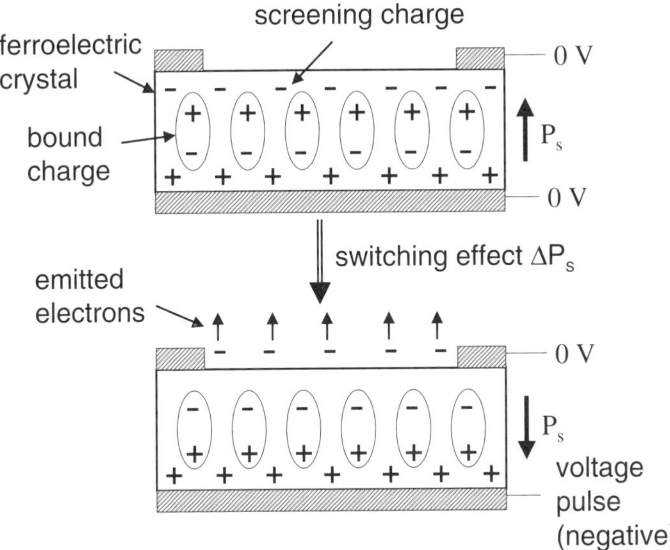

FIGURE 6.21. Schematic illustrating the ferroelectric emission mechanism.

Many of the ferroelectric emitting materials belong to the class of perovskite ceramics including barium titanate ($BaTiO_3$), barium–lead–titanate (BPT), and lead–zirconium–titanate (PZT) [178–180]. Other reported materials include $Pb_5Ge_3O_{11}$ [181], TGS (triglycine sulfate) [182], $Gd_2(MoO_4)_3$ [183], $LiNbO_3$ [184] and $LiTaO_3$ [185]. Emission current densities are typically low ($<10^{-7}$ A/cm^2) for ferroelectrics purturbed thermally, mechanically, and optically. However, recent studies have observed current densities in excess of 100 A/cm^2 from PZT and PLZT (lead-lanthanum-zirconium-titanate) cathodes [179,180]. This strong ferroelectric emission might be based on a different mechanism, because it occurs regardless of the phase state (ferroelectric, antiferroelectric, relaxor, or paraelectric), while true ferrolelectric emission by polarization reversal is only observed from the ferroelectric phase [177,186]. A typical approach to excite the strong ferroelectric emission is to apply a pulsing (several kHz) electric field (1–10 V/μm) to appropriately patterned electrodes (e.g., a blank electrode on the back surface and a grated electrode on the front surface) on the ferrolelectric cathode [187]. It has been experimentally shown that a plasma is generated at the cathode surface during these experiments, and this plasma is possibly responsible for the observed large current densities. The lifetimes of these strong ferroelectric emitters are typically only 10^6–10^7 pulses. By using conductive oxide instead of platinum as electrodes, the lifetime of PZT cathode has been increased up to 10^{12} switching cycles [188].

Ferroelectric cathodes have very robust surfaces that may be exposed to air and operate in poor vacuum conditions (up to 10^{-2} torr) or even in a plasma [189]. In particular, the high current density achievable under fast polarization reversal conditions could make ferroelectric emitters suitable for applications such as triggers for gas switches that could be used in particle accelerators or pulsed power systems [190] and microwave power generation and amplification [191]. A scheme of local domain switching to generate ferroelectric emission from switched regions (i.e. pixels) has also been proposed for use in flat panel displays [192]. However, many issues in ferroelectric emission, such as polarization fatigue during multiple fast switching, emission current stability and domain structure aging, need to be clarified before those envisioned devices could be realized.

6.4.3. Metal-Insulator-Metal Emitters

Metal-insulator-metal (MIM) cathodes are a type of thin film tunneling structure that was first proposed by Mead [193] in the early 1960s and has been experimentally studied by many researchers [194–198]. It is an interesting device, because the whole structure is fabricated as a stack. The cathode is less susceptible to surface contamination due to the fact that the emitting material is buried inside, and electrons tunnel through interfacial Schottky barriers instead of surface barriers. Typically, the device consists of a thin insulating film (e.g., Al_2O_3) sandwiched between two metal electrodes (e.g., Al and Au) as shown schematically in Fig. 6.22(a). The insulating layer is so thin (several nanometers) that an electron can tunnel through it when an electric field is applied across the layer. As the energy diagram in Fig. 6.22(b) shows, the

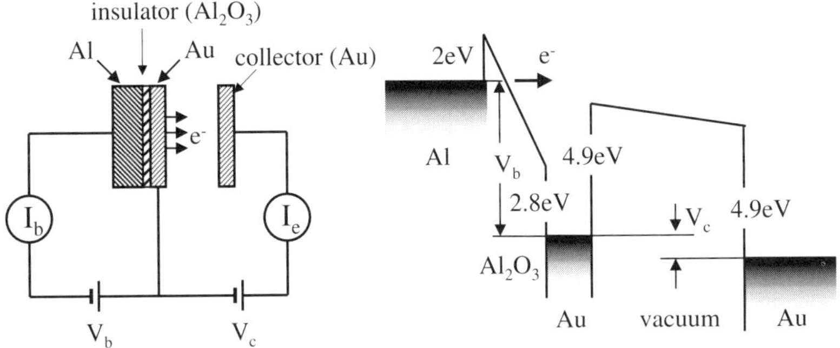

FIGURE 6.22. Schematic depicting the MIM emission mechanism [198].

tunneling electrons are injected from the negative electrode (Al, the emitter) through the insulator into the positive electrode (Au, the gate) as hot electrons and are detected as a diode current I_b. A portion of the injected electrons that have kinetic energies larger than the work function of the Au surface can go through Au and emit to vacuum, which is collected as an emission current I_e. However, a majority of the tunneling electrons lose their kinetic energies while they pass through the structure because of scattering events in both the insulator and the gate metal, resulting in a very inefficient emission process with low current transfer ratios ($I_e/I_b < 10^{-3}$). Reducing the thickness of both the insulator and the gate electrode would increase I_e/I_b, but for the insulator, it needs to have a certain thickness to withstand the applied voltage that must be higher than the work function of the gate electrode. The use of low work function materials for the gate electrode would also help to enhance I_e/I_b. In addition, very high current transfer ratios (0.7%) has been reported from the metal-oxide-semiconductor (MOS) structure, a variant of the MIM cathode [199].

It is important to note that stable and reproducible electron emission from MIM and MOS cathodes requires the fabrication of ultrathin (a few nanometers thick) insulating and metal or semiconductor layers, the structures of which need to be sufficiently controlled on an atomic scale. Roughness at the interfaces or the presence of defects and structural inhomogeneity in these thin layers will cause significant fluctuations in emission current and emission nonuniformity. The emission current density from MIM cathodes is also generally too low ($\sim$50 μA/cm^2) for practical applications. Although higher currents can be obtained by increasing the electric field applied across the insulating layer, the insulator would degrade quickly under such high fields, and electroforming of the MIM cathodes can occur [200]. This would further result in unstable emission and poor emission uniformity. However, recent work by Kusunoki and Suzuki [201] indicated that by using a multilayer (Ir–Pt–Au) gate electrode over Al/Al$_2$O$_3$, the emission current density could be increased by 2 orders of magnitude. This makes it possible to build displays based on MIM cathodes. Researchers at Matsushita demonstrated a 4×4 array of triode MIM elements with spatially uniform field emission [202], and researchers at Hitachi reported a 30×30 MIM cathode array with a peak current density of 5.8 mA/cm^2 operating at a gate voltage of 10 V [201]. However, the limited lifetime of MIM cathodes operating at useful emission current densities remains the greatest obstacle.

6.5. CONCLUSION

We have reviewed the status of carbon-based cold cathodes and three other types of noncarbon-based field emitters. Compared to the microtip FEAs, these novel cathode materials potentially offer the important advantage of ease in fabrication and low-cost manufacturing. They have unique and interesting field emission characteristics, and device applications based on these materials are being actively explored. However, our understanding of the fundamental physics governing the electron emission process from these materials is still poor, and our ability in controlling and reproducing their physical properties is still weak. As a result, issues such as emission uniformity, reliability, and reproducibility remain to be overcome for these cathodes materials to be eventually implemented in working devices.

REFERENCES

1. F. S. Baker, A. R. Osborn, and J. Williams, *Nature* **239**, 96 (1972).
2. C. Lea, *J. Phys. D. Appl. Phys.* **6**, 1105 (1973).
3. Http://cnst.rice.edu/pics.html.
4. F. J. Himpsel, J. A. Knapp, J. A. van Vechten, and D. E. Eastman, *Phys. Rev. B* **20**, 624 (1979).
5. A. Zangwill, *Physics at Surfaces*, Cambridge University Press: Cambridge, England, 1988.
6. P. K. Baumann and R. J. Nemanich, in *Low Pressure Synthetic Diamond* (B. Dischler and C. Wild, Eds.), Springer Verlag: Berlin, 1998.
7. R. L. Bell, *Negative Electron Affinity Devices*, Oxford: Clarendon Press, 1973.
8. E. S. Kohn, *IEEE Trans. Electron Dev.* **20**, 321 (1973).
9. R. U. Martinelli, *J. Appl. Phys.* **45**, 1183 (1974).
10. R. U. Martinelli, *J. Appl. Phys.* **45**, 3203 (1974).
11. R. U. Martinelli, *J. Appl. Phys.* **45**, 3896 (1974).
12. J. van der Weide and R. J. Nemanich, *Appl. Phys. Lett.* **62**, 1878 (1993).
13. B. B. Pate, P. M. Stefan, C. Binns, P. J. Jupiter, M. L. Shek, I. Lindau, and W. E. Spicer, *J. Vac. Sci. Technol.* **19**, 349 (1981).
14. B. B. Pate, *Surf. Sci.* **165**, 83 (1986).
15. R. U. Martinelli and D. G. Fisher, *Proc. IEEE* **62**, 1339 (1974).
16. R. J. Nemanich, P. K. Baumann, M. C. Benjamin, S. W. King, J. van der Weide, and R. F. Davis, *Diam. Relat. Mater.* **5**, 790 (1996).

17. J. van der Weide and R. J. Nemanich, *Phys. Rev. B* **49**, 13629 (1994).

18. P. K. Baumann and R. J. Nemanich, *Surf. Sci.* **409**, 320 (1998).

19. D. P. Malta, J. B. Posthill, T. P. Humphreys, R. E. Thomas, G. G. Fountain, R. A. Rudder, G. C. Hudson, M. J. Mantini, and R. J. Markun, *Appl. Phys. Lett.* **64**, 1929 (1994).

20. G. T. Mearini, I. L. Krainsky, J. Dayton, J. A., Y. Wang, C. A. Zorman, J. C. Angus, and R. W. Hoffman, *Appl. Phys. Lett.* **65**, 2702 (1994).

21. J. B. Cui, J. Ristein, and L. Ley, *Phys. Rev. B* **60**, 16135 (1999).

22. M. J. Rutter and J. Robertson, *Phys. Rev. B* **57**, 9241 (1998).

23. W. E. Pickett, *Phys. Rev. Lett.* **73**, 1664 (1994).

24. M. W. Geis, J. C. Twichell, J. Macaulay, and K. Okano, *Appl. Phys. Lett.* **67**, 1328 (1995).

25. J. van der Weide and R. J. Nemanich, *J. Vac. Sci. Technol. B* **10**, 1940 (1992).

26. P. K. Baumann, and R. J. Nemanich, *Phys. Rev. B* **58**, 1643 (1998).

27. P. K. Baumann and R. J. Nemanich, *Mat. Res. Soc. Symp. Proc.* **416**, 157 (1996).

28. P. K. Baumann and R. J. Nemanich, *Appl. Surf. Sci* **104/105**, 267 (1996).

29. P. K. Baumann and R. J. Nemanich, *J. Vac. Sci. Technol. B* **15**, 1236 (1997).

30. P. K. Baumann and R. J. Nemanich, *J. Appl. Phys.* **83**, 2072 (1998).

31. C. Bandis, D. Haggerty, and B. B. Pate, *Mat. Res. Soc. Symp. Proc.* **339**, 75 (1994).

32. C. Wang, A. Garcia, D. C. Ingram, M. Lake, and M. E. Kordesch, *Electron Lett.* **27**, 1459 (1991).

33. M. W. Geis, N. Efremow, J. Woodhouse, M. Mcalese, M. Marchywka, D. Socker, and J. Hochedez, *IEEE Elect. Dev. Lett.* **12**, 456 (1991).

34. B. C. Djubua and N. N. Chubun, *IEEE Trans. Electron Dev.* **38**, 2314 (1991).

35. W. Zhu, G. P. Kochanski, S. Jin, and L. Seibles, *J. Vac. Sci. Technol. B* **14**, 2011 (1996).

36. W. Zhu, G. P. Kochanski, and S. Jin, *Science* **282**, 1471 (1998).

37. R. H. Fowler and L. Nordheim, *Proc. R. Soc. London, Ser. A* **119**, 173 (1928).

38. W. Zhu, C. Bower, and S. Jin, *Solid State Electron.* **45**, 921 (2001).

39. J. Prins, *Thin Solid Films* **212**, 11 (1992).

40. J. Prins, *Mater. Sci. Rep.* **7**, 271 (1992).

41. K. K. Das, in *Diamond Films and Coatings* (R. F. Davis, Ed.), Noyes Publications: Park Ridge, NJ, p. 381, 1993.

42. M. W. Geis, J. C. Twichell, N. N. Efremow, K. Krohn, and T. M. Lyszczarz, *Appl. Phys. Lett.* **68**, 2294 (1996).

43. K. Okano and K. K. Gleason, *Electron Lett.* **31**, 74 (1995).

44. K. Okano, S. Koizumi, S. R. P. Silva, and G. A. J. Amaratunga, *Nature* **381**, 140 (1996).

45. K. Okano, T. Yamada, H. Ishihara, S. Koizumi, and J. Itoh, *Appl. Phys. Lett.* **70**, 2201 (1997).

46. A. T. Sowers, B. L. Ward, S. E. English, and R. J. Nemanich, *J. Appl. Phys.* **86**, 3973 (1999).

47. W. Zhu, G. P. Kochanski, S. Jin, and L. Seibles, *J. Appl. Phys.* **78**, 2707 (1995).

48. D. Zhou, A. R. Krauss, L. C. Qin, T. G. McCauley, D. M. Greun, T. D. Corrigan, R. P. H. Chang, and H. Gnaser, *J. Appl. Phys.* **82**, 4546 (1997).

49. D. M. Gruen, *Annu. Rev. Mater. Sci.* **29**, 211 (1999).

50. T. G. McCauley, T. D. Corrigan, A. R. Krauss, O. Auciello, and D. Zhou, *Mater. Res. Soc. Symp. Proc.* **498**, 227 (1998).

51. A. R. Krauss, D. M. Gruen, D. Zhou, T. G. McCauley, and L. C. Lin, *Mater. Res. Soc. Symp. Proc.* **495**, 299 (1998).

52. O. Groning, O. M. Kuttel, P. Groning, and L. Schlapbach, *J. Vac. Sci. Technol. B* **17**, 1970 (1999).

53. N. S. Xu, Y. Tzeng, and R. V. Latham, *J. Phys. D: Appl. Phys.* **26**, 1776 (1993).

54. N. S. Xu, R. V. Latham, and Y. Tzeng, *Electron Lett.* **28**, 1596 (1993).

55. N. S. Xu, Y. Tzeng, and R. V. Latham, *J. Phys. D: Appl. Phys.* **27**, 1988 (1994).

56. R. V. Latham, *High Voltage Vacuum Insulation*, Academic Press: San Diego, 1995.

57. A. Talin, L. S. Pan, K. F. McCarthy, T. E. Felter, H. J. Doerr, and R. F. Bushah, *Appl. Phys. Lett.* **69**, 3842 (1996).

58. J. Liu, V. V. Zhirnov, G. J. Wojak, A. F. Myers, W. B. Choi, J. J. Hren, S. D. Wolter, M. T. McClure, B. R. Stoner, and J. T. Glass, *Appl. Phys. Lett.* **65**, 2842 (1994).

59. W. B. Choi, J. J. Cuomo, V. V. Zhirnov, A. F. Myers, and J. J. Hren, *Appl. Phys. Lett.* **68**, 720 (1996).

60. W. B. Choi, J. Liu, M. T. McClure, A. F. Myers, V. V. Zhirnov, J. J. Cuomo, and J. J. Hren, *J. Vac. Sci. Technol. B* **14**, 2050 (1996).

61. V. Raiko, R. Spitzl, B. Aschermann, D. Theirich, J. Engemann, N. Puteter, T. Habermann, and G. Müller, *Thin Solid Films* **290/291**, 190 (1996).

62. E. I. Givargizov, V. V. Zhirnov, A. N. Stepanova, E. V. Rakova, A. N. Kieselev, and P. S. Plekhanov, *Appl. Surf. Sci.* **87/88**, 24 (1995).

63. V. V. Zhirnov, *J. de Physique IV* **6**, C5 (1996).

64. A. R. Krauss, O. Auciello, M. Q. Ding, D. M. Gruen, Y. Huang, V. V. Zhirnov, E. I. Givargizov, A. Breskin, R. Chechen, E. Shefer, V. Konov, S. Pimenov, A. Karabutov, A. Rakhimov, and N. Suetin, *J. Appl. Phys.* **89**, 2958 (2001).

65. M. W. Geis, J. C. Twichell, J. Macaulay, and K. Okano, *Science* **67**, 1328 (1995).

66. W. Zhu, G. P. Kochanski, S. Jin, L. Seibles, D. Jacobson, M. McCormack, and A. White, *Appl. Phys. Lett.* **67**, 1157 (1995).

67. Z. H. Huang, P. H. Cutler, N. M. Miskovsky, and T. E. Sullivan, *Appl. Phys. Lett.* **65**, 2562 (1994).

68. R. M. Hill, *Phil. Mag.* **23**, 59 (1971).

69. Y. Muto, T. Sugino, J. Shirafuji, and K. Kobachi, *Appl. Phys. Lett.* **59**, 843 (1991).

70. J. E. Henderson and R. E. Badgley, *Phys. Rev.* **38**, 5401 (1931).

71. R. Schlesser et. al., *Appl. Phys. Lett.* **70**, 1596 (1997).

72. O. Groning, O. M. Kuttel, P. Groning, and L. Schlapbach, *Appl. Phys. Lett.* **71**, 2253 (1997).

73. O. Groning, O. M. Kuttel, P. Groning, and L. Schlapbach, *Appl. Surf. Sci.* **111**, 135 (1997).

74. C. Bandis and B. B. Pate, *Appl. Phys. Lett.* **69**, 366 (1996).

75. M. L. Yu, H. S. Kim, B. W. Hussey, T. H. Chung, and W. A. Mackie, *J. Vac. Sci. Technol. B* **14**, 3797 (1996).

76. C. S. Athwal, K. H. Bayliss, R. Calder, and R. V. Latham, *IEEE Trans. Plasma Sci.* **13**, 225 (1985).

77. N. S. Xu and R. V. Latham, *J. Phys. D: Appl. Phys.* **19**, 477 (1986).

78. O. M. Kuttel, O. Groening, and L. Schlapbach, *J. Vac. Sci. Technol. A* **16**, 3464 (1998).

79. K. H. Bayliss and R. V. Latham, *Proc. R. Soc. A* **403**, 285 (1986).

80. J. D. Shovlin and M. E. Kordesch, *Appl. Phys. Lett.* **65**, 863 (1994).

81. S. Bajic and R. V. Latham, *J. Phys. D: Appl. Phys.* **21**, 200 (1998).

82. Y. D. Kim, W. Choi, H. Wakimoto, S. Usami, H. Tomokage, and T. Ando, *Appl. Phys. Lett.* **75**, 3219 (1999).

83. B. Fiegl, R. Kuhnert, M. Ben-Chorin, and F. Koch, *Appl. Phys. Lett.* **65**, 371 (1994).

84. I. Andrienko, A. Cimmino, D. Hoxley, S. Prawer, and R. Kalish, *Appl. Phys. Lett.* **77**, 1221 (2000).

85. M. W. Geis, J. C. Twitchell, and T. M. Lyszczarz, *J. Vac. Sci. Technol. B* **14**, 2060 (1996).

86. G. A. J. Amaratunga and S. R. P. Silva, *Appl. Phys. Lett.* **68**, 2529 (1996).

87. V. S. Veersamy, J. Yuan, G. A. J. Amaratunga, W. I. Milne, K. W. R. Gilkes, M. Weilner, and L. M. Brown, *Phys. Rev. B* **48**, 17954 (1993).

88. S. R. P. Silva, G. A. Amaratunga, and K. Okano, *J. Vac. Sci. Technol. B* **17**, 557 (1999).

89. W. B. Nottingham, *Phys. Rev.* **59**, 907 (1941).

90. N. M. Miskovsky and P. H. Cutler, *Appl. Phys. Lett.* **75**, 2147 (1999).

91. J. Robertson, *J. Vac. Sci. Technol. B* **17** (1999).

92. O. Groning, O. M. Kuttel, E. Schaller, P. Groning, and L. Schlapbach, *Appl. Phys. Lett.* **69**, 476 (1996).

93. P. W. May, S. Hohn, M. N. R. Ashfold, W. N. Wang, N. A. Fox, T. J. Davis, and J. W. Steeds, *J. Appl. Phys.* **84**, 1618 (1998).

94. M. W. Geis, N. N. Efrenow, K. E. Krohn, J. C. Twichell, T. M. Lyszcarz, R. Kalish, J. A. Greer, and M. D. Tabat, *Nature* **393**, 431 (1998).

95. B. S. Satyanarayana, A. Hart, W. I. Milne, and J. Robertson, *J. Appl. Phys.* **71**, 1430 (1997).

96. N. Missert, T. A. Friedmann, J. P. Sullivan, and R. G. Copeland, *Appl. Phys. Lett.* **70**, 1995 (1997).

97. S. Lee, B. K. Ju, Y. H. Lee, D. Jeon, and M. H. Oh, *J. Vac. Sci. Technol. B* **15**, 425 (1997).

98. K. L. Park, J. H. Moon, S. J. Chung, J. Jang, M. H. Oh, and W. I. Milne, *Appl. Phys. Lett.* **70**, 1381 (1997).

99. F. Y. Chuang, C. Y. Sun, T. T. Chen, and I. N. Lin, *Appl. Phys. Lett.* **69**, 3504 (1996).

100. F. S. Baker, A. R. Osborn, and J. Williams, *J. Phys. D Appl. Phys.* **7**, 2105 (1974).

101. E. Braun, J. F. Smith, and D. E. Sykes, *Vacuum* **25**, 425 (1975).

102. R. V. Latham and D. A. Wilson, *J. Phys. E:Sci. Instrum.* **15**, 1083 (1982).

103. R. V. Latham and M. A. Salim, *J. Phys. E:Sci. Instrum.* **20**, 181 (1987).

104. A. N. Obraztsov, A. P. Volkov, and I. Y. Pavlovskii, *JEPT Lett.* **68**, 59 (1998).

105. Y. Chen, S. Patel, Y. Ye, D. T. Shaw, and L. Guo, *Appl. Phys. Lett.* **73**, 2119 (1998).

106. B. F. Coll, J. E. Jaskie, J. L. Markham, E. P. Menu, A. A. Talin, and P. von Allmen, *Mater. Res. Soc. Symp. Proc.* **498**, 185 (1998).

107. I. Musa, D. A. I. Munindrasdasa, G. A. J. Amaratunga, and W. Eccleston, *Nature* **395**, 362 (1998).

108. J. Robertson, *Phys. Rev. B* **53**, 16302 (1996).

109. J. Rinstein, J. Schafer, and L. Ley, *Diam. Relat. Mater.* **4**, 508 (1995).

110. R. S. Becker, G. S. Higashi, Y. Chabal, and A. J. Becker, *Phys. Rev. Lett.* **65**, 197 (1990).

111. D. R. Tallant, J. E. Parmeter, M. P. Siegal, and R. L. Simpson, *Diam. Relat. Mater.* **4**, 191 (1995).

112. R. L. Fink, Z. Li Tolt, and Z. Yaniv, *Surf. Coat. Technol.* **108/109**, 570 (1998).

113. N. Kumar, H. K. Schmidt, and C. Xie, *Solid State Technol.* **38**, 71 (1995).

114. Z. Li Tolt, R. L. Fink, R. C. Robinder, and Z. Yaniv, in *Technical Digest of Asia Display*, Seoul, Korea, 1998.

115. R. C. Robinder, Z. Yaniv, R. L. Fink, Z. Li Tolt, and L. Thuesen, in *Technical Digest of Asia Display*, Seoul, Korea, 1998.

116. R. L. Fink, L. H. Thuesen, T. Visel, and Z. Yaniv, in *Proc. Int. Display Res. Conf.*, 2000, p. 304.

117. S. Ijima, *Nature* **354**, 56 (1991).

118. T. W. Ebbesen (Ed.), *Carbon Nanotubes: Preparation and Properties*, CRC: Cleveland, OH, 1997.

119. R. Saito, G. Dresselhaus, and M. S. Dresselhaus, *Physical Properties of Carbon Nanotubes*, Imperial College Press: London, 1998.

120. C. Dekker, *Physics Today*, May 1999.

121. B. I. Yakobson and R. E. Smalley, *Am. Sci.* **85**, 324 (1997).

122. J. W. G. Wildoer, L. C. Venema, A. G. Rinzler, R. E. Smalley, and C. Dekker, *Nature* **391**, 59 (1998).

123. T. W. Odum, J. L. Huang, P. Kim, and C. M. Lieber, *Nature* **391**, 62 (1998).

124. S. Frank, P. Poncharal, Z. L. Wang, and W. A. de Heer, *Science* **280**, 1744 (1998).

125. M. S. Dresselhaus, G. Dresselhaus, and P. C. Eklund, *Science of Fullerenes and Carbon Nanotubes*, Academic: New York, 1996.

126. S. Suzuki, C. Bower, Y. Watanabe, and O. Zhou, *Appl. Phys. Lett.* **76**, 4007 (2000).

127. H. Ago, T. Kugler, F. Cacialli, W. R. Salaneck, M. S. P. Shaffer, A. H. Windle, and R. H. Friend, *J. Phys. Chem. B* **103**, 8116 (1999).

128. A. G. Rinzler, J. H. Hafner, P. Nikolaev, L. Lou, S. G. Kim, D. Tomanek, P. Nordlander, D. T. Colbert, and R. E. Smalley, *Science* **269**, 1550 (1995).

129. W. A. de Heer, A. Chatelain, and D. Ugarte, *Science* **270**, 1179 (1995).

130. Q. H. Wang, T. D. Corrigan, J. Y. Dai, R. P. H. Chang, and A. R. Krauss, *Appl. Phys. Lett.* **70**, 3308 (1997).

131. P. G. Collins and A. Zettl, *Appl. Phys. Lett.* **69**, 1969 (1996).

132. C. Bower, O. Zhou, W. Zhu, A. G. Ramirez, G. P. Kochanski, and S. Jin, *Mater. Res. Soc. Symp. Proc.* **593**, 215 (1999).

133. X. Xu and G. R. Brandes, *Appl. Phys. Lett.* **74**, 2549 (1999).

134. O. M. Kuttel, O. Groening, C. Emmenegger, and L. Schlapbach, *Appl. Phys. Lett.* **73**, 2113 (1998).

135. Z. F. Ren, Z. P. Huang, J. W. Xu, J. H. Wang, P. Bush, M. P. Siegal, and P. N. Provencio, *Science* **282**, 1105 (1998).

136. S. Fan, M. G. Chapline, N. R. Franklin, T. W. Tombler, A. M. Cassell, and H. Dai, *Science* **283**, 512 (1999).

137. C. Bower, W. Zhu, S. Jin, and O. Zhou, *Appl. Phys. Lett.* **77**, 830 (2000).

138. C. Bower, W. Zhu, D. J. Werder, S. Jin, and O. Zhou, *Appl. Phys. Lett.* **77**, 2767 (2000).

139. H. Murakami, M. Hirakawa, C. Tanaka, and H. Yamakawa, *Appl. Phys. Lett.* **76**, 1776 (2000).

140. L. Nilsson, O. Groening, C. Emmenegger, O. Kuettel, E. Schaller, L. Schlapbach, H. Kind, J. M. Bonard, and K. Kern, *Appl. Phys. Lett.* **76**, 2071 (2000).

141. K. A. Dean and B. R. Chalamala, *Appl. Phys. Lett.* **76**, 375 (2000).

142. W. Zhu, C. Bower, O. Zhou, G. Kochanski, and S. Jin, *Appl. Phys. Lett.* **75**, 873 (1999).

143. J. Bonard, J. Salvetat, T. Stockli, W. A. de Heer, L. Forro, and A. Chatelain, *Appl. Phys. Lett.* **73**, 918 (1998).

144. P. G. Collins and A. Zettl, *Phys. Rev. B* **55**, 9391 (1997).

145. Y. Saito, K. Hamaguchi, S. Uemura, K. Uchida, Y. Tasaka, F. Ikazaki, M. Yumura, A. Kasuya, and Y. Nishina, *Appl. Phys. A* **67**, 95 (1998).

146. J. P. Barbour, W. W. Dolan, J. K. Trolan, E. E. Martin, and W. P. Dyke, *Phys. Rev.* **92**, 45 (1953).

147. W. A. Anderson, *J. Vac. Sci. Technol. B* **11**, 383 (1993).

148. K. A. Dean, P. von Allmen, and B. R. Chalamala, *J. Vac. Sci. Technol. B* **17**, 1959 (1999).

149. K. A. Dean and B. R. Chalamala, *J. Appl. Phys.* **85**, 3832 (1999).

150. C. D. Child, *Phys. Rev.* **32**, 492 (1911).

151. I. Langmuir, *Phys. Rev.* **2**, 450 (1913).

152. J. W. Gadzuk and E. W. Plummer, *Rev. Mod. Phys.* **45**, 487 (1973).

153. H. Morikawa, K. Okamoto, Y. Yoshino, F. Iwatsu, and T. Terao, *Jpn. J. Appl. Phys.* **36**, 583 (1997).

154. A. P. Komar and A. A. Komar, *Sov. Phys. Tech. Phys.* **7**, 634 (1963).

155. Y. H. Lee, S. G. Kim, and D. Tomanek, *Chem. Phys. Lett.* **265**, 667 (1997).

156. Y. Saito, K. Hamaguchi, K. Hata, K. Uchida, Y. Tasaka, F. Ikazaki, M. Yumura, A. Kasuya, and Y. Nishina, *Nature* **389**, 554 (1997).

157. Y. Saito, K. Hata, and T. Murata, *Jpn. J. Appl. Phys.* **39**, L271 (2000).

158. J. Bonard, T. Stockli, F. Maier, W. A. de Heer, A. Chatelain, J. Salvetat, and L. Forro, *Phys. Rev. Lett.* **81**, 1441 (1998).

159. R. Tamura and M. Tsukada, *Phys. Rev. B* **52**, 6015 (1995).

160. D. L. Carroll, P. Redlich, P. M. Ajayan, J. C. Charlier, X. Blasé, A. De Vita, and R. Car, *Phys. Rev. Lett.* **78**, 2811 (1997).

161. K. A. Dean, O. Groening, O. M. Kuttel, and L. Schlapbach, *Appl. Phys. Lett.* **75**, 2773 (1999).

162. W. P. Dyke and W. W. Dolan, Field Emission, in *Adv. Electron Phys.* vol. 8, 1956, pp. 89.

163. J. Y. Cavaille and M. Dechsler, *Surf. Sci.* **75**, 342 (1978).

164. K. A. Dean and B. R. Chalamala, *Appl. Phys. Lett.* **75**, 3017 (1999).

165. I. Brodie and C. A. Spindt, Vacuum microelectronics, in *Advances in Electronics and Electron Physics, Vol. 83* (P. W. Hawkes, Ed.), Academic: New York, p. 1, 1992.

166. C. Chinnock, *Laser Focus World*, Feb. 1996.

167. Y. Saito, S. Uemura, and K. Hamaguchi, *Jpn. J. Appl. Phys.* **37**, L346 (1998).

168. Q. H. Wang, A. A. Setlur, J. M. Lauerhaas, J. Y. Dai, E. W. Seelig, and R. P. H. Chang, *Appl. Phys. Lett.* **72**, 2912 (1998).

169. W. B. Choi, D. S. Chung, J. H. Kang, H. Y. Kim, Y. W. Jin, I. T. Han, Y. H. Lee, J. E. Jung, N. S. Lee, G. S. Park, and J. M. Kim, *Appl. Phys. Lett.* **75**, 3129 (1999).

170. E. Yamaguchi, K. Sakai, I. Nomura, T. Ono, M. Yamanobe, N. Abe, T. Hara, K. Hatanaka, Y. Osada, H. Yamamoto, and T. Nakagiri, *J. SID* **5**, 345 (1997).

171. M. I. Elinson, A. G. Zhdan, G. A. Kudintseva, and M. E. Chugunova, *Radio Eng. Electron Phys.* **10**, 1290 (1965).

172. G. Dittmer, *Thin Solid Films* **9**, 317 (1972).

173. M. Hartwell and C. G. Fonstad, in *Tech. Dig. IEDM*, 1975, p. 519.

174. H. Araki, O. Hirabaru, and T. Hanawa, *Vacuum* **26**, 22 (1983).

175. R. C. Miller and A. Savage, *J. Appl. Phys.* **31**, 662 (1960).

176. A. Koller and M. Beranek, *Czech. J. Phys.* **9**, 402 (1959).

177. G. Rosenman, D. Shur, Y. E. Krasik, and A. Dunaevsky, *J. Appl. Phys.* **88**, 6109 (2000).

178. K. Biedrzycki, H. Aboura, and R. Le Bihan, *Phys. Stat. Sol.* **140**, 257 (1993).

179. H. Gundel, J. Handarek, and H. Riege, *J. Appl. Phys.* **69**(2), 975 (1991).

180. A. Airapetov, I. Ivanchik, A. Lebedev, I. Levshin, and N. Tikhomirova, *Sov. Phys. Dokl.* **35**, 267 (1990).

181. G. Rosenman, V. Okhapkin, Y. L. Chepelev, and V. Shur, *Sov. Phys.—JEPT* **39**, 477 (1984).

182. K. Biedrzycki, *Phys. Status Solidi A* **93**, 503 (1986).

183. G. Rosenman, V. Letuchev, Y. Chepelev, O. Malyshkina, V. Shur, and V. Kuminov, *Appl. Phys. Lett.* **56**, 689 (1990).

184. B. Rosenblum, P. Braunlich, and J. Carrico, *Appl. Phys. Lett.* **25**, 17 (1974).

185. I. Rez, G. Rosenman, Y. Chepelev, and N. Angert, *Sov. Tech. Phys. Lett.* **5**, 1352 (1979).

186. H. Riege, I. Boscolo, J. Handerek, and U. Herleb, *J. Appl. Phys.* **84**, 1602 (1998).

187. L. Schachter, J. D. Ivers, J. A. Nation, and G. S. Kerslick, *J. Appl. Phys.* **73**, 8097 (1993).

188. M. Angadi, O. Auciello, A. R. Krauss, and H. Gundel, *Appl. Phys. Lett*, **77**, 2659 (2000).

189. H. Gundel, H. Riege, E. J. N. Wilson, J. Handerek, and K. Zioutas, *Ferroelectrics* **100**, 1 (1989).

190. H. Gundel, H. Riege, J. Handerek, and K. Zioutas, *Appl. Phys. Lett.* **54**, 2071 (1989).

191. R. Drori, M. Einat, D. Shur, E. Jerby, G. Rosenman, R. Advani, R. J. Temkin, and C. Pralong, *Appl. Phys. Lett.* **74**, 335 (1999).

192. G. Rosenman, D. Shur, and A. Skliar, *J. Appl. Phys.* **79**, 7401 (1996).

193. C. A. Mead, *J. Appl. Phys.* **32**, 646 (1961).

194. Y. Kumagai, K. Kawarada, and Y. Shibata, *Jpn. J. Appl. Phys.* **6**, 290 (1967).

195. J. G. Simmons, *J. Appl. Phys.* **34**, 1793 (1963).

196. K. Ohta, J. Nishida, and T. Hayashi, *Jpn. J. Appl. Phys.* **7**, 784 (1968).

197. T. Kusunoki, M. Suzuki, S. Sasaki, T. Yaguchi, and T. Aida, *Jpn. J. Appl. Phys.* **32**, L1695 (1993).

198. H. Adachi, *J. Vac. Sci. Technol. B* **14**, 2093 (1996).

199. K. Yokoo, H. Tanaka, S. Sato, J. Murota, and S. Ono, *J. Vac. Sci. Technol. B* **11**, 429 (1993).

200. T. W. Hickmott, *Thin Solid Films* **9**, 431 (1972).

201. T. Kusunoki and M. Suzuki, *IEEE Trans. Electron Devices* **47**, 1667 (2000).

202. T. Komoda, Y. Honda, T. Hatai, Y. Watabe, T. Ichihara, K. Aizawa, and Y. Kondo, in 6th Int. Display Workshop, Sendai, Japan, 1999, p. 939.

CHAPTER 7

Field Emission Flat Panel Displays

HEINZ H. BUSTA

Sarnoff Corporation, Princeton, New Jersey 08543-5300

7.1. INTRODUCTION

A field emission display (FED) is, in principle, a simple device. It consists of a baseplate containing an array of addressable gated field emitters whose electrons are projected towards a phosphor plate that is positioned in close proximity (0.2–2 mm) to the array. In contrast to the cathode ray tube (CRT) where three hot filament electron guns address all of the red, green, and blue phosphor spots on the screen, in the FED, each color pixel or a small group of color pixels is addressed by a field emitter source. The virtues of the FED over other flat panel devices are higher brightness, better viewing angles, lower power consumption, and a larger range of operating temperatures. Table 7.1 compares some of the characteristics of FEDs with those of active matrix liquid crystal displays (AMLCDs), electroluminescent, and plasma displays. The table was adapted from Ref. [1] with an added column for organic light emitting diode (OLED) from Ref. [2]. The table is intended as a guide and will change as these display technologies evolve.

The FED was first proposed by Crost, Shoulders, and Zinn and is described in US Patent 3,500,102 issued in 1970. The first working monochrome prototype was demonstrated at Japan Display in 1986 by a group from LETI, France [3]. The first color display (6 in.) was demonstrated by Pixel/LETI in 1993 [4]. Initially, it was conjectured that, by using standard CRT phosphor and encapsulation technologies, FEDs could be manufactured at lower cost than LCDs because of simpler processing and proven CRT technology. In reality, FED manufacturers encounter tremendous difficulties, mostly caused by the transition from the low surface-to-volume ratio CRT to a very high surface-to-volume ratio device. Issues of spacer breakdown, vacuum flash over, gettering, placement of getters, device outgassing, uniformity of emission, emitter array yield, emitter positioning, and others — all of which are of minor concern in CRT processing — are becoming major problems that have to be resolved in FEDs.

The FED has been the driving engine for the continuous development of vacuum microelectronics (VME), similar to microprocessor and memory chips in microelectronics and pressure sensors and accelerometers in microelectromechanical systems

Vacuum Microelectronics, Edited by Wei Zhu.
ISBN 0-471-32244-X. 2001 John Wiley & Sons, Inc.

TABLE 7.1. Comparison of FED with Other Flat Panel Displays

Feature	Thin Film Transistor LCD	Electro-luminescent Display	FED	Plasma Display Panels	OLED Display
Brightness $(\mathrm{cd/m^2})$	200	100	150 (low-V) >600 (high-V)	300	300
Viewing angle (degrees)	±40	±80	±80	±80	±80
Emission efficacy (lm/W)	3–4	0.5–2	1.5–3 (low-V) 10–15 (high-V)	1.0	10–15
Response time (ms)	30–60	<1	0.01–0.03	1–10	<0.001
Contrast ratio (intrinsic)	>100:1	50:1	300:1	100:1	100:1
Number of colors (millions)	16	16	16	16	16
Number of pixels	1024 × 768	640 × 480	800 × 600	852 × 480	640 × 480
Resolution (mm in pitch)	0.31	0.31	0.27	1.08	0.012
Power consumption (W)	3 $(25.4)^a$	6 (25.4)	2 (25.4)	200 (106.7)	6 (15.2)
Maximum screen size in diagonal (cm)	55.9 $(22)^b$	25.4 (10)	35.6 (14)	106.7 (42)	15.2 (6)
Panel thickness (mm)	8	10	10	75–100	3
Operating temperature range (°C)	0–50	−5 – +85	−5 – +85	−20 – +55	−25 – +65

^{a} Values in parentheses in this row indicate panel size in centimeters.
^{b} Values in parentheses in this row indicate panel size in inches.

(MEMS). This chapter will describe in detail the basic FED devices that are intended for desktop applications. This will be followed by profiles of other display applications, such as electronic billboards, wall television, and infrared-to-visible light converters.

7.2. FIELD EMISSION DISPLAYS

7.2.1. Principle of Operation

7.2.1.1. Triode Structure. FEDs can be operated as either diode or triode devices with either low-voltage (<2000 V) or high-voltage phosphors (2000–8000 V).

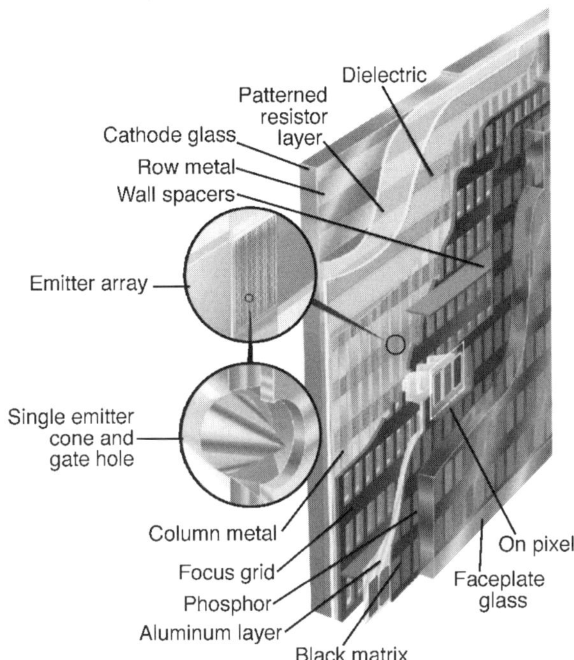

FIGURE 7.1. Perspective view of a high-voltage phosphor FED architecture, including cut-away enlargements of a subpixel emitter array and a single gated tip [5].

Diode devices are usually fabricated using planar carbon-type emitters, and triode devices use Spindt-type gated emitters fabricated in molybdenum or silicon. Other triode approaches such as surface conduction electron emitters and gated edge emitters have been proposed and used.

At present, the most mature device is the Spindt-type triode device using high-voltage aluminized phosphors. Figure 7.1 shows a perspective cutaway drawing of such a device [5]. The glass baseplate consists of row metal lines upon which a patterned resistor layer is fabricated. This layer is necessary for obtaining uniform emission per subpixel array by providing a voltage drop to the emitters that is proportional to the emission current. Thus, the gate voltage is reduced by the largest amount for the dominant emitters. On top of the resistor film, the gate dielectric is deposited, followed by the deposition of the gate material. Gate column lines are then formed, followed by the processing of gate openings and tip formation. A gated tip is shown in the enlargement on the left side of the drawing. For obtaining uniform emission, it is desirable to fabricate as many emitters per subpixel as possible (ranging from hundreds to several thousands).

The faceplate that contains the red, green, and blue phosphor regions is separated from the baseplate by spacer rails. These are needed to avoid collapse of the device by the atmospheric pressure once it is evacuated. By aluminizing the phosphors, higher brightness can be achieved, since light that would reflect back into the display from the glass/air interface is reflected towards the viewer by the aluminum layer. A black matrix surrounding the rectangular phosphor pixels enhances the contrast of the display.

By applying a positive voltage to the gate columns with respect to the emitter rows, selective pixels can be energized. The electrons generated at the tips traverse through the vacuum space, penetrate through the thin aluminum layer, and cause the phosphor pixels to light up via the electroluminescent process. If the distance between the phosphor plate and the emitters is short, the electrons will not spill over to adjacent pixels. This situation is called proximity focusing. For larger distances that are generally needed for high-voltage phosphor operation to avoid flashover of the spacers, the electron beam has to be focused to avoid illumination of adjacent pixels. In Fig. 7.1, this is accomplished by surrounding each subpixel emitter area by a conductive thick film "waffle" structure that is negatively biased with respect to the gate. Not shown in this cutaway drawing are the ancillary items such as the device seal, pumping ports, getters, electrical feedthroughs, and placement of the driver chips.

To reduce the number of emitter arrays of a device such as the one shown in Fig. 7.1 by a factor of 3 and therefore, increase the processing yield, one array can be shared with the corresponding red, green, and blue subpixels. To accomplish this, the phosphor lines must be switched, which limits the device to low-voltage phosphor operation. The first color FED prototype from Pixel/LETI utilized this scheme [4]. Figure 7.2 shows a perspective view of the switched phosphor triode device [6].

7.2.1.2. Diode Structure. The devices shown in both Figs. 7.1 and 7.2 are of the triode type, in which the phosphor voltage is constant. A diode device is shown in perspective view in Fig. 7.3 [7]. It consists of a glass baseplate onto which thin film emitter lines and metal spacers are fabricated. The emitters are fabricated from diamond-like carbon (DLC). Because of the fragile nature of these films and the possibility of contamination from subsequent processing steps, the emitters are deposited at the end of the processing sequence using a shadow-mask technique. The top plate contains groups of red, green, and blue nonmetalized phosphor stripes. In operation, the phosphor lines have to be switched to enable selection of a given subpixel. This limits the device to low-voltage phosphor technology, since achieving line-to-line isolation at high voltages (>2 kV) is difficult. In addition, the device would also require high-voltage addressing circuitry. In order to obtain low-voltage operation, the distance between the two plates is only about 50–100 μm. This is dictated by the turn-on field of carbon films which range from 5 to 10 V/μm. The advantage of the diode FED is that a much simpler device structure is used; the disadvantage is the need for higher voltage driver circuitry.

7.2.2. Emitter Technologies

7.2.2.1. Spindt Emitters. The most prominent and widely used emitter technology for FEDs is the Spindt emitter using e-beam evaporated molybdenum cones

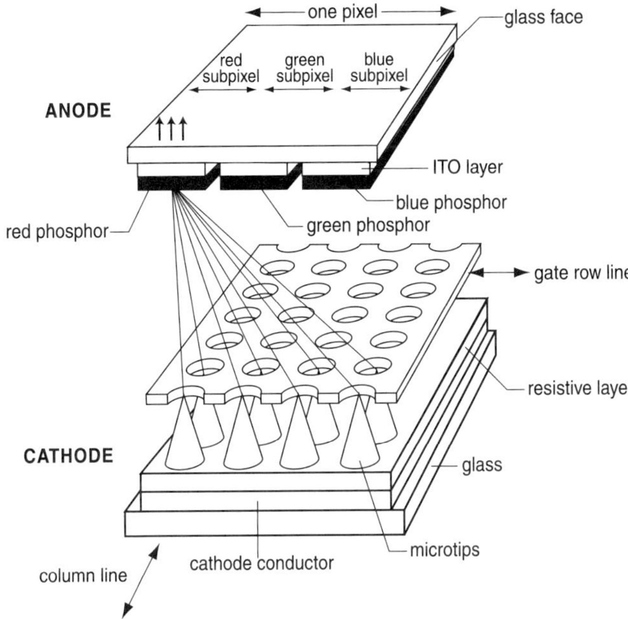

FIGURE 7.2. Perspective view of an FED in which one emitter array is shared by a red, green, and blue subpixel [6]. Selection of the subpixel is accomplished by biasing the appropriate phosphor line. The drawing is not to scale. The distance between the emitters is at least 100 times larger than the tip height, meaning that the electron paths are not as distorted as shown.

ranging in heights from about one to several micrometers, and gate diameters ranging from 0.15 to 1.5 μm. Other materials, such as tungsten, tantalum, hafnium, metal silicides, carbides and nitrides, have been proposed, along with thin layers (nanometer thick) of insulators, semiconductors, metals, and others on the tips. Emission results have been reported in the literature, notably in proceedings of the International

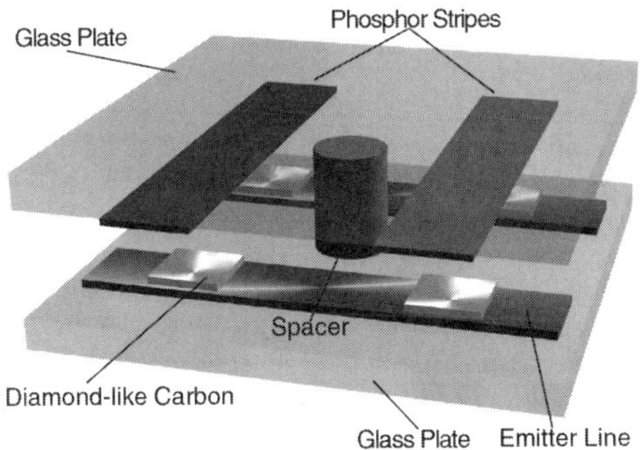

FIGURE 7.3. Perspective view of a diode device using DLC emitters [7].

Vacuum Microelectronics Conferences (IVMCs) that have been held annually since 1988 (Williamsburg, VA.) Even though some improvements in emission behavior over the molybdenum devices have been reported, none of the companies pursuing product development are using any of these materials besides molybdenum at this point. One reason is historical, since at the time when most companies launched their product development efforts, the only device for which sufficient data were available was the Spindt device. Another reason is that the molybdenum cone device works well in comparison to other refractory metals and silicon. For instance, initial conditioning is not that critical as compared to silicon, which is attributed to the more conductive molybdenum oxide as compared to the native silicon oxide. Detailed discussions on the processing, array formation, and performance of the Spindt device can be found in Chapter 4.

7.2.2.2. *Silicon Emitters.* Next in popularity is the gated silicon emitter. By using the entire arsenal of VLSI processing, an impressive body of work has been developed, ranging from pyramid to wedges, to cone, and to tip-on-post emitters. The advantage of silicon technology is that new device concepts can be proven in a relatively short period of time and in most academic institutions. By combining field emitters with novel current-limiting circuits on a silicon substrate, low-noise emitter arrays have been demonstrated. Monocrystalline silicon, however, limits the size of FEDs to applications such as helmet displays. For larger products, in which all the emitters are fabricated on one substrate, amorphous or polycrystalline silicon cone devices are used. Silicon field emitter arrays (FEAs) are described in detail in Chapter 5.

7.2.2.3. *Thin Film Edge Emitters.* Thin film edge emitter is, in principle, much simpler to fabricate, since the cone deposition process (either by etching or collimated deposition through a diminishing hole diameter) is eliminated. Some of the basic device structures are revealed in US patents 5,382,185 by Gray and Hsu and 5,874,808 by Busta and Pogemiller, and some of the initial devices were discussed in Ref. [8]. Figure 7.4 shows the perspective view of a single gated lateral device with comb-shaped thin film emitters [9]. Figure 7.5 shows the SEM micrograph of a 0.2 μm tungsten device with a p/a ratio (defined in Fig. 7.4) of 3.3 and an emitter-to-gate distance of 0.5 μm. Figure 7.6 shows the I–V curves of two 150-comb devices with different p/a ratios.

By folding the combs and gates into cylindrical structures and rotating them by 90°, gated edge emitters can be fabricated. The cross section of a stepped oxide device is shown in Fig. 7.7, and the SEM micrograph of a TiW–Au emitter is shown in Fig. 7.8 [10]. By stepping the oxide near the gate plateau, the turn-on voltage can be reduced without increasing the gate-to-emitter capacitance significantly. The device in Fig. 7.8 was processed by an etch process, taking advantage of the pile-up of photoresist near the gate posts and the thinning on top of the plateau. For large-area implementation, a chemical mechanical planarization (CMP) process is preferred. Sputtered SiC emitters fabricated by CMP were described in Ref. [11]. Other edge emitters, such as a dispenser-type, vertical multilayer emitter using a low work function lithium film sandwiched between two noble metals, were described in Ref. [12]. Double-gated edge emitters, which are formed by the inclusion of another insulating

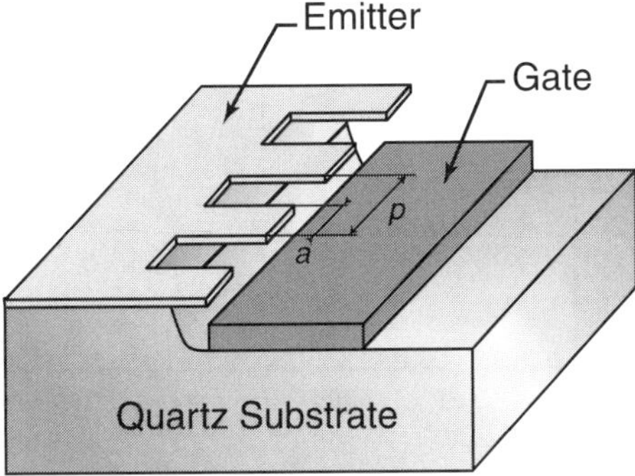

FIGURE 7.4. Perspective drawing of a lateral, self-aligned comb-shaped field emitter device [9].

and conductive layer, are shown schematically in Fig. 7.9. Such devices can avoid the excessive gate currents recorded in devices with single or asymetrically positioned gates, thus reducing the power consumption of FEDs.

Emission from these structures originates from sharp features at the thin film edges and is not very controlled at present. Although this development appears promising, no FED prototypes currently use thin film edge emitters. Considerable work must be done to raise the level of performance of these edge emitters to that of cone devices.

7.2.2.4. Surface Conduction Electron Emitters.

A very intriguing technology based on surface conduction electron emitters has entered the field in recent years [13,14]. The basic phenomenon was described in 1965, but had not been identified

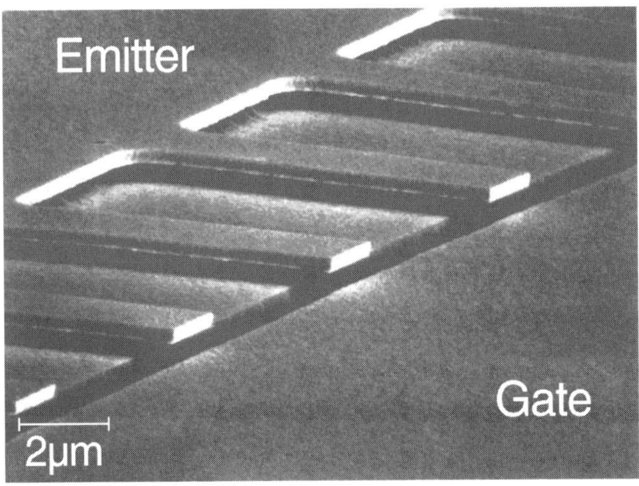

FIGURE 7.5. SEM micrograph of a lateral edge emitter with a pitch-to-edge width ratio (p/a) of 3.3 [9].

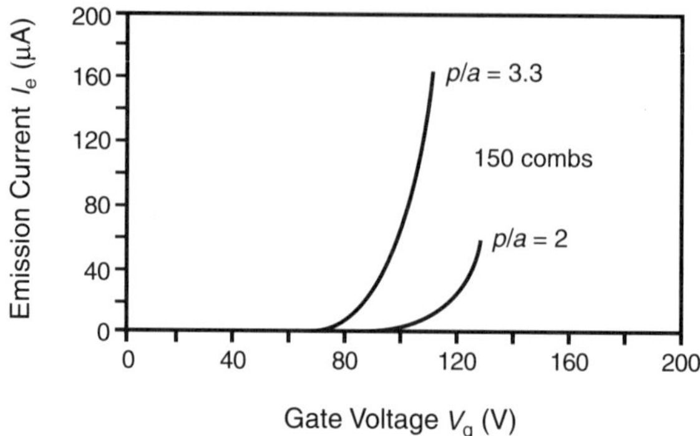

FIGURE 7.6. *I–V* characteristics of two 0.2-μm thick tungsten edge emitters with different *p/a* ratios [9].

as a possible candidate for display applications until recently. The device is based on a fine particle film of PdO (5–10 nm) that is deposited between two Pt electrodes. A forming process generates a nanofissure parallel to the edges of the two electrodes that acts as an electron emission source. Electrons emitting from the fissure are scattered towards the positive side of the electrode and some eventually reach the anode. The efficiency, i.e., the ratio of anode current to electrode current, is below 1%, but still high enough for a low-power display. It is estimated that the power consumption for a 40-in. display is only 53 W.

For FED applications, the electrodes are screen-printed in air, and the PdO film is deposited via an ink jet printer. This eliminates thin film processing completely and brings this technology into the arena of plasma display manufacturing. When keeping the electrode voltage below 20 V, the electrons focus sufficiently so that no extra focusing electrodes are needed even for an emitter-to-anode spacing of 3–5 mm. A 10-in. diagonal prototype with 240 × 240 × 3 pixels and 8.6-mm panel spacing has been demonstrated using stripe-patterned P-22 (R/G/B) phosphors and a

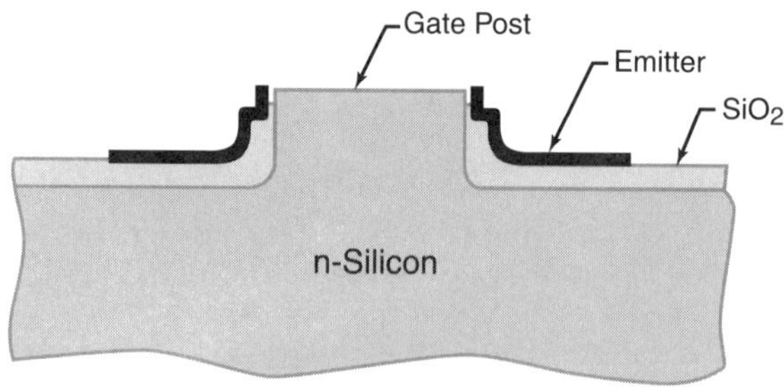

FIGURE 7.7. Cross section of a single gated edge emitter using a stepped oxide approach to reduce the gate-to-emitter distance near the emission area without significantly increasing device capacitance [10].

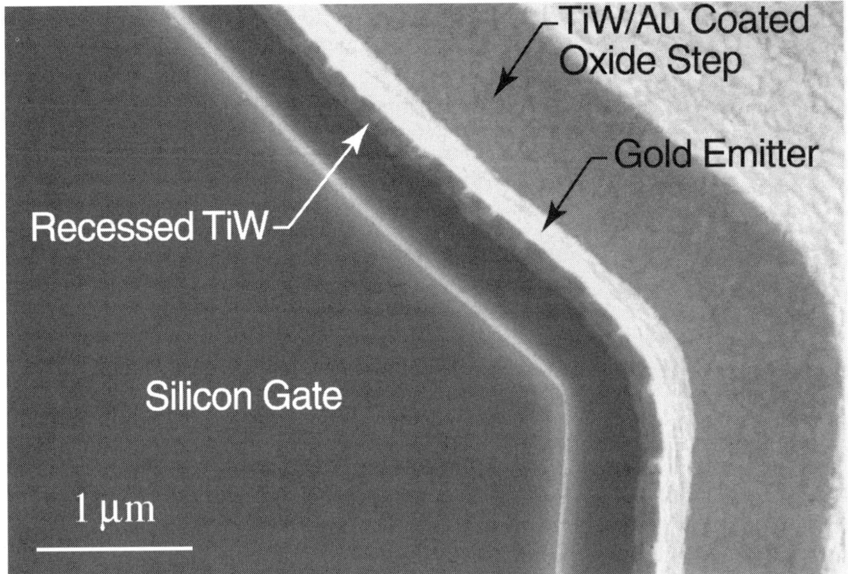

FIGURE 7.8. SEM micrograph of a stepped oxide TiW–Au vertical edge emitter [10].

black matrix [13]. Little information is available at present about the reliability of this display. Since there is only one fissure formed per subpixel, there is no redundancy, as is the case for multiple cone emitters.

7.2.2.5. Other Emitters. Other emitters gaining popularity in recent years, such as diamond, DLC, and carbon nanotubes, are described in detail in Chapter 6. The main objective in developing alternative cold electron sources for display applications is low-cost manufacturability over large areas. One such alternative might exist in printable inks consisting of semiconductive and conducting particles dispersed in an insulator-forming precursor. Such inks need to be vacuum compatible at both the

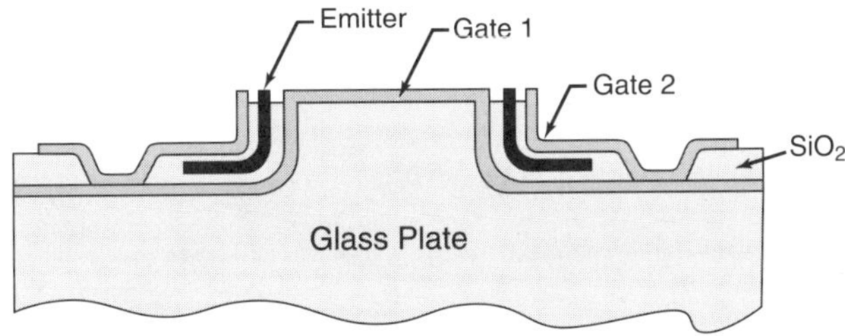

FIGURE 7.9. Cross section of a vertical edge emitter with symmetrically placed gate [H. Busta, unpublished].

fabrication stage and throughout the life of the display. Initial results yielding an emission site density of about 1×10^5 cm^{-2} at an extraction field of about 25 V/μm were described in Ref. [15]. Field emission from polymers has also been recently demonstrated [16]. By embedding transistors fabricated from the same semiconducting polymer, polyalkylthiophenes, inexpensive FEDs are being proposed (Bill Eccleston, University of Liverpool, UK).

7.2.3. Uniformity of Emission Enhancement

When refractive and semiconductive cone emitters are fabricated on conductive substrates such as thin film metal emitter lines, they typically show poor emission uniformity within an array. This is caused by the extreme sensitivity of the Fowler–Nordheim (F–N) tunneling process to small variations in tip radii, cone angles and heights, gate diameters, position of the gate holes, and the work function of the tips. A key contribution in the development of FEDs was made by introducing resistive layers between the tips and the emitter lines. Vertical resistors were proposed in Ref. [17] and resulted in significant improvements in emission uniformity. Additional improvements, especially in cases where some of the tips cause shorts between the emitter and gate metals, were obtained by the introduction of a lateral resistor mesh [18]. For silicon tips and tips fabricated on silicon substrates, active resistors in the form of on-substrate integrated field effect transistors (FETs) have been reported with excellent results (see Chapter 5).

Figure 7.10 shows the cross section of two Spindt emitters fabricated on a resistive layer. To demonstrate the effect of the resistive layer, we assume that the second tip has a field enhancement factor β that is 20% higher than the first tip. This could be caused, for instance, by a sharper tip. The current can be calculated using the F–N equation

$$I = \left(\frac{1.54 \times 10^{-6} \alpha E^2}{\Phi} \right) \exp \left(\frac{-6.87 \times 10^7 \Phi^{3/2} v(y)}{E} \right) \tag{1}$$

where $y = 3.79 \times 10^{-4} E^{1/2}/\Phi$, $v(y) = 0.95 - y^2$, $E = \beta V_g$, $\alpha = 4 \times 10^{-15}$ cm^2, $\beta = 8 \times 10^5$ cm^{-1}, and $\Phi = 4.5$ eV.

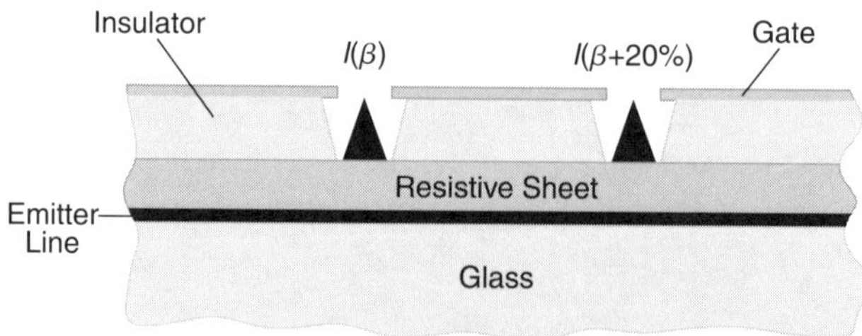

FIGURE 7.10. Cross section of two tips fabricated on a resistive sheet.

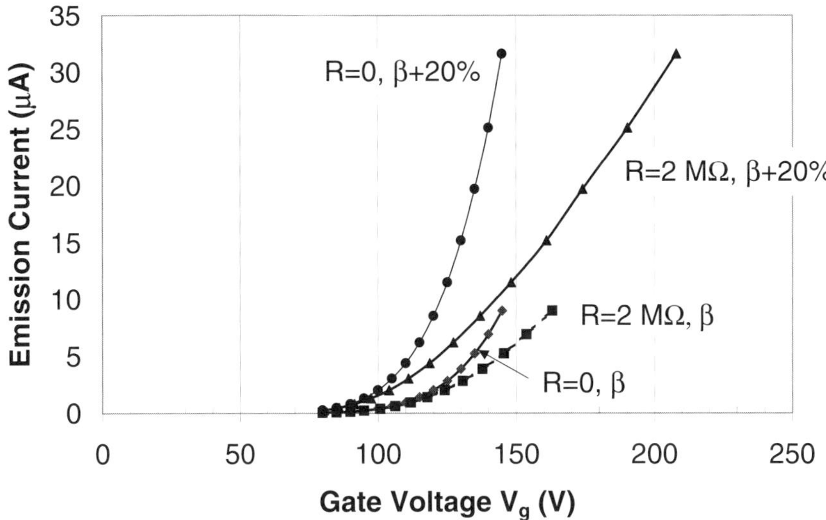

FIGURE 7.11. Calculated emission currents of two single-tip devices with and without a vertical resistor of 2 MΩ, assuming that one of the devices has a 20% higher field enhancement factor compared to the other one.

This is shown in Fig. 7.11 as the line designated with $(R = 0, \beta)$. Assuming that the resistance between the base of the tip and the emitter line is 2 MΩ, the effect of the voltage drop IR is calculated by substituting V_g with $V_g - IR$ and solving the F–N equation numerically. The result is shown as the line designated with $(R = 2\,\text{M}\Omega, \beta)$. It can be seen that, at a gate voltage of 145 V, the emission current is reduced from 9 μA without the resistive layer to 5 μA with the resistive layer. Now the field enhancement factor β of the second tip is 20% higher, i.e., 9.6×10^5 cm^{-1}, the calculated I–V curves with and without the resistive layer are shown by the lines designated with $(R = 2\,\text{M}\Omega, \beta + 20\%)$ and $(R = 0, \beta + 20\%)$, respectively. At 145 V gate bias, the emission current of the dominant second tip reduced from 32 μA to 11 μA by including the resistance.

Assuming that the base radii of these two emitters and the film thickness of the resistive layer are all 1 μm, the resistivity of the film can be calculated to be 6.3×10^2 Ω cm, and the sheet resistance is 6.3×10^6 Ω/□. This is a relatively high resistivity, and very little information is available concerning actual materials that are used by the FED manufacturers. Sputtered amorphous silicon, Cr cermets, and SiC are some of the candidates [19]. Care has to be taken that the voltage drops across the resistive layer do not reach magnitudes that can cause film breakdown.

To avoid the large voltage drop, and particularly to reduce the tip shorting problem, the lateral resistor mesh was introduced. In addition to increasing the emission uniformity, the lateral resistor mesh allows faster tip conditioning and results in a more rugged array [20]. However, the lateral film consumes more real estate at the expense of tip density. Figure 7.12 shows the top view and cross section of an FEA

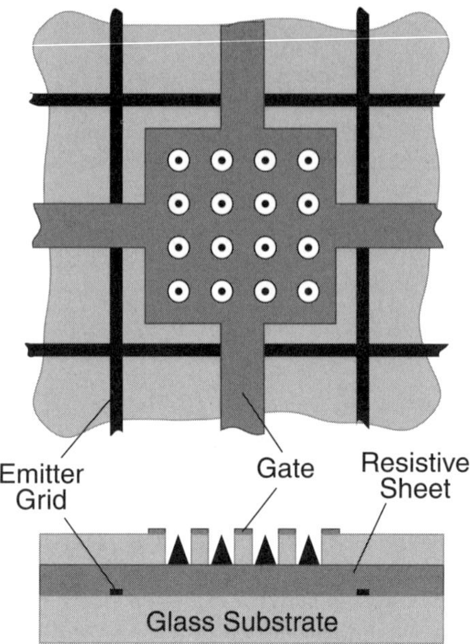

FIGURE 7.12. Design of a 4 × 4 emitter array with a lateral resistor film [21].

with a lateral resistor mesh onto which a 4 × 4 microtip array is positioned [21]. The bias to the emitters is supplied by a thin film metal grid, on top of which the resistive film is deposited. The gate lines are positioned perpendicular to the emitter grid. The performance of this 4 × 4 array was calculated in Ref. [22] by assuming an emission current of 1 μA from each emitter at a gate voltage of 100 V and a sheet resistance of 10^7 Ω/□ for the film.

Figure 7.13 shows the voltage drops and currents for each of the emitters. The inner 2 × 2 matrix shows identical results which, for reasons of symmetry, is easily understandable. Some nonuniformity exists for the outer 12 emitters. The resistive layer causes a reduction of the total current from 16 μA to 4.53 μA.

The effect of one tip shorting to the gate is shown in Fig. 7.14. Shorting can be caused during processing or due to the tip exploding during operation, leaving a conductive film on the sidewall of the insulator between the gate and the emitter line [23]. It is assumed that the upper left emitter in the 4 × 4 array is shorted to the gate. The voltage drop is then 100 V, drawing a non-emissive current of 9.8 μA. The total emitted current from the other 15 tips is now 1.75 μA as compared to the 4.53 μA for the ideal case of identical emitters with the resistive layer. If the short is assumed at a different position, slightly different results are obtained.

A combination of vertical and lateral resistors was described in Ref. [24], and the cross section of such a configuration is shown in Fig. 7.15. It consists of a mesh structure and a conductive plate underneath the emitters. A comparison study was performed with a 21 × 21 array of mesh and mesh with plate structures, with the

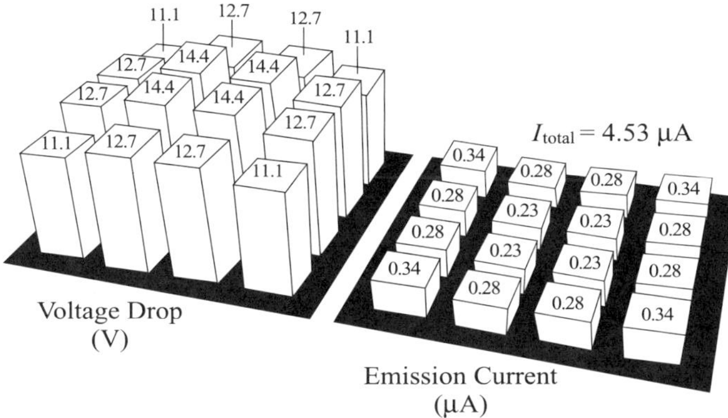

FIGURE 7.13. Voltage drop and current distribution of a 4 × 4 emitter array embedded in a lateral rasistor mesh with a sheet resistance of $10^7\ \Omega/\square$ and in which each emitter emits 1 μA at 100 V [22].

same distance between the cathode grid and the nearest emitter in the array and the same 0.5-μm thick amorphous Si resistive film. With the mesh structure, a current density of about 0.1 A/cm^2 was reached and 1.5 A/cm^2 when the conductive plate structure was included. The authors attributed this result to higher gate voltages per tip and improved uniformity of the voltage drops within the array. Postmortem inspection of failed arrays showed significant damage at the periphery of the array for the mesh structure and a more uniform distribution of failed devices for the mesh with conductive plate array.

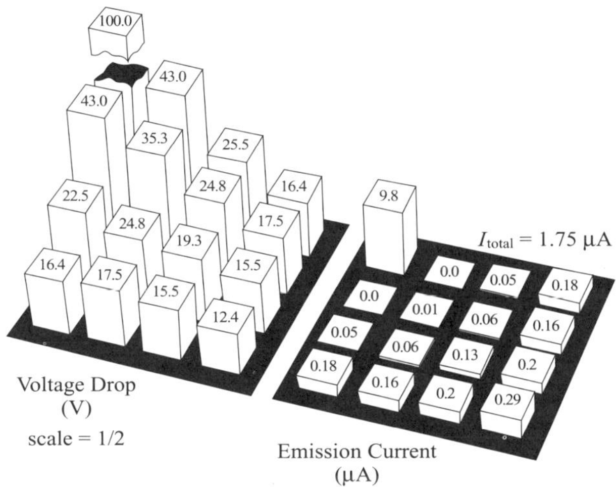

FIGURE 7.14. Voltage drop and current distribution of a 4 × 4 emitter array embedded in a lateral resistor mesh with a sheet resistance of $10^7\ \Omega/\square$ in which the upper right emitter forms a short to the gate. Each emitter is capable of delivering 1 μA at 100 V [22].

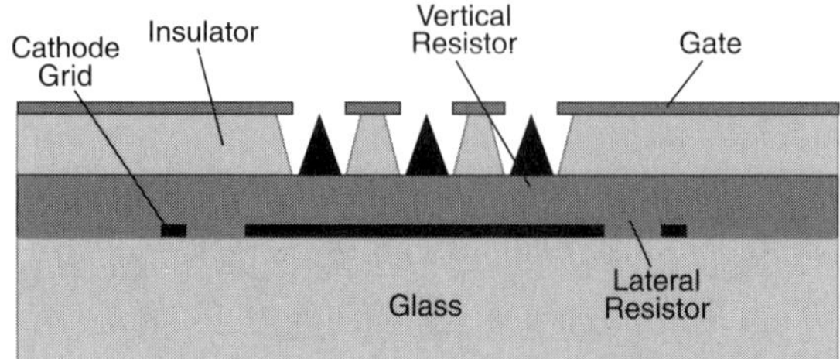

FIGURE 7.15. Cross section of an emitter array combining the vertical and lateral resistor concepts [24].

7.2.4. Emitter Array Fabrication

The objective of array fabrication is to produce low-cost, high-yield bottom plates that can be easily mated with the phosphor plate and also survive the sealing and evacuation process without imposing any detrimental effects to the emitters. In addition to fabricating uniform arrays containing millions of microtips, it is necessary to properly design the periphery of the bottom plate so that sealing can be performed without damaging the interconnect lines, and the thin film getter material can be incorporated without interfering with the performance of the arrays.

The array formation using the classical Spindt process is described in Chapter 4. For optimized tip formation, the incoming molybdenum beam should be collimated, which means a large source-to-substrate distance; one-to-two story high evaporators are on the market to satisfy this demand. In addition to this approach, other processing strategies such as random ion beam damage, laser interferometric lithography, and chemical mechanical polishing have been developed.

7.2.4.1. Ion Tracking Lithography.

Ion tracking is used by one of the FED houses and was described in Ref. [5]. After the emitter lines, resistive material, gate dielectric, and gate rows are photolithographically defined, a polymer is applied and exposed to a low current density ion beam that is scanned across the emitter plate. Every ion induces a damage track in the resist. These tracks are then exposed to an etchant with a large (100:1) etch selectivity between the damaged and undamaged polymer. This causes the removal of polymer at the damaged sites, resulting in gate holes with diameters of about 0.15 μm. Other etchants are then used to remove the gate electrode material and to form the cavities in the interelectrode dielectric layer. After that, tip processing proceeds in a modified Spindt-type fashion similar to that described in Chapter 4. Figure 7.16 shows several gated tips in a small region of a subpixel that have been formed by this method. Tip densities exceeding 10^8 tips/cm^2 have been achieved for substrate sizes exceeding 14-in. diagonal.

7.2.4.2. Laser Interferometric Lithography.

Large area arrays of dots or holes with submicron diameters and submicron-to-micron size spacings can be fabricated

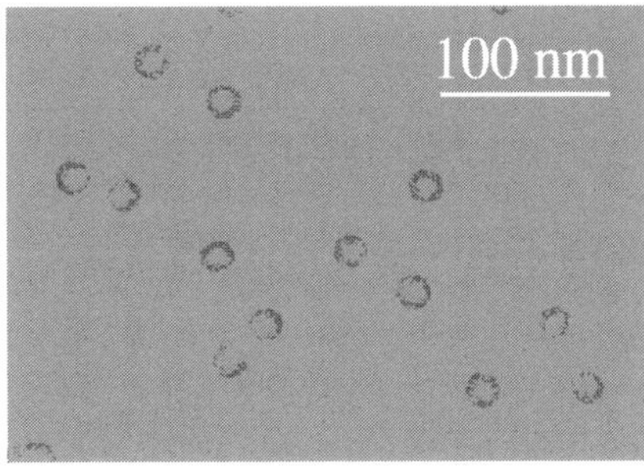

FIGURE 7.16. Top view SEM micrograph of several gated tips fabricated by the ion beam tracking method [5].

using laser interferometry [25], and working FED prototypes using this technique have been reported [26]. The technique, in some of its implementations, offers essentially unlimited depth of field, large-area (100 cm × 100 cm) and submicrometer patterning with the use of existing lasers, and relatively inexpensive optics. A two-beam exposure system can form a dense array of patterns as shown in Fig. 7.17(a) with the distance between the dots or holes given by

$$d = \lambda/2\sin(\theta) \tag{2}$$

where λ is the wavelength of the laser, and θ is the angle of incidence.

For obtaining sparser arrays, three-, four- and five-beam exposure strategies, each with its own subtlety, were described in Ref. [27], and the corresponding patterns are shown in Figs. 7.17(b) and (c). Sparser arrays are desirable from a reliability point of view, since failure propagation from tip-to-tip is reduced as the tip distance is increased. The trade-off of the three- and five-beam methods compared to the two- and four-beam systems is that the advantage of near infinite depths of focus is lost, necessitating the use of extremely flat substrates. Since arrays of holes are formed over the entire substrate in both the ion tracking and laser interferometric lithographies, an

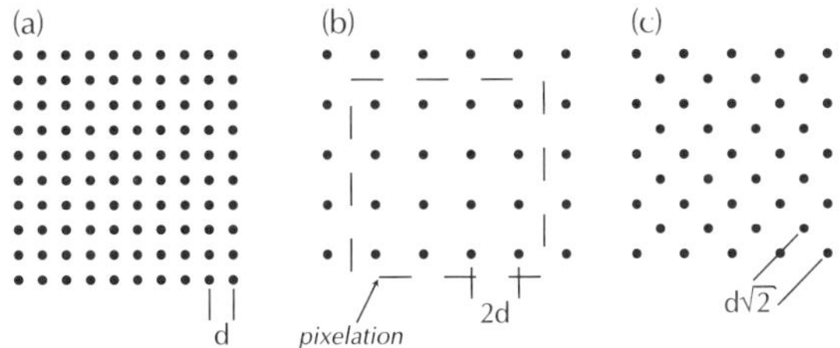

FIGURE 7.17. Hole pattern formation for (a) two-beam double exposure, (b) three-beam double exposure and five-beam single exposure, and (c) four-beam single exposure [27].

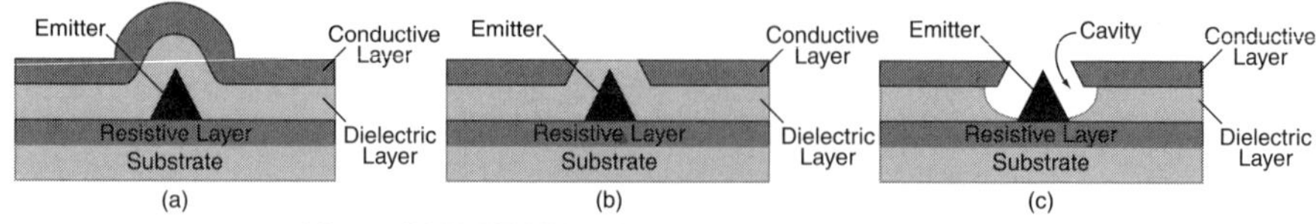

Micron CMP-FED™ self-aligned grid formation

FIGURE 7.18. Cross sections of a gated device at different stages of the CMP process [28].

additional processing step is needed to block formation of tips in unwanted regions. This is indicated in Fig. 7.17 as pixelation.

7.2.4.3. Chemical Mechanical Polishing.

Another attractive, large-area tip forming process is that of chemical mechanical polishing/planarization (CMP). Figure 7.18 shows the basic processing steps in forming gated tips [28]. After tip etching, the gate dielectric is deposited, followed by the gate material (Fig. 7.18(a)). At this point, the CMP process is performed, which consists of polishing off the protruding gate metal and insulator portions over the tips in a slurry that is chemically altered to facilitate enhanced removal (Fig. 7.18(b)). Once the caps are polished off, the pressure of the polishing pad increases, because the substrate is planarized. The point of increased pressure serves as an endpoint indicator of the process, and the polishing is subsequently stopped. The gate lines are then photolithographically defined, and the insulator near the tips is removed by an etching process (Fig. 7.18(c)). Commercial CMP machines capable of polishing 100 cm × 100 cm plates are readily available.

7.2.5. Phosphors and Phosphor Plate Fabrication

The color quality and brightness of FEDs should be at least as good or better than LCDs for desktop monitors or as good as the CRT for video image applications. When most of the efforts concentrated on the development of the emitter array during the early days of FED development, it was thought that standard CRT phosphors, which were developed to operate at 20–30 kV for optimum performance, could be employed (perhaps with minor modifications). It was soon learned that the maximum phosphor voltage for FEDs had to be limited below 10 kV dictated by the short distance between the two plates. Much lower voltages are desired to eliminate flashover problems that can occur at those high fields.

7.2.5.1. Definition of Terms.

Phosphors are characterized by their chromaticity, efficacy, and luminance. For FED applications, it is also important to understand the degradation mechanisms under e-beam bombardment and interactions of the phosphors with the emitters. The color of a given phosphor is defined by its *chromaticity coordinates*, x and y, and is displayed within the *chromaticity diagram* or the so called

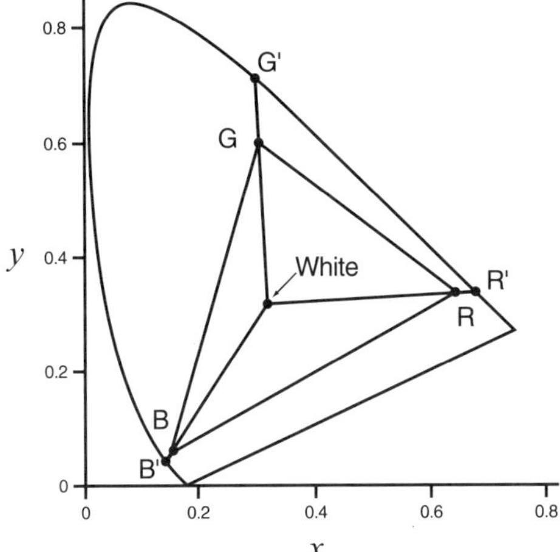

FIGURE 7.19. The CIE chromaticity diagram including x and y coordinates for P22 red, green, and blue CRT phosphors.

CIE (Commission Internationale de l'Eclairage) diagram [29] as shown in Fig. 7.19 (including the chromaticity coordinates for the P22 red, green, and blue phosphors that are currently used in CRTs, see Table 7.2).

The *color gamut* is the range of colors that can be obtained with a group of phosphors and is defined by the triangle that connects the three primary colors. The triangle (RGB) should be as large as possible, and the R, G, B coordinates should lie close to the envelope (R'G'B') of the diagram. By extending lines from the white point through the R, G, B coordinates to the envelope, the degree of *color saturation* can be obtained by determining the ratio of distances. For instance, the green in Fig.7.19 is the least saturated color amongst the three phosphors, since the ratio of the distances, $W - G/W - G'$, is the smallest. A good phosphor should be color saturated.

The chromaticity can shift as a function of excitation current density. This is shown in Fig. 7.20 for three ZnS phosphors [31], where increased current density follows the arrows. The gamut of colors decreases because of the relatively large shift of the red and green phosphors. Prior to 1964, ZnCdS:Ag was used for P22 red, but

TABLE 7.2. Chromaticity Coordinates for P22 CRT Phosphors [30]

Phosphor	Color	x	y	Efficacy (lm/W at 25 kV)
ZnS:Ag:Cl	Blue	0.150	0.0060	10
ZnS:Cu:Au:Al	Green	0.305	0.605	65
Y_2O_2S:Eu	Red	0.650	0.340	20

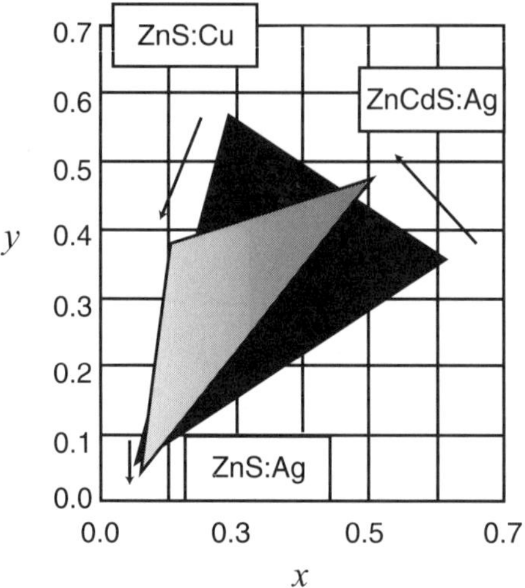

FIGURE 7.20. Chromaticities of red, green, and blue zinc sulfide phosphors for different current excitation densities at 5 kV anode voltage [31].

has since been replaced by Y_2O_2S:Eu due to the fact that ZnCdS:Ag exhibits a large current density shift, is hard to screen, and has a low efficacy.

There are other CRT phosphors listed in Table 7.3 that are used for monitor and projection TV applications. The degree of current saturation in these phosphors is less than that of the P22 phosphors, and their Coulomb maintenance is better. However, these improvements are counterbalanced by their inferior chromaticities.

The *efficacy* of a phosphor under electron beam excitation is defined as [32]

$$\varepsilon = \pi LA/P \qquad (\text{lm/W}) \tag{3}$$

where L is the luminance in cd/m^2, A is the area of the electron beam spot in m^2, and P is the power of the incident electron beam in watts obtainable by multiplying the anode voltage by the current. Note that the dimension on the left side of the equation

TABLE 7.3. Other Popular Phosphors for CRT Applications [30]

		x		y		
Phosphor	Color	Low Current	High Current	Low Current	High Current	Efficacy (lm/W at 25 kV)
P31 – ZnS:Cu:Cl	Green	0.226	0.193	0.528	0.420	60
P43 – Gd_2O_2S:Tb	Green		0.333		0.556	35–40
P53 – $Y_3Al_5O_{12}$:Tb	Green		0.368		0.539	30
P56 – Y_2O_3:Eu	Red		0.640		0.335	18

is in lm/W and on the right side in cd/W. *Luminance* is a measure of the total energy output of the light source emitted in the visible region of the spectrum. The subjective sensation produced in the eye by this energy is known as brightness. Several units are used for luminance: 1 nit = 1 cd/m^2 = 0.292 fL (foot-Lambert). An objective measure of phosphor efficacy is defined by the integrated photon energy divided by the input power and is measured in W/W.

The *intrinsic luminous* efficacy is the efficacy of a powder sample of the phosphor. The *screen luminous efficacy* is the efficacy of a thin layer of phosphor powder deposited onto a substrate. Screen efficacies are typically lower than intrinsic efficacies because of the presence of binders in thin film phosphors that can absorb fractions of the excitation electrons. The binders may also chemically react with the phosphor. Screen efficacies can be measured in back reflection mode or transmission mode [32]. For FEDs to compete with LCDs on a power consumption level, the phosphors must have screen efficacies of 11, 22, and 3 lm/W for the red, green, and blue components, respectively [33]. To produce a white pixel of a given luminance, about 60% has to be generated by the green, 30% by the red, and 10% by the blue phosphor [30].

Phosphors developed for the CRT industry are most efficient in the 20–30 kV range. In general, for a given current, luminance increases with anode voltage, and for a given anode voltage, it increases with current. An example is shown in Fig. 7.21 for picture element tubes developed at Sarnoff using P56(Y$_2$O$_3$:Eu), P53(Y$_3$Al$_2$(AlO$_4$)$_3$:Tb), and P55(ZnS:Ag) phosphors for R, G, B, and gated Si emitters for the excitation source. The phosphor voltage was 10 kV.

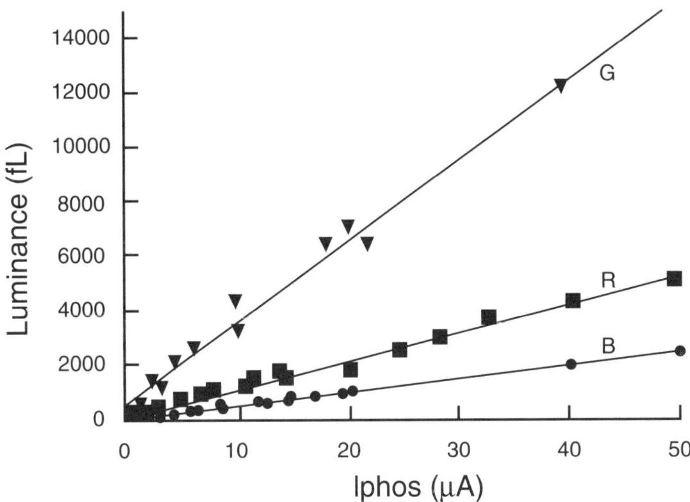

FIGURE 7.21. Luminance for P56 (red), P53 (green), and P55 (blue) phosphors as a function of phosphor current at a phosphor voltage of 10 kV. Spindt-type arrays were used as the excitation source.

At lower anode voltages, higher current densities are required in order to obtain the same brightness. Unfortunately, phosphors start to saturate at higher current densities due to the generation of higher defect densities that are caused by the decreasing penetration depth, and thus smaller excited phosphor volume. In addition to current saturation, phosphors decrease in efficacy under prolonged electron bombardment. This is known as *Coulomb aging.* Since low-voltage phosphors are excited at much higher current densities, the time at which the initial luminance is reduced to one-half becomes much shorter. Coulomb aging is attributed to the formation of color centers (point defects that act as traps for electron–hole pairs) and to surface damage. Electron-stimulated chemical reactions between the phosphor and constituents of the residual atmosphere in vacuum (H_2, CO, CO_2, H_2O) are another potential Coulomb aging mechanism. Typical values for Coulomb aging when the luminance decreases to half its initial value are 100–200 C/cm^2 [34].

The *dwell time* is the time that the electron beam spends to excite the phosphor. For CRTs, it is several tens of nanoseconds. After the beam scans to the next phosphor dot, the luminance decays. Typical decay times for the P22 phosphors are 800 μs for red, 100 μs for green, and 60 μs for blue. As the beam scans the screen, it can select the same phosphor dot 1/30 s later. During this time interval, the luminance of the dot has sufficiently decayed and is ready for new excitement. FEDs are generally addressed in a one-line-at-a-time mode. This means that, with a repeat time of 1/30 s and a display of 480 lines, the dwell time is 0.033/480 or 69 μs. Exciting a pixel during this time period at increased current densities and reduced phosphor voltages causes severe current saturation with P22 phosphors. It has been shown that the current saturation can be reduced with phosphors in which the decay time (defined as the time to reach 10% of the initial luminance) is lower than the dwell time [35,36]. This means that the phosphor can be excited, relaxed by luminescing, and re-activated several times during the dwell time.

7.2.5.2. Low and High-Voltage Phosphors.

Phosphors can be divided as low-voltage (<2 kV) and high-voltage phosphors (>2 kV). Most FED companies currently use high-voltage phosphors. To improve contrast, a black matrix is used between the color stripes (triade) or the rectangular subpixels (quad). A thin aluminum (Al) layer, 50–100 nm, is also used to increase luminance, since the light that would be radiated towards the emitter plate is reflected by the Al screen towards the viewer. The layer also minimizes outgassing of the phosphor. Depending on the thickness, anode voltages of 800–2000 V are needed for the electron beam to penetrate the Al layer.

Since the low-voltage phosphors are usually not metallized, resistive effects can severely limit their performance [37]. Mixing of conductive particles such as In_2O_3 with the phosphor powder is common [38]. ZnO:Zn is a low-voltage phosphor with the highest efficacy (10 lm/W at 100 V). Unfortunately, its chromaticity does not meet color TV standards. One reason for its high efficacy is that it has the highest conductivity. Since the electron penetration depth is reduced at low voltages, it is important to synthesize phosphors with long diffusion length and also investigate surface coatings using wide bandgap dielectrics such as WO_3 and V_2O_5 for increased efficacy and stability [39].

TABLE 7.4. Low-Voltage Phosphors for FED Applications [30]

In Use	Alternate Phosphors	Experimental
Y_2SiO_5:Ce – Blue (.5)	$Y_3(Al, Ga)_5O_{12}$:Tb – Green (YAGG) (7)	$SrGa_2S_4$:Ce – Blue
Gd_2O_2S:Tb – Green (14)	$ZnGa_2O_4$:Mn – Green (3)	$CaMgSi_2O_6$:Eu –Blue
Y_2O_3:Eu – Red (4.5)	ZnO:Zn – Bluish green (8)	$Sr_3MgSi_2O_8$:Eu – Blue (.5)
		$ZnGa_2O_4$ – Blue (.3)
		$SrGa_2S_4$:Eu – Green

The Phosphor Technology Center of Excellence (PTCOE) at the Georgia Institute of Technology, Atlanta, Georgia (C.J. Summers) is maintaining a database on phosphors, and Sandia National Labs, Albuquerque, New Mexico (L.E. Shea) is establishing characterization protocols for the proper evaluation of phosphors. Several good reviews about the status of phosphor development were given in special sessions on FED phosphors in the 1997–1999 proceedings of the IVMCs. Tables 7.4 and 7.5 summarize some of the phosphors under consideration for low voltage (Table 7.4) and high voltage (Table 7.5) FEDs. The numbers in the parenthesis are efficacies in lm/W at 1000 V for Table 7.4 and at 10000 V for Table 7.5. Most phosphors used at this point are fabricated from powders and have particle sizes of several micrometers. Thin film phosphors deposited by sputtering and e-beam evaporation are also considered. They have better heat sinking properties and lower outgassing rates. However, they are less efficient because of lower optical scattering and increased light piping.

7.2.5.3. Deposition Methods. Deposition of phosphors from powders can be performed by the slurry process, screen printing, electrophoresis, dusting, and several other methods. As shown in Fig. 7.22 for the entire steps in the slurry process [40], the starting glass substrate (7059 Corning or equivalent) is first coated with indium tin oxide (ITO) (transmission 94–95%) with a sheet resistance of 20–25 $\Omega/\square$. After cleaning the substrate, the black matrix pattern is formed. This can be accomplished by using silicate-based oxide powders (0.1–0.5-μm size particles) through suitable screens by screen printing. The substrate is then baked, and loose particles are blown off. The phosphors, in the form of a slurry containing sodium or ammonium dichromate, are coated onto the substrate with polyvinyl alcohol (PVA) by spin coating, dried, and exposed to light through a shadowmask. After removal

TABLE 7.5. High-Voltage Phosphors for FED Applications [30]

In Use	Alternate Phosphors	Experimental
ZnS:Ag:Cl – Blue (10)	Gd_2O_2S:Tb – Green (35)	$SrGa_2S_4$:Ce – Blue (4)
ZnS:Cu:Au:Al – Green (65)	Y_2O_3:Eu – Red (9)	$SrGa_2S_4$:Eu – Green (45)
Y_2O_2S:Eu – Red (20)		

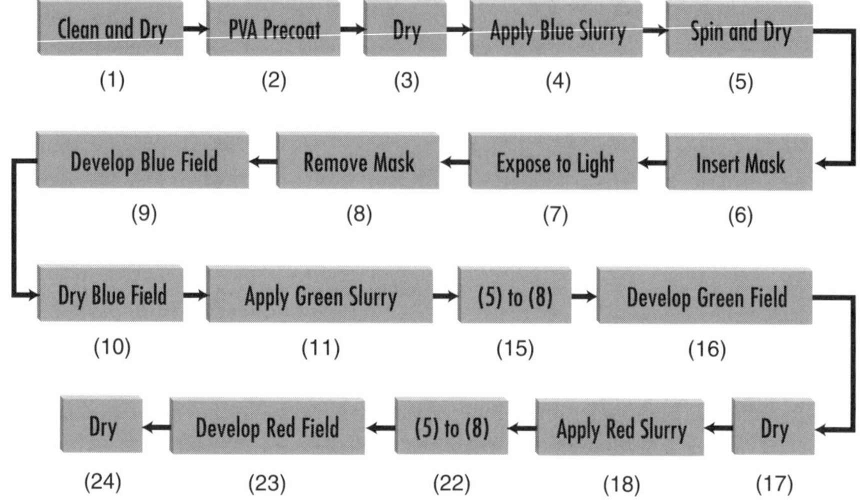

FIGURE 7.22. Steps in forming a phosphor plate by the slurry process [40].

of the mask, the film is developed by washing away unwanted portions of the photo-resist/phosphor pattern and dried. This sequence has to be repeated for the other two colors.

After the R, G, B phosphor patterns are defined, a thin film of lacquer (acrylic polymer) is deposited, followed by aluminum deposition. The lacquer acts as a planarizer to obtain highly reflective Al coverage and is subsequently evaporated (with the PVA) in air at 440°C by escaping through pinholes in the Al. For high-voltage phosphors, the initial ITO layer is not necessary, since the Al layer is conductive. For low-voltage phosphor plates, the ITO layer is essential, since no Al is deposited. The cross section of the finished phosphor plate is shown in Fig. 7.23. An application of this method for fine-line (25 μm) phosphor deposition was given in Ref. [41].

For the screen printing process shown in Fig. 7.24, the phosphor slurry is forced through the opening in a silk or metal screen with a rubber squeegee and deposited

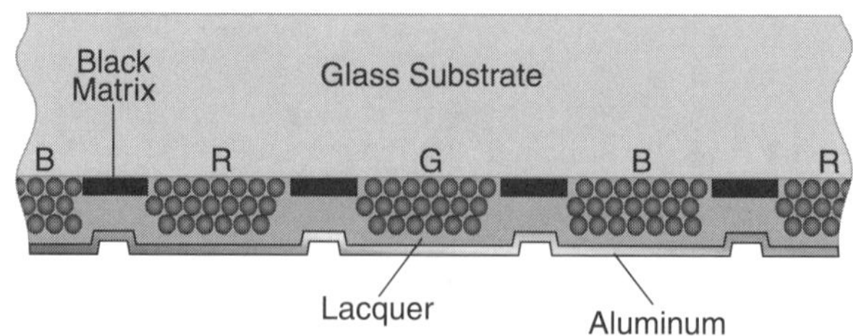

FIGURE 7.23. Cross section of a high-voltage FED phosphor plate. At least three layers of phosphor powder are needed per subpixel.

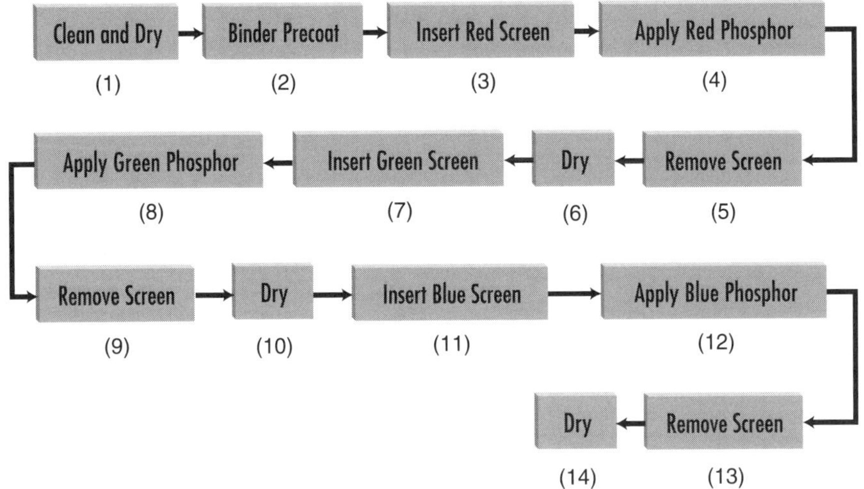

FIGURE 7.24. Steps in forming a phosphor plate by screen printing [40].

onto the glass substrate. The screen is then lifted off, and the plate is dried. This process needs to be performed with all three colors. Similar to the slurry process, both the ITO layer and black matrix are formed prior to phosphor screening, and both the lacquer and Al are deposited later.

Electrophoresis is a transport process in which conducting or nonconducting particles in a suspension move under the influence of an applied field and deposit on substrate surfaces. A phosphor deposition process based on electrophoresis was described in Ref. [42] and in additional references given therein.

7.2.5.4. Phosphor Summary.

It is clearly desirable to use low voltage phosphors in FEDs for reduced voltage operation and simplified device designs. However, efficient low-voltage phosphors are still under development. To be efficient, they should possess high electrical and thermal conductivities. Since the screens will not be aluminized, these phosphors can become potential emitter contaminants, and it is, therefore, important to minimize the outgassing from these phosphor surfaces. The degradation by total charge injection also needs to be minimized, and no saturation or emission wavelength shifts at high current densities should take place. The emission spectra should be narrow and lie near the boundaries of the CIE diagram.

High voltage phosphors, on the other hand, do exist and are currently being used at voltages ranging from 5–7 kV. They are efficient and are usually aluminized. The thin aluminum layer provides electrical conductivity and additional viewer intensity. It also reduces the contamination of the emitters from phosphor and binder byproducts. Because of the lower current densities required, the degradation due to charge loading is minimized. Disadvantages include the need for increased panel spacings and high voltage spacer technology. The larger distance between the

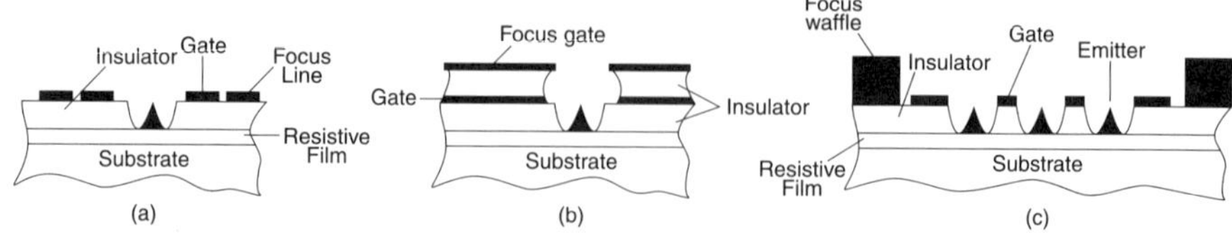

FIGURE 7.25. Methods for generating focused electron beams: (a) in-plane, thin film focusing of each emitter [43]; (b) double-gate, thin film focusing of each emitter [43]; and (c) thick film focusing of a group of emitters [26,45].

emitters and phosphor screen takes the display out of the realm of proximity focusing, thus requiring a more complex scheme that includes the addition of focusing electrodes.

7.2.6. Beam Focusing

When the subpixel size is less than about one-tenth of the anode-to-cathode spacing, the electron beam is generally required to be focused. For an 80 μm subpixel size, the maximum distance for proximity focusing is about 0.8 mm. This is far too short compared to the 1.5 mm spacing that is typically required for high-voltage operation. Figure 7.25 shows several focusing or beam confinement schemes, and theoretical analysis of beam focusing was described in Ref. [43].

Applying a focusing electrode to each tip as shown in Fig. 7.25(a), which would be the most effective means, is not very practical for FED applications, since it complicates manufacturing considerably. For the configuration shown in Fig. 7.25(b), experimental results indicated that, at a gate voltage of 80 V and anode voltage of 1000 V, focusing can be achieved at voltages between 10 and 20 V [44]. Below 10 V, the anode current drops drastically. At present, the method of choice in the FED industry is that of Fig. 7.25(c), which is more of a beam confinement method rather than true individual tip focusing [26,45]. Modeling results have shown that the best focusing is obtained with the double gate structure, followed by the in-plane structure, and then by the waffle structure [46]. A microphotograph taken with a digital camera of light emitted from a display using a 40-μm thick film focusing "waffle" is shown in Fig. 7.26 [45]. The column pixel pitch is 94 μm.

7.2.7. Encapsulation

In addition to the fabrication of emitter arrays on the baseplate and phosphor subpixels on the faceplate, spacers need to be inserted or fabricated on one of the plates, the plates have to be brought together and aligned, and the periphery of the plates must be sealed to form a vacuum envelope. Getters, which are placed at the periphery of the device (near or inside the pumping stem), are activated as the final step to provide

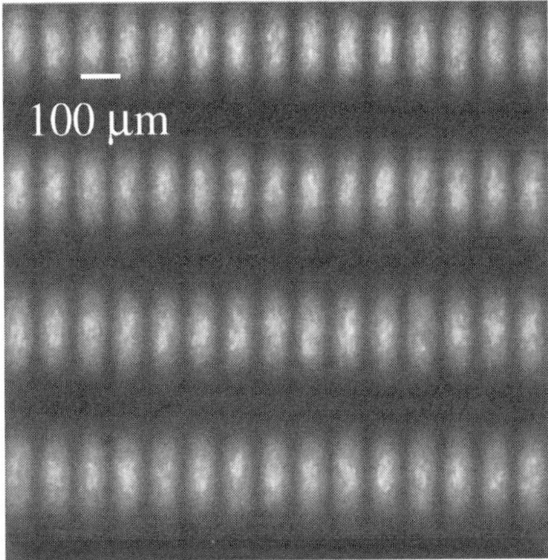

FIGURE 7.26. Microphotograph of light originating from a display with a pixel pitch of 94 µm using a thick film "waffle" around each subpixel for focusing. The waffle is biased at a few volts (dc) negative with respect to the gate voltage [45].

chemical pumping throughout the life of the display. Figure 7.27 shows some details of an FED near the periphery [47]. The row and column address leads of the baseplate have to be extended beyond the FEA, and the lower lying emitter leads must have the gate insulator material removed to form pads that can be connected to the driver electronics via solder or pressure connectors. Provisions also need to be made for the anode pad(s), the pump port(s), and the glass frit seal.

7.2.7.1. Spacers.

Spacers are the "orphans" of the FED technology. It is interesting to observe that during the first 10 years of the IVMC, no papers on spacers were presented. Only recently breakdown results on small gaps in vacuum without spacers [48] and with spacers [49–51] have been seen in publications. Nevertheless, none of the companies involved in FED product development are sharing performance results of their spacer technologies.

An ideal spacer should possess the following properties: (i) good mechanical strength to support the baseplate and faceplate under the influence of the atmospheric pressure, (ii) low cost and easy to manufacture and to position, (iii) good dielectric strength (>5 V/µm), (iv) low coefficient for secondary electron emission, (v) high vacuum flashover resistance, (vi) slightly conductive to avoid charging, (vii) transparent to the viewer, and (viii) matched thermal expansion coefficient with glass. Spacers can take the forms of spheres, rails, or posts. Materials that have been investigated include glass, ceramic, polyimide, and silicon nitride coated polyimide. Resistive coatings that have been investigated to reduce charging include metal oxides (Fe_2O_3, Cu_2O, Cr_2O_3), DLC films, amorphous silicon, and SiC [52]. Spacer fabrication can be an integral part of the faceplate or baseplate manufacturing process. Alternatively,

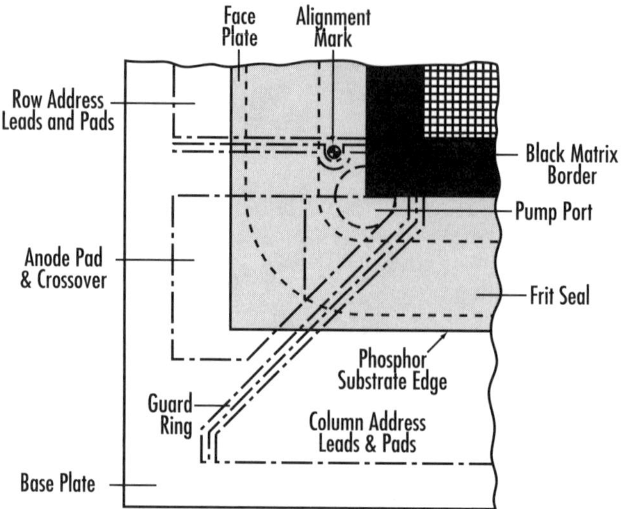

FIGURE 7.27. Segment of the lower left edge of an FED showing conductor pads, pumping port, and glass seal. Not shown are the getters [47].

individually processed spacers can be attached to either one of these plates. The maximum dimensional requirement for a spacer is that it fits within the black matrix. It should be somewhat smaller to allow for alignment tolerances in placing the spacer (if it has not been lithographically defined) and in aligning of the two plates. For instance, for a 17-in. diagonal super extended graphics array (SXGA) (1280 × 1024) FED, the black matrix is 60 μm between the rows of the display and smaller between the columns. By placing the 2.4-mm tall glass fiber posts only in the rows of the black matrix, 40-μm diameter spacers were chosen [53]. This corresponds to an aspect ratio of 2400/40 or 60. Spacer geometries and placement need to be determined very carefully for each display application. The spacer has to be electrically invisible, meaning that no electrons should strike it. Nevertheless, stray electrons and secondary electrons from the faceplate can still end up striking the spacer, which causes charging and distortion of the image around the spacer. This problem must be eliminated by the proper choice of the spacer materials and technology.

The breakdown voltage of thin film electrodes deposited on glassplates and measured in a relatively poor vacuum of 1×10^{-6} torr was reported to be larger than 18 kV for a 750-μm gap using Ni/Cr electrodes [48], corresponding to a breakdown field of 24 V/μm. By inserting teflon disks and wall spacers, breakdown voltages ranged from 13 to 16 kV and 15 to 20 kV, respectively, for the same 750-μm gap [49]. By replacing the teflon spacers with thin-wall machinable glass ceramic (MACOR), Coors ceramic, Coors YTZP Zirconia, and AMZIROX Zirconia, breakdown fields ranging from 6.4 to 24.5, 5.9 to 8.8, 8.0 to 16.7, and 4.0 to 10.4 V/μm, respectively, were obtained [49]. These values were obtained when the

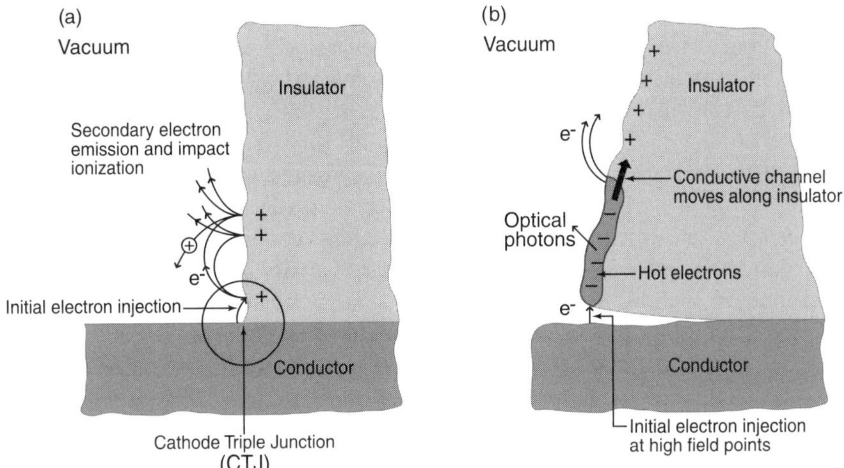

FIGURE 7.28. Proposed spacer breakdown mechanisms: (a) electron injection, hopping, secondary electron emission, and impact ionization [52] and (b) electron injection and the formation of a surface conducting channel. An essential prerequisite for growth of the conductive channel is the emission of electrons from the surface of the insulator [54].

spacers were inserted between upper and lower thin film electrodes. It is anticipated that, if these experiments are performed using FED faceplate and baseplate, the breakdown fields will be further reduced, because the spacers are now subjected to bombardment from primary electrons from the emitters and secondary electrons from the phosphors.

Several breakdown mechanisms have been proposed to explain spacer flashover. Figure 7.28 shows two of the proposed spacer breakdown mechanisms: (a) electron injection, hopping, secondary electron emission, and impact ionization, and (b) electron injection and the formation of a surface conducting channel. Most of these explanations relate to macrosize spacers with diameters much larger than the spacers used in FEDs. Breakdown generally occurs in two stages: prebreakdown electron injection, followed by vacuum flashover. The role of the triple junction, which is defined as the intersection of conductor, insulator, and vacuum, plays a critical role in spacer breakdown. Triple junctions can be formed at both the cathode and anode. In Fig. 7.1, the cathode triple junction (CTJ) is formed at the intersection of the conductive thick film focusing grid, the wall spacer, and the vacuum. The anode triple junction (ATJ) is formed at the intersection of the Al-coated phosphor plate, the spacer, and the vacuum. It is conjectured that high fields established at the conductor–insulator interface can cause electron injection from the conductor into the insulator. For macrospacers, this injection is also aided by microparticles and rough surface features near the spacer–conductor periphery. Early theories in the 1980s [52,54] proposed that the electrons injected into the insulator "hopped" along the surface of the insulator, generating more electrons by secondary electron emission at the electron–insulator impact points, thereby causing an electron avalanche and positive charging of the spacer. In parallel with the charging process, electron-stimulated desorption of neutral or positively ionized atoms and molecules from the insulator surface can also take place. The positive ions drift towards the cathode (in the case of Fig. 7.1, the focus grid) and, with the positively charged insulator from impact ionization, give rise to an enhanced field at the triple junction. As a result, through a combination of gas desorption, regenerative secondary electron emission and impact ionization, the necessary conditions can be established for striking an avalanche-based plasma discharge across the surface of the spacer. Other models, such as hot-electron processes within the surface layers of the insulator that could give rise to ionization sufficient to enable a solid-state avalanche discharge [55] and explosive relaxation of trapped charges [56], have also been proposed.

For high-voltage macrospacers, it has been shown that suppression of spacer breakdown can be achieved by specially shaping the insulator near the triple junction [57,58] or reducing secondary emission by coating the spacer with materials exhibiting low secondary electron yield, ideally with the yield $\delta_e \leq 1$. These coatings, which should also be slightly conductive, reduce the build-up of the positive charge. Truncated cone-shaped spacers with one end larger than the other end (the larger end should be positioned on the cathode electrode) have also been shown to increase the breakdown fields by six-fold (from 5 to 30 V/μm) when the truncated cone angle is greater than 45° [58]. This approach might be suitable for low resolution FEDs, because real estate considerations preclude such an approach in high-resolution displays. So far, the most prominent method is to coat the spacers with a low secondary electron yield material with its conductivity just enough to inhibit charge build-up.

Recent studies of 2-mm tall, conically shaped alumina macrospacers have yielded additional information about spacer breakdown mechanism [54]. It is shown that the pre-breakdown conduction process is "nucleated" at a highly localized site on the CTJ with simultaneous emission of optical photons. This process is believed to be associated with the injection of electrons from a particulate-based emission site located under the bounding edge of the insulator. These sites are switched on by the enhanced fields in the range of 5–10 V/μm. Because of the high field within the insulator, the injected electrons will be ballistically "heated" and, as a result of internal ionization processes, give rise to a point source of light at the CTJ. Depending on the microscopic configuration of the emission site, the switch-on process may result in an energetically favorable regime for the subsequent growth of a conducting channel along the surface of the insulator. Some of the field-emitted electrons may bombard the insulator surface in front of the ATJ and give rise to an anode conducting channel. Typically, it can take several seconds for the channels to form. The final breakdown or flashover of the gap is a very rapid process and generally occurs at some random times after the growth of a conduction channel.

It is yet to be reported how these findings relate to microspacers and what breakdown mechanisms are the most relevant. In addition to spacer breakdown, vacuum breakdown can also occur at sharp points on the baseplate and from dislodged particles (such as phosphors, aluminum) derived from the faceplate. It is critical for high voltage robustness that all components and assembly be handled in a clean room environment to minimize particles and contamination issues.

There are some promising developments in the search of innovative spacer technologies. The MEMS approach might offer some interesting solutions. Spacers with features that inhibit charge initiation and propagation could be embedded into the cathode or anode manufacturing process and then released and actuated as a final step. Actuation could be achieved via built-in stresses in the spacer or other means. One such approach, in which a Ti–Ni alloy film was used for activation, was described in Ref. [59].

7.2.7.2. Alignment/Sealing. After fabrication or placements of the spacers, the two plates must be aligned and sealed. There are two basic methods of sealing, namely at atmospheric pressure or in vacuum. The atmospheric method is similar in many ways to the standard CRT sealing procedure. However, due to the sensitivity of the emitters towards oxidation and poisoning, special precautions should be taken. If air is the sealing ambient, the tips must be coated to prevent oxidation or specially treated after sealing. Sealing in nonoxidizing atmospheres such as nitrogen and argon has also been used [60]. Depending on the glass used for the manufacture of the baseplate and faceplate (borosilicate, float, sodalime, etc.), it is important to work with the glass manufacturers to select the appropriate frits for sealing and follow the recommended temperature cycles. Typical frit firing temperatures range from 350 to 450°C. Shrinkage of the glass during sealing also needs to be considered. Additionally, one must consider the CTE (coefficient of thermal expansion) of the spacers.

For a 17-in. SXGA display, an alignment tolerance of about 5 μm is required [53], and the appropriate alignment/sealing fixturing systems have to be designed. Generally, fixturing is less expensive when aligned and sealed at atmospheric pressure as compared to alignment/sealing in vacuum. The basic atmospheric sealing process involves dispensing of the glass frit on either the faceplate or the baseplate, followed by pre-firing. The two plates are then aligned and heated together. As the frit softens, the plates are squeezed together, and the frits are extruded. The plates are then cooled, and a vacuum seal is established. The package is then evacuated through an exhaust tube. Evacuation is usually performed at elevated temperatures to desorb moisture and outgas the envelope prior to pinching off the exhaust tube. Pinch-off is done by heating the tube until it softens and then pulling on it until a seal is formed. A chemical pump (getter) is usually placed around the exhaust tube and activated by rf heating. In most cases, evaporable getters such as barium (as they are used in CRTs) are employed. Care has to be taken that none of the barium is deposited onto the emitter tips.

In the vacuum alignment/sealing process, the two plates are aligned and placed into a high vacuum station. A gap exists between the two plates for the efficient removal of air. The temperature in the vacuum chamber is raised to allow moisture to be desorbed and outgassing to take place. The temperature is then raised to the point where the glass frit begins to soften, and the two plates are forced together, causing the glass frit to extrude. The glass frits used for vacuum sealing must be specifically processed to minimize outgassing. At this point, the envelope is sealed, and the temperature is slowly lowered to prevent thermal stressing of the glass plates. A thermally activated getter (e.g., by the temperature of the sealing process) can be used in the vacuum process. This vacuum seal process is better suited for FEDs, since the baseplate is not contaminated by air. In principle, the two plates can be outgassed prior to alignment and joining, and the faceplate can be exposed to an electron beam to condition and pre-age the phosphor prior to sealing. Disadvantages include a more expensive system and potentially reduced throughput.

By replacing the air sealing process with sealing in nonoxidizing atmospheres, some of the advantages of the vacuum sealing process can be retained in the less expensive atmospheric pressure sealing method. In Ref. [60], Spindt-type Mo tips were mounted on glass substrates that contain pre-baked frits. The substrates were then placed into the sealing furnaces filled with N_2, Ar, and air. Severe degradation of the $I–V$ characteristics was observed in the air-sealed devices. Reduced degradation was seen with the nitrogen-sealed devices, and the best results were obtained when Ar was used. For example, for a constant current, the gate voltages shifted upwards by 250 V in air, 30 V in N_2, and 15 V in Ar. The surface morphologies of Mo tips as measured by atomic force microscopy (AFM) also roughened significantly, by an average roughness of 39 Å (air), 36 Å (N_2), and only 9.7 Å (Ar), respectively.

In Ref. [61], the authors showed that Mo Spindt emitters that were treated at 470°C for 20 min in highly purified N_2 and Ar gases displayed no degradation in their $I–V$ characteristics, similar to those vacuum-sealed (10^{-6} torr) devices. The authors concluded that if the partial pressure of oxygen in Ar or N_2 was kept below 7×10^{-6} torr, no detrimental effects upon the $I–V$ characteristics would be observed. In Ref. [62], the authors observed a 20-V upward shift in the $I–V$ characteristics of Mo tips when they were sealed in an Ar ambient, while no shift was observed when using Ar with 1.5% H_2. They claimed that the small amount of hydrogen keeps the Mo tips from oxidizing. Further, the Ar/H_2 treatment did not result in any detrimental changes in the luminance and chromaticity of the devices.

7.2.7.3. Vacuum Requirements.

7.2.7.3. Vacuum Requirements. Although CRTs can operate at pressure of about 10^{-6} torr, Spindt-type emitters fabricated from refractory metals and silicon should be operated below 5×10^{-7} torr. Above these pressure, noise increases and life expectancy decreases due to ion bombardment effects. Some of the planar emitters such as BN [63] and carbon-type emitters [64] were reported to be capable of operating at 10^{-4} torr. However, no good lifetime data exist at these pressures.

The basic panel structure consists of the faceplate and baseplate joined together with a glass frit as shown in Fig. 7.1 as well as in Fig. 7.29 in a cross-sectional form. To increase pumping efficiency, a panel structure with an auxiliary tank as shown in Fig. 7.30 was investigated [65]. Several pumping holes are fabricated in the baseplate, and the device is evacuated through an exhaust tube that is positioned at the back of the auxiliary tank. The advantages of this structure are that placement of the getters

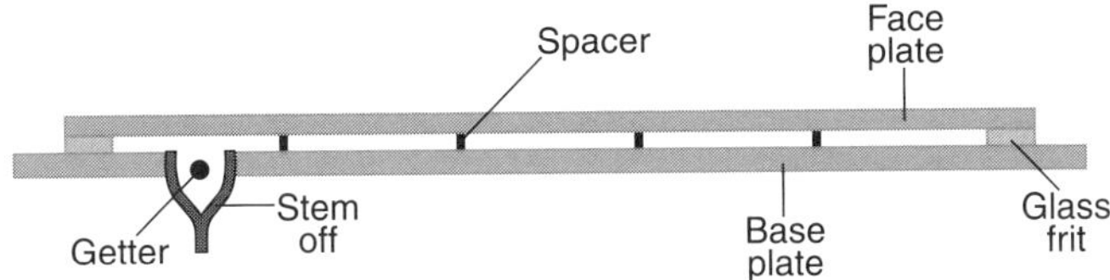

FIGURE 7.29. A conventional FED panel consisting of two glass plates [65].

is more convenient, and that sufficient getter volume can be used. The disadvantages are a relatively thicker display panel and the need for additional spacers as indicated in the figure.

The evacuation stems are a nuisance during the final packaging of the displays and can cause outgassing problems during heating prior to pinch-off. A stemless process that employs an anodic bonding technique was described in Ref. [66]. As shown in Fig. 7.31, the anodic bonding takes place at the sputtered silicon–glass interface at an elevated temperature of 280°C under dc bias. By applying a negative bias to the glass seal-off plate via the ITO layer, positive metal ions (Na, Al, K, Mg, ...) migrate to the ITO layer, leaving a space-charge layer near the silicon–glass interface that supports most of the applied bias. The resulting large electrostatic force pulls the plates together, resulting in Si–O–Si bonds. The advantages of this method are that no frit is needed, and that the sealing temperature is only 280°C. The disadvantage is that the bonded surfaces have to be extremely smooth, with peak-to-valley roughness not exceeding 200 Å.

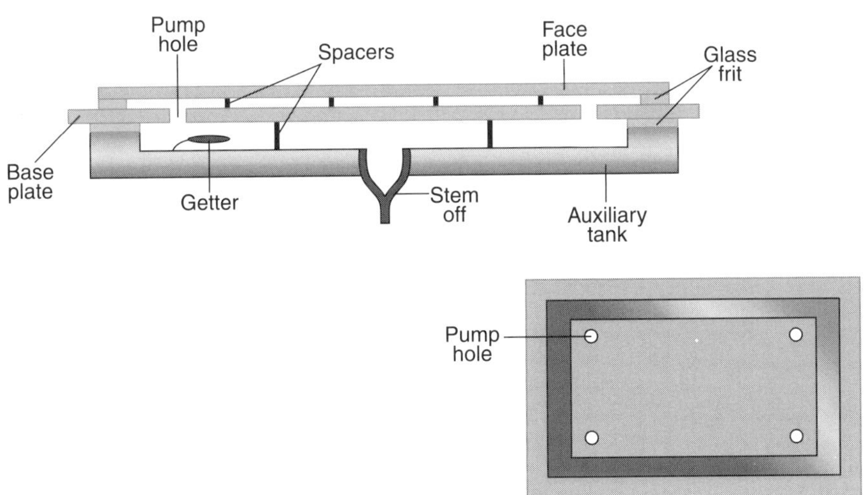

FIGURE 7.30. Cross section and top view showing four pump holes of a panel structure with an auxiliary tank [65].

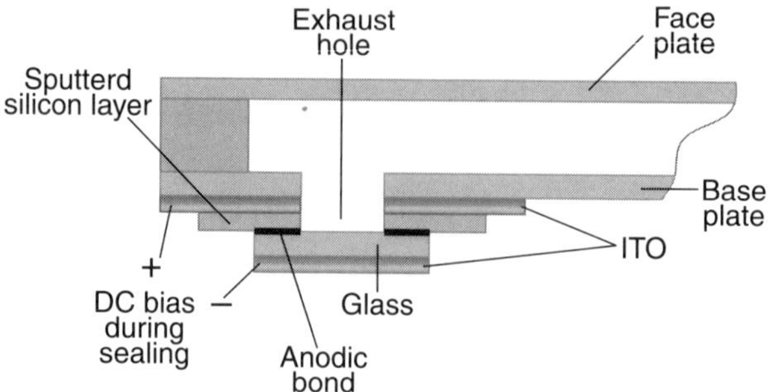

FIGURE 7.31. Cross section of a stemless panel in which the tipped-off exhaust tube is replaced by an anodically bonded glass plate [66].

Some of the pumping results from the conventional structure shown in Fig. 7.29 and the structure shown in Fig. 7.30 are presented in Fig. 7.32. The 5.7-in. diagonal faceplate and baseplate are fabricated from sodalime glass with a panel separation of 200 μm. The diameter of the pump hole is 5 mm, and that of the exhaust opening is 5.8 mm. The results show that after 250 min of pumping, the pressure in the conventional panel has still not reached its lowest value. The pumping speed is similar when one pump hole in the baseplate is used. The condition improves quite considerably only when four pump holes in the baseplate are used.

It is interesting to observe that the equilibrium pressure is rather high at about 6×10^{-5} torr. This is caused by the low conductance of the 200-μm spaced panels and by the outgassing of the two plates. By assuming a panel that is closed at one

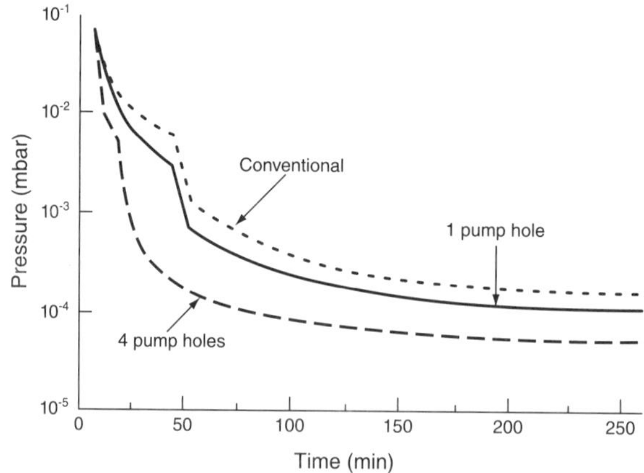

FIGURE 7.32. Pressure vs. pump time for a conventional 5.7″ diagonal panel and a panel with an auxiliary tank in which the baseplate has one and four pump holes of 5-mm in diameter [65]. Note that the pressure unit of mbar is used. As a refresher: 1 Pa = 1 Newton/m²; 1 torr = 133.32 Pa; 1 mbar = 100 Pa; 1 atmosphere = 1.013×10^5 Pa; and 1 torr = 1 mm Hg.

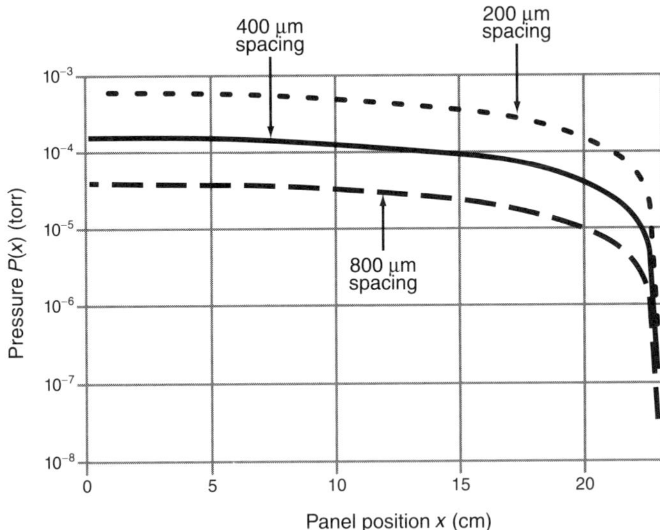

FIGURE 7.33. Pressure vs. distance along a 23-cm long panel with panel spacings of 200, 400, and 800 μm. The panel is closed on the left side ($x = 0$) and pumped on the right side ($x = 23$ cm) with infinite pumping speed [67].

end and pumped at the other with infinite pumping speed, the pressure at a distance x along the panel for $t > 3600$ s can be calculated as [67]:

$$P(x) = \frac{64}{\pi^3}\left(\frac{L}{a}\right)^2 Q\left(\frac{mk_BT}{2\pi}\right)^{1/2} \sum_{n=0}^{\infty}\left[\frac{(-1)^n}{(2n+1)^3}\right]\cos\left(\frac{\pi(2n+1)x}{2L}\right) \quad (4)$$

where L is the length of the panel, a is the panel spacing, Q the outgassing rate in number of molecules/m² s, m the gas mass (g/mole), k_B the Boltzmann constant, T the absolute temperature, and x the distance along the panel. Results for a 23-cm long panel with 200, 400, and 800 μm spacings are shown in Fig. 7.33. It is assumed that the sticking coefficient is zero, i.e., no entrainment of gas to the surface takes place once it is released by outgassing. In the calculations, Q is assumed to be 6.4×10^{15} molecules/m² s for the outgassing rate of stainless steel, and m is 29 g/mole for air. These data show that spacing and outgassing rates are the most important parameters in obtaining low pressure inside the panel. Unfortunately, they also show that if flashover events occur away from the getter pump, the getter might not be able to reduce these local pressure bursts fast enough to avoid localized damage.

7.2.7.4. Chemical Pumping or Gettering.

Once a panel is sealed, changes in pressure over its life can be caused by leaks, permeation, and outgassing. Assuming that materials and processes are chosen to eliminate the first two causes, outgassing remains. Outgassing is caused by thermal fluctuations in the materials used in panel production, electron bombardment of spacers and phosphors, and electron-induced desorption of species at the tips. To combat the pressure increase, getters are installed inside the panel and activated. A good review concerning getters was provided in Ref. [68].

Gettering is routinely used in CRTs, X-ray tubes, particle accelerators, and many other vacuum devices. The getter material chemically interacts with active gas species, forming very stable chemical compounds both at the surface and in the bulk. At room temperature, getters can chemically absorb hydrogen, oxygen, water, nitrogen, and carbon monoxide and dioxide in an irreversible manner. The sorption capacity of getters at room temperature is related to the available effective surface area, while at elevated temperatures (above a threshold), it is close to the theoretical stoichiometric capacity for the chemical compounds that are formed such as oxides and nitrides. Sorption of hydrocarbons (e.g. methane) only occur at high temperatures. Noble gases such as Ar and He cannot be gettered.

Getters are available in two forms: evaporable and nonevaporable. The evaporable getters are commonly based on a mixture of Ba–Al$_4$ alloy and Ni powders compressed together into ring- or wire-shaped containers. These containers are then heated, by rf coils, to 800–900°C to promote an exothermic reaction between the two components. These getters are mainly used in CRTs, since large areas are available for the deposition of a porous gettering film. Although they are also used in FED prototypes, their effectiveness is limited, because large surface areas for getter deposition are not available.

Over the last few years, getter manufacturers have been increasingly working on the development of non-evaporable getters based on Zr–Al and Zr–V–Fe alloys. Of great interest for FED applications are high porosity (60–65%), screen printable getters based on Zr–V–Fe and Ti. Sorption capacities of about 1 (cm^3 · torr/cm^2) have been measured at room temperature for carbon monoxide. The capacities are 10 times higher for O$_2$ and H$_2$O, and 100 times higher for H$_2$. These getters are typically screen-printed with 100, 200, and 400-μm thick lines and are activated at 500°C for about 10 min. Figure 7.34 shows the sorption characteristics of a screen printable HPTF (high porosity thin film) NEG (non evaporable getter).

To design an appropriate gettering system, it is important to measure the electron-induced outgassing rate of different gaseous species in the displays. The outgassing rate typically decreases as a function of time t in the form of t^{-1} for most gases and $t^{-1/2}$ for hydrogen. By knowing the outgassing rate, the total gas load and the corresponding increase in pressure over the life of the displays can be calculated. The amount of getter material can, in turn, be estimated from the total gas load. For example, a 20 cm × 20 cm display area has an estimated gas load of 1.5 (cm^3 · mbar) after 10000 h of operation [34,68]. This corresponds to a pressure build-up of 10^{-2} mbar, and only an area of several square centimeters of HPTF NEG would be needed to absorb this gas load.

The real devices are more complicated. Getters are typically placed at the periphery of the panel, near or in the evacuation stem. This limits their effectiveness in terms of absorbing pockets of outgassing deep inside the panel. Distributed getters that consist of a network of getter material uniformly distributed across the entire panel

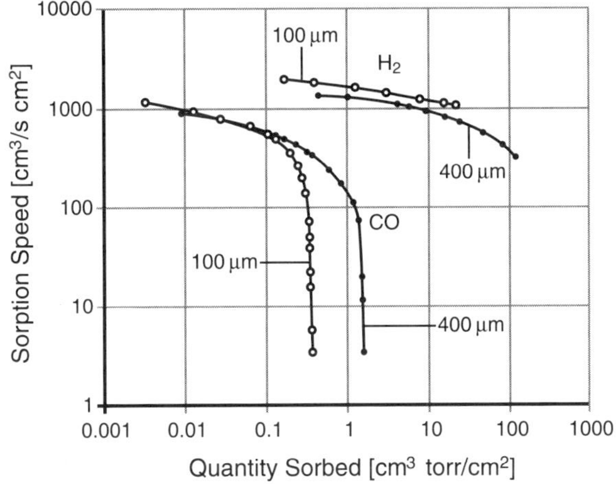

FIGURE 7.34. Sorption speed vs. quantity absorbed for CO and H_2 of a screen printable getter with thicknesses of 100 and 400 μm [68].

are, therefore, considered more effective. However, for high resolution panels, this is difficult to achieve and would further increase the complexity of the display structure. One possibility that is worth investigating is that the focusing grid shown in Fig. 7.1 can be fabricated from the gettering material.

7.2.8. Driver Electronics

Figure 7.35 shows the basic dimensions of a 10-in. diagonal video graphics array (VGA) display that consists of 480 row grid lines and 640 × 3 column emitter lines [69]. The row lines are 300-μm wide with a 30-μm spacing, and the column lines are 92 μm wide with a 18-μm spacing. Thus, the size of a white pixel is 330 μm × 330 μm. Each 92 μm × 300 μm color pixel is partitioned into 4 × 13 = 52 subpixels of 23 μm × 23 μm dimensions, and each subpixel contains four tips. The total number of tips per color pixel is thus 208 (only three tips are shown in Fig. 7.35).

Both emitter and gate drivers must be able to supply enough charging current I

$$I = \frac{C\,dV}{dt} \tag{5}$$

to support the required scan rates [70]. Charging and discharging the matrix conductors dissipates power P in the resistance of the circuit, which is given by

$$P = CV^2 f \tag{6}$$

where C is the capacitance, V the switching voltage, and f the frequency. Although

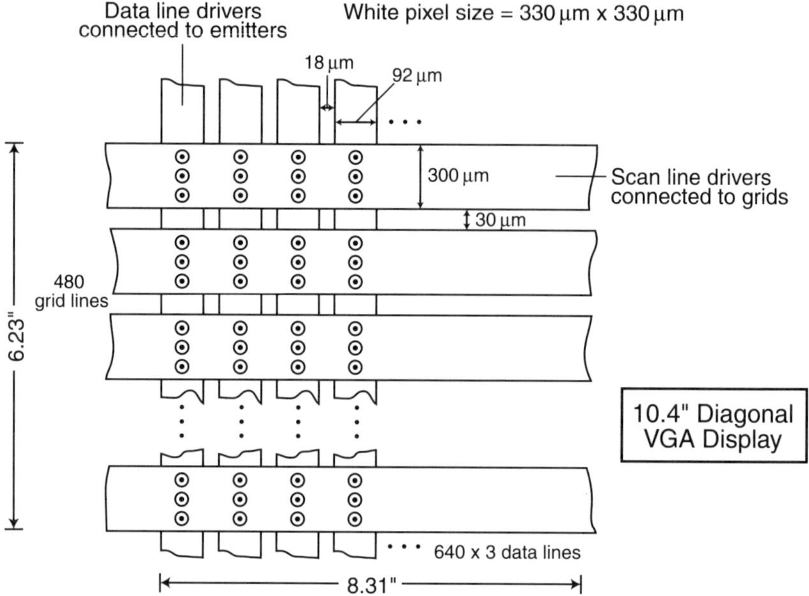

FIGURE 7.35. The row and column arrangement of a $10''$ VGA FED prototype display [69].

the transient power is dissipated in the resistance of the circuit, the magnitude of the lost power is independent of the value of the resistance. Nevertheless, the resistance should be kept low in order to avoid excessive voltage drops along the emitter lines when most of the pixels are excited. In addition, the line resistance, capacitance, and switching voltage should all be kept low in order to minimize the driver circuit power requirements.

The line resistance R of a 6.23-in. (15.82 cm) long and 92-μm wide line in Fig. 7.35 can be estimated by

$$R = l\rho/A \tag{7}$$

where l is the line length, ρ the resistivity ($\sim 20 \times 10^{-6}$ Ωcm), and A the cross section of the line. Assuming a 1-μm thick line, R is estimated as

$$R = \frac{15.82 \times 20 \times 10^{-6}}{92 \times 10^{-4} \times 1 \times 10^{-4}} = 344 \ \Omega$$

The gate-to-emitter capacitance is given by

$$C = \varepsilon_0 \varepsilon A/d \tag{8}$$

where $\varepsilon_0 = 8.86 \times 10^{-14}$ F/cm, $\varepsilon = 3.9$ for SiO_2, A is the area of the gate line to emitter line overlap (300 μm $\times$ 92 μm), and d is the thickness of the dielectric. Assuming $d = 1$ μm, one obtains

$$C = \frac{8.86 \times 10^{-14} \times 3.9 \times 300 \times 10^{-4} \times 92 \times 10^{-4}}{1 \times 10^{-4}}$$

$$= 9.54 \times 10^{-13} \quad \text{or} \quad 0.95 \text{ pF/color pixel overlap}$$

There are 480 overlaps along one emitter line, which means that the line capacitance is 480 $\times$ 0.95 pF $=$ 450 pF. In addition to the parallel plate capacitance, the tip-to-gate stray capacitance should be included in the calculations. The value of this capacitance depends strongly on both the gate diameter and tip geometry and is usually below 0.5 fF per tip [71]. Thus for 208 tips, the additional capacitance is less than 0.1 pF/color pixel, which would increase the total capacitance by less than 10%.

The required pixel current can be estimated from the equation for luminance, which is given by [23]

$$L = \frac{1}{\pi A} V I D_c \varepsilon \tag{9}$$

where L is the luminance in cd/m^2, A is the faceplate area in m^2, V the anode voltage in volts, I the emission current in amps, D_c the duty cycle, and ε the luminous efficacy of the phosphor in lm/W. The current can then be calculated as

$$I = L\pi A / V D_c \varepsilon \tag{10}$$

We assume that the display in Fig. 7.35 yields an overall luminance of 200 cd/m^2 at an anode voltage of 4000 V and a screen luminous efficacy of 10 lm/W. With a duty cycle of 1/480 (for sequential line addressing) and an area of 6.23 in. $\times$ 8.31 in. $=$ 51.77 in.^{2} (0.0334 m^2), a total current of 0.252 A can be derived. The individual pixel current can be obtained by dividing this value by the number of pixels, which yields an average current per pixel of 0.252/480 $\times$ 640 $\times$ 3 $\cong$ 0.3 μA. Operating this display at 400 V increases the current to about 30 μA due to a 10-fold decrease in both voltage and luminous efficacy. This estimated range of 0.3–30 μA agrees well with the quoted values of 1–10 μA in Ref. [70] for the currents needed to excite a white pixel.

With the knowledge of current/color pixel, line capacitance, line resistance, and I–V_g characteristics of a subpixel, the driver circuitry can then be designed. Typically, the row connections are made to the gates and the column connections to the emitter lines. The rows are scanned sequentially from top to bottom. During each row select time, the column connections are used to imprint intensity information to the pixels of the selected row.

There are three basic methods to modulate the pixel intensity with the column drivers [70]: pulse height (amplitude) modulation (PHM), pulse width modulation (PWM), and hybrid approach (PHM/PWM). Within each group, either the voltage or current drive methods can be selected. By choosing the PHM voltage drive method, the column driver is a voltage source with sufficient current to support the requirements for slew rate. The drive capability based on this PHM voltage method is good. The circuit is relatively simple, but must maintain high accuracy at low emission levels. At low currents, the effect of ballast resistors is reduced due to insufficient voltage drops across the resistors. As a result, pixel emission tends to become nonuniform. Transient power is low, though, because the voltage excursions from one row select time to the next are minimal.

In the PHM current drive scheme, a current proportional to the pixel intensity is applied to the column conductor. The column voltage floats to a value corresponding to the applied current. Circuit implementation is easier in this current mode than the voltage drive, because the applied current is linear with brightness and requires less accuracy at low brightness levels. Transient power is also minimal, since there are little changes in column voltage from row to row. However, the current approach suffers from two significant disadvantages. First, since the current is limited to the emission level (which is much smaller than the charging currents), the voltage changes on the column conductors are too slow to support typical scan rates. Second, emitter defects that result in pixel leakages will distort the pixel brightness. This does not occur in the voltage drive method.

In the PWM, the pixel intensity is modulated in time. The pixels are operated at a constant current, but the pixel on-time is varied as a function of the pixel intensity. This can be accomplished with either voltage or current drive, or a combination of both. In the PWM voltage drive, during the row select time, a voltage pulse to the column turns the pixel on for a fraction of the time. Dark pixels are not turned on at all, while white pixels remain on for the entire row select time. Gray scale is achieved by varying the on-time between these two extremes. Since the driver is digital, the output voltage accuracy is much relaxed as compared to the PHM approaches. The transient power is large, since the same voltage transitions occur whether the pixel is at a low or high brightness level. Depending upon the number of gray scales, the slew rate can be quite high, resulting in 100 ns pulse width. This leads to high transient currents. The image uniformity is very good, since the pixel is always at a current level where significant voltage drops can develop at the ballast resistors.

In the PWM current drive, a constant current is forced to the column conductors with the on-time proportional to the pixel intensity, and again, the column voltage floats to a value corresponding to the applied current. This can potentially produce images with very high quality, except that limiting the currents also restricts the possible slew rate and, therefore, the number of possible gray levels. The pixel leakage will result in brightness distortion in the pixel. Transient power is much reduced from the PWM voltage method, but so is the ability to support the required bandwidth and gray scales. The PWM can also be operated in a combined voltage and current drive mode. This offers advantages of fast slew rates of the voltage drive and uniformity of the current drive. When the pixel is turned on, a voltage source is used to precharge the line to a voltage near that needed for the required pixel current. The voltage source is then switched to a high impedance state, and a current source is switched on for emission control. The main disadvantage of PWM is the requirement for very short pulse widths for low brightness pixels. Short pulses require fast rise and fall times, and result in very high charge currents.

The hybrid PHM and PWM method has distinct advantages over the PWM approaches, because it does not require a very high charging current at low pixel brightness. The PHM/PWM approach can also be implemented in either the voltage or current drive mode. Typically, the video data are split into two nibbles, and multiple current/voltage sources are used for the implementation.

In addition to the technical considerations, the overall driver electronics cost, driver size, IC technology, and chip processing yield have to be taken into account in order to select an appropriate display drive method. Based on the current cathode technology, the PWM voltage drive appears to be the better solution, since cathode nonuniformity and leakage prevents the use of either the PHM approach or the current drive. As the technology advances, other drive approaches will become possible. In the long run, if the problem of pixel leakage can be resolved, the current drive technique will offer the best solution for display uniformity.

Two custom chip implementations of data line drivers were described in Refs. [69] and [72]. In Ref. [69], a PHM voltage scheme was considered for driving the display shown in Fig. 7.35. The functional block diagram of the data line driver is presented in Fig. 7.36. The features of this data driver include 120 driver outputs, the support of interlaced and noninterlaced video signals, the support of interleaved and noninterleaved panel connections, the support of three types of color filter arrangements: vertical stripe, diagonal mosaic (top-left to bottom-right), and bicolor triangular patterns, and three phase clock inputs with frequencies up to

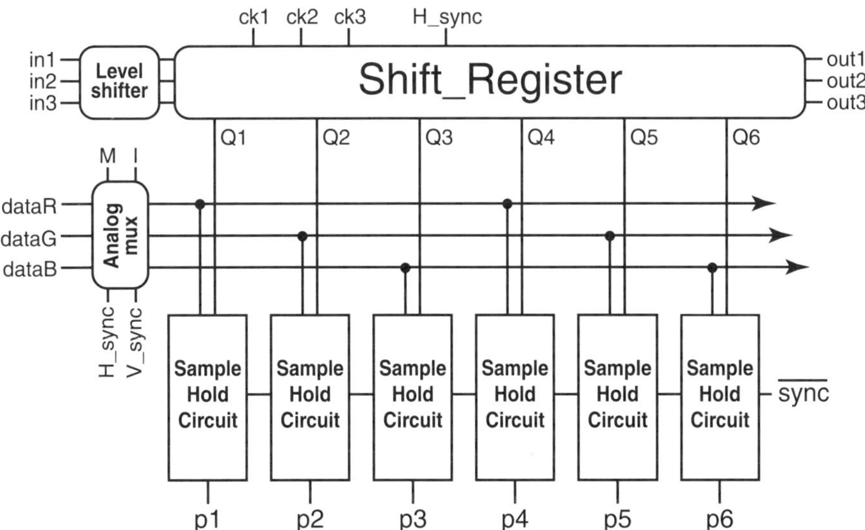

FIGURE 7.36. Functional block diagram of a pulse height modulation (PHM) data line driver [69].

20 MHz. The major functional blocks are bidirectional shift register, level shifter, analog multiplexer, and sample-and-hold circuits.

The operation of the data line is repetitive per scan line and is synchronized by the horizontal synchronization signal (H_sync). At the beginning of each scanning line, input pins In1, In2, and In3 are held high for one clock cycle. Thus, the outputs of the first three shift registers are high. These outputs are used to put the first three sample-and-hold circuits in the sampling mode. After one clock cycle, the outputs of the first three shift registers become low, and the outputs of the next three shift registers are high. Thus, the first three sample-and-hold circuits are in the hold mode, and the next three are in the sampling mode. As the clock goes on, the three logic 1's in the shift register are shifted to the right by three steps, and each sample-and-hold circuit samples the video input in sequence. With 120 stages, the data line driver can handle 40×3 data lines. A panel of 640×3 data lines requires 16 driver chips. The driver is designed to support direct cascade connects between chips. The output pins Out1, Out2, and Out3 are connected to In1, In2, and In3 pins, respectively. At the end of a scan line, all 640×3 sample-and-hold circuits are holding the appropriate video signals. At the beginning of the next scan line, these values are output through the output buffer during the entire scan line period, while the other set of sample-and-hold circuits are sampling the current scan line video signals.

By using three phase clocks that are offset by $120°$, the maximum operating frequency is reduced by a factor of three. Every input signal (except the R, G, B video signals) is connected to a level shifter. These shifters change the 0 to -5 V input signal swing to -19 to -5 V internal swing, which is needed to obtain the appropriate brightness on the selected rows. A MOS (metal-oxide-semiconductor) switch and a MOS capacitor are used to perform the sample-and-hold function. Two sets of sample-and-hold were used, thus allowing the unity gain buffer the entire scan line period to drive the data. To reduce power, a class AB operational amplifier was used as the output buffer. These buffers are able to supply large currents when the outputs are changing, and draw little or no current when the outputs reach their steady state. The dimensions of the chip are 7.62 mm $\times$ 17.50 mm.

In Ref. [72], the authors described a PHM current method and claimed that the resistive layer was not needed, and nonuniform pixel I–V characteristics could be allowed (probably assuming that there was no gate leakage). Figure 7.37 shows the schematic diagram of this driver scheme. The chip consists of both high- and low-voltage devices. The high-voltage devices protect the low-voltage n-channel current sources when capacitively coupled high voltages develop at the cathode lines during the scan time. T1, T2, T3, T4 are low-voltage n-channel devices used for current sources. Their channel widths increase from W, $2W$, $4W$ to $8W$, which accommodate 16 grey scales for this experimental chip. When V_{gate} goes from ground to high (>60 V) during the scan time, the cathode lines will reach relatively high voltages due to the gate-to-cathode capacitive coupling. The induced high voltage is protected by T5 (high-voltage n-channel LDMOS (laterally diffused metal-oxide-semiconductor) with a breakdown voltage of 120 V). When the scan pulse (V_{gate}) goes from ground to high, node (B) is grounded by circuit block (A) for preventing floating the source of T5, and $V_{control}$ is low. After $V_{control}$ is high (T5 turns on), the digitized four-bit

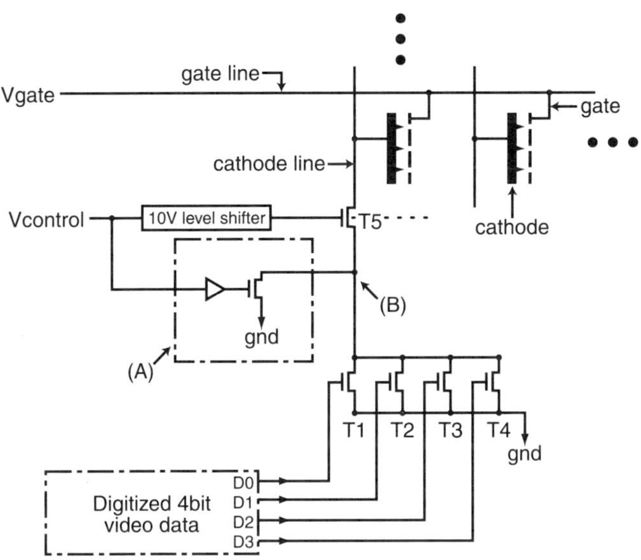

FIGURE 7.37. Schematic diagram of a constant-current data driver scheme [72].

video data are applied to the gates of current sources (T1 through T4), and the proper current level is supplied to the cathode line for a given gray scale.

The methods described above are primarily for passive matrix addressing. Active matrix addressing can be achieved by including thin film transistors (TFT) per subpixel in a similar scheme to the AMLCD display [73].

7.2.9. Aging/Surface Chemistry

7.2.9.1. Coulomb Load. In contrast to thermionic emitters where chemical impurities are burned off, field emitters are extremely sensitive to changes in surface chemistry. Submonolayer coverage of certain species can change the work function of field emitters significantly. In addition, sputtering events can change the surface morphologies of emitters. Not only do the emitter arrays have to maintain the performance during the life of the display, which should last at least 10 000 h, but the phosphors also need to operate satisfactorily and should not reach their Coulomb limit of 100–200 C/cm^2 during the projected operation period. It is interesting to observe that a phosphor dot in a CRT TV set is only excited by the electron beam for a total of 1.6 min during 10 000 h of operation [74]. If we assume a TV signal of 525 lines per frame and a typical 4:3 aspect ratio, there are about 700 addressable points in each horizontal line. Thus, the number of phosphor dots is $525 \times 700 = 367{,}500$. The dwell time is the frame time divided by the number of dots. With a frame time of 1/30 s, the dwell time is 90.7 ns. The number of times a phosphor dot gets excited (hit) in 10 000 h is $10\,000 \times (3600$ s/h)/frame time (sec), and the total time that the dot is irradiated is the number of hits ($=1.08 \times 10^9$) times the dwell time. This turns out to be 100 s or 1.6 min. The Coulomb loading that this phosphor dot experiences is $Q = It/A = 1 \times 10^{-3} \times 100/10^{-2} = 10$ C/cm^2. Here we assume an average current of 1 mA/dot and an area of about 1 mm^2 per phosphor dot.

For a passive addressing scheme as discussed in relation to Fig. 7.35, the dwell time is $(1/30)/480$ (number of lines) or 69 μs. Since the number of hits one phosphor pixel sees during 10 000 h of operation is the same as in the CRT, i.e., life time/frame time, the total time one pixel is energized is $1.08 \times 10^9 \times 69 \times 10^{-6} = 74.52 \times 10^3$ s or 20.7 h. The average current per pixel for the same brightness can be estimated by equating the energy densities/pixel for the two cases. For the CRT, assuming 1 mA/phosphor dot, 1 mm^2 area, 30 kV phosphor voltage and a dwell time of 90 ns, the energy density is $IVt_{\text{Dwell}}/A = 1 \times 10^{-3} \times 30000 \times 90 \times 10^{-9}/0.01 = 2.7 \times 10^{-4}$ W s/cm^2. For the FED shown in Fig. 7.35, the area is 92 μm $\times$ 300 μm, and the dwell time is 69 μs. Assuming an anode voltage of 4 kV, the current is then calculated to be 0.27 μA/color pixel. This agrees well with the current of 0.3 μA obtained from Eq. (10). This then corresponds to a Coulomb load of

$$Q = \frac{It}{A} = \frac{0.27 \times 10^{-6} \times 20.7(3600 \text{ s/h})}{92 \times 10^{-4} \times 300 \times 10^{-4}} = 72.9 \text{ C/cm}^2$$

The increase in the Coulomb load is only proportional to the ratio of the two operating voltages of 30 kV/4 kV. One can immediately see that the situation becomes critical when low-voltage FEDs are involved. Going from 4 kV operation to 400 V, for instance, will increase the Coulomb loading to 729 C/cm^2, assuming that the luminous efficiency remains the same. The efficiency, however, also decreases by almost 1 order of magnitude as compared to the high-voltage phosphors. Expanding this argument further, one can conclude that an ideal application for FEAs lies in high brightness video billboards fabricated from individual picture element modules. These elements are usually operated at 10 kV, which brings the Coulomb loading to 30 C/cm^2. Thus, a factor of six in brightness can be achieved by increasing the pixel current without reaching the Coulomb loading limit. Since the average TV set operates at about 200 cd/m^2, a billboard of 1200 cd/m^2 can be built by operating it at 10 kV. By increasing the operating voltage to 20–30 kV, true sunlight readable billboards can be built with luminance exceeding 3000 cd/m^2 and operating lifes of longer than 10 000 h. Table 7.6 summarizes these comparative results.

7.2.9.2. Aging of Mo Emitters. In a typical packaged and sealed FED, the main gases present are oxygen, water, carbon dioxide, carbon monoxide, methane, and hydrogen. Gas evolution can take place by outgassing during sealing, electron-stimulated desorption from the phosphor and other components, and desorption of process residues. Figure 7.38 shows some of the results obtained from a controlled experiment in an UHV (ultra-high vacuum) analytical chamber using a platinum-coated silicon wafer as the anode. This avoided any potential outgassing contribution from a phosphor anode. In the figure, the percentage changes in emission current are plotted against gas exposure ranging from 1 to almost 10 000 Langmuir (1 Langmuir = 1×10^{-6} torr s).

TABLE 7.6. Summary of Coulomb Loads from Different Displays

Display	Display Operating Time (h)	Operating Time of Emitters (h)	Operating Time of Phosphor (h)	Anode (Phosphor) Voltage V_{ph} (V)	Luminance L (cd/m^2)	Coulomb Load Q (C/cm^2)
CRT	10 000	10 000	0.027	30 000	200	10
FED (high voltage)	10 000	20[a]	20	4000	200	70
FED (low voltage)	10 000	20[a]	20	400	200	>700
Video billboard[b]	10 000	20	20	30 000	3000	150

[a]The 20 hours correspond to the emitter on-time. In most cases the column drivers have a continuous dc offset that could stress the gate dielectric and thus reduce product life.
[b]We assume a $480 \times 640 \times 3$ pixel billboard that is a-line-at-a-time addressed.

It can be seen that both methane and hydrogen increase the emission current, helium has no effect up to 1000 Langmuir, and carbon dioxide, oxygen, and water decrease the performance. The authors conjectured that in the case of water, the degradation mechanism is the interaction of oxygen with molybdenum, forming surface molybdenum oxides that have a higher work function than clean molybdenum. Similar situations exist for oxygen and carbon dioxide where the tips become oxidized, thus resulting in higher local work function and decreased emission. Methane undergoes dissociation and ionization by the field-emitted electrons, in which the energetic hydrogen species react with Mo to form volatile Mo hydrides, which are pumped away

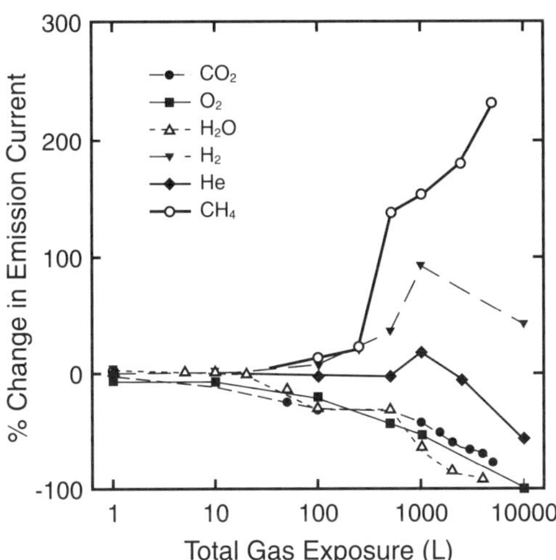

FIGURE 7.38. Effects of gas exposure on the emission characteristics of a 2.8×10^5 Mo Spindt emitter array [75].

TABLE 7.7. Carbon and Oxygen Concentrations on Mo Tip Surfaces in Test and Control Areas as Measured by Micro-Auger Spectroscopy

Impurity	Tips in Test Area (Average of 9 Tips)	Tips in Control Area (Average of 9 Tips)
Carbon	10%	28%
Oxygen	47%	31%

by the vacuum system, leaving clean tip surfaces; the carbon free radicals interact with Mo to form stable Mo carbides. Hydrogen undergoes similar dissociation and ionization upon impact with electrons and forms volatile Mo hydrides. It was found that the effects of gas exposure are similar in both dc and pulsed mode operation of the device. Thus, dc mode testing can be used as an effective acceleration method in establishing the device lifetimes under various vacuum conditions.

Although the above measurements show some trends in tip behavior upon exposure to certain gases, different results might be obtained in an enclosed FED vacuum envelope, where volatile species cannot be pumped away as readily as in a high vacuum test system. By using micro-Auger and X-ray photoelectron spectroscopy (XPS) measurements, tips that were part of a 320×240 matrix addressable array were investigated [76]. The panel was divided into two sections. One section served as a control area, and in the other the tips were operated at 50 μs pulses with an interval of 13 ms for 1800 h when the emission current had decreased by 50%. The anode phosphor used was ZnO:Zn. The vacuum was then broken, and analytical measurements were performed. Table 7.7 summarizes the micro-Auger results, in which the values quoted are the averages of nine tips measured, both in the test and control areas.

From the Auger data on tips and XPS measurements performed over larger areas, it was concluded that the tips oxidized during operation in the vacuum envelope, and that the reduction in current was caused by the increased work function of the Mo oxide. Zinc from the phosphor anode was also found on the emitters with its XPS intensity three times higher in the test area as compared to the control area. No explanation was given for the reduction in carbon in the test area and how that might contribute to the change in emission behavior.

7.2.9.3. Aging of Carbon-Based Emitters.

High temperature CVD deposited carbon thin film emitters are under intense investigation for display applications [77,78]. Experiments similar to the ones described here showed severe degradation in emission currents when the carbon cathodes were operated in both water and oxygen at pressures of about 10^{-6} torr, and showed almost no changes when operated in hydrogen [79]. It is believed that both the oxygen and water molecules were ionized by the electrons, and they could subsequently modify and etch the surface of the carbon

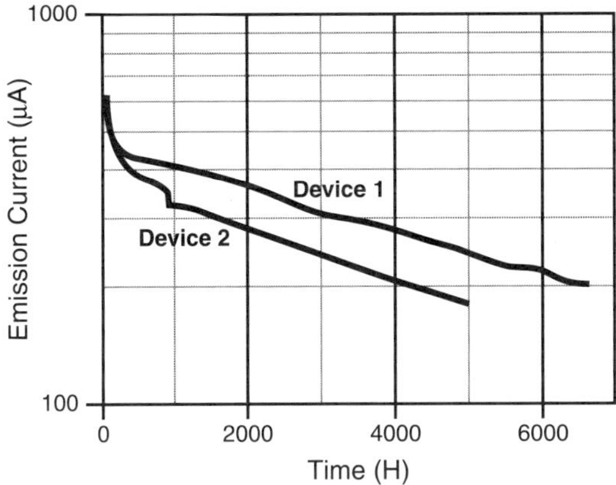

FIGURE 7.39. Cathode emission as a function of time for two frit sealed PETs [79]. The measurements were performed at an anode voltage of 7 kV using aluminized phosphor.

films. The fact that hydrogen has little effect on cathode lifetime might indicate that the surface termination of the carbon films by hydrogen was crucial. Figure 7.39 shows the emission current as a function of time for two vacuum frit sealed picture element tubes (PETs). It can be seen that the current decreases monotonically with time. Similar results of decreasing emission current as a function of time were obtained from the operation of PETs using carbon nanotubes as cathodes [80].

When graphite emitters were operated with a copper anode (instead of phosphors used in the experiments described above), no significant current decrease was observed as shown in Fig. 7.40 [81]. Initially, the tube was operated under continuous pumping for about 500 h, and a slight decrease in current was observed. After pinch-off,

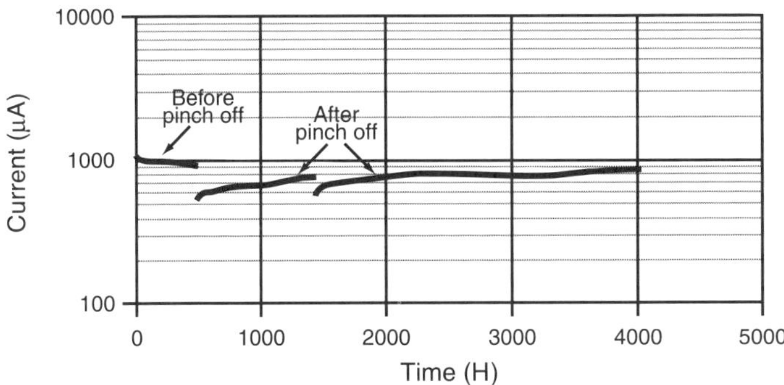

FIGURE 7.40. Emission aging of a nanocrystalline graphite emitter as a function of time. The tube was operated in a dc diode mode using a copper anode of 2 mm in diameter [81].

the current increased monotonically until it showed signs of saturation at about 4000 h (the discontinuity at 1500 h was due to power supply failure during the measurement).

From the different aging behaviors of these carbon-based emitters and the Mo emitters, it becomes apparent that the interplay between emitters, phosphors, and sealing conditions must be both understood and controlled in order to manufacture a reliable and reproducible display. Some of the peculiarities of the $I–V$ characteristics of carbon emitters when operated in a diode mode with different phosphors have been reported in Ref. [82]. The authors observed different $I–V$ characteristics with ZnS:Cu,Al and $Y_3Al_5O_{12}$:Tb,Ge phosphors, as compared to an ITO-covered glass anode and an yttrium phosphor. Similar results were obtained in PETs using graphite emitters operated in the triode mode [81]. The triode PETs were formed by placing a metal mesh about 100 μm above the emitter surface. The phosphor anode was placed about 60 mm away from the grid. The area of the openings in the grid was 25% of the total area. Sets of R, G, B aluminized P22 phosphor tubes were fabricated using CRT frit sealing techniques. It was found that the electron transmission coefficient, which is defined as

$$T = \frac{I_{\text{phosphor}}}{I_{\text{grid}} + I_{\text{phosphor}}} \times 100 \tag{11}$$

was 25% before the sealing for all three tubes, consistent with the grid area consideration. After frit sealing, the green tubes had emission coefficients ranging from 25 to 50%, the red tubes from 5 to 24%, and the blue tubes from 0.6 to 5%. This suggested that during the sealing process, changes on the emitter surfaces occurred, possible related to the outgassing from the phosphors. These changes apparently affected either the work function of the emitters, the number of active emission sites, or emission areas. Figure 7.41 shows the voltage dependencies of the grid and phosphor currents of green, red, and blue tubes that are shown in the upper left corner in Fig. 7.44 later in Section 7.3.2. For the green tube, the phosphor current was about half of the grid current, resulting in a transmission coefficient of 33%. For the red tube, the grid current was larger than that of the green tube, but the phosphor current was reduced by an order of magnitude at $V_{\text{grid}} = 800$ V. For the blue tube, the transmission coefficient is only about 6%. The offset in operating voltage could be caused by a larger grid-to-emitter distance or by surface changes during the sealing.

7.3. OTHER DISPLAY TECHNOLOGIES

So far, the main emphasis has been placed on FEDs intended for laptop applications. There are a couple of companies that have demonstrated 13- and 15-in. prototypes and are probably investigating the possibility of scaling the technology to 40 in. for true, very thin, TV-on-the-wall applications. This is likely a very costly endeavor, since up-scaling to 40-in. requires cost-effective solutions to several technological

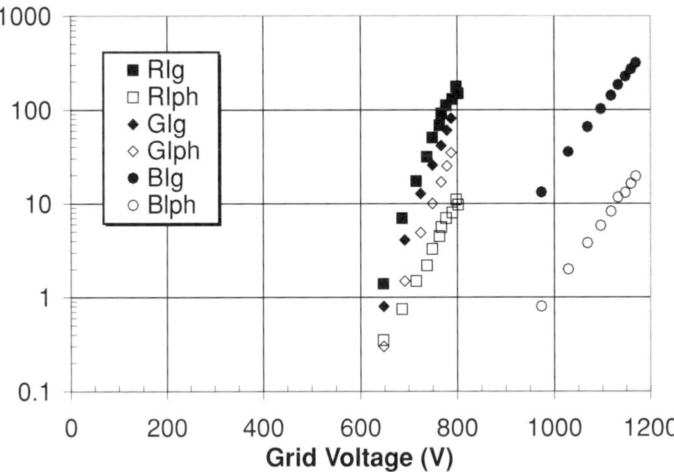

FIGURE 7.41. I–V characteristics of green, red, and blue PETs using carbon emitters. These tubes are shown in the upper left corner of Fig. 7.44. (RI_g = grid current of the red tube, RI_{ph} = phosphor current of the red tube, GI_g = grid current of the green tube, ... etc.).

hurdles. A potentially interesting solution was presented by SI Diamond Technology (SIDT) in its hybrid FED concept. Other applications for field emitters include high brightness and large area video billboards and a direct view IR image converter, in which IR-activated gates modulate the field at the tips and the optical readout is performed by a phosphor plate. There are other potential products such as X-ray fluorescence spectrometer, e-beam lithography, etc., but these are beyond the scope of this chapter.

7.3.1. Hybrid FED

In the hybrid FED (HyFED) structure, the three electron guns of a CRT are replaced by a matrix of $N \times M$ field emitting electron guns. Each gun rasters a small section of the screen; thus, the distance between the guns and the phosphor plate can be reduced. No spacers are needed in HyFEDs, and standard CRT faceplates can be used in manufacturing. But the device is still relatively bulky and heavy as compared to 40-in. plasma displays with 1-mm thick glass plates and the proposed TV-on-the-wall FEDs using spacers. Implementation of the HyFED technology is not simple, since several metal grids are needed to provide focusing and deflection capabilities. Also, compensation schemes have to be developed to obtain seamless images near the boundaries of the individual electron guns. Figure 7.42 illustrates the reduction in display thickness from L (conventional CRT) to approximately L/N for the HyFED, where N is the larger number in the (N,M) matrix. Figure 7.43 shows the cross section of one of the $N \times M$ gun elements. Note that, in addition to the field emission extraction grid (B), there are eight more grids involved with a total thickness of the

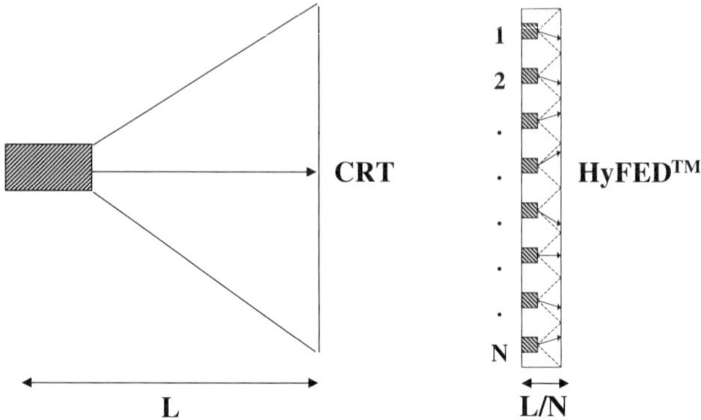

FIGURE 7.42. Cross sections of conventional CRT (left) and HyFED (right) devices.

grid assembly at 3 mm (excluding the thickness of the emitter panel and the phosphor panel). This HyFED concept has been demonstrated in a 4-in. diagonal display with a 0.45 mm pixel pitch [83].

7.3.2. Picture Element Tubes

Picture element tubes (PETs) with hot filament emitters are used in giant screen displays [84]. They are available in either individual tube form with each tube having one color or in the modular form with each module having at least four white pixels. An exaggerated image of a giant display with the Statue of Liberty as foreground is

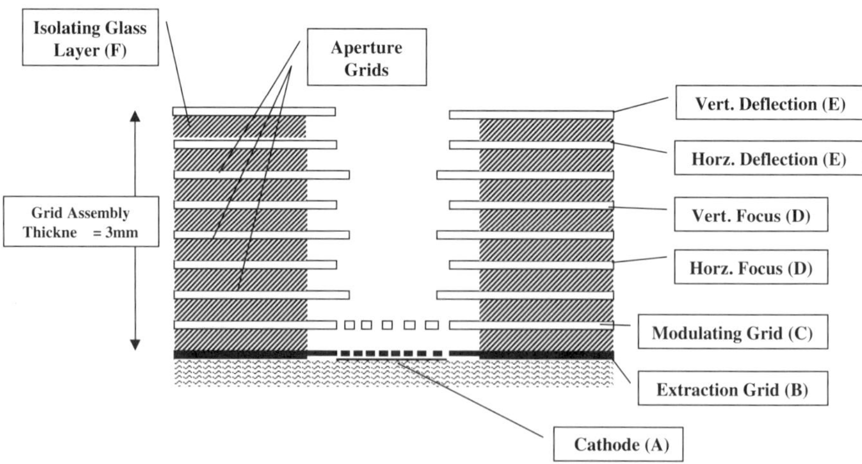

FIGURE 7.43. Cross section of one of the individual electron guns in a $N \times M$ HyFED gun array.

FIGURE 7.44. A conceptual giant video billboard composed of individual, high luminance, cold cathode PETs. Prototypes of these tubes, in single color elements and in a four white pixel module, are shown on the left (developed at Sarnoff under partial funding by Applied Photonics Technology, San Jose, CA.)

shown in Fig. 7.44, and single color PETs and a four-white pixel module are shown in the upper and lower left corners, respectively. Implementing a large array of hot wires into one envelope is not an easy manufacturing task. Replacing these hot wires with the cold emitter technology is thus a logical choice in improving manufacturability and reducing power consumption. A typical hot cathode in a PET is operated at 45 mA and 0.45 V, which corresponds to 0.02 W. The same tube at a 10 kV phosphor voltage and an anode current of 100 μA dissipates a total of 1 W. Thus, the filament power is 2% of the maximum brightness (about 20000 fL) power of the tube. If the hot cathode is replaced with field emitters, this would translate into a 24.5 kW reduction in standby power for a $480 \times 640 \times 4$ white pixel billboard (60×80 feet).

Figure 7.45 shows the cross section of a PET module that is operated with hot cathode filaments [85]. These modules are usually 3 in. $\times$ 3 in. in area and about 1 in. in depth. Sixteen of these modules are usually packaged into individually serviceable 1-ft $\times$ 1-ft units. Billboards of any size can then be assembled using these components. The metal anode prevents spillover of electrons to adjacent pixels. High-voltage contact to the phosphors is provided by a spring-loaded contact placed inside the evacuation tube. The center of the plug is used for the outside high-voltage connection. There is plenty of space available for effective gettering, and the peripheral glass envelope of the module functions as the spacer. The front shield grid is a defocusing lens to spread the electrons as uniformly as possible across a given color subpixel. The processing yield is high, since the number of emitters per module is low.

Replacing the hot filaments with cold cathodes can be accomplished by several means. Using area emitters such as thin film carbon or carbon nanotubes, diode or triode structures can be formed. Triode structures are preferred, since addressing

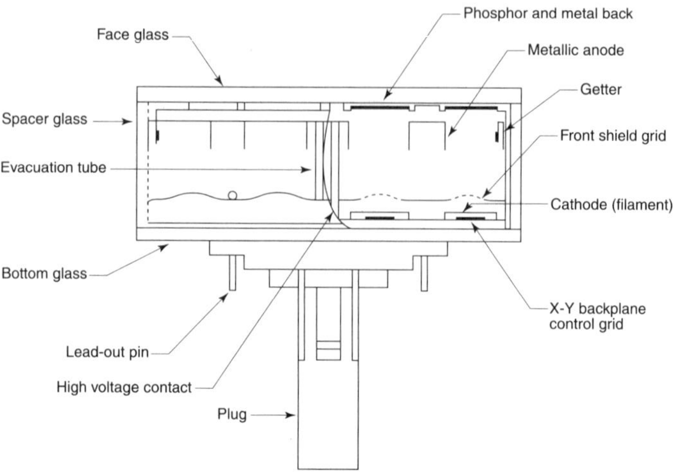

FIGURE 7.45. Cross section of a PET module that is operated with hot filament emitters [85].

voltages are lower. At present, metal grids are placed about 10–100 μm above the emitters. Gated metal or Si emitter arrays can also be used, but only as point sources, since area emitters would be too expensive. All the advantages of built-in ballasting and flashover protection of silicon chips as described in Chapter 5 can be used in this approach. Silicon technology in the single chip form is only economical for low resolution displays, since pick-and-place operations and the cost per chip prohibit manufacturing of billboards with a large number of silicon chips.

Some of the implementation techniques are shown in Fig. 7.46 with corresponding prototype devices shown in Figs. 7.44 and 7.47. Method (a) was jointly developed by SIDT and ISE Electronics and was implemented in the device shown on the left side in Fig. 7.47. It shows a 64 white pixel module energized by carbon emitters and operated at 10 kV. For comparison, a 16-pixel conventional module with the same cross section shown in Fig. 7.45 was placed on the right side in Fig. 7.47. The reduction in module thickness was achieved by the use of proximity focusing. Method (c) was implemented in the device shown in the lower left corner in Fig. 7.44. It is a four white pixel (G, G, R, B), 3-in. × 3-in. module energized by individual gated silicon tip arrays, one array per color pixel. The size of the chip is about 1 mm × 1 mm, containing about 6000 tips. Method (d) was used in fabricating the individual picture element tubes shown in the upper left corner in Fig. 7.44. Spreading of the electrons exiting the grid takes place due to the lensing action of the extraction grid. The distance between the grid and the phosphor in these tubes is 2.5 in.

7.3.3. Direct View Infrared FED

An interesting application that combines MEMS with FEAs for infrared (IR) imaging has been proposed [86]. The basic operation is illustrated in Fig. 7.48. The gate plate

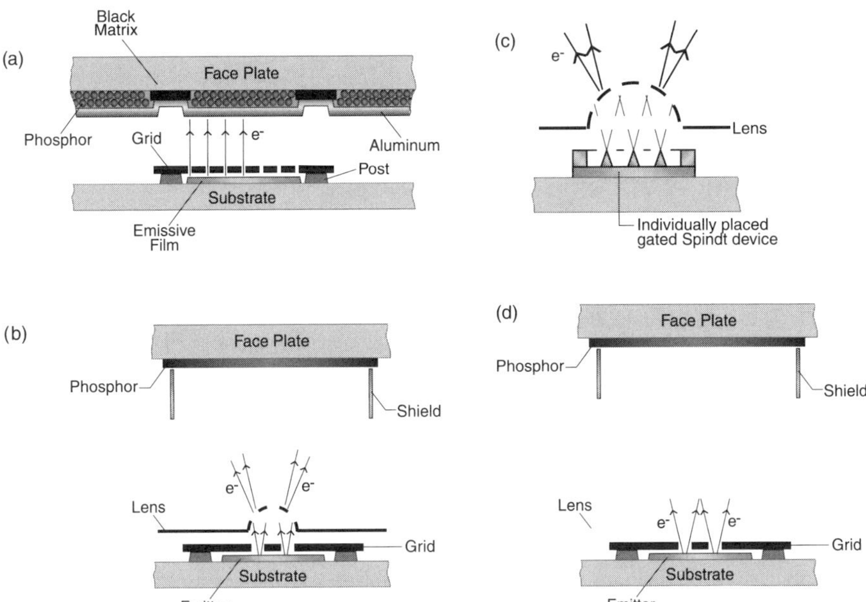

FIGURE 7.46. Emitter implementation schemes for replacing thermionic emitters with field emitters: (a) large area planar emitter triode operated in proximity mode; (b) "point" source planar emitter triode with defocusing lens; (c) "point" source gated Spindt emitter with defocusing lens; and (d) planar emitter "point" source without defocusing lens (some defocusing is obtained by the grid) [81].

FIGURE 7.47. PET modules. Left, a 64 white pixel array using method (a) depicted in Fig. 7.46 (joint development of SIDT and ISE Electronics). Right, a conventional hot filament cathode module with its cross section shown in Fig. 7.45 [85].

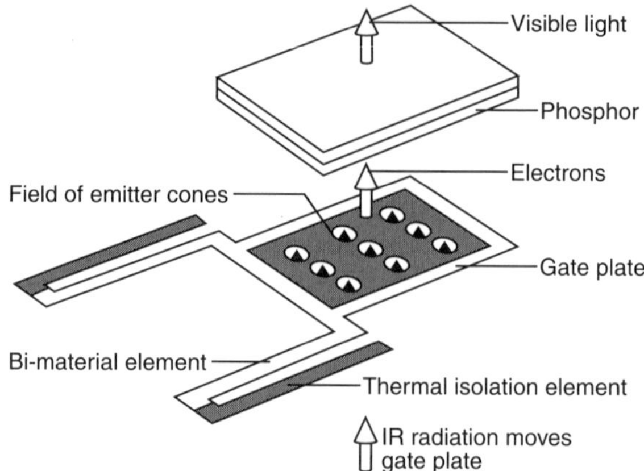

FIGURE 7.48. Schematic of a direct view IR FED [86].

of an FEA is released with respect to the tips and is supported by two double-layer (SiC/Al) arms that end in thermal isolation elements anchored to the substrate (anchors are not shown). The infrared radiation that enters the silicon chip from the back is absorbed in the gate plate, causing the double-layer arms to move with respect to the tips. Thus, for a given gate voltage, the field at the tips is modulated, which causes the emission current to vary as a function of incoming IR power. Readout is accomplished by a monochrome FED phosphor. Sensitivity calculations predict that a 1K change in temperature should move a gate plate that is suspended by a 50-μm long cantilever arm by 0.2 μm, which causes the emission current to be modulated by a factor of 2–10, depending on the tip geometry.

However, when a gate voltage is applied, an electrostatic pull-in between the gate and the emitter substrate occurs. For a 0.85 μm gate-to-substrate separation, the pull-in voltage is only 2–5 V for a 50-μm long cantilever for the chosen double-layer arm materials and thicknesses. One way to overcome this problem is to include a gate shield. By applying the gate voltage also to the shield, a field-free region (except near the tips) is created, allowing emission to occur without pulling in the gate. The shield can be processed in the region enveloping the tips, which avoids potential IR loss through the shield.

Figure 7.49 shows a micrograph of a single pixel element of a 16 $\times$ 16 array device, with each single pixel element containing 30 field emitters. The IR absorber plate is shield free and electrically isolated from the Ti–W gate. An additional contact is provided for electrically moving the structure by a small amount from its equilibrium position for compensation purposes. The cross section of the prototype package for such an IR-to-visible display is shown in Fig. 7.50. The intended application of this device is for night vision operation in an IR wavelength regime of 8–12 μm.

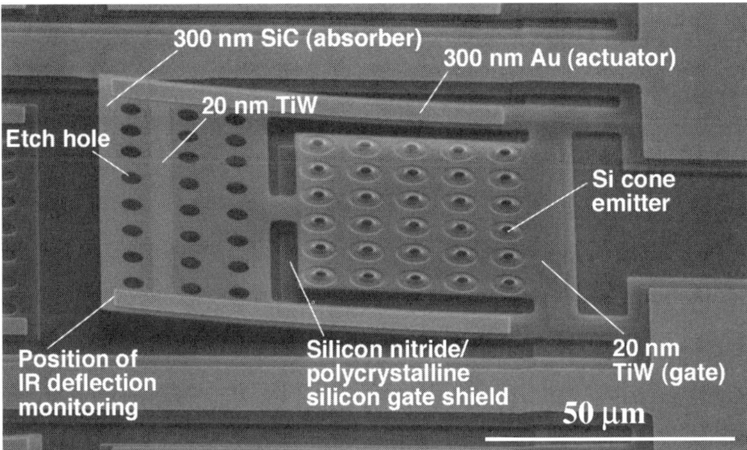

FIGURE 7.49. SEM micrograph of a 30 emitter IR sensitive pixel. The fully released cantilevered gate moves toward the tips when IR is absorbed from the back of the chip. IR-induced motion of about 1 µm has been demonstrated with this structure.

7.4. SUMMARY

In this chapter, we have attempted to profile the excitement that is associated with the development of FED technologies. The excitement lies in the interdisciplinary approach of combining large-area FEAs, high-voltage phosphors or the still developing low-voltage phosphor technology, and large-area, small-gap vacuum technology for the production of high resolution, medium-size displays. Material science, surface chemistry, and physics all become intimately involved in the development of

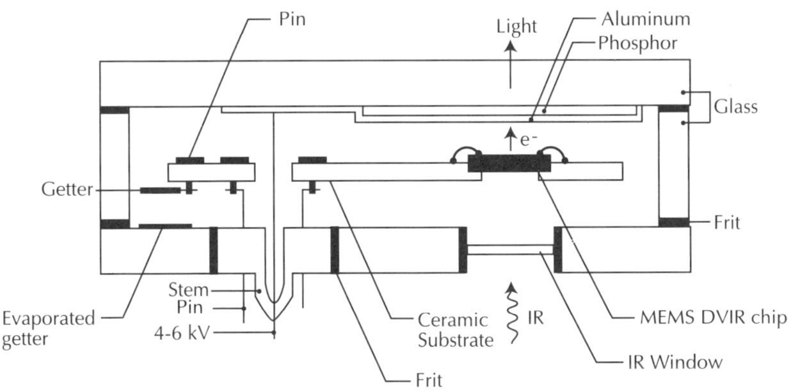

FIGURE 7.50. Cross section of a prototype direct view IR FED [86].

key display components such as stable emitters, efficient phosphors, and reliable flashover-resistant spacers.

The main portion of the chapter relates to FEDs using cone-shaped FEAs that are brought in close vicinity to a phosphor plate. The rest of the chapter focuses on encapsulation and aging issues and methods of addressing the panel. There is currently great motivation in exploring non-FEA-based emitter technologies, particularly carbon-based emitters, with the aim of lowering the production cost and enhancing the device reliability. For example, researchers from Samsung recently presented display prototypes based on two different gated versions (undergate triode and mesh triode) of carbon nanotube emitters as low cost alternatives to Spindt-type emitters [87]. The same group also incorporated electron multiplying microchannel plates (MCPs) into the display [88,89]. By generating secondary electrons inside of SiO_2 covered pores, the brightness of the display increased by a factor of three to four for the same emission current from the tips. By properly designing the MCPs, focusing electrodes can be eliminated for high voltage phosphor displays, because the MCPs can act as a focusing element.

FEDs are still a nascent technology that requires a fair amount of development for the devices to reach the marketplace. The most exciting prototypes today are the 13-in. $600 \times 800 \times 3$ Candescent display operating at 5000 V and the 15-in. PixTech display operating at 6000 V. These prototypes are currently being refined to meet the stringent requirements for laptop applications and groomed for mass production so that they can compete effectively with other mature display technologies, most notably LCDs. For video billboard displays of sizes 2×4 ft^2 and larger, the specifications are not as stringent as lap top FEDs. Here, the conventional hot filament approach can be adequately replaced by the use of more efficient field emitters, because spacers are no longer needed, very efficient high-voltage phosphors can be used, and enough space exists within the vacuum envelope to provide for sufficient gettering. Other applications such as TV-on-the-wall using a matrix of field emitting electron guns and an IR imaging display are also being contemplated.

INSTITUTIONS INVOLVED IN FED COMMERCIALIZATION

Information about FED technologies, products, and products under development can be found on the following websites:

Candescent Technology Corporation: www.candescent.com
European Network of Field Emission Research: www.cmp-cientifica.com/Eurofe
Futaba: www.futaba.com
PixTech: www.pixtech.com
Printable Field Emitters Ltd.: www.pfe-ltd.com
Samsung: www.samsungsdi.com
SI Diamond Technology: www.carbontech.net
Society for Information Displays: www.sid.org

ACKNOWLEDGMENTS

I would like to acknowledge the contributions, suggestions, and assistance of the following individuals: Dr. Theodore Fahlen of Candescent for technical discussion and assistance, reviewing the manuscript, and providing Figure 1; Dr. Capp Spindt of SRI for many useful conversations; Dr. Niel Yocom of Sarnoff for discussion on phosphor-related issues and for reviewing the phosphor section of the manuscript; Dr. Aris Silzars, president of SID 2000, for discussion on systems-related issues and for critical review of the manuscript; Charlie Kasano of ISE Electronics for assistance on picture element tubes; Drs. Z. Tolt, R. Fink, and Z. Yaniv of SIDT for assistance on carbon field emitters; Dr. Bill Taylor of Printable Field Emitters for discussion on spacer-related issues; the technical and management teams of the former Coloray Display Corp. (B. Cantos, J. Pogemiller, Drs. G. Gammie, J. Hubacek, R. Nowicki, R. Rao, S. Skala, D. Devine, and R. Young); Dr. Junji Itoh of Electrotechnical Laboratory of Japan for information on lateral emitters; Dr. Jongmin Kim of Samsung for discussion on general FED issues; Dr. Takao Kishino of Futaba for information about resistor ballasting; Dr. Bruce Gnade of University of Northern Texas and Prof. Wiley Kirk of Texas AMU for discussions on pressure-related issues; Robert Espinosa, president of Microwave Power Technology, for providing reliability data on carbon emitters; LouAnn Wingerter, Sarah Paris, Nicole Luczak, and Lauren Shuke of the Sarnoff Creative Services Department for creating most of the figures; and E. Amos of Sarnoff for typing the initial version of the manuscript. Special thanks go to Dawn Coppola, Nicole Turner and Chris Barbieri of Sarnoff for revising and editing the manuscript.

REFERENCES

1. B. R. Chalamala, B. E. Gnade, and Y. Wei, *IEEE Spectrum*, **35**, 42 (1998).

2. R. Dawson, Y. Fukuda, S. Miyaguchi, S. Ishizuka, T. Wakimoto, J. Funaki, H. Kubota, T. Watanabe, H. Ochi, T. Sakamto, M. Tsuchida, I. Ohshita, H. Nakada, and T. Tohma, Organic LED full-color passive-matrix display, in *SID Digest*, 1999, p. 430.

3. R. Meyer, Microtips fluorescent display, in *Workshop Digest, Japan Display*, 1986.

4. R. Meyer, in *Workshop Digest, Eurodisplay*, 1993, p. 189.

5. T. S. Fahlen, Performance advantages and fabrication of small gate openings in Candescent's thin CRT, in *Proceedings IVMC*, 1999, p. 56.

6. K. Werner, *IEEE Spectrum* **34**, 40 (1997).

7. N. Kumar, H. K. Schmidt, M. H. Clark, A. Ross, B. Lin, L. Fredin, B. Baker, D. Patterson, W. Brookover, C. Xie, C. Hilbert, R. L. Fink, C. N. Potter, A. Krishnan, and D. Eichman, Development of nano-crystalline diamond-based field-emission displays, in *SID Digest*, 1994, p. 43.

8. H. H. Busta, *J. Micromech. Microeng.* **2**, 43 (1992).

9. J. Itoh, K. Tsuburaya, S. Kanemaru, T. Watanabe, and S. Itoh, *Jpn. J. Appl. Phys.* **32**, 1221 (1993).

10. H. Busta, G. Gammie, S. Skala, J. Pogemiller, R. Nowicki, J. Hubacek, D. Devine, R. Rao, and W. Urbanek, Volcano-shaped field emitters for large area displays, in *Technical Digest IEDM*, 1995, p. 405.

11. H. H. Busta, *J. Micromech. Microeng.* **7**, 37 (1997). Also US Patent 5,930,590, Fabrication of volcano-shaped field emitters by chemical-mechanical polishing, (1999).

12. D. S. Y. Hsu and H. F. Gray, A low-voltage, low-capacitance, vertical multi-layer thin-film-edge dispenser field emitter array electron source, in *Proceedings IVMC*, 1998, p. 82.

13. E. Yamaguchi, K. Sakai, I. Nomura, T. Ono, M. Yamanobe, N. Abe, T. Hara, K. Hatanaka, Y. Osada, H. Yamamoto, and T. Nakagiri, A 10-in. surface-conduction electron-emitter display, in *Technical Digest SID*, 1997, p. 52.

14. M. Okuda, S. Matsutani, A. Asai, A. Yamano, K. Hatanaka, T. Hara, and T. Nakagiri, Electron trajectory analysis of surface conduction electron emitter displays (SEDs), in *SID Digest*, 1998, p. 185.

15. A. P. Burden, H. E. Bishop, M. Brierley, C. Hood, P. G. A. Jones, W. Lee, R. J. Riggs, V. L. Shaw, and R. A. Tuck, Field emitting inks for consumer-priced broad-area flat-panel displays, in *Proc. IVMC*, 1999, p. 62.

16. I. Musa, D. A. I. Munindrasdasa, G. A. J. Amaratunga, and W. Eccleston, *Nature*, **395**, 362 (1998).

17. M. Borel, J. F. Borenat, R. Meyer, and P. Rambbaud, Electron source with micropoint emissive cathodes and display means by cathodoluminescence excited by field emission using said source, US Patent 4,940,916 (1990).

18. R. Meyer, Electron source with microtip emissive cathodes, US Patent 5,194,780 (1993).

19. R. S. Nowicki, H. H. Busta, J. E. Pogemiller, A. R. Forouhi, I. Bloomer, W. M. Clift, J. L. Yio, and T. Felter, Studies of rf-sputtered $CrSiO_x$ and SiC as series resistor films in FEDs, in *SID Digest*, 1996, p. 456.

20. J. D. Levine, R. Meyer, R. Baptist, T. E. Felter, and A. A. Talin, *J. Vac. Sci. Technol. B* **13**, 474 (1995).

21. C. Constancias and R. Baptist, *J. Vac. Sci. Technol. B* **16**, 841 (1998).

22. R. Baptist, F. Bachelet, and C. Constancias, *J. Vac. Sci. Technol. B* **15**, 385 (1997).

23. J. D. Levine, *J. Vac. Sci. Technol. B* **14**, 2008 (1996).

24. S. Itoh, T. Niiyama, M. Taniguchi, and T. Watanabe, A new structure of field emitter arrays, in *Proc. IVMC*, 1995, p. 99. Also US Patent 5,786,659, Field emission type electron source, (1998).

25. J. P. Spallas, R. D. Boyd, J. A. Britten, A. Fernandez, A. M. Hawryluk, M. D. Perry, and D. R. Kania, *J. Vac. Sci. Technol. B* **14**, 2005 (1996).

26. M. Anandan, G. Jones, and A. P. Gosh, Field emission display for commercial avionics applications, in *Workshop Digest of Technical Papers, Asia Display*, 1998, p. 131.

27. X. Chen, S. H. Zaidi, and S. R. J. Brueck, *J. Vac. Sci. Technol. B.* **14**, 3339 (1996).

28. T. T. Doan, J. B. Rolfson, T. A. Lowrey, and D. A. Cathey, Method to form self-aligned gate structures around cold cathode emitter tips using chemical mechanical polishing technology, US Patent 5,229,331 (1993).

29. L. E. Tannas, Jr. (Ed.), *Flat-Panel Displays and CRTs*, Van Nostrand Reinhold Co.: Ch. 3, New York, 1985.

30. N. Yocom, Sarnoff Corporation, private communications.

31. H. Bechtel and H. Nikol, Applicability of zinc sulfide phosphors in low voltage cathode ray displays, in *Workshop Digest, Asia Display*, 1998, p. 283.

32. L. E. Shea, The Electrochemical Society *Interface* **7**, 24 (Summer 1998).

33. S. S. Chadha, in *3rd Int. Conf. on the Science and Tech. of Display Phosphors*, Huntington Beach, CA, Nov. 1997.

34. P. H. Holloway, J. Sebastian, T. Trottier, H. Swart, and R. O. Peterson, *Solid State Technol.* **38**, 47 (1995).

35. S. Jacobsen, *J. SID* **4**, 331 (1996).

36. C. Stoffers, S. Yang, S. M. Jacobsen, and C. J. Summers, *J. SID* **4**, 337 (1996).

37. S. S. Kim, S. H. Cho, J. S. Yoo, S. H. Jo, and J. D. Lee, The effect of the resistivity of $ZnGa_2O_4$:Mn phosphor screens on the emission characteristics of field emitter array, in *Proc. IVMC*, 1997, p. 676.

38. S. A. Bukesov, N. V. Nikishin, A. O. Dmitrienko, S. L. Shmakov, and J. M. Kim, Electrophysical characteristics and low-energy cathodoluminescence of VFD and FED screens, in *Proc. IVMC*, 1997, p. 663.

39. C. J. Summers and B. K. Wagner, The role of surface modifications and coating in improving phosphor performance, in *Workshop Digest, Asia Display*, 1998, p. 261.

40. R. Ravi, Plasmaco, Private communications.

41. J. E. Jang, J. H. Gwak, Y. W. Jin, N. S. Lee, J. E. Jung, and J. M. Kim, Polyvinyl alcohol-slurry screening of phosphors for full-color SVGA FED applications, in *Proc. IVMC*, 1999, p. 402.

42. M. J. Shane, J. B. Talbot, B. G. Kinney, E. Sluzky, and K. R. Hess, *J. Electrochem. Soc.* **165**, 334 (1994).

43. W. D. Kesling and C. E. Hunt, *IEEE Trans. Electronic Devices* **42**, 193 (1995).

44. J. Itoh, Y. Tohma, K. Morikawa, S. Kanemaru, and K. Shimizu, *J. Vac. Sci. Technol. B* **13**, 1968 (1995).

45. T. S. Fahlen, Focusing properties of thin CRT™ display, in *Technical Digest SID*, 1999, p. 830.

46. Y. C. Lan, J. T. Lai, S. H. Chen, W. C. Wang, C. H. Tsai, K. L. Tsai, and C. Y. Shen, Simulation of focusing field emission devices, in *Proc. IVMC*, 1999, p. 72.

47. Notes from Technology Forum, Coloray Display Corporation, November 1994.

48. X. Ma, P. G. Muzykov, and T. S. Sudarshan, *J. Vac. Sci. Technol. B* **17**, 769 (1999).

49. X. Ma and T. S. Sudarshan, *J. Vac. Sci. Technol. B* **17**, 1 (1999).

50. X. Ma and T. S. Sudarshan, Flashover performance of thin-wall spacers in field emission displays, in *IS&T/SPIE Conference on Flat Panel Display Technology*, Vol. 3636 0277-786/99, 1999, p. 143.

51. W. Taylor, N. S. Xu, R. V. Latham, B. J. Goddard, W. Kalbreier, J. Tan, H. J. Hopman, J. Verhoeven, and D. J. Chivers, The effects of surface treatments on the high voltage vacuum performance of alumina based insulators, Poster presented at *IVMC*, Asheville, NC, 1998.

52. R. V. Latham, Flashover across solid insulator in a vacuum environment [Appendix], in *High Voltage Vacuum Insulation: The Physical Basis*, Academic Press: New York, 1981, p. 229.

53. J. Browning, C. Watkins, D. Dynka, D. Zimlich, and J. Alwan, Field emission displays for desktop monitors, in *Workshop Digest of the 18th Int. Display Research Conference, Asia Display*, 1998, p. 149.

54. R. V. Latham, J. A. Mir, B. Goddard, J. Tan, and W. Taylor, The electro-optical signature of pre-breakdown electron injection from "electron pin-holes" at the cathode triple junction of a bridged vacuum gap, in *Proceedings of the 3rd International Conference on Space Charge in Solid Insulators*, Tours, France, June 1988.

55. N. C. Jaitly and T. S. Sudarshan, *IEEE Trans. on Elec. Insul.* **23**, 261 (1988).

56. G. Blaize and C. LeGressus, Charge trapping-detrapping processes and related breakdown phenomena, in *High Voltage Vacuum Insulation: Basic Concepts and Technological Practice* (R. V. Latham, Ed.), Academic Press: London, UK, 1995, Ch. 9, p. 329.

57. M. J. Kofoid, *Trans. AIEE (USA), Pt. III*, **79**, 991 (1960).

58. R. A. Anderson, Time-resolved measurements of surface flashover of conical insulators, Annual Report CEIDP, p. 475, 1978.

59. A. D. Johnson, H. H. Busta, R. S. Nowicki, Fabrication system, method and apparatus for microelectromechanical devices, US Patent 5,903,099 (1999).

60. H. Kim, B. K. Ju, K. B. Lee, M. S. Kang, J. Jang, and M. H. Oh, Influences of ambient gases upon emission characteristics of Mo-FEAs during frit sealing process, in *Proc. IVMC*, 1998, p. 67.

61. S. J. Jung, K. J. Woo, K. S. Kim, G. J. Moon, M. S. Kim, N. Y. Lee, and S. Ahn, The effect of ambient gas in sealing process on field emission characteristics, in *Proc. IVMC*, 1999, p. 40.

62. J. M. Kim, N. S. Lee, J. E. Jung, H. W. Lee, J. Kim, Y. J. Park, S. N. Cha, J. W. Kim, J. H. Choi, N. S. Park, J. E. Jang, S. S. Hong, Y. W. Jin, I. T. Han, and S. Y. Jung, Development of high voltage operated 5.2″ full color field emission displays, in *Proc. IVMC*, 1999, p. 54.

63. H. H. Busta and R. W. Pryor, *J. Vac. Sci. Technol. B* **16**, 1207 (1998).

64. H. H. Busta and R. J. Espinosa, The need for standardization of field emitter testing, in *Workshop Digest, Asia Display*, 1998, p. 113.

65. Y. R. Cho, J. Y. Oh, H. S. Kim, J. D. Mum, and H. S. Jeong, A new panel structure for field emission displays, in *Proc. IVMC*, 1997, p. 271.

66. W. B. Choi, B. K. Ju, Y. H. Lee, S. J. Jeong, N. Y. Lee, M. Y. Sung, and M. H. Oh, *J. Electrochem. Soc.*, **146**, 400 (1999).

67. B. Gnade and J. Levine, SID Seminar Lecture Note, Vol. 1, M-8/3, 1995. Eq. 4 was derived by Prof. Wiley P. Kirk of Texas AMU.

68. S. Tominetti, Chemical pumping in flat CRTs, in *Workshop Digest, Asia Display*, 1998, p. 101.

69. C. C. Wang, J. C. Wu, and C. M. Huang, Data line driver design for a 10″ 480 × 640 × 3 color FED, in *Proc. IVMC*, 1996, p. 557.

70. R. T. Smith, Drive approaches for FED displays, in *Proc. Int. Display Research Conference*, 1997, p. F-35.

71. H. H. Busta, J. E. Pogemiller, W. Chan, and G. Warren, *J. Vac. Sci. Technol. B* **11**, 445 (1993).

72. Y. S. Na, O. K. Kwon, C. H. Hyun, and J. W. Choi, A new driver circuit for field emission display, in *Workshop Digest, Asia Display*, 1998, p. 137.

73. Y. D. Lee, Seoul National University, Private communication.

74. A. Silzars, Society for Information Display, Private communication.

75. B. R. Chalamala, R. M. Wallace, and B. E. Gnade, Residual gas effects on the emission characteristics of active Mo field-emission cathode arrays, in *SID Digest*, 1998, p. 107.

76. Y. Wei, B. R. Chalamala, B. G. Smith, and C. W. Penn, Surface chemical changes on field emitter arrays due to device aging, in *Proc. IVMC*, 1998, p. 46.

77. Z. Li Tolt, R. L. Fink, and Z. Yaniv, Carbon thin film cathode for large area displays, in *Proceedings 5th Int. Display Workshops*, Kobe, Japan, 1998, p. 651.

78. V. M. Polushkin, S. N. Polyakov, A. T. Rakhimov, N. V. Suetin, M. A. Timofeyev, and V. A. Tugarev, *Diamond Relat. Mater.* **3**, 531 (1994).

79. Z. Li Tolt, L. H. Thuesen, R. L. Fink, and Z. Yaniv, Field emission carbon thin film and its life time and stability, in *Proc. IVMC*, 1999, p. 160.

80. J. Yotani, S. Uemura, T. Nagasako, Y. Saito, and M. Yumura, Proceedings, High-luminance triode panel using carbon nanotube field emitters, in *Proc. 6th Int. Display Workshop*, Sendai, Japan, 1999, p. 971.

81. H. Busta, R. Espinosa, A. Rakhimov, N. V. Suetin, M. A. Timofeyev, M. E. Kordesch, P. Bressler, M. Schramme, J. R. Fields, and A. Silzars, *Solid State Electron.* **45**, 1039 (2001).

82. B. V. Seleznev, A. V. Kandidov, A. T. Rakhimov and N. P. Sostchin, Peculiarities of emission current flow in diode-structured FEDs, in *Workshop Digest, Eurodisplay*, 1996, p. 207.

83. R. C. Robinder, Z. Yaniv, R. L. Fink, Z. Li Tolt, and L. Thueseu, Application of diamond thin film cathodes to the hybrid field emission display/CRT (HyFED), in *Proceedings 5th International Display Workshops*, Kobe, Japan, 1998, p. 655.

84. M. Morikawa, Y. Seko, H. Kamogawa, S. Uemura, and T. Shimojo, Study to improve the flood-beam CRT for giant screen display, in *Workshop Digest, Japan Display*, 1992, p. 385.

85. N. Hasegawa, *Display Devices*, Dempa Publications, 1998, p. 33.

86. H. Busta, R. Amantea, D. Furst, C. Ling, L. White, and J. M. Chen, MEMS direct view infrared vision system (DVIR) with FED readout, in *Proc. IVMC*, 2000, p. 232.

87. N. S. Lee, J. H. Kang, W. B. Choi, Y. S. Choi, Y. J. Park, H. Y. Kim, Y. J. Lee, D. S. Chung, J. E. Jung, C. J. Lee, J. H. Kim, J. H. Yon, S. H. Jo, C. G. Lee, and J. M. Kim, Triode structure field emission displays using carbon nanotube emitters, in *Proc. IVMC*, 2000, p. 193.

88. W. Yi, S. Jin, T. Jeong, J. Lee, S. G. Yu, and Jongmin Kim, New electron multiplying channel for high efficiency field emission display, in *Proc. IVMC*, 2000, p. 9.

89. T. Jeong, J. Lee, S. G. Yu, S. Jin, J. Heo, W. Yi, and Jongmin Kim, The secondary electron emission characteristics for sol-gel based SiO_2 thin films, in *Proc. IVMC*, 2000, p. 20.

CHAPTER 8

Cold Cathode Microwave Devices

R. ALLEN MURPHY

Lincoln Laboratory, Massachusetts Institute of Technology, 244 Wood Street, Lexington, Massachusetts 02420-9108

MARY ANNE KODIS

Jet Propulsion Laboratory, 4800 Oak Grove Drive, Pasadena, California 91109-8099

8.1. INTRODUCTION

Historically, field emission cathodes were conceived as a substitute for the thermionic cathodes of conventional microwave tubes [1]. Subsequently, this application has driven much of the development of field emitter arrays (FEAs), and has spawned the field of vacuum microelectronics, in which microlithography is used to fabricate FEA cathodes for such amplifiers, as well as for flat panel displays [2–12].

This chapter is adapted and updated from an early publication in John Webster's *Encyclopedia of Electrical and Electronics Engineering* by the authors [13]. It will focus upon the realization of cold cathode microwave amplifiers using FEAs and describe the development and status of FEAs in this context. Variants of the TWT (traveling wave tube) and klystron will be discussed in detail, and the potential advantages of such devices over other tubes and solid-state devices will be described. Both the physical principles that underlie the issues and simple mathematical analyses that describe them will be presented.

Although this chapter will not discuss specifically some newly developed cold cathode approaches, encouraging results have been obtained and are worth mentioning. Using carbon nanotubes, high current densities and stable emission have been obtained [14], and a gated cathode has been demonstrated [15]. A self-aligned gated diamond microtip array has been fabricated using a silicon wafer molding technique [16]. In diode form, these microtips have demonstrated emission currents of 7 μA/tip and have operated well at 10^{-5} torr, a relatively poor vacuum. Selective growth techniques have been used to fabricate arrays of GaN [17,18] and diamond [19] emitters. Electron emission from diamond and III–V nitrides is reviewed by Nemanich [20]. Good emission and excellent resistance against vacuum degradation has been obtained from carbon-doped boron nitride emitters [21].

Vacuum Microelectronics, Edited by Wei Zhu.
ISBN 0-471-32244-X. 2001 John Wiley & Sons, Inc.

8.2. MICROWAVE AMPLIFIERS

It can be shown [22,23] that, for triode- and transistor-like three-terminal devices, the maximum power P_m that can be delivered to a load is

$$P_m = \frac{E_m^2 v_s^2}{X_o (2\pi f_T)^2} \tag{1}$$

In Eq. (1), E_m is the critical field at which electrical breakdown occurs, v_s is the electron velocity, X_o is output impedance level, and f_T is the cutoff frequency. Eq. (1) can be used to understand the difference between solid-state and vacuum devices. In a solid-state device, f_T can be quite high because device dimensions are small. However, the electron velocity in a solid-state device cannot exceed approximately 10^7 cm/s because of electronic collisions with the semiconductor lattice, whereas for vacuum tubes even relativistic velocities can be attained. The breakdown process in a semiconductor is initiated by valence- to conduction-band transitions, which typically require energies only of the order of 1 eV. In contrast, secondary emission processes, which can be minimized by proper choice of materials and geometry, determine breakdown in tubes. Furthermore, the heat dissipated in the semiconductor is more problematic in a solid-state device and often limits the output power. In general, semiconductors have much lower thermal conductivity than metals, so a properly designed microwave tube can provide better thermal paths to dissipate heat. Consequently, for quite fundamental reasons, the output power provided by a microwave tube can be much higher than that provided by a solid-state device.

Lee de Forest invented the triode, the first microwave tube, in 1906 [24]. In a triode, the electron beam is modulated by a grid and collected by a plate (anode), which is connected through a load to ground. The classical high-frequency vacuum triode reached maturity in the late 1940s with the "lighthouse" family of cavity-driven, gridded tubes [25–28]. In the "lighthouse" design, all high-frequency connections were made radially through disk leads, which minimized parasitic losses. The 416A triode operated to 4 GHz. Its grid was 90 mil (2.3 mm) in diameter, was fabricated from a 0.3 mil (7.6 μm) tungsten wire wrapped at 1000 turns/in., and was mounted 0.6 mil (15 μm) above the cathode surface. The variation in wire spacing and grid–cathode spacing was less than 10%, although the cathode diameter was 150 times the grid-cathode spacing. Measurements indicated that the performance of the 416A triode was within a factor of 5 of the theoretical maximum performance imposed by the thermal velocity spread of electrons emitted by a thermionic cathode. The power-handling capability of the grid structure posed the most severe limitation to further extensions of the frequency range. The stiffness of the grid wires limited the diameter of the cathode (and so the diameter of the beam), and interception of beam current by the grid limited the maximum current density to about 180 mA/cm². At 4 GHz, the 416A triode achieved a gain of 10 dB with an instantaneous bandwidth of 2.5%. Transit-time effects also limited the frequency response of these triodes.

As discussed later, the transit time of electrons through the region of a triode (or pentode) in which they interact with the grid fields must be less than the period of the electromagnetic radiation. Further advances in the gain-bandwidth product of the 416A tube would have required the grid-cathode spacing to be smaller than 0.2 mil (5 μm). The fabrication of such a tube would be quite challenging, and even if it were feasible, the high operating temperatures of the cathode (of the order of 700–800°C) would create thermal expansion and reliability problems on the nearby grid. Since a reduction in grid-cathode spacing increases the grid-cathode capacitance, the lateral dimensions of the cathode must be reduced as well, which further reduces the peak power. Subsequent development shifted to linear-beam velocity-modulated (klystrons, TWTs) and crossed-field devices.

For frequencies exceeding about 500 MHz, the wavelength of the signal becomes comparable to the dimensions of the circuit, so that circuit elements in conventional microwave tubes become distributed. No longer are capacitors purely capacitive and inductors purely inductive. In addition, transit-time effects occur. Only by reducing the physical size of the elements can such distributed effects be eliminated. In no element is this size reduction more profitable than the part of the tube that produces the electrons, i.e., the cathode. In contrast, miniaturization of the output circuit tends to reduce its power-handling ability and thus limit the average power of the amplifier. In a field emission cathode, a cathode heater is absent, and the input signal and emitted beam dissipate only a moderate amount of heat in the cathode region.

In contrast to the triodes discussed previously, inductive output amplifiers (IOAs), such as klystrons and TWTs, utilize inductive circuits to exchange energy with the electron beam [29]. Electron transit times that are long compared to the period are not problematic because the high-frequency coupler is distinct and separate from the beam emission and collection circuit. The electron beam does not strike the rf output circuit as it does in a resistive output circuit; rather, it is decelerated by passing through a traveling or standing electromagnetic wave that is developed in an output circuit. Such an inductive output circuit can be many wavelengths long, allowing a cumulative interaction that converts part of the kinetic energy of the beam into electromagnetic energy in each period. The spent beam is dumped into a collector only after the complete extraction of high-frequency power. As shown in Fig. 8.1(a), the cathode emits an unmodulated electron beam (a "dc beam"). In the first section of the circuit, the rf input signal imposes a small velocity modulation ($v = v^{dc} + v^{rf}$) on the electron beam, which launches longitudinal space-charge waves. As the electrons drift through the microwave tube, the beam modulation cycles between kinetic-energy modulation, i.e., velocity modulation, and potential-energy modulation, i.e., density modulation ($n = n^{dc} + n^{rf}$). If the initial modulation is small, it is increased by passing the beam through intermediate interaction regions where rf modulation of the beam current exchanges energy with an electromagnetic wave. In the last interaction region (the output region), the relative phase of the plasma wave and the electromagnetic wave is adjusted to maximize energy transfer from the beam to the electromagnetic wave.

Appropriate interaction circuits include resonant cavities and structures that will support a slow electromagnetic wave with a phase velocity close to the beam velocity.

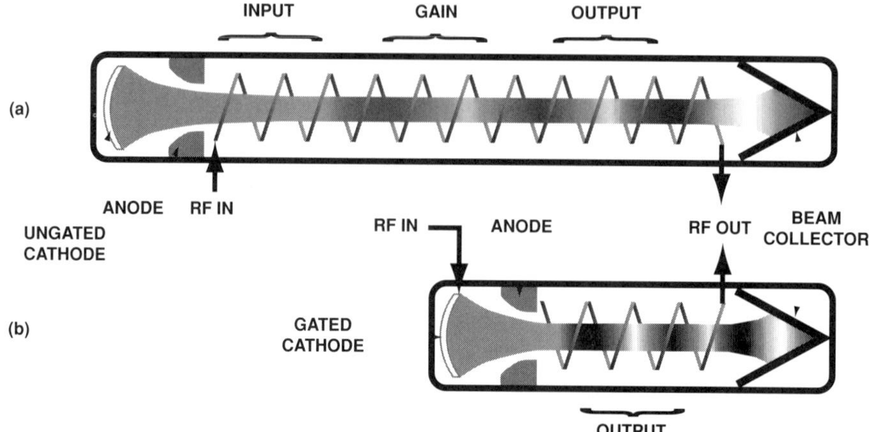

FIGURE 8.1. (a) Traveling wave tube (TWT) and (b) twystrode. The twystrode's gated cathode replaces the input and gain sections of the TWT by modulating the emitted beam current.

Two extreme cases are the pillbox cavity and the helical coil; the amplifiers that use them are the narrowband, high-gain klystron and the wideband, low-gain helix TWT, as will be discussed later. Amplifiers with intermediate gain and bandwidth use circuits such as the various coupled-cavity and ring-bar structures. In all of these velocity-modulated tubes, approximately the first two-thirds of the length of the circuit is employed in achieving a strong modulation of the beam, with the last one-third allocated to extracting output power. If higher gain is desired, the circuit must be lengthened to convert a very small input signal into a large rf modulation of the electron-beam current. Since the electron beam must be magnetically focused over the whole length of the interaction circuit, velocity modulation can be an expensive approach in terms of size and weight.

In variants of these tubes, the klystrode and twystrode, velocity modulation is replaced by density modulation from a gated cathode that emits a directly modulated electron beam. A twystrode, as illustrated in Fig. 8.1(b), combines a gated cathode, in which the input signal modulates the beam density, with a wideband output circuit where the modulated beam interacts with a synchronous electromagnetic wave. The device is analogous to a TWT, with the inductive input circuit replaced by a gated cathode. The gated cathode must be embedded in a broadband input circuit to realize the full bandwidth potential of the broadband output circuit. A klystrode, as illustrated in Fig. 8.2, is a narrowband, high-gain amplifier consisting of a gated cathode modulated by the input signal, followed by a resonant-cavity output circuit. Klystrodes and twystrodes thus combine the best features of triodes and velocity-modulated tubes. A gridded cathode imposes a strong initial density modulation on the beam current, eliminating approximately two-thirds of the size and weight of the tube, while the inductively coupled output circuit provides high power-handling capability and high efficiency.

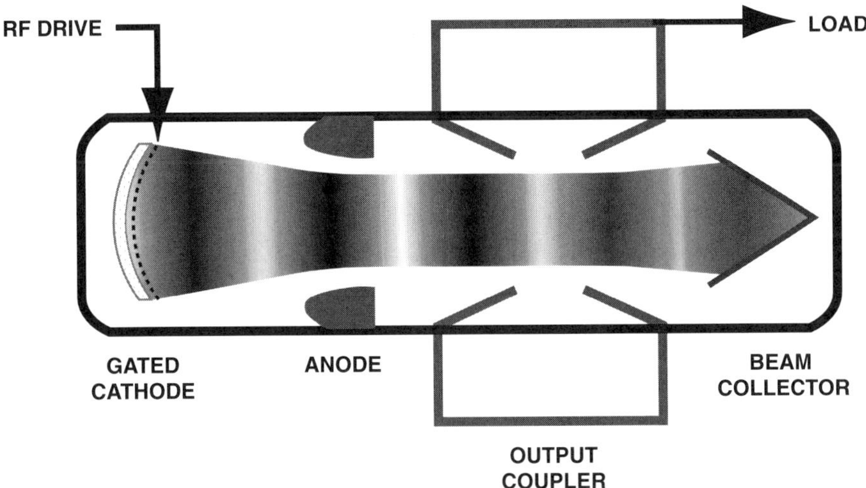

FIGURE 8.2. Klystrode. A field emission cathode provides a density-modulated beam. A rf energy is extracted from the accelerated beam by the output coupler, which is a resonant cavity, and the spent beam is recovered in a beam collector.

A grid-controlled thermionic cathode is limited in high-frequency response by transit-time effects, high grid-cathode capacitance, and low transconductance. The spacing of the grid-cathode gap and the transparency of the grid are limited by fabrication technology and material properties. At common operating voltages and at frequencies above 1 GHz, structures smaller than 10 μm must be fabricated to tolerances of 10% or less. Grid materials must possess good thermal and electrical conductivity, excellent mechanical stability at temperatures above 1000°C, and low secondary electron emission coefficient. The most common materials meeting these requirements are tungsten or molybdenum, possibly coated with noble metals. Graphite, an early contender that was dropped because of excessive fragility, has returned in the form of pyrolytic graphite.

Microelectronic FEAs can modulate the beam density at high frequency and with good spatial localization, extending the frequency range of density-modulated amplifiers by orders of magnitude. In FEA structures, the grid (or gate) is fabricated in nearly the same plane as the emitting surface, dramatically reducing interception current and increasing transconductance. A microtriode using an FEA cathode is illustrated in Fig. 8.3. The use of field emission also eliminates the need to dispense a continuous supply of low-work function material as is often done in thermionic cathodes. This material, which vaporizes in the tube, can coat the grids and grid-cathode insulators, resulting in secondary emission and shorts.

FEA cathodes require no heater power, offer 'instant on' capability because the cathode is not heated, can provide extremely high current densities, and can be operated at high pulse repetition rates. Consequently, incremental performance improvements can be obtained in conventional velocity-modulated amplifiers by replacing an

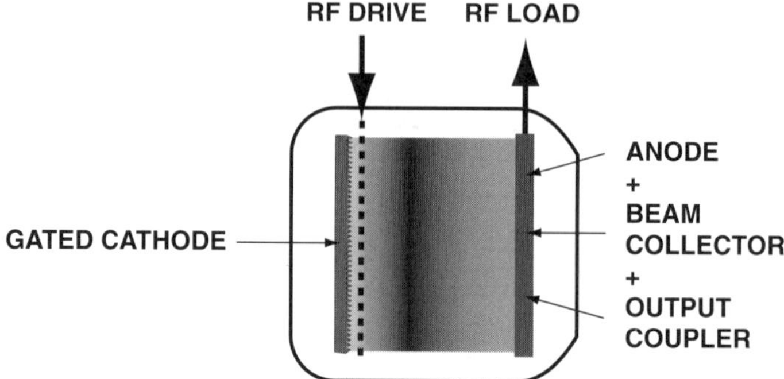

FIGURE 8.3. Vacuum microtriode. A rf signal is applied to the gate electrode of the FEA, providing a density-modulated beam. The anode both collects the beam and delivers amplified current to the rf load.

ungridded thermionic cathode with an FEA cathode that produces an unmodulated electron beam. FEAs have been used as cathodes in velocity-modulated fast-wave [30] and slow-wave [31–35] devices. However, a gated FEA cathode that provides a density-modulated beam current at the input of an amplifier enables amplifiers with substantial qualitative advantages over both velocity-modulated and conventional gated-cathode amplifiers. Because of improvements in gated FEA technology, such cathodes are feasible alternatives to thermionic electron beam sources [35–39] for emission gating at frequencies above UHF. This new opportunity, however, depends critically upon the ability to integrate the gated FEA technology into the vacuum tube environment.

To summarize, IOAs use a gated cathode to modulate the current and an inductive output circuit, which uses a resonance or synchronous electromagnetic wave, to couple power from the beam. In this class of devices, the beam is fully modulated before the anode accelerates it, no drift space is needed to convert velocity modulation into density modulation, and the rf output electrodes are separated from the beam-collecting electrodes. The absence of an inductive input circuit and its associated length of magnet make IOAs potentially more compact and higher in specific power (W/g) than their analogous velocity-modulated amplifiers. Additionally, in contrast to velocity-modulated tubes, IOAs can operate as power amplifiers in Class B or C. In an IOA, only the modulated beam conveys information about the input signal to the output circuit. This results in a physical isolation of the input circuit, which dominates the gain, from the output circuit, which controls the efficiency. The design criteria for the cathode and the output circuit are thus clearly distinguished. The role of the gated cathode and its impedance-matching circuit is to modulate the electron beam with a minimum of input signal power over the desired frequency band. The role of the output interaction circuit is to efficiently convert beam energy to electromagnetic energy at the desired frequency in as short a circuit length as possible. Inductive output amplifiers are classified according to the type of output interaction: klystrodes use a standing-wave cavity and twystrodes use a traveling-wave circuit.

8.3. FIELD EMITTER ARRAYS

8.3.1. Operation and Fabrication

The key to the performance advantages of IOAs is the emission gating of the electron beam at the cathode surface before acceleration to anode potential. The cathode assembly that performs this modulation is usually an old technology pushed to its fundamental limitations (i.e., gridded thermionic cathodes) or a new technology pushed to its present limits of performance (i.e., FEAs or laser-driven photo-cathodes). The critical measures of the performance of any emission gated cathode are low transit time, high transconductance, and low capacitance. The current density must be sufficiently high for good performance, but not too high for good beam optics. Each of these factors is reviewed in the following.

In order to extract electrons from a metal or semiconductor into vacuum, the potential energy barrier that confines electron to the metal (with the work function ϕ) must be overcome by external means. This situation, in the absence of any such physical means, is illustrated in Fig. 8.4(a). In a thermionic cathode, the host cathode metal is heated until electrons can escape, as is illustrated in Fig. 8.4(b). Very high temperatures (of the order of 700–800°C) are required for this purpose and emission from such thermal cathodes cannot be modulated at microwave frequencies.

In field emission cathodes, the electrons are extracted by high electric fields, as illustrated in Fig. 8.4(c). In this case, the width and height of the confining potential barrier is reduced by the external electric field, which allows a significant fraction of the electrons to escape by quantum-mechanical tunneling. The emission process can be described by the Fowler–Nordheim (F–N) relationship [40], which gives the emitted current density J as a function of an electric field E_{tip} applied normal to the emitting surface.

In view of the variety of expressions and units that appear in the literature, an outline of the derivation of F–N expression seems useful. For the most part, the derivation of Good and Müller [41] will be followed, but SI units will be used. The current density J of the emitted electrons is given by

$$J = e \int_{-\infty}^{\infty} P(W)\,\mathrm{d}W \tag{2}$$

In this expression, e is the electronic charge (1.6×10^{-19} C), and $P(W)\,\mathrm{d}W$ is the number electrons per second per area that emerge from the emitting material at energy W. The quantity $P(W)$ is given by the product of a supply function $N(W)$, which describes the availability of electrons in the emitting material, and a transmission coefficient $D(W)$, which describes the probability of transmission through the barrier, so that

$$J = e \int_{-\infty}^{\infty} D(W)N(W)\,\mathrm{d}W \tag{3}$$

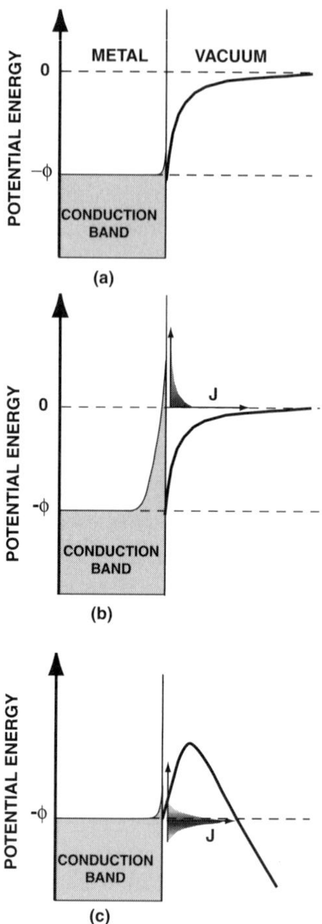

FIGURE 8.4. Emission processes. (a) Potential energy of electrons at a metal/vacuum interface in the absence of any external stimuli. The electrons are confined in the metal by a potential barrier Φ that must be overcome to extract electrons into vacuum. (b) Thermionic emission. The cathode is heated to temperatures exceeding $700°C$, allowing thermally excited electrons to escape from the metal. (c) Field emission. The application of a high external electric field diminishes the width of the potential barrier that confines the electrons, which allows them to escape by quantum mechanical tunneling. This process is usually well described by the F–N relationship.

According to Fermi–Dirac statistics, $N(W)$ is given by

$$N(W) = \frac{4\pi mkT}{h^3} \ln\left[1 + \exp\left(-\frac{W - W_F}{kT}\right)\right] \qquad (4)$$

In this expression, m is the electronic mass (9.1×10^{-31} kg), h is Planck's constant (6.626×10^{-34} J s), k is Boltzmann's constant (1.38×10^{-23} J/K), W_F is the

Fermi energy of the electrons in the emitter, and T is the absolute temperature. In the WKB (Wentzel-Kramers-Brillouin) approximation, $D(W)$ is given by $D(W) = \exp[-2\kappa(W)]$, where

$$\kappa(W) = \int_{x_1}^{x_2} \sqrt{\frac{2m}{\hbar}[V(x) - W]}\, dx \tag{5}$$

In Eq. (5), the integration is performed along the coordinate normal to the emitting surface, $\hbar$ is the reduced Planck constant, and $V(x)$ is the potential experienced by an electron of energy W. The limits of integration, x_1 and x_2, are the positions at which the integrand vanishes. Including the effects of image-force lowering [42], $V(x)$ is given by

$$V(x) = -eE_{\text{tip}}x - \frac{e^2}{16\pi\varepsilon_0 x} \tag{6}$$

In Eq. (6), ε_0 is the permittivity of free space (8.85×10^{-12} F/m). Substituting Eq. (6) in Eq. (5), solving for x_1 and x_2, and changing the variable of integration results in [41]:

$$\kappa(W) = \frac{2\sqrt{2m|W|^3}}{3\hbar e E_{\text{tip}}} v(y) \tag{7}$$

where

$$v(y) = \frac{3}{2\sqrt{2}} \int_a^b \sqrt{(\eta^2 - a^2)(b^2 - \eta^2)}\, d\eta$$

$$a^2 = 1 - \sqrt{1 - y^2}$$

$$b^2 = 1 + \sqrt{1 - y^2}$$

$$y^2 = \frac{e^3 E_{\text{tip}}}{4\pi\varepsilon_0 |W|^2}$$

The function $v(y)$ is a correction factor that results from image-force lowering, and is given by [43–45]

$$v(y) = (1 + Y)^{1/2}[E(m) - yK(m)] \cong 0.95 - y^2 \tag{8}$$

In Eq. (8), $K(m)$ and $E(m)$ are the complete elliptic integrals of the first and second kinds, respectively, and $m = (1 - y)/(1 + y)$. As is evident from an inspection of Eqs. (3), (4), and (7), emission arises from electrons near the Fermi level. Expanding

$\kappa(W)$ about $W = W_F$ gives

$$D(W) \cong \exp\left[-\frac{4\sqrt{2m|W_F|^3}}{3\hbar e E_{\text{tip}}}v(y_F)\right]\exp\left[\frac{2\sqrt{2m|W_F|}}{\hbar e E_{\text{tip}}}t(y_F)(W - W_F)\right] \qquad (9)$$

In Eq. (9), $y_F^2 = e^3 E_{\text{tip}}/4\pi\varepsilon_0|W_F|^2$, and $t(y)$ is given by [43–45]

$$t(y) = v(y) - \frac{2}{3}y\frac{dv(y)}{dy} = \frac{(1 + y)E(m) - yK(m)}{\sqrt{1 + y}} \cong 1.1 \qquad (10)$$

Inserting Eqs. (4) and (9) into Eq. (3) yields a formula that can be integrated analytically. The result is

$$J = \frac{e^3 E_{\text{tip}}^2}{16\pi^2\hbar W_F t^2(y_F)}\exp\left[-\frac{4\sqrt{2m W_F}W_F}{3\hbar e E_{\text{tip}}}v(y_F)\right]\frac{\dfrac{2\pi\sqrt{2m W_F}kT t(y_F)}{\hbar e E_{\text{tip}}}}{\sin\left(\dfrac{2\pi\sqrt{2m W_F}kT t(y_F)}{\hbar e E_{\text{tip}}}\right)} \qquad (11)$$

Eq. (11) can be written in a somewhat more accessible form,

$$J = \frac{\alpha E^2}{t^2(y_F)}\exp\left[-\frac{4\phi}{3\phi_E}v(y_F)\right]\frac{\theta}{\sin\theta} \qquad (12)$$

where

$$\theta = \frac{2\pi(kT/e)t(y_F)}{\phi_E}$$

$$\alpha = \frac{e^2}{16\pi^2\hbar\phi} = \frac{1.541}{\phi}\mu A/V^2$$

$$\phi_E = \frac{\hbar E_{\text{tip}}}{\sqrt{2me\phi}} = 1.959\frac{E(V/\text{\AA})}{\sqrt{\phi}}\,eV$$

It is evident from Eq. (12) that a very high electric field, of the order of 1 V/Å, is required for emission. For $\phi = 4.5$ eV and $E = 1$ V/Å, $\phi_E = 0.92$ eV and $\theta \cong 0.04$ at room temperature ($kT/e = 0.026$ eV). The temperature-correction factor, $\theta/\sin\theta$, departs appreciably from unity at very high temperatures and is usually ignored. The transition from field emission to thermionic emission is discussed in detail by Murphy and Good [46]. In addition, density-gradient theory has been applied to the analysis of field emission [47], the effect of electric field penetration into the emitting material has been calculated [48], field emission in a microwave field has been discussed [49], and the theory has been expanded to semiconductor emitters [50–52].

Impracticably high voltages are required to attain emission fields between simple parallel-plane electrodes. For example, a voltage of 1000 kV is required to achieve a field of 1 V/Å between two planar electrodes spaced apart by 100 μm, a value that

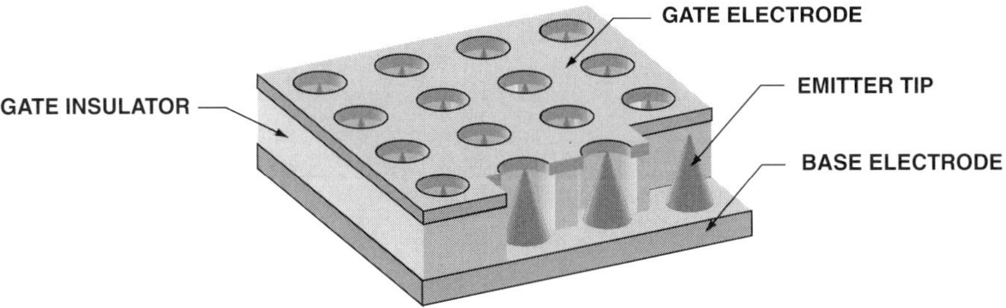

FIGURE 8.5. Drawing of an FEA. Using microelectronic fabrication techniques, an array of sharply pointed tips with a gate electrode in close proximity can be fabricated. Typically, the tips are spaced 0.3–2 μm apart and are 0.2–1 μm high. Radii of curvature at the tip range from 100 to 1000 Å. High tip fields are achieved by applying a positive voltage difference between the gate electrode and the tips, typically in the range of 50–150 V. After the electrons are extracted from the cathode, these electrons can be formed into a useful beam by the anode electric field and an external magnetic field.

would cause arcing between the electrodes. Consequently, the field enhancement that occurs at a sharpened metallic tip must be utilized to achieve the required electric fields. A typical gated FEA is shown in Fig. 8.5. Semiconductor fabrication techniques are used to form conical emitter tips of semiconductor or metal materials on an appropriate substrate, e.g., doped or undoped silicon. A metal gating electrode is fabricated in close proximity to the tips, typically supported by an insulating layer such as SiO_2. Electrons are extracted from the emitter surface by the extremely high fields that are created at the emission tip by voltage on the gate electrode. The electrons' momenta carry them quickly out of the strong field region where they are captured by the anode field and accelerated away from the field emission structure. In most cases, the anode is much farther away from the tips than the gate electrode, so that the electric field at each tip is primarily determined by the gate-tip voltage V_g. The small size and high initial accelerating field leads to insignificant transit-time effects, as discussed later.

The F–N relation of Eq. (11) applies strictly to only a perfectly flat planar emitter. The effects of the sharply pointed tips in an FEA have been discussed in several publications [53–56]. Nevertheless, Eq. (11) is often used as a basis for characterizing experimental emitters. If the space charge of the emitted electrons is neglected, the field at the tip, E_{tip}, is proportional to V_g, i.e., $E_{tip} = \beta_{tip} V_g$. The field-enhancement factor β_{tip} can be approximated by [57]

$$\beta_{tip} = \frac{R}{kr(R-r)} \tag{13}$$

In Eq. (13), r is the radius of curvature of the tip, R is the tip-gate distance, and k is a constant that typically ranges from 1 to 5, depending on the tip geometry. Neglecting thermal effects, the current I_b emitted from an array is

$$I_b = A V_g^2 \exp(-B/V_g) \tag{14}$$

If the array contains N_{tip} identical tips and the effective area of emission for each tip is A_{tip}, the parameters A and B are given by

$$A = N_{tip} A_{tip} \left(\frac{e^2}{16\pi^2\hbar} \right) \frac{\beta_{tip}^2}{\phi}$$

$$B = \frac{4\sqrt{2me}}{3\hbar} \frac{\phi^{3/2}}{\beta_{tip}} \tag{15}$$

The parameters A and B are usually determined experimentally by the slope and intercept of a plot of $\ln (I/V_g^2)$ vs. $1/V_g$, a plot referred to as an F–N plot [58,59]. The expression for the parameter A in Eq. (15) is of limited value in interpreting experimental data. Theoretically, A_{tip} can depend on V_g [58–60], and individual emitting tips are rarely uniform across an array, as discussed later. However, since the dominant voltage dependence of Eq. (14) is the exponential term, the parameter B, as extracted from an F–N plot, is often used to compare the quality of fabricated FEAs.

Planar fabrication processes have been used to fabricate a variety of FEAs, including arrays with Mo [36–39,44,45,61–65], Si [66–68], and GaAs [69] tips. Arrays having a tip density as large as 10^9 tips/cm^2 have been fabricated [62]. A scanning electron micrograph of such an array is shown in Fig. 8.6. Several research groups

FIGURE 8.6. Scanning electron micrograph of an FEA. The tips are arrayed with a 0.32-μm periodicity, corresponding to a density of 10^9 tips/cm^2. Interferometric laser lithography was used to achieve the small tip sizes. The conical Mo tips are 2000 Å high with tip radii of approximately 100 Å.

have reported gated FEA emission currents that are sufficiently large for application to microwave tubes [36,37,63,64].

The physical emission processes involved in recently fabricated emitters are far more complex than the simple model described earlier. The characteristics of field emitters depend upon a number of environmental factors, such as vacuum quality and the cleanliness of neighboring structures [70–78]. Unfortunately, high-temperature desorption of contaminants is often the only technique that can completely clean the field emitter tips in an array, but this is precluded by the diverse materials used to fabricate an FEA. In addition, field emitter surfaces are not perfectly smooth, as assumed by Eq. (2), but are populated with atomic-scale "nanotips" that further increase the local electric field [79]. A body of evidence [79–84] indicates that emission occurs primarily at a subset of these sites. It appears that the time variation of the constituents and character of this subset can often account for the complex behavior observed in experimental FEAs. These include the lengthy conditioning procedures that are necessary to stabilize and increase emission [82,83], improve noise properties of the emitters [84], and eliminate premature burnout [45,82]. Studies of how emission current and noise scale with array size show that only a small fraction of the tips in an array participates in electron emission [73,85]. Consequently, the actual business of operating emitters involves a number of empirical procedures that are not well understood. The development and adaptation of conditioning procedures for the tube environment is a major challenge in the application of FEAs to IOAs.

8.3.2. Noise Characteristics

A thermionic cathode is usually operated in a space-charge-limited mode, i.e., the emission current is limited by the field associated with the charge in the cathode–anode region, rather than by the ability of the cathode to supply electrons. Consequently, noise waveforms inherent to the cathode emission process do not strongly appear in the electron beam. In contrast, an FEA cathode is not usually operated in a space-charge-limited mode, because the presence of appreciable space charge causes emitted electrons to be reflected to the microfabricated grid. The reflected electrons induce gate current, which degrades both the reliability and operating characteristics of the FEA. In the absence of the stabilizing effect of the space charge, fluctuations in FEA emission current can adversely affect tube performance.

The emission current from FEAs is dominated by burst noise [86,87], which is also called random-telegraph or popcorn noise. Other forms of the ubiquitous "$1/f$ noise" have also been observed [73], as well as shot noise and thermal noise, but they are typically negligible compared to burst noise for frequencies less than ~ 1 kHz. Burst noise consists of current pulses of nearly equal amplitudes (occasionally, the pulse amplitudes are distributed between several discrete levels) that have randomly distributed pulse lengths and times. The physical explanation for burst noise in field emitters is not known with certainty, but it can be caused by any affect that randomly modulates the emission current [86]. Examples of such effects are the appearance and disappearance of nanoprotrusions, field-aided migration of impurities, or adsorption/desorption of gases. If the burst-noise waveform is bistable, and the transitions between levels are assumed to follow Poisson statistics, the noise power spectral density $S_I(\omega)$ is given by [86]

$$S_I(\omega) = \frac{8\nu\Delta I^2}{(4\nu^2 + \omega^2)} \tag{16}$$

In Eq. (16), ΔI is the magnitude of the current pulses and ν is the mean number of transitions per second. Experimentally, the low-frequency noise can be described by a power spectral density of

$$S(f) = \frac{\text{Constant}}{f^\gamma} \tag{17}$$

The quantity γ is referred to as the spectral density index, and is usually between 1 and 2 [44,73,86,88] for field emitters. Baseband flicker noise may get up-converted by the nonlinearities of the tube to appear as skirts at the microwave signal frequency, producing phase noise and adversely affecting its spectral purity. The effects of these noise sources on TWT operation have not been studied experimentally because of the limited life of previous FEA TWT amplifiers. Consequently, the conversion efficiency of the low-frequency flicker noise to the microwave frequency of the amplifier has not been measured.

Workers at Lincoln Laboratory, Massachusetts Institute of Technology, have measured the baseband spectra from FEAs that are comparable to those used in the klystrode tests [85]. Initial tests indicate that γ is typically between 1.7 and 1.9, and that over 99% of the noise power exists below 20 Hz with over 90% concentrated below 2 Hz. Similar results have also been found in another study using Si emitters [73]. The low-frequency nature of the noise power bodes well for the use of FEAs in microwave power tubes. For example, a 10-GHz carrier signal should not be broadened by more than 20 Hz, which should not interfere with any practical voice-communication or radar application (including most Doppler systems).

8.3.3. Modulation of Gated FEAs

To achieve acceptable gain in an IOA, it must be possible to modulate the emission from a gated cathode with a low power input signal. This section will discuss a number of important factors that influence the suitability of a gated cathode for high-frequency modulation. The requirements for tube operation, the physical and practical considerations that limit the emission current, the beam quality that can be achieved with such emitters, and the reliability of such emitters will be discussed.

8.3.3.1. Transit Time. The performance of a gated cathode will degrade if the rf fields experienced by an electron change appreciably during its transit from the emitting surface to the gate–anode region. In this context, "transit time" refers to the time that an electron spends under the influence of the electric field between the cathode and gating structure. In gridded thermionic cathodes, it is the time for an electron to reach the plane of the grid, while in FEAs it is the time for an electron to reach the gate potential. In a gridded thermionic cathode, the dc bias voltage on the grid is usually negative with respect to the cathode in order to suppress the extraction of thermally emitted electrons by the anode electric field. The grid polarity is rarely positive because emitted current will then be intercepted by the grid, which would unacceptably load the input circuit and damage the grid at high power densities. Consequently, the electric field that accelerates electrons away from a thermionic cathode is relatively small, and in fact must be negative for part of each rf cycle in Class C operation. In contrast, the strong electric fields at the emitting surface of an FEA accelerate emitted electrons to high velocity immediately upon emission. In addition, the gate electrode is approximately co-planar with the emitting tip, so that the electron passes from the influence of the oscillating gate potential into that of the anode static field in a short distance.

Emission-gated cathodes offer the most dramatic performance advantages in Class C operation. Under these conditions, the accurate determination of the limitations imposed by transit-time effects requires simulations of 2-D electron trajectories that include time-varying space charge and electrons returning to the emitting surface. However, transit-time effects in thermionic and field emission cathodes can be roughly compared by focusing on the gross distinctions between the two structures. The gate voltage of a field emitter, V_g, modulates the current by causing electron emission, while the grid voltage of a thermionic emitter, V_{gr}, modulates the current by suppressing electron extraction from the thermally emitted cloud on the cathode surface. For a space-charge-limited thermionic cathode with an ideal grid (an ideal grid is a thin, perfectly conducting sheet that intercepts no current), the limiting current density J_{CL} is determined by the Child–Langmuir law [89,90]:

$$J_{CL} = \frac{4\varepsilon_0}{9}\left(\frac{2e}{m}\right)^{1/2}\frac{V_{gr}^{3/2}}{d^2} \tag{18}$$

In Eq. (18), d is the cathode-to-grid separation. Consequently, the ratio of the full-on voltage V_{gr}^+ to cut-off voltage V_{gr}^- required for a ratio of full-on current I^+ to cut-off current I^- of $I^+/I^- = 1000$ is

$$\frac{V_{gr}^+}{V_{gr}^-} = \left(\frac{I^+}{I^-}\right)^{2/3} = 100 \tag{19}$$

Thus, the electrons emitted near cutoff depart the cathode surface with only 1% of the acceleration of electrons emitted near full-on conditions.

In contrast, for a field emitter in which the current is given by Eq. (14), the ratio of the currents is

$$\frac{I^+}{I^-} = \left(\frac{V_g^+}{V_g^-}\right)^2 \exp\left[-\frac{B}{V_g^+}\left(1-\frac{V_g^+}{V_g^-}\right)\right] \tag{20}$$

For $B = 750$ V and $V_g^+ = 75$ V, the reasonable values for today's field emitters, Eq. (20) yields $V_g^+/V_g^- = 1.6$. The field that accelerates electrons that are emitted near cutoff is over 60% of that at full-on conditions. This simple example shows why field emission is inherently better adapted to Class C amplifiers than thermionic emission; no field emitted electron can linger in the time-varying electric field of the gate.

In addition to the differing cut-off conditions of thermionic and field emission cathodes, the transit time under full-on conditions differs substantially as well. The field between the gate (grid) and the emitting surface is approximately constant and equal to the potential change divided by the gate (grid)-cathode distance. For the thermionic cathode, the electric field in the cathode-to-grid region is sufficient to extract the required current density, as determined by the Child-Langmuir law of Eq. (18). Using $E = V_{gr}/d$ in Eq. (18) and solving for E gives

$$E = \left[\left(\frac{9}{4} \frac{J_{CL}}{\varepsilon_0} \right)^2 \frac{dm}{2e} \right]^{1/3} \tag{21}$$

For $J_{CL} = 2\,\text{A/cm}^2$ and $d = 250\,\mu\text{m}$, Eq. (21) predicts that $E = 264\,\text{kV/cm}$. The resulting transit time τ is

$$\tau = \int_0^d \frac{dz}{v(z)} = \sqrt{\frac{m}{2e}} \int_0^d \frac{dz}{\sqrt{\phi(z)}} = \sqrt{\frac{2dm}{eE}} = 100\,\text{ps} \tag{22}$$

This corresponds to a cutoff frequency $f_c = 1/2\pi\tau \cong 1.6\,\text{GHz}$. As the gate voltage declines toward cut-off, the transit time approaches infinity, resulting in the return of some electrons to the cathode.

Jensen [91] derived the potential on the axis of symmetry for a gated field emitter with an anode:

$$V(z) = V_g \frac{E_{tip}z}{V_g + E_{tip}z} \left(1 + \frac{E_0 z}{V_g} \right) \tag{23}$$

In Eq. (23), E_{tip} is the field at the emitter tip on its center axis, and E_0 is the background field due to the anode. An emitted electron can be significantly influenced by the gate when $V(z) < V_g$. Solving Eq. (23) for $V(z) = V_g$ yields an upper bound to the extent of the control region, $z_g = V_g/\sqrt{E_{tip}E_0}$. Since collisions can be neglected, the electron velocity $v(z)$ is determined by the electrostatic potential $\phi(z)$ as

$$\frac{1}{2}mv^2(z) = e\phi(z) = eV_g \frac{E_{tip}z}{V_g + E_{tip}z} \left(1 + \frac{E_0 z}{V_g} \right) \tag{24}$$

The electron velocity is

$$v(z) = \sqrt{\frac{2eV_g}{m}} \sqrt{\left(\frac{E_{tip}z}{V_g + E_{tip}z} \right) \left(1 + \frac{E_0 z}{V_g} \right)} \tag{25}$$

Then, the transit time is

$$\tau = \int_0^{z_g} \frac{dz}{v(z)} = \sqrt{\frac{m V_g}{2e E_{\text{tip}} E_0}} \int_0^1 \sqrt{\frac{u + \sqrt{E_0/E_{\text{tip}}}}{u(1 + \sqrt{E_0/E_{\text{tip}}}u)}}\, du \cong \sqrt{\frac{m V_g}{2e E_{\text{tip}} E_0}} \tag{26}$$

In Eq. (26), $\sqrt{E_0/E_{\text{tip}}} \ll 1$ has been used in approximating the integral. In a typical FEA, a gate voltage of 75 V produces a tip field of 0.5 V/Å. The anode field must be large enough to draw all of the field emitted current away from the grid, yet small enough to avoid arc breakdown. A value of 20 kV/cm is reasonable for moderate emission currents; this is much higher than for thermionic emission because of the very high local current densities obtained from FEAs. The transit time is then $\tau = 0.15$ ps, which is nearly 3 orders of magnitude shorter than the thermionic case and corresponds to $f_c \cong 1000$ GHz.

8.3.3.2. Input Impedance.

Although the close spacing of the gate and cathode diminishes the transit time, it increases the grid-cathode capacitance. Further, the gate-cathode region constitutes a distributed transmission line, as depicted in Fig. 8.7. Calame [92] provided a detailed analysis of the voltage distribution within the FEA and the input impedance presented by the FEA. A simplified versions is given here.

The array is assumed to be composed of cells that repeat with periodicity a. The gate capacitance of each repeat cell arises from the capacitance through the gate insulator, C_{pc}, and the gate-tip capacitance C_{tc}, as shown in Fig. 8.7(a). If the extent of the array in the direction of propagation (the z direction, hereafter called the length) is l and the array width is w, the capacitance per unit length, C, is

$$C = \frac{w(C_{\text{tc}} + C_{\text{pc}})}{a^2} \tag{27}$$

If the effects of the gate-tip holes are neglected (Calame includes these effects and shows them to be small), the resistance per unit length is $R = \rho_g/wt$, where ρ_g and t are the resistivity and thickness of the gate metal, respectively. Using a TEM transmission-line approximation [93], the inductance per unit length is $L = \mu_0 h/w$, where μ_0 is the permeability of free space and h is the gate-insulator thickness.

Consider the gate to be excited by the superposition of a dc voltage V_g^{dc} and a sinusoidal rf gate voltage V_g^{rf}. Using the equivalent transmission line of Fig. 8.7(b), the rf gate current $I_g^{\text{rf}}(z, t) = \text{Re}[\tilde{I}_g^{\text{rf}}(z) e^{j\omega t}]$ and the rf gate voltage $V_g^{\text{rf}}(z, t) = \text{Re}[\tilde{V}_g^{\text{rf}}(z) e^{j\omega t}]$ are determined by the transmission-line equations:

$$\frac{\partial \tilde{I}_g^{\text{rf}}(z)}{\partial z} = -j\omega C \tilde{V}_g^{\text{rf}}(z)$$

$$\frac{\partial \tilde{V}_g^{\text{rf}}(z)}{\partial x} = -(R + j\omega L)\tilde{I}_g^{\text{rf}}(z) \tag{28}$$

Solving Eq. (28) subject to the boundary condition that an open circuit exists at

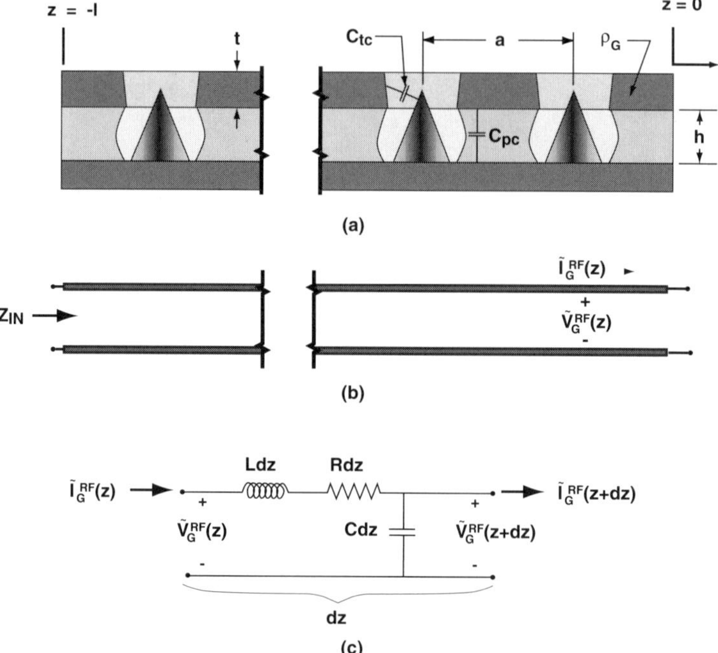

FIGURE 8.7. Transmission-line effects for a gated FEA. (a) The relevant parameters of a gated-FEA input circuit; (b) the equivalent transmission line; (c) the incremental transmission line used to calculate the gate-voltage distribution.

$z = 0 \, [\tilde{I}_{\mathrm{g}}^{\mathrm{rf}}(0) = 0]$ gives

$$\tilde{V}_{\mathrm{g}}^{\mathrm{rf}}(z) = V_0 \, \cos(\beta z)$$

$$\tilde{I}_{\mathrm{g}}^{\mathrm{rf}}(z) = \frac{V_0}{j Z_0} \, \sin(\beta z) \tag{29}$$

where

$$\beta = \omega\sqrt{LC}\left(1 - j\frac{R}{\omega L}\right)^{1/2} = \frac{\omega}{a}\sqrt{\mu_0 h (C_{\mathrm{pc}} + C_{\mathrm{tc}})}\left(1 - \frac{j}{\omega}\frac{\rho_{\mathrm{g}}}{\mu_0 h}\right)^{1/2}$$

$$Z_0 = \omega\sqrt{\frac{L}{C}}\left(1 - j\frac{R}{\omega L}\right)^{1/2} = \frac{a}{w}\sqrt{\frac{\mu_0 h}{C_{\mathrm{pc}} + C_{\mathrm{tc}}}}\left(1 - \frac{j}{\omega}\frac{\rho_{\mathrm{g}}}{\mu_0 h}\right)^{1/2}$$

Thus, the rf input impedance of the gate Z_{in}, is

$$Z_{\mathrm{in}} = \frac{\tilde{V}_{\mathrm{g}}^{\mathrm{rf}}(-l)}{\tilde{I}_{\mathrm{g}}^{\mathrm{rf}}(-l)} = -j Z_0 \, \cot(\beta l) \tag{30}$$

For $|\beta l| \ll 1$, the cotangent function can be expanded, and

$$Z_{\text{in}} = \frac{Rl}{3} + \frac{1}{j\omega(Cl)} + j\omega\frac{Ll}{3} = \left(\frac{\rho_g}{3t}\right)\left(\frac{l}{w}\right)$$

$$+ \frac{1}{j\omega N_{\text{tip}}(C_{\text{tc}} + C_{\text{pc}})} + j\omega\left(\frac{\mu_0 h}{3}\right)\left(\frac{l}{w}\right) \tag{31}$$

Here, the quantity N_{tip} is the total number of tips in the array.

Eq. (29) shows that each tip does not experience the same gate–tip voltage. To examine the effects of the gate-voltage distribution on the emission current, consider the beam current to be composed of a dc component I_b^{dc} and a small rf component $I_b^{\text{rf}}(t)$, so that $I_b(t) = I_b^{\text{dc}} + I_b^{\text{rf}}(t)$. If an individual tip emits current $I_{\text{tip}}(V_g)$ at a gate voltage V_g, the current emitted per unit length, $K_b(z, t)$, is given by

$$K_b(z, t) = \frac{w}{a^2} I_{\text{tip}}\left[V_g^{\text{dc}} + V_g^{\text{rf}}(z, t)\right] \cong \frac{w}{a^2}\left[I_{\text{tip}}\left(V_g^{\text{dc}}\right) + g_{\text{m tip}} V_g^{\text{rf}}(z, t)\right] \tag{32}$$

On the right-hand side of Eq. (32), the approximation is valid only in small-signal conditions, and the transconductance per tip, $g_{\text{m tip}}$, is given by

$$g_{\text{m tip}} = \left.\frac{\partial I_{\text{tip}}}{\partial V_g}\right|_{V_g^{\text{dc}}} \tag{33}$$

For small-signal sinusoidal excitation, the rf beam current is

$$I_b^{\text{rf}}(t) = \int_{-l}^{0} K_b^{\text{rf}}(z, t)\, dz = \frac{w g_{\text{m tip}}}{a^2} \text{Re}\left[e^{j\omega t}\int_{-l}^{0} \tilde{V}_g^{\text{rf}}(z)\, dz\right] \tag{34}$$

Using Eq. (29) and writing $I_b^{\text{rf}}(t) = \text{Re}(\tilde{I}_b^{\text{rf}} e^{j\omega t})$,

$$\tilde{I}_b^{\text{rf}} = \frac{w g_{\text{m tip}}}{a^2} \int_{-l}^{0} \tilde{V}_g^{\text{rf}}(z)\, dz = N_{\text{tip}}\, g_{\text{m tip}}\frac{\sin(\beta l)}{\beta l} \tag{35}$$

Thus, the reduction of the transconductance by the nonuniform gate voltage is expressed by the term $\sin(\beta l)/\beta l$, implying that the entire array will not be effectively modulated unless $|\beta l| \ll 1$.

The small dimensions of an FEA, together with the high emission current required by a microwave tube, often result in input impedance much lower than 50 Ω. Within a factor of 2, the total capacitance per cell, $C = C_{\text{pc}} + C_{\text{tc}}$, may be estimated as the parallel-plate capacitance $C = \varepsilon a^2 / h$. Suppose that a total emission current of 100 mA is required from an FEA, for which $a = 1\,\mu\text{m}$, $h = 1\,\mu\text{m}$, and $I_{\text{tip}} = 1\,\mu\text{A}$. Neglecting resistive losses, $\beta = 4.2\,\text{cm}^{-1}$, which implies that l must be less than 250 μm for $\beta l < 0.1$. In order to emit 100 mA at 1 μA/tip, 10^5 tips ($N_{\text{tip}} = 10^5$) must be used. For $l = 250\,\mu\text{m}$, the array width $w = N_{\text{tip}}a^2/l$ must be at least $400\,\mu\text{m}$. Using Eq. (30), the input reactance is then approximately $5\,\Omega$ at 10 GHz. Consequently, an impedance-matching network must be inserted between the power source and FEA to efficiently couple the power to the FEA, as shown in Fig. 8.8(a).

Impedance-matching considerations are important because they can affect the FEA design and packaging techniques. In the equivalent circuit of Fig. 8.8(b), the source is represented by a conductance $g_s = 1/r_s$, and the FEA is represented by a series connection of a resistor r_L and capacitor c_L. By Poynting's Theorem [94], the input admittance of the matching circuit, Y_1, is given by

$$Y_1 = \frac{2P_d + 4j\omega(\langle W_e\rangle - \langle W_m\rangle)}{\left|\tilde{V}_1^{\text{rf}}\right|^2} \tag{36}$$

In Eq. (36), P_d is the power dissipation, and $\langle W_m\rangle$ and $\langle W_e\rangle$ are the average magnetic and electric energies, respectively. For optimum power transfer, $Y_1 = g_s$, so that $\langle W_m\rangle = \langle W_e\rangle$, i.e., the circuit is resonant. Near resonance, the circuit can be

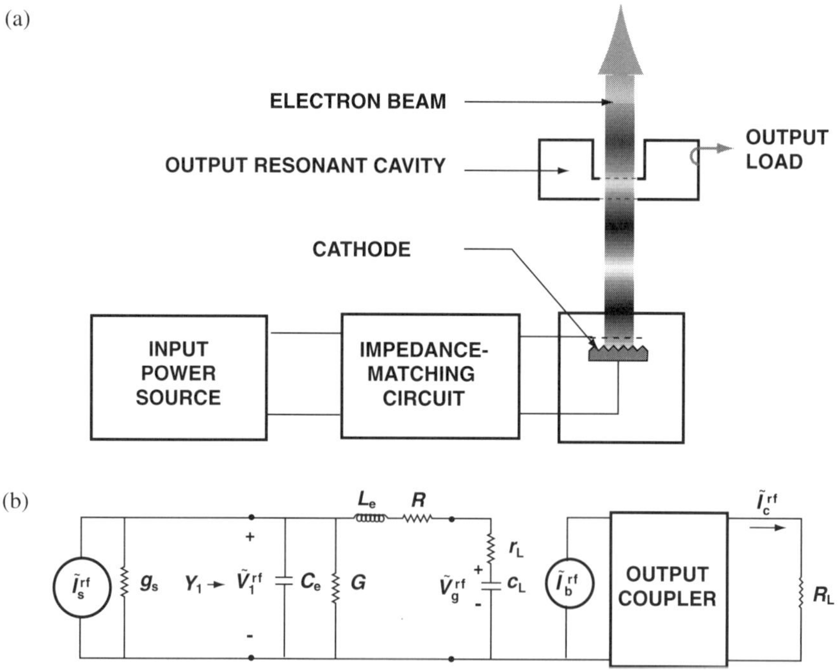

FIGURE 8.8. Input circuit of an FEA. (a) A power source is coupled to the gate of an FEA by an impedance-matching network; (b) an equivalent circuit for the input networks. The impedance-matching network is a resonant circuit that is resonant at the operating frequency. The resonant circuit is characterized by circuit elements C_e, G, L_e, and R.

approximated by an effective inductance L_e and capacitance C_e, as shown in Fig. 8.8(b). The values of L_e and C_e are chosen so that resonance occurs at the design frequency ω, presenting a parallel resonance at port 1 and a series resonance at port 2. A parallel conductance G and a series resistance R are added to L_e and C_e, respectively, to represent losses in the matching circuit. For simplicity, it will be assumed that

$$\frac{\omega L_e}{R} = \frac{\omega C_e}{G} = Q \tag{37}$$

The quantity Q is the quality factor of the matching circuit. Circuit analysis yields the ratio of the rf output power P_O^{rf} to the rf power available from the source, P_A^{rf}

$$\frac{P_O^{rf}}{P_A^{rf}} = \frac{4 r_L g_s}{[1-\chi_c + (g_s + G)]^2 + \chi_c \left[\dfrac{(r_L + R)}{\eta} + \eta(g_s + G) \right]^2} \tag{38}$$

In Eq. (38),

$$\chi_c = \omega^2 L_e C_e \left(1 - \frac{1}{\omega^2 L_e C_L} \right)$$

$$\eta = \sqrt{\frac{L_e}{C_e} \left(1 - \frac{1}{\omega^2 L_e C_L} \right)} \tag{39}$$

The maximum value of P_O^{rf} as a function of ωL_e is

$$\left(\frac{P_O}{P_A} \right)_{max} = \left(\frac{r_L}{r_L + R} \right) \left(\frac{g_s}{g_s + G} \right) \tag{40}$$

and occurs when

$$\omega L_e = \sqrt{\frac{r_L + R}{g_s + G}} \sqrt{1 - (r_L + R)(g_s + G)} + \frac{1}{\omega C_L} \cong \sqrt{\frac{r_L}{g_s}} + \frac{1}{\omega C_L}$$

$$\omega C_e = \sqrt{\frac{g_s + G}{r_L + R}} \sqrt{1 - (r_L + R)(g_s + G)} \cong \sqrt{\frac{g_s}{r_L}} \tag{41}$$

It is clear from Eq. (40) that r_L must be much larger than R to avoid power loss by the matching circuit. An estimate of R and G can be obtained by using Eqs. (37)

and (41), which gives

$$R = \frac{\omega L_e}{Q} \cong \frac{1}{Q}\left(\sqrt{\frac{r_L}{g_s}} + \frac{1}{\omega c_L}\right)$$

$$G = \frac{\omega C_e}{Q} \cong \frac{1}{Q}\sqrt{\frac{g_s}{r_L}} \tag{42}$$

Defining the load quality factor $Q_L = 1/\omega r_L c_L$, Eq. (41) gives a condition for efficient impedance matching

$$r_L \gg R \cong \frac{1}{Q}\left(\sqrt{r_L r_s} + \frac{1}{\omega c_L}\right) \quad \text{or} \quad Q \gg \sqrt{\frac{r_s}{r_L}} + Q_L \tag{43}$$

The matching circuit must be designed so that Eq. (43) is not violated. Alternately, the FEA designer, faced with unavoidable circuit losses, must design both the FEA and the FEA packaging with Eq. (43) in mind. The matching circuit can be realized in a variety of ways. Stub transmission lines near the emitting area can add the shunt inductance needed to match the capacitance of the FEA. For narrowband operation, a quarter-wave impedance transformer [93] can be used. Lumped-element circuits are often more compact, but suffer from low quality factors.

The emission current is modulated by the rf voltage that is applied to the gate $\tilde{V}_g^{\mathrm{rf}}$, as shown in Fig. 8.8(b). This voltage is given by

$$\left|\tilde{V}_g^{\mathrm{rf}}\right|^2 = \frac{2 P_O^{\mathrm{rf}}}{r_L(\omega c_L)^2} = \frac{2 g_s P_A^{\mathrm{rf}}}{(\omega c_L)^2 (g_s + G)(r_L + R)} \cong \frac{2 Q_L}{\omega c_L} P_A^{\mathrm{rf}} \tag{44}$$

As an example, at 10 GHz each quadrant of the Lincoln Laboratory FEA cathode [64] has input impedance $Z_{\mathrm{FEA}} = 2.5 - j12\ \Omega$ and requires a peak-to-peak voltage of approximately 20 V ($|\tilde{V}_g^{\mathrm{rf}}| = 10$ V) to modulate the emission current. The required power, assuming lossless matching, is

$$P_A^{\mathrm{rf}} = \frac{\omega c_L \left|\tilde{V}_g^{\mathrm{rf}}\right|^2}{2 Q_L} = \frac{100}{2 \times 12.5 \times (12.5/2)} = 0.6\ \mathrm{W}$$

8.3.3.3. Beam-Current Modulation.

In the absence of transit time delays, the modulated beam current $I_b(t)$ of a gated cathode is given by substituting the gate voltage $V_g(t)$ into the current–voltage relation of the cathode. In the case of a gated FEA, the voltage modulation is usually sinusoidal, and the current–voltage relation of an FEA is taken to be the F–N relation of Eq. (14). Because the characteristic curve is nonlinear, the resulting beam-current waveform will include harmonic frequencies. Computer simulations must be used to exactly obtain the emission current modulation that results from a given gate voltage modulation. However, an approximate analysis, coupled with Eq. (44), can be used to estimate the beam-current modulation produced by the FEA. The gate voltage is assumed to be

$$V_g(t) = V_g^{dc} - V_g^{rf} \cos(\omega t) \tag{45}$$

Then, defining $\chi = V_g^{rf}/V_g^{dc}$ and using Eq. (14), the emission current is

$$I_b(t) = A\left[V_g^{dc} - V_g^{rf} \cos(\omega t)\right]^2 \exp\left[-\frac{B}{V_g^{dc} - V_g^{rf} \cos(\omega t)}\right]$$

$$\cong A\left(V_g^{dc}\right)^2 \exp\left[-\frac{B}{V_g^{dc}[1 - \chi \cos(\omega t)]}\right] \tag{46}$$

By Fourier analysis,

$$\frac{1}{1 - \chi \cos(\omega t)} \cong 1 + \frac{2\chi}{1 + \sqrt{1-\chi^2}} \cos(\omega t) \tag{47}$$

If Eq. (47) is inserted into Eq. (46), the emission current can be expressed in terms of the fundamental and harmonic frequencies as

$$I_b(t) \cong A\left(V_g^{dc}\right)^2\left[I_0(\delta) + 2\sum_{k=1}^{\infty} I_k(\delta) \cos(\omega t)\right] \tag{48}$$

In Eq. (48),

$$\delta = \frac{2B}{V_g^{dc}\sqrt{1-\chi^2}}\left(\frac{\chi}{1 + \sqrt{1-\chi^2}}\right),$$

and $I_k(z)$ is the modified Bessel function of the first kind. The identity

$$e^{z\cos(\theta)} = I_0(z) + 2\sum_{k=1}^{\infty} I_k(z) \cos(k\theta)$$

has been used [95]. The appropriate modulation will depend on the application: a frequency multiplier will require a more strongly modulated beam than a linear amplifier. Eq. (48) is a good indicator of the fraction of the beam energy that can be converted to electromagnetic energy in the fundamental frequency, provided that the inductive output circuit only extracts power from the beam to the circuit. If the output circuit is lengthened to increase the modulation of the beam before extraction begins, space-charge effects and nonlinear interactions between the beam and the inductive output can result in conversion of power betwen the harmonics [96–99]. Writing $I_b(t) = I_b^{dc} + \text{Re}[\tilde{I}_b^{rf} e^{j\omega t}]$, the dc and rf components of the beam current are given by

$$I_{\mathrm{b}}^{\mathrm{dc}} = A\left(V_{\mathrm{g}}^{\mathrm{dc}}\right)^2 \exp\left[\frac{B}{V_{\mathrm{g}}^{\mathrm{dc}}\sqrt{1-\chi^2}}\right] I_0(\delta)$$

$$\left|\tilde{I}_{\mathrm{b}}^{\mathrm{rf}}\right| \cong 2A\left(V_{\mathrm{g}}^{\mathrm{dc}}\right)^2 \exp\left[\frac{B}{V_{\mathrm{g}}^{\mathrm{dc}}\sqrt{1-\chi^2}}\right] I_1(\delta) = 2I_{\mathrm{b}}^{\mathrm{dc}}\frac{I_1(\delta)}{I_0(\delta)}$$

$$(49)$$

The values of χ and δ can be estimated for optimal impedance matching, using Eq. (44).

The transconductance of a voltage-controlled current source is another indicator of the efficiency by which gate-voltage rf modulation is converted to emission-current rf modulation [99]. It is defined as the incremental change in beam current divided by the incremental change in gate potential, $g_m = \partial I_{\mathrm{b}}/\partial V_{\mathrm{g}}$. In the absence of transit-time effects, the transconductance is the slope of the characteristic curve $I_{\mathrm{b}}(V_{\mathrm{g}})$; if the characteristic curve is nonlinear, the transconductance will depend upon V_{g}. The transconductance of a gated FEA is thus

$$g_m = \frac{\partial I_{\mathrm{b}}}{\partial V_{\mathrm{g}}} = AV_{\mathrm{g}}\left(2 + \frac{B}{V_{\mathrm{g}}}\right)\exp\left(-\frac{B}{V_{\mathrm{g}}}\right) = \left(2 + \frac{B}{V_{\mathrm{g}}}\right)\frac{I_{\mathrm{B}}}{V_{\mathrm{g}}} \qquad (50)$$

This transconductance, like the current, is exponentially sensitive to the F–N B parameter. To relate cathode performance to the gain of an IOA, a generalized transconductance α may be defined as the incremental rf current that results for an increment in rf gate-drive power, i.e., $\alpha = \partial|\tilde{I}_{\mathrm{b}}^{\mathrm{rf}}|/\partial P_{\mathrm{g}}^{\mathrm{rf}}$. Since the application of an oscillating potential to the gate performs the emission gating, α is related to g_m as

$$\alpha = \frac{\partial\left|\tilde{I}_{\mathrm{b}}^{\mathrm{rf}}\right|}{\partial P_{\mathrm{g}}^{\mathrm{rf}}} = \frac{\partial\left|\tilde{I}_{\mathrm{b}}^{\mathrm{rf}}\right|}{\partial V_{\mathrm{g}}} \times \frac{\partial V_{\mathrm{g}}}{\partial V_{\mathrm{g}}^{\mathrm{rf}}} \times \frac{\partial V_{\mathrm{g}}^{\mathrm{rf}}}{\partial P_{\mathrm{g}}^{\mathrm{rf}}} = g_m \times 1 \times \frac{\partial V_{\mathrm{g}}^{\mathrm{rf}}}{\partial P_{\mathrm{g}}^{\mathrm{rf}}} = \frac{2Q_{\mathrm{L}}}{\omega c_{\mathrm{L}}}P_{\mathrm{A}}^{\mathrm{rf}}g_m \qquad (51)$$

In Eq. (51), Eq. (44) has been used to relate $V_{\mathrm{g}}^{\mathrm{rf}}$ to $P_{\mathrm{A}}^{\mathrm{rf}}$. This, of course, assumes optimal impedance matching. More generally, the relation between the drive power and the rf voltage at the gate depends upon the input circuit as discussed earlier.

8.3.4. Current Density

In most cases, both emission current and current density are limited by reliability considerations. These are discussed below. However, fundamental limits apply to the current that can be obtained from field emitter cathodes.

8.3.4.1. Space Charge. As emission current from the cathode increases, the reduction of the field near the cathode by the space charge of the emitted electrons can no longer be neglected. For an FEA diode, the space-charge-limited current density is described by the Langmuir-Child law of Eq. (18) and is determined by the reduction of the extraction field at the tip by the space charge of the emitted electrons [100,101]. However, for a gated field emitter, space charge does not greatly diminish the tip field,

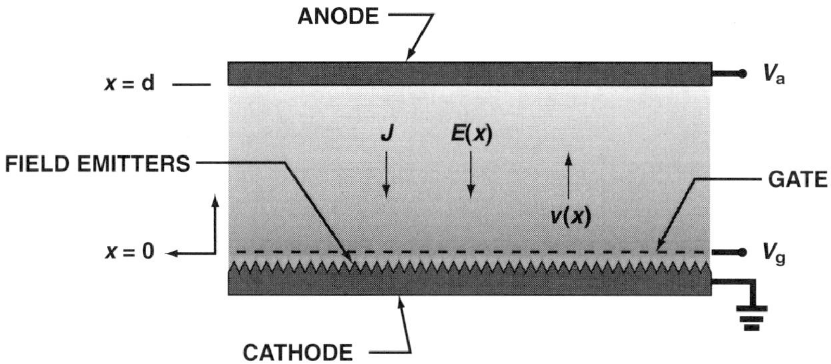

FIGURE 8.9. Simple model used to estimate space-charge effects in a gated-FEA cathode.

but rather, gives rise to large gate current. Because of the high current density that is required, such space-charge effects must be considered in any microwave tube design.

The one-dimensional analysis of Lau et al. [102] can be extended to provide some insight into the nature of these effects. A gated FEA with dc voltages V_g and V_a applied to the gate and anode, respectively, is depicted in Fig. 8.9. To minimize confusion, the polarities of the current density J, electric field $E(x)$, and electron velocity $v(x)$ are defined to be positive for electron flow from the cathode to the anode. In the gate-anode region ($0 < x < d$), Poisson's equation relates the electrostatic potential $\phi(x)$ to the charge density $\rho(x)$:

$$\frac{d^2\phi(x)}{dx^2} = \frac{\rho(x)}{\varepsilon_0} \tag{52}$$

For static conditions, the current density J is independent of x and is given by

$$J = \rho(x)v(x) \tag{53}$$

Because field emitters operate in UHV, any electron collisions with gaseous molecules can be neglected, so that $v(x)$ is given by

$$\frac{m}{2}v^2(x) = e\phi(x) \tag{54}$$

Using the polarity definitions of Fig. 8.9, the electric field $E(x)$ is given as

$$E(x) = \frac{d\phi(x)}{dx} = \frac{m}{e}\frac{dv}{dt} \tag{55}$$

Differentiating Eq. (55),

$$\frac{d^2\phi}{dx^2} = \frac{m}{e}\frac{d}{dx}\left(\frac{dv}{dt}\right) = \frac{m}{e}\frac{d}{dt}\left(\frac{dv}{dt}\right)\frac{dt}{dx} = \frac{m}{ev}\frac{d^2v}{dt^2} \tag{56}$$

In view of Eqs. (52) and (53), Eq. (56) becomes

$$\frac{d^2v}{dt^2} = \frac{eJ}{\varepsilon_0 m} \tag{57}$$

The emitted electrons are assumed incident upon the gate-anode region with a velocity derived from the gate voltage, i.e.,

$$v(0) = \sqrt{\frac{2eV_g}{m}} \tag{58}$$

Defining $t = 0$ at $x = 0$, and solving Eq. (58) for $v(t)$ and $x(t)$,

$$
\begin{aligned}
v(t) &= \frac{eJ}{2m\varepsilon_0}t^2 + \frac{eE_s}{m}t + \sqrt{\frac{2eV_g}{m}} \\
x(t) &= \frac{eJ}{6m\varepsilon_0}t^3 + \frac{eE_s}{2m}t^2 + \sqrt{\frac{2eV_g}{m}}t
\end{aligned}
\tag{59}
$$

In Eq. (59), E_s is the electric field at $x = 0$. If the emitted electrons reach the anode at time T, then at $t = T$ Eq. (59) becomes

$$
\begin{aligned}
\sqrt{\frac{2eV_a}{m}} &= \frac{eJ}{2m\varepsilon_0}T^2 + \frac{eE_s}{m}T + \sqrt{\frac{2eV_g}{m}} \\
d &= \frac{eJ}{6m\varepsilon_0}T^3 + \frac{eE_s}{m}T^2 + \sqrt{\frac{2eV_g}{m}}T
\end{aligned}
\tag{60}
$$

In a gated FEA, J is determined by V_g through Eq. (14) and is thus a given quantity. As a result, Eq. (60) determines E_s and T as a function of V_g, V_a, J, and d. This allows $x(t)$ and $\phi[x(t)]$ to be determined from Eqs. (59) and (54). Figure 8.10 displays the dependence of ϕ upon x for several values of emission current, using the FEA parameters of Fig. 8.11. As J increases, $E_s = d\phi/dx$ diminishes until, analogous to the Child–Langmuir law, $E_s = 0$ at current density J_L given by

$$J_L = J_{CL}\left(1 - \sqrt{\frac{V_g}{V_a}}\right)\left(1 + 2\sqrt{\frac{V_g}{V_a}}\right)^2 \tag{61}$$

The quantity J_L only roughly estimates the upper limit to the current density. A number of important factors have been neglected in this simple analysis, including the two-dimensional (2-D) geometry of the FEA and the spreading of the emitted-electron beam. In most cases, numerical simulations [103] must be used to accurately determine these effects. More importantly, the redirection of the emission current from the anode to the gate electrode occurs at current densities lower than J_L. The resulting high gate current degrades FEA performance and enhances failure probability, as discussed later. Figure 8.11 shows experimental data from a 6100-tip array that was tested in a

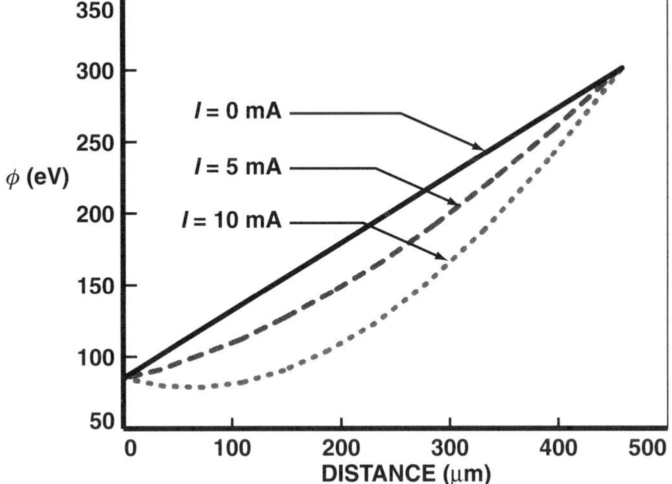

FIGURE 8.10. Calculated dependence of the electrostatic potential in the gate–anode region for several values of emitted current. As the emission current increases, the electric field near the cathode surface diminishes. At sufficiently high emission current, increased gate current results.

UHV probing apparatus [91]. The FEA was approximately 25×25 μm in area and the probe anode was spaced about 18 mil (0.46 mm) from the FEA. Figure 8.11(a) shows how anode current saturates as a result of space-charge effects. As the anode voltage is increased, higher values of emitted current can be achieved, as is indicated by Eq. (61). Figure 8.11(b) shows F–N plots of the same data and includes the gate

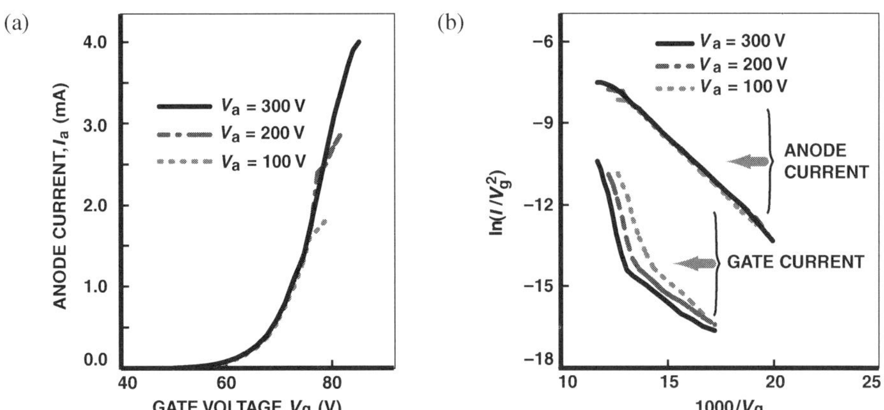

FIGURE 8.11. Experimental manifestation of space-charge effects. (a) Anode current vs. gate voltage, showing the saturation of anode current at high emission levels because of the space charge. (b) F–N plots of gate and anode current, showing the increase of gate current that accompanies anode-current saturation [91].

current. The departure of the anode current from a F–N dependence and the accompanying gate-current increase is evident.

Since gain depends so strongly upon minimizing the gate-to-cathode capacitance, a small-area source that is operating near peak intensity will generally provide the best simultaneous gain and efficiency. Since efficiency is also improved when the bulk of the beam passes close to the output circuit electrodes, the optimum electron beam geometry is a thin annulus. Because such compact annular beams improve the performance of rf output couplers, it is advantageous to draw the maximum current density consistent with a reasonable cathode lifetime. This raises issues in electron gun design, including initial velocity effects, beam spreading, axial demodulation, and beam stability. All are of concern in a design context. Electron guns for inductive output amplifiers should be designed to exploit cathodes such as FEAs that are capable of emitting hundreds of amperes per square centimeter.

8.3.4.2. Beam Quality. FEAs emit current from sharply pointed cones or pyramids. Although the emitting tips are sharp, the radius of curvature is finite, typically ranging from 10 to 100 nm. Electrons are emitted from the sides as well as the tops of the emitter tips, resulting in an angular distribution of the emitted current. Jensen [104] has applied the F–N equation for the current density to a field emitter that is approximated by hyperbolic surfaces and surrounded by a co-planar anode, as shown in Fig. 8.12. The electron distribution as a function of emission angle θ_g was calculated, and the rms average angle of emission from a single tip, θ_{rms}, was approximated by

$$\theta_{rms} \equiv \sqrt{\langle \theta_g^2 \rangle} = \sqrt{\frac{1}{2}\left[1 - \frac{I_1(\chi_g)}{I_0(\chi_g)}\right]} \tag{62}$$

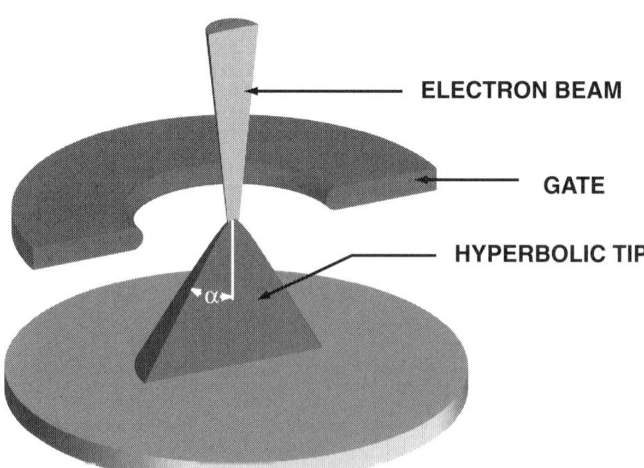

FIGURE 8.12. Field emitter modeled using the surfaces of a hyperbolic coordinate system. Both calculations and experiments show that the angular spread of the emitted electron beam is approximately 20°.

In Eq. (62), $I_0(x)$ and $I_1(x)$ are modified Bessel functions of the first kind, and

$$\chi_g = 1 + \frac{B}{2V_g} + \frac{1}{2}\sin^2 \alpha \tag{63}$$

where B and V_g are defined by Eq. (14), and α is the half-angle of the conical emitter tips.

Experimental measurements of single-tip emitters with $\theta_{rms} = 20°$ correlated well with the theory. The mean transverse energy is given in terms of θ_{rms} and the gate potential by

$$E_\perp = eV_g \frac{\sin^2 \theta_{rms}}{1 + \sin^2 \theta_{rms}} \tag{64}$$

Thus, for example, electrons emitted from a tip having a mean angle of emission of $20°$ have a mean transverse energy of $0.1V_g$, which implies $E_\perp \cong 5\,\mathrm{eV}$ for contemporary submicron FEAs. This transverse energy is irreducibly introduced into the electron beam [105].

For the linear-beam amplifiers considered here, the angular emission of the FEA becomes more problematic at frequencies above X band. At these wavelengths, the small size of the output circuit demands a small beam diameter, and low beam voltage is required because small circuits are not able to dissipate as much power. Therefore, low gate voltages are required to achieve acceptable beam quality. For some applications, such as gyroamplifiers and free-electron lasers, maintaining a high-quality beam is of paramount importance. Focusing grids have been suggested as a means of collimating the emission from single tips [106,107].

8.3.5. Lifetime and Failure Mechanisms

The greatest limitation to the utility of FEAs in applications requiring high emission currents, such as the microwave-tube application, is the precocious and seemingly random failure of the FEAs at high emission currents. The environment and procedures used in testing field emitters have proved to be quite critical. Hydrocarbon-free UHV, lengthy in situ conditioning procedures, and electrostatic safeguards are necessary for FEA longevity. In ultraclean conditions, several studies of single-tip field emitters have shown that tip failure is predictable and occurs at tip currents in the multi-milliampere range [108–111]. For example, resistive heating was identified as the failure mechanism of single-tip emitters fabricated out of single-crystal tungsten [108,109]. In this study, repeatable precursors of failure were identified that enabled tips to be reversibly cycled near burnout conditions. Unfortunately, trace amounts of contamination invariably remain because of the fabrication processing of an FEA, and the environment near the FEA cathode in a microwave tube invariably contains absorbed impurities. Figure 8.13 shows data taken on Lincoln Laboratory emitters [64]. In this case, the conical molybdenum tips were approximately 2000 Å high, with a conical half-angle of $30°$ and a radius of curvature of

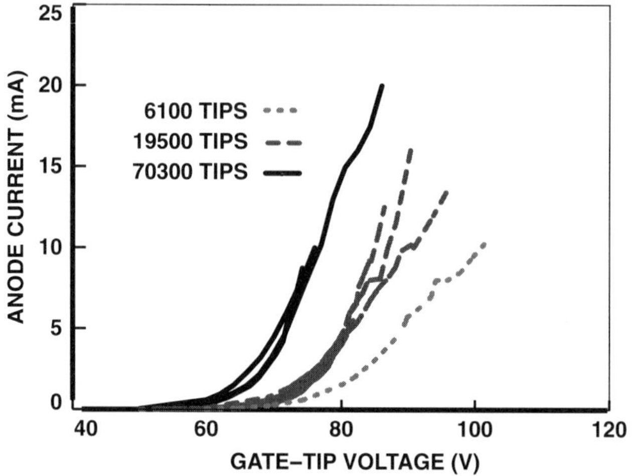

FIGURE 8.13. Tip failure data for Lincoln Laboratory FEAs. Emission currents exceeding 20 mA and tip currents exceeding 1.7 μA were attained. The incidence of failure does not correlate well with the average current per tip.

approximately 100 Å. Arrays of different size were tested to destruction in a UHV probing station. Assuming a uniform distribution of current across the array, failure occurred for the 6100-tip array at 1.7 μA/tip, while the best 70 300-tip array failed at 0.3 μA/tip.

A simple analysis can be used to show that average tip currents cannot heat the tips to unacceptable temperatures. The tip is approximated by a conical section of a sphere having inner radius a, outer radius b, and conical half-angle α. In operation, tip heating arises from two sources. The first of these is resistive heating by the emission current, and the second is Nottingham heating [112]. Nottingham heating is described by a thermal heat flux $\varphi_0 = J_{\text{tip}}(E_F - \bar{E})$ at the emitting surface, where J_{tip} is the tip current density, E_F is the Fermi-level energy, and $\bar{E}$ is the average energy of the emitter electrons. The energy deposited per emitted electron, $E_d = E_F - \bar{E}$, is approximately 0.25 eV [113]. Neglecting any angular variations and using spherical coordinates, the tip temperature $T(r, t)$ is given by the equation [114]

$$\frac{1}{\kappa}\frac{\partial T(r,t)}{\partial t} = \frac{1}{r^2}\frac{\partial}{\partial r}\left(r^2\frac{\partial T(r,t)}{\partial r}\right) + \frac{A(r)}{K_{\text{th}}} \tag{65}$$

In Eq. (65), $\kappa = K_{\text{th}}/\rho_m c_p$, and K_{th}, ρ_m, and c_p are the tip-metal thermal conductivity, mass density, and specific heat, respectively. If I_{tip} is the tip current and ρ_e is the electrical resistivity of the tip metal, the resistive power dissipation $A(r)$ is

$$A(r) = \rho_e\left[\frac{I_{\text{tip}}}{2\pi(1-\cos\alpha)r^2}\right]^2 \tag{66}$$

If a thermal flux ϕ_0 is applied at $r = a$, $T = T_0$ at $r = b$, and $a \ll b$, the steady-state solution for the tip temperature is

$$T(a) - T_0 = \frac{1}{2\pi a(1 - \cos\alpha)K_{\text{th}}}\left[\frac{\rho_e I_{\text{tip}}^2}{\pi a(1 - \cos\alpha)} + I_{\text{tip}}E_{\text{d}}\right] \tag{67}$$

If resistive heating is neglected, the transient temperature response is

$$T(a, t) - T_0 = \frac{I_{\text{tip}}E_{\text{d}}}{2\pi a(1 - \cos\alpha)K_{\text{th}}}\left[\frac{1 - e^{\kappa t}}{a^2}\,\text{erfc}\left(\frac{\sqrt{\kappa t}}{a}\right)\right] \tag{68}$$

The tip temperatures predicted by Eq. (68) for molybdenum and tungsten tips are plotted in Fig. 8.14. Only modest temperatures are predicted for the average tip currents of Fig. 8.13. Ancona [115,116], using detailed simulations, also concludes that experimentally observed FEA failures cannot be explained by tip heating if a uniform distribution of tip currents across the array is assumed.

Eq. (67), however, does predict that the tip temperature would be much higher if the emission current of the array were concentrated in one or several tips. This is quite possible because of imperfections in the FEA fabrication process. Further, random processes associated with migration of surface contaminants or changes in surface morphology can give rise to tip-destroying bursts of emission or gate current

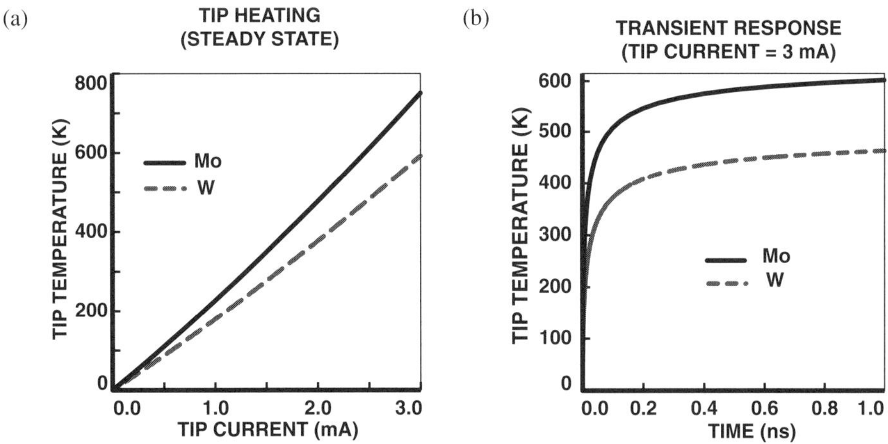

FIGURE 8.14. Calculated tip temperature during emission. A simple analytical model that includes resistive heating and Nottingham heating is used. W and Mo conical tips are considered, with tip height = 2000 Å, tip radius = 100 Å, and conical half-angle = 30°. (a) Steady-state temperature vs. tip current; (b) transient temperature response for an initial tip-current step of 3 mA.

[110,111,113,117]. This also is quite plausible since the thermal time constants predicted by Eq. (68) imply that a large excursion in tip temperature will result from short ($\sim$1 ns) current pulses.

Current limiting by external circuit elements has proved effective in reducing tip burnout. In the simplest such scheme, which is used in field emitters for display applications, a resistor is incorporated in series with the tips [8,118,119]. Because of such resistive stablization, cathode arcing is no longer considered an issue for displays, and lifetimes exceeding 1000 h are routinely achieved. However, the resistance that is introduced into the FEA equivalent circuit can limit the ability to modulate emission at GHz frequencies. FETs (field-effect transistors) have also been integrated with emitters [6,120] to provide current limiting in display applications. NEC Corporation has described the VECTL (Vertical Current Limiter) approach [31–34], which has enabled much higher currents, stability, and longevity in a tube environment. In the VECTL scheme, depicted in Fig. 8.15, the pinch-off of a FET-like structure beneath the emitters limits the current, thereby preventing any dramatic rises in current. Under conditions of an arc, the bias across the VECTL channels substantially increases, which causes the conducting channel to constrict. This greatly increases the effective resistance and limits the current. In normal operation, the resistance of the VECTL structure had a negligible effect on the current. Stable pulsed dc emission at current levels sufficient for meaningful levels of gain and power output was obtained for 5000 h without FEA failure, an unprecedented achievement.

The current-limiting structures described previously do not address the fundamental causes of tip failure. These causes are quite diverse [117,121], including vacuum conditions [72–78,122,123], improper anode design [70], insulator breakdown [124–126], contamination [75,76], and surface irregularities [127]. It is likely that the elimination of the causes of failure will involve stringent cleaning procedures and/or the use of ultraclean and stable tip materials. Both approaches have been used, but

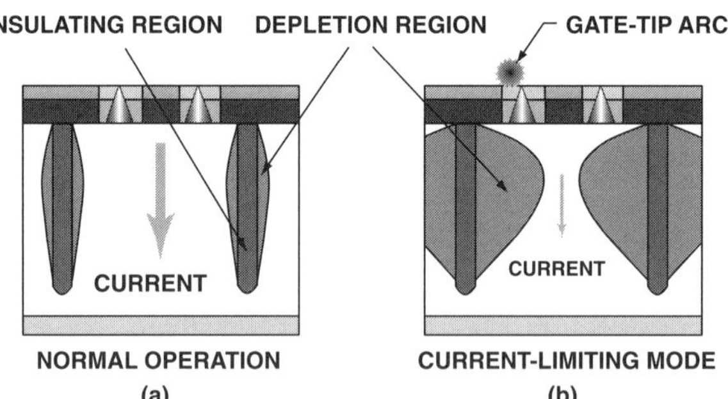

FIGURE 8.15. Schematic diagram of VECTL-stabilized FEA. An arc causes the tip potential to rise to the gate-voltage level. Depletion regions (shaded) are induced by the increased bias across the VECTL channel, which increases the resistance of the VECTL structure, thereby limiting emission current. (a) Normal operation. The VECTL structure presents a low series resistance. (b) Response to an event that would increase the tip current. The VECTL channel constricts, presenting high impedance that limits current flow.

often the only effective cleaning procedure involves high-temperature bake-outs that are impractical for many applications. A number of new materials for tips are under development that promise to improve upon the characteristics of molybdenum, silicon, or tungsten tips. Encouraging results have been obtained from Mo and Si FEAs that have been overcoated with transition metal carbides [128–135], titanium nitride [136], thin metallic layers [137], and diamond [138].

8.4. CHARACTERISTICS OF FEA-CATHODE MICROWAVE TUBES

When a gated cathode modulates the electron beam, the primary function of the output circuit is to couple rf power from the beam to an electromagnetic wave. The choice of output circuit, whether cavity, coupled cavity, ring-bar, or helix, depends chiefly on the bandwidth, size, and output power required of the amplifier. The modulated current [see Eq. (49)] that is available from the gated FEA cathode also influences the choice of circuit.

8.4.1. Efficiency

Although superior efficiency might be expected from a more strongly modulated beam, this is not always the case. In a broadband output circuit, a strongly modulated beam can drive higher frequency parasitic modes, which reduce the efficiency. In a narrowband output circuit, any high-frequency modes are filtered out. However, they increase peak electric fields, which limits the power handling capability of the output cavity. In addition, strong modulation severely reduces the gain, which must be included in any performance optimization.

To optimize the tube for total efficiency, both the beam power converted to rf and the beam power recovered in the collector must be considered. The total beam power P_b entering the tube leaves as output power P_o, as power recovered in the collector P_c and as waste heat in the collector. The net efficiency of the amplifier, η_N, is given by

$$\eta_N = \frac{\eta_e}{[1-\eta_c(1-\eta_e)]} \tag{69}$$

In Eq. (69), $\eta_e = P_o/P_b$ is the electronic efficiency and $\eta_c = P_c/(P_b-P_o)$ is the collector efficiency. The collector efficiency depends upon both the "quality" of the spent beam and the design of the collector itself. If the velocity distribution of the spent beam that enters the collector is broadly smeared, much less energy will be recovered than from a more monoenergetic beam. Therefore, in optimizing an output coupler for an IOA, equal attention must be given to the rf power that is coupled out and the velocity distribution of the spent beam.

8.4.2. Klystrode Output Power

A klystrode extracts rf power from the beam by passing the beam through a capacitive gap in a resonant cavity, as illustrated in Fig. 8.2. The operation of a klystrode output cavity is fundamentally similar to the final cavity of a klystron. The constraints are the same, with the exception that the quality factor of the output cavity, Q, must be appropriate to the current ratio of the injected beam.

According to a theorem first derived by Shockley [139], the current induced in the plates of a capacitive gap by an electron beam having a current $I_b(x, t)$ is

$$I_c(t) = \frac{1}{d} \int_0^d I_b(x, t) \, dx \tag{70}$$

In Eq. (70), the capacitive gap extends from $x = 0$ to $x = d$. In klystrode operation, the modulated electron beam that is incident upon the cavity induces a rf current at the input terminals of the output cavity in accordance with Eq. (70). Since the klystrode operates at the resonant frequency of the cavity, the cavity presents a high resistance R_c at the operating frequency and presents much lower impedance at other frequencies. Consequently, a rf voltage $V_c^{rf}(t) = \text{Re}[\tilde{V}_c^{rf} e^{j\omega t}]$ is developed across the gap that, together with the space charge, modifies the electrostatic potential and electron velocity within the gap. The power available from a cavity output circuit is obtained from the conservation of energy [140]:

$$\frac{dW}{dt} = -\frac{\omega_0}{Q} W + \langle E \cdot J \rangle \tag{71}$$

The maximum energy in the cavity is limited by arcing in regions of strong electric field, usually in the cavity gap near the beam. The rate of power extraction from the beam $\langle E \cdot J \rangle$ is limited by the deceleration of electrons by the electric fields induced in the coupling gap. The harmonic content of the beam current and the Q of the cavity determine the harmonic content of $\langle E \cdot J \rangle$. This is often an important design factor. Higher order harmonics contribute no usable output power, but increase the electric field in the cavity, reducing the maximum output power.

The interaction between the electron beam and output cavity is quite complex and can only be treated accurately using computer-aided techniques. However, some important characteristics of the output coupling can be displayed by a simple analysis. As the beam enters the gap, at $x = 0$, the beam current can be written as

$$I_b(0, t) = I_b^{dc} + \text{Re}\left(\tilde{I}_b^{rf} e^{j\omega t}\right) \tag{72}$$

In Eq. (72), I_b^{dc} is the dc component of the electron beam and $\tilde{I}_b^{rf}$ is a complex quantity that describes the emission-current modulation at the rf frequency ω. The electronic charge density in the absence of electric field in the coupling gap, $\rho_b^0(x, t)$, is given by the current-continuity equation as

$$\rho_b^0(x, t) = \frac{I_b^{dc}}{v_0 A} + \text{Re}\left[\frac{\tilde{I}_b^{rf}}{v_0 A} \exp\left(-j\omega x / v_0\right) \exp\left(j\omega t\right)\right] \tag{73}$$

The velocity of the electrons, $v_e(x, t)$, is determined by the electric field within the gap and the anode voltage V_a according to

$$\frac{m}{2} v_e^2(x, t) = eV_a + e\phi(x, t) \tag{74}$$

If space-charge effects are neglected, and $|V_c(t)| \ll V_a$,

$$v_e(x, t) = v_0 \sqrt{1 + \frac{\phi(x, t)}{V_a}} \cong v_0 + \frac{v_0}{2V_a} \phi(x, t) \cong v_0 + \frac{v_0 x}{2V_a d} V_c(t) \tag{75}$$

In Eq. (75), $v_0 = \sqrt{2eV_a/m}$ is the velocity of the electrons at $x = 0$ as they enter the coupling gap. If the rf modulation of the electron beam is small compared to the dc current, the induced current can be approximated, using Eq. (70), as

$$I_c(t) = \frac{A}{d} \int\limits_0^d \rho(x, t) v(x, t) \, dx$$

$$\cong \int\limits_0^d \left\{ \frac{I_b^{dc}}{d} + \mathrm{Re}\left[\left(\frac{\tilde{I}_b^{rf}}{d} \exp(-j\omega x/v_0) + \frac{I_b^{dc} x}{2V_a d^2} \tilde{V}_c \right) \exp(j\omega t) \right] \right\} dx \tag{76}$$

If the induced current is described by $I_c(t) = I_c^{dc} + \mathrm{Re}[\tilde{I}_c^{rf} e^{j\omega t}]$, Eq. (76) gives

$$\tilde{I}_c^{rf} = \tilde{I}_b^{rf} \exp(-j\omega d/2v_0) \left[\frac{\sin(\omega d/2v_0)}{\omega d/2v_0} \right] + \frac{I_b^{dc}}{4V_a} \tilde{V}_c^{rf} \tag{77}$$

Using $\tilde{V}_c = R_c \tilde{I}_c^{rf}$ in Eq. (77) and solving for $\tilde{I}_c^{rf}$,

$$\tilde{I}_c^{rf} \cong \frac{\dfrac{\sin(\omega d/2v_0)}{\omega d/2v_0}}{1 + \dfrac{R_c |I_b^{dc}|}{4V_a}} \tilde{I}_b^{rf} \tag{78}$$

The rf output power P_O^{rf} is then

$$P_O^{rf} = \frac{R_c}{2} |\tilde{I}_c^{rf}|^2 = \frac{R_c}{2} \frac{\left[\dfrac{\sin(\omega d/2v_0)}{\omega d/2v_0} \right]^2}{\left[1 + \dfrac{R_c |I_b^{dc}|}{4V_a} \right]^2} |\tilde{I}_b^{rf}|^2 \tag{79}$$

Eq. (79) shows that the quantity $\omega d/2v_0$ must be small for efficient operation. Consequently, the output cavity must be designed so that $d \ll (1/\pi f)\sqrt{2eV_a/m}$. At $f = 10$ GHz with $V_a = 5$ kV, this implies that $d \ll 1$ mm. Eq. (79) also displays first-order effects of the deceleration of the electron beam by the rf voltage and indicates that there are design trade-offs involving the beam modulation, output-cavity quality factor, and anode voltage.

Output circuit efficiency for the klystrode can be predicted using techniques ranging from basic analytical theory to detailed electromagnetic particle-in-cell (PIC) simulation. Resonator saturation theory (RST) [140] is an analytical approach that predicts the power in the output cavity from startup through saturation. It is based upon the conservation of energy, Eq. (71); the power loss term $\langle E \cdot J \rangle$ is calculated by integrating the velocity of the electrons crossing the gap with sinusoidal voltage. For simple assumptions, an analytical result can be obtained; otherwise, the power loss term must be integrated numerically. In this manner, details such as arbitrary bunch and interaction field shape can be included.

When space charge is important, multi-dimensional, electromagnetic PIC techniques such as MAGIC [141] can be used to obtain a fully self-consistent calculation of the beam interaction with the circuit. In the PIC code, the circuit can be modeled with a full-cavity transient simulation or with a "port approximation"—a transmission line. The predictive accuracy of these methods has been well substantiated, most notably for the 487-MHz klystrode designed by Varian, for which the code predicted 71% efficiency in excellent agreement with the experimental data [142].

A klystrode designed by workers at Communications and Power Industries (CPI) is intended to provide an output power of 50 W at 10 GHz using a gated FEA cathode. It requires a peak current of 112 mA from a ring cathode with inner and outer diameters of 550 and 610 μm, respectively [143,144]. Twystrodes, investigated at NRL [145], require similar (but potentially higher) currents. For example, if a gate voltage $V_g(t) = V_{pk} - V_{rf}[1 - \cos(\omega t)]$ is applied to a reduced-geometry FEA as described earlier, V_{pk} of 39.3 V and V_{rf} of 11.9 V will produce the required $I_{ave}/I_{pk} = 0.2$ and correspond to a peak current density of 100 A/cm^2. In separate measurements of field emitters, currents of 180 mA have been obtained, and SRI and MIT's Lincoln Laboratory have obtained current densities in excess of 2000 A/cm^2. Emission currents as high as 22 mA/quadrant have been obtained in test stations for four-quadrant ring cathodes [64], but only 2.6 mA/quadrant has been obtained in the klystrode vehicle. In the tube, cathode arc failure typically occurs just beyond the 2-mA/quadrant level. The reasons for the premature failure are presumed to relate to environmental factors with the tube, such as contaminants and backscattering from tube surfaces. Other factors, of course, affect the appropriateness of FEAs in a TWT/Twystrode; they shall be discussed in the following section.

8.4.3. Twystrode Output Power

A twystrode extracts power from the electron beam by passing the beam through the fields of an electromagnetic wave propagating with a phase velocity slightly lower than the beam velocity [145], as shown in Fig. 8.1(b). On casual inspection, a twystrode circuit closely resembles a TWT output section: optimizing the circuit impedance leads to the same specifications for beam and circuit radii. When the electron beam is density modulated, special consideration must be given to the gradual reduction of the phase velocity along the length of the circuit ("tapering") in order to optimize power extraction and the quality of the spent beam. In tapering, the phase velocity of the electromagnetic wave is reduced as the electron beam slows in order to maintain tight coupling between the traveling wave and the slowing beam. The maximum useful reduction is determined by loss of coherence in the electron bunches, which results in the re-acceleration of some electrons to high energy, thereby degrading the efficiency of the collector. Tightly bunched beams permit the use of greater total velocity tapers, as do longer circuits in which the taper occurs more gradually. Therefore, increasing the efficiency of a twystrode output circuit requires compromises with the size and the gain of the amplifier.

Small-signal theory does not adequately describe the interaction of a density-modulated electron beam with a traveling-wave circuit. The modulation is typically too strong to allow a linearization of the beam current, and the beam–circuit interaction is sufficiently strong to materially alter the beam-current waveform within one wave period. The efficiency of twystrodes can be analyzed with PIC computations covering the full length of the circuit. In two-dimensional calculations, the electromagnetic fields can be represented either by polarized boundary conditions [146] or by mode decomposition [147,148].

The polarizer model [146] uses a sheath approximation in which the finite-wire helix is represented as a cylindrical sheet with infinite conductance when parallel to the helix wire and zero conductance in the perpendicular direction. This representation, which is realized as a boundary condition on the fields, enables accurate modeling of a helical circuit in a two-dimensional PIC simulation. The model is implemented in MAGIC as a projection operator that constrains axial and azimuthal fields at the helix radius. Special diagnostics have been developed to analyze fundamental mode power as a function of axial distance. This model has been in use for several years and has been successfully tested against a series of experiments on emission gated amplifiers [149].

Investigations of FEA twystrodes are underway at laboratories in the United States [35] and Japan. Although no report of a density-modulated electron beam in a traveling-wave output circuit has yet appeared, NEC Corporation has reported a velocity-modulated TWT with an FEA cathode [31–34].

8.5. FUTURE WORK

Gated FEA cathodes have operational characteristics (e.g., small size, density modulation, high cut-off frequency, and instant-on capability) that should enable superior performance in microwave IOAs. Further, the modulation of FEAs at microwave frequencies is possible, and sufficiently high emission currents have been experimentally demonstrated in clean and well-controlled environments. Future research in this area must center on the development of FEAs that will perform reliably in a tube environment. This endeavor can be divided into two thrusts, improving the processes and materials with which FEAs are fabricated and accommodating the instabilities that remain with the best available near-term solutions.

A proper approach to the first of these areas requires a systematic and scientifically supportable study of tip and gate materials. It is not clear that currently available physical diagnostic techniques are capable of such a task. The dimensions of the structures that provide field emission are at least as small as 100 Å, and may indeed be even smaller, in view of the possibility that nanoprotrusions provide the true emission centers on microtips. Studies of advanced materials, such as the carbides, need to continue, and such investigations must be constantly mindful of the requirements for FEA cathodes, e.g., low gate current and high packing density. FEA stabilization may well be the best near-term solution to the problem. Current-limiting techniques using both resistive and active devices, along the lines of the VECTL approach of NEC, should be investigated.

The application of gated FEAs to the microwave tube spawned the field of vacuum microelectronics, but this application has yet to be convincingly demonstrated. It is evident that the operation of real-life FEAs is quite complex, and this complexity has thus far thwarted several concerted efforts to insert a gated FEA cathode into a microwave tube. Nevertheless, much progress has been made. It is clear that the insertion of a gated FEA cathode into a microwave amplifier tube is within the grasp of the technical community.

REFERENCES

1. F. M. Charbonnier, J. P. Barbour, L. F. Garrett, and W. P. Dyke, *Proc. IEEE* **51**, 991 (1963).

2. I. Brodie and C. A. Spindt, *Appl. Sur. Sci.* **2**, 149 (1979).

3. C. A. Spindt, C. E. Holland, I. Brodie, J. B. Mooney, and E. R. Westerburg, *IEEE Trans. Electron Devices* **36**, 225 (1989).

4. C. A. Spindt, C. E. Holland, A. Rosengreen, and I. Brodie, *IEEE Trans. Electron Devices* **38**, 2355 (1991).

5. J. D. Levine, R. Meyer, R. Baptist, T. E. Felter, and A. A. Talin, *J. Vac. Sci. Technol. B* **13**, 474–477 (1995).

6. K. Yokoo, M. Arai, M. Mori, J. Bae, and S. Ono, *J. Vac. Sci. Technol. B* **13**, 491–493 (1995).

7. J. D. Levine, *J. Vac. Sci. Technol. B* **13**, 553–557 (1995).

8. J. D. Levine, *J. Vac. Sci. Technol. B* **14**, 2008–2010 (1996).

9. J. M. Kim, J. P. Hong, J. H. Choi, Y. S. Ryu, and S. S. Hong, *J. Vac. Sci. Technol. B* **16**, 736–740 (1998).

10. J. O. Choi, H. S. Jeong, D. G. Plug, and H. I. Smith, *Appl. Phys. Lett.* **74**, 3050 3052 (1999).

11. A. A. Talin, B. Chalamala, B. F. Coll, J. E. Jaskie, R. Petersen, and L. Dworsky, *Mat. Res. Soc. Symp. Proc.* **509**, 21–22 (1998).

12. D. Temple, *Mat. Sci. Eng. R* **24**, 185–239 (1999).

13. R. A. Murphy and M. A. Kodis, Vacuum microelectronics, in *Wiley Encyclopedia of Electrical and Electronics Engineering*, Supplement 1 (John G. Webster, Ed.), John Wiley and Sons: New York, NY, p. 761, 2000.

14. W. Zhu, C. Bower, O. Zhou, G. Kochanski, and S. Jin, *Appl. Phys. Lett.* **75**, 873–875 (1999).

15. X. Xu and G. R. Brandes, *Mat. Res. Soc. Symp. Proc.* **509**, 107–112 (1998).

16. D. E. Patterson, K. D. Jamison, M. L. Kempel, and S. Freeman, *Mat. Res. Soc. Symp. Proc.* **509**, 65–75 (1998).

17. T. Kozawa, M. Suzuki, Y. Taga, Y. Gotoh, and J. Ishikawa, *J. Vac. Sci. Technol. B* **16**, 833–835 (1998).

18. R. D. Underwood, S. Keller, U. K. Mishra, D. Kapolnek, B. P. Keller, and S. P. Denbaars, *J. Vac. Sci. Technol. B* **16**, 822–825 (1998).

19. S.-C. Ha, D.-H. Kang, B.-S. Kim, S.-H. Min, and K.-B. Kim, *Mat. Res. Soc. Symp. Proc.* **509**, 77–82 (1998).

20. R. J. Nemanich, P. K. Baumann, M. J. Benjamin, S. L. English, J. D. Hartman, A. T. Sowers, B. L. Ward, and P. C. Yang, *Mat. Res. Soc. Symp. Proc.* **509**, 35–46 (1998).

21. R. W. Pryor, *Appl. Phys. Lett.* **68**, 1802–1804 (1996).

22. J. M. Early, *IRE Trans. Electronic Devices* **6**, 322–325 (1959).

23. E. O. Johnson, *RCA Rev.* **26**, 163–177 (1965).

24. T. S. W. Lewis, *Empire of the Air: The Men Who Made Radio*, Harper Collins Publishers: New York, NY, 1991.

25. D. B. Hamilton, J. K. Knipp, and J. B. H. Kuper, *Rad. Lab. Ser., Vol 7: Klystrons and Microwave Triodes*, Chs. 1, 2, and 6.

26. J. A. Morton, *Bell Labs Res.* **27**, 166–170 (1949).

27. J. A. Morton and R. M. Ryder, *Bell Syst. Tech. J.* **29**, 496–530 (1950).

28. A. E. Bowen and W. W. Mumford, *Bell Syst. Tech. J.* **29**, 531–552 (1950).

29. A. V. Haeff and L. S. Nergaard, *Proc. IRE* **28**, 126–130 (1940).

30. M. Garven, A. D. R. Phelps, S. N. Spark, D. F. Howell, and N. Cade, *Vacuum* **45**, 513–517 (1994).

31. H. Takemura, Y. Tomihari, N. Furutake, F. Matsuno, M. Yoshiki, N. Takada, A. Okamoto, and S. Miyano, A novel vertical current limiter fabricated with a deep trench forming technology for highly reliable field emitter arrays, in *IEDM Tech. Dig.*, 1997, p. 709.

32. H. Imura, S. Tsuida, M. Takahasi, A. Okamoto, H. Makishima, and S. Miyano, Electron gun design for traveling wave tubes (TWTs) using a field emitter array (FEA) cathode, in *IEDM Tech. Dig.*, 1997, p. 721.

33. H. Makishima, H. Imura, M. Takahashi, H. Fukui, and A. Okamoto, Remarkable improvements of microwave electron tubes through the development of the cathode materials, in *Proc. 10th IVMC*, Kyongju, Korea, 1997, p. 194.

34. H. Makishima, S. Miyano, H. Imura, J. Matsuoka, H. Takemura, and A. Okamoto, *Appl. Sur. Sci.* **146**, 230–233 (1999).

35. D. R. Whaley, C. M. Armstrong, M. A. Basten, B. Gannon, P. Mukhopadhyay-Phillips, C. A. Spindt, and C. E. Holland, PPM focused TWT using a field emitter array cold cathode, in *Proc. IEEE Int. Conf. Plasma Sci.*, Raleigh, NC, 1998, p. 125.

36. C. A. Spindt, C. E. Holland, P. R. Schwoebel, and I. Brodie, *J. Vac. Sci. Technol. B* **14**, 1986–1989 (1996).

37. C. A. Spindt, C. E. Holland, P. R. Schwoebel, and I. Brodie, *J. Vac. Sci. Technol. B* **16**, 758–761 (1998).

38. M. A. Kodis, K. L. Jensen, E. G. Zaidman, B. Goplen, and D. N. Smithe, *J. Vac. Sci. Technol. B* **14**, 1990–1993 (1996).

39. C. A. Spindt, C. E. Holland, A. Rosengreen, and I. Brodie, *J. Vac. Sci. Technol. B* **11**, 468–473 (1993).

40. R. H. Fowler and L. W. Nordheim, *Proc. R. Soc. (London) A* **119**, 173–181(1928).

41. R. H. Good, Jr. and E. W. Müller, Field Emission, in *Handbuch der Physik, Vol. 21*, Springer Verlag: Berlin, Germany, p. 176, 1956.

42. S. M. Sze, *Physics of Semiconductor Devices*, Wiley: New York, 1981.

43. R. G. Forbes, *J. Vac. Sci. Technol. B* **17**, 516–521 (1998).

44. I. Brodie and C. A. Spindt, *Adv. Electron. Electron Phys.* **83**, 1–106 (1992).

45. C. A. Spindt, I. Brodie, L. Humphrey, and E. R. Westerberg, *J. Appl. Phys.* **47**, 5248–5263 (1976).

46. E. L. Murphy and R. H. Good, Jr., *Phys. Rev.* **102**, 1464–1473 (1956).

47. M. G. Ancona, *Phys. Rev. B* **46**, 4874–4883 (1992).

48. L. G. ll'chenko and Yu. V. Kryuchenko, *J. Vac. Sci. Technol. B* **13**, 566–570 (1995).

49. G. N. Fursey, *J. Vac. Sci. Technol. B* **13**, 558–565 (1995).

50. K. L. Jensen, *J. Vac. Sci. Technol. B* **13**, 566–570 (1995).

51. K. L. Jensen, *J. Vac. Sci. Technol. B* **13**, 516–521 (1995).

52. D. Nicolaescu, V. Filip, J. Itoh, and F. Okuyama, *J. Vac. Sci. Technol. B* **17**, 542–546 (1999).

53. P. H. Cutler, Jun He, N. M. Miskovsky, T. E. Sullivan, and B. Weis, *J. Vac. Sci. Technol. B* **11**, 387–391 (1993).

54. K. L. Jensen and E. G. Zaidman, *J. Vac. Sci. Technol. B* **12**, 776–780 (1994).

55. K. L. Jensen and E. G. Zaidman, *J. Vac. Sci. Technol. B* **13**, 511–515 (1995).

56. G. N. Fursey and D. V. Glazanov, *J. Vac. Sci. Technol. B* **16**, 910–915 (1998).

57. I. Brodie and P. R. Schwoebel, *Proc. IEEE* **82**, 1006–1034 (1994).

58. D. Nicolaescu, *J. Vac. Sci. Technol. B* **11**, 392–395 (1993).

59. R. G. Forbes, *J. Vac. Sci. Technol. B* **17**, 526–533 (1998).

60. K. L. Jensen, E. G. Zaidman, M. A. Kodis, B. Goplen, and D. N. Smithe, *J. Vac. Sci. Technol. B* **14**, 1942–1946 (1996).

61. C. A. Spindt, *J. Appl. Phys.* **39**, 3504–3505 (1968).

62. C. O. Bozler, C. T. Harris, S. Rabe, D. D. Rathman, and M. A. Hollis, *J. Vac. Sci. Technol. B* **12**, 629–632 (1994).

63. C. O. Bozler, D. D. Rathman, C. T. Harris, G. A. Lincoln, C. A. Graves, R. H. Mathews, S. Rabe, R. A. Murphy, and M. A. Hollis, Field-emitter arrays for microwave power tubes, in *Proc. IEEE Int. Conf. Plasma Science*, San Diego, CA, 1996, p. V-13.

64. R. A. Murphy, C. T. Harris, R. H. Mathews, C. A. Graves, M. A. Hollis, M. A. Kodis, J. Shaw, M. Garven, M. T. Ngo, K. L. Jensen, M. Green, J. Legarra, J. Atkinson, L. Garbini, and S. Bandy, Fabrication of field-emitter arrays for inductive output amplifiers, in *Proc. IEEE Int. Conf. Plasma Sci.*, San Diego, CA, 1997, p. 127.

65. S. Itoh, T. Watanabe, K. Ohtsu, M. Taniguchi, S. Uzawa, and N. Nishimua, *J. Vac. Sci. Technol. B* **13**, 487–490 (1995).

66. C. T. Sune, G. W. Jones, and D. G. Velenga, *J. Vac. Sci. Technol. B* **10**, 2984–2988 (1992).

67. W. D. Palmer, J. E. Mancusi, C. A. Ball, W. J. Joines, G. E. McGuire, D. Temple, D. G. Vellenga, and L. Yadon, *IEEE Trans. Electron Devices* **41**, 1867–1870 (1994).

68. W. D. Palmer, H. F. Gray, J. Mancusi, D. Temple, C. Ball, J. L. Shaw, and G. E. McGuire, *J. Vac. Sci. Technol.* **13**, 576–579 (1995).

69. F. Ducroquet, P. Kropfeld, O. Yaradou, and A. Vanoverschelde, *J. Vac. Sci. Technol.* **16**, 787–789 (1998).

70. C. E. Holland, A. Rosengreen, and C. A. Spindt, *IEEE Trans. Electron Devices* **38**, 2368–2372 (1991).

71. M. S. Mousa, P. R. Schwoebel, I. Brodie, and C. A. Spindt, *Appl. Surf. Sci.* **67**, 56–58 (1993).

72. S. Meassick and H. Champaign, *J. Vac. Sci. Technol.* **14**, 1914–1917 (1996).

73. J. T. Trujillo, A. G. Chakhovski, and C. E. Hunt, *J. Vac. Sci. Technol. B* **15**, 401–404 (1997).

74. B. R. Chalamala, R. M. Wallace, and B. E. Gnade, *J. Vac. Sci. Technol. B* **16**, 2859–2865 (1998).

75. B. R. Chalamala, R. M. Wallace, and B. E. Gnade, *J. Vac. Sci. Technol. B* **16**, 2866–2870 (1998).

76. B. R. Chalamala, R. M. Wallace, and B. E. Gnade, *J. Vac. Sci. Technol. B* **16**, 3073–3076 (1998).

77. B. R. Chalamala, R. M. Wallace, and B. E. Gnade, *J. Vac. Sci. Technol. B* **17**, 303–304 (1999).

78. Y. Gotoh, K. Utsumi, M. Nagao, H. Tsuji, J. Ishikawa, T. Nakatani, T. Salashita, and K. Betsui, *J. Vac. Sci. Technol. B* **17**, 604–607 (1999).

79. Vu Thien Binh, N. Garcia, and S. T. Purcell, Electron field emission from atom-sources: fabrication, properties, and applications of nanotips, in *Advances in Imaging and Electron Physics, Vol. 95* (P. W. Hawkes, Ed.), Academic Press: New York, p. 63, 1996.

80. Vu Thien Binh, S. T. Purcell, G. Gardet, and N. Garcia, *Surf. Sci. Lett.* **279**, L197–L201 (1992).

81. Vu Thien Binh, S. T. Purcell, N. Garcia, and J. Doglioni, *Phys. Rev. Lett.* **69**, 2527–2530 (1992).

82. S. T. Purcell, Vu Thien Binh, and R. Baptist, *J. Vac. Sci. Technol. B* **15**, 1666–1677 (1997).

83. F. Ito, K. Konuma, A. Okamoto, and A. Yano, *J. Vac. Sci. Technol. B* **16**, 783–786 (1998).

84. N. Garcia, M. J. Marques, A. Asemjo, and A. Correia, *J. Vac. Sci. Technol. B* **16**, 654–664 (1998).

85. L. Parameswaran, Private communication, 1999.

86. M. J. Buckingham, *Noise in Electronic Devices and Systems*, Wiley: New York, 1983.

87. M. J. Kirton and M. J. Uren, Noise in solid-state microstructures: A new perspective on individual defects, interface state and low frequency (1/f) noise, in *Advances in Electron Physics, Vol. 38* (P. W. Hawkes, Ed.), Academic Press: New York, p. 367, 1989.

88. R. N. Amirkhanov, S. S. Ghots, and R. Z. Bakhtizin, *J. Vac. Sci. Technol. B* **14**, 2135–2137 (1996).

89. C. D. Child, *Phys. Rev.* **32** (Series 1), 492–511 (1911).

90. I. Langmuir, *Phys. Rev.* **21** (Series 2), 419–431 (1923).

91. K. L. Jensen, M. A. Kodis, R. A. Murphy, and E. G. Zaidman, *J. App. Phys.* **82**, 845–854 (1997).

92. J. P. Calame, H. F. Gray, and J. L. Shaw, *J. Appl. Phys.* **73**, 1485–1503 (1993).

93. S. Ramo, J. R. Whinnery, and T. van Duzer, *Fields and Waves in Communication Electronics*, Wiley: New York, 1965.

94. R. E. Collin, *Foundations for Microwave Engineering*, McGraw-Hill: New York, 1966.

95. M. Abramowitz and I. A. Stegun, *Handbook of Mathematical Functions*, Dover: New York, 1965.

96. J. E. Rowe, *IRE Trans. Electron Devices* **3**, 39–56 (1956).

97. Y. Y. Lau and D. Chernin, *Phys. Fluids B: Plasma Phys.* **4**, 3473–3497 (1992).

98. H. P. Freund, E. G. Zaidman, and T. M. Antonsen, *Phys. Plasmas* **3**, 3145–3161 (1996).

99. D. N. Smithe, B. Goplen, M. A. Kodis, and N. R. Vanderplaats, Predicting twystrode output performance, in *Proc. IEEE Int. Conf. Plasma Science*, Madison, WI, 1995.

100. J. P. Barbour, W. W. Dolan, J. K. Trolan, E. E. Martin, and W. P. Dyke, *Phys. Rev.* **92**, 45–51 (1953).

101. W. A. Anderson, *J. Vac. Sci. Technol. B* **11**, 383–386 (1993).

102. Y. Y. Lau, Y. Liu, and R. K. Parker, *Phys. Plasmas* **1**, 2082–2085 (1994).

103. L. Yun-Peng and L. Enze, *Appl. Surf. Sci.* **76/77**, 7–10 (1994).

104. K. L. Jensen, P. Mukhopadhyay-Phillips, E. G. Zaidman, K. Nguyen, M. A. Kodis, L. Malsawma, and C. Hor, *Appl. Surf. Sci.* **111**, 204–212 (1997).

105. Y. Liu and Y. Y. Lau, *J. Vac. Sci. Technol. B* **14**, 2126–2129 (1996).

106. C.-M. Tang, T. A. Swyden, and A. C. Ting, *J. Vac. Sci. Technol. B* **13**, 571–575 (1995).

107. A. Hosono, S. Kawabuchi, S. Horibata, S. Okuda, H. Harada, and M. Takai, *J. Vac. Sci. Technol. B* **17**, 575–579 (1999).

108. W. P. Dyke, J. K. Trolan, E. E. Martin, and J. P. Barbour, *Phys. Rev.* **91**, 1043–1054 (1953).

109. W. W. Dolan, W. P. Dyke, and J. K. Trolan, *Phys. Rev.* **91**, 1054–1057 (1953).

110. G. N. Fursey, *IEEE Trans. Electr. Insul.* **20**, 659–670 (1985).

111. G. N. Fursey and D. V. Glazanov, *J. Vac. Sci. Technol. B* **13**, 1044–1049 (1995).

112. W. B. Nottingham, *Phys. Rev.* **59**, 906–43 (1941).

113. F. Charbonnier, *Appl. Surf. Sci.* **94/95**, 26–43 (1996).

114. H. S. Carslaw and J. C. Jeager, *Conduction of Heat in Solids*, 2nd ed., Oxford University Press: London, 1959.

115. M. G. Ancona, *J. Vac. Sci. Technol. B* **13**, 2206–2214 (1996).

116. M. G. Ancona, *J. Vac. Sci. Technol. B* **14**, 1918–1923 (1996).

117. F. Charbonnier, *J. Vac. Sci. Technol. B* **16**, 880–887 (1998).

118. R. Baptist, F. Bachelet, and C. Constancias, *J. Vac. Sci. Technol. B* **15**, 385–390 (1997).

119. L. Parameswaran, C. T. Harris, C. A. Graves, R. A. Murphy, and M. A. Hollis, *J. Vac. Sci. Technol. B* **17**, 773–777 (1999).

120. S. Kanemaru, T. Hirano, K. Honda, and J. Itoh, *Appl. Surf. Sci.* **146**, 198–202 (1999).

121. P. R. Schwoebel and I. Brodie, *J. Vac. Sci. Technol. B* **13**, 1391–1410 (1995).

122. M. Takai, H. Morimoto, A. Hosono, and S. Kawabuchi, *J. Vac. Sci. Technol. B* **16**, 799–802 (1998).

123. B. R. Chalamala, R. M. Wallace, and B. E. Gnade, *J. Vac. Sci. Technol. B* **16**, 2855–2858 (1998).

124. H. C. Miller, *IEEE Trans. Electr. Insul.* **24**, 765–786 (1989).

125. X. Ma and T. S. Sudarshan, *J. Vac. Sci. Technol. B* **16**, 745–748 (1998).

126. V. A. Nevrosky, Analysis of elementary processes leading to surface flashover in vacuum, in *Proc.* 18th *IEEE Int. Symp. On Discharges and electrical Insulation in Vacuum*, 1998, p. 159.

127. Y. Wei, B. R. Chalamala, B. G. Smith, and C. W. Penn, *J. Vac. Sci. Technol. B* **17**, 233–236 (1999).

128. W. A. Mackie, R. L. Harman, and P. R. Davis, *Appl. Surf. Sci.* **67**, 29–35 (1993).

129. W. A. Mackie, T. Xie, and P. R. Davis, *J. Vac. Sci. Technol. B* **13**, 2459–2463 (1995).

130. M. L. Yu, W. A. Mackie et al., *J. Vac. Sci. Technol. B* **13**, 2436–2440 (1995).

131. T. Xie, W. A. Mackie, and P. R. Davis, *J. Vac. Sci. Technol. B* **14**, 2090–2092 (1996).

132. W. A. Mackie, T. Xie, J. E. Blackwood, S. C. Williams, and P. R. Davis, *J. Vac. Sci. Technol. B* **16**, 1215–1218 (1998).

133. W. A. Mackie, T. Xie, M. R. Matthews, B. P. Routh, Jr., and P. R. Davis, *J. Vac. Sci. Technol. B* **16**, 2057–2062 (1998).

134. W. A. Mackie, T. Xie, and P. R. Davis, *J. Vac. Sci. Technol. B* **17**, 613–619 (1998).

135. W. A. Mackie, T. Xie, M. R. Matthews, and P. R. Davis, *Mat. Res. Soc. Symp. Proc.* **509**, 173–178 (1998).

136. S.-Y. Kang, J. H. Lee, Y.-H. Song, Y. T. Kim, K. I. Cho, and H. J. Yoo, *J. Vac. Sci. Technol. B* **16**, 871–874 (1998).

137. P. R. Schwoebel, C. A. Spindt, and I. Brodie, *J. Vac. Sci. Technol. B* **13**, 338–343 (1995).

138. Z. X. Yu, S. S. Wu, and N. S. Xu, *J. Vac. Sci. Technol. B* **17**, 562–566 (1999).

139. W. Shockley, *J. Appl. Phys.* **9**, 635–636 (1938).

140. K. Nguyen, G. D. Warren, L. Ludeking, and B. Goplen, *IEEE Trans. Electron Devices* **38**, 2212–2220 (1991).

141. B. Goplen, L. Ludeking, D. Smithe, and G. Warren, *Computer Phys. Commun.* **87**, 54–86 (1995).

142. D. H. Preist and M. B. Shrader, *IEEE Trans. Electron Devices* **38**, 2205–2211 (1991).

143. S. G. Bandy, M. C. Green, C. A. Spindt, M. A. Hollis, W. D. Palmer, B. Goplen, and E. G. Wintucky, Application of gated field emitter arrays in microwave amplifier tubes, in *Proc. IEEE Int. Conf. Plasma Science*, San Diego, CA, 1997, p. 132.

144. K. L. Jensen, R. H. Abrams, and R. K. Parker, *J. Vac. Sci. Technol. B* **16**, 749–753 (1998).

145. K. L. Jensen, *J. Appl. Phys* **83**, 7982–7992 (1998).

146. J. R. Pierce, *Traveling Wave Tubes*, Van Nostrand: Princeton, NJ, Ch. III, Appendix II, 1950.

147. D. Smithe, G. Warren, and B. Goplen, *Bull. Am. Phys. Soc.* **37**(6), 1992.

148. H. P. Freund and E. G. Zaidman, *Phy. Plasmas* **4**, 2292–2301 (1997).

149. M. A. Kodis, N. R. Vanderplaats, H. P. Freund, B. Goplen, and D. N. Smithe, *Bull. Am. Phys. Soc.* **38**(10), 1993.